AF530726

VERTEBRATE PESTS IN AGRICULTURE

"Man has lot of choices even if he happens to be a subsistence farmer living in villages around the protected area but the herbivores of the forest pushed into or restricted to fragmented and degraded habitat have absolutely no choice but to feed on alternate, available food in their range. It also happens that cultivated crops become an easily accessible resource"

— Shakunthala Sridhara

Other Related Books

A Manual of Practical Entomology 3rd ed.	– *Trigunayat, M.M.*
Agricultural Insect Pests and Their Control	– *Awasthi, V.B.*
Beekeeping : A Compressive Guide to Bees and Beekeeping	– *Abrol, D.P.*
Biological and Molecular Approaches in Pest Management	– *Singh, B.*
Biopesticides in Environment and Food Security : Issues and Strategies	– *Koul, O.*
Biopesticides in Sustainable Agriculture Progress and Potential	– *Koul, O.*
Botanical Pesticides for Pest Management	– *Dodia, D.A.*
Dynamics of Insect Behaviour	– *Ananthakrishnan, T.N.*
Ecodynamics of Insect Communities	– *Ananthakrishnan, T.N.*
Entomology Ecology & Biodiversity	– *Tyagi, B.K.*
Handbook of Practical Nematology	– *Bajaj, H.K.*
Integrated Pest Management	– *Dhawan, A.K*
Introduction to General and Applied Entomology, 3rd Edition	– *Awasthi, V.B.*
Perspectives in Entomological Research	– *Agrawal, O.P.*
Pests of Forest Importance & their Management	– *Tyagi, B.K.*
Principles of Insect Behaviour, 2nd Ed.	– *Awasthi, V.B.*
The Invincible Deadly Mosquitoes	– *Tyagi, B.K.*
Zoology At A Glance	– *Pandey, B.D.*
बागवानी फसलों में समेकित रोग कीट प्रबंधन	– *Singh, S.K.*

VERTEBRATE PESTS IN AGRICULTURE

Shakunthala Sridhara

Professor & Head,
AICRP on Rodent Control,
G.K.V.K.
Bangalore - 560 065 (Karnataka)

Published by:

SCIENTIFIC PUBLISHERS (INDIA)

Jodhpur – *5 A, New Pali Road, P.O. Box 91, Jodhpur - 342 001, INDIA*

Delhi – *4806/24, Ansari Road, Daryaganj, New Delhi - 110 002, INDIA*

E-mail: info@scientificpub.com
Website: www.scientificpub.com

Print: 2016

ISBN: 978-81-7233-980-7 (HB)
978-81-7233-981-4 (PB)

Printed in India

Dedicated to

late Dr. Ishwar Prakash

FOREWORD

The most urgent need today is to reduce all losses to crops due to pests and diseases. Rodents take a heavy toll in both the field and in the threshing ground. Later they do damage at the storage level. Our country has to produce more from less land in the coming years. This book provides detailed guidelines on methods of reducing and even eliminating damage by rodents to crops. Dr. Shakunthala Sridhara, the author of the book has been working on rodent control from 1978. She is thus, very eminently qualified to write a book on this topic. I hope, it will be read widely by all concerned with improving the profitability and stability of agricultural production in our country. We owe Dr. Sridhara a deep debt of gratitude for this labour of love.

M.S. Swaminathan
Chairman, National Commission on Farmers, Govt. of India
President, Purwash Conferences on Science and World Affairs
Chairman, M.S. Swaminathan Research Foundation
Third Cross Street, Taramani Institutional Area
Chennai - 600 113 (India)

PREFACE

In our endeavour to increase agricultural yields no stone has been left unturned. Not long ago, it was fine when pesticides were applied liberally to kill the tiny insects, other arthropods and the ugly, scary destructive rodents to save the cultivated food. Not many of us felt guilty of homicide till Rachel Carson (Silent Spring, 1962) drew attention to our lack of sensitivity to the disappeared, destroyed, threatened and mutilated – mutated animals and environments around us. The battle was carried forward with the issues of toxic residues in the food chain and the concurrent heath hazards. For sometime, the silent cry of anguish of the subsistence farmer living close to protected areas went unheard whenever his entire crop was trampled and devoured by animals much bigger than him. He was a helpless witness to the devastation caused by crop raiding elephants, ungulates, monkeys and even some granivore birds. Almost all of them, with the exception of birds were not only bigger than him and as intelligent as him if not more, but were protected by the law of the land. He was also bogged down by the divinity associated with them. The problem of crop raiding vertebrate pests is not only economical but also ethical, social, legal and emotional one. In this scenario it is difficult to judge who is suffering more – the unfortunate, farmer or the more unfortunate animal who was forced to seek his nutritional requirements away from his home which was so much degraded and depleted that the resources available were highly inadequate. Thus, our rich wildlife is as much a problem for its conflict with humans for depleting resources as for its conservation priorities. In spite of regular occurrence of crop raiding by vertebrate pests, there does not appear to be a consolidated source of critical evaluations of the problem and its management. The information generated mostly by field biologists and forestry departments are scattered in scientific journals and departmental reports respectively. Inspite of being a serious agricultural and conservation problem, we don't have a quick, detailed reference manual on the problem of wildlife damage to crops with the exception of "Handbook of Vertebrate Pest Control" by Fitzwater and Ishwar Prakash (1989).

Thus, there is a need to have detailed information on not only the damage caused by vertebrate pests but also the contributory factors as well as the options available to us to mitigate the problem. This book, conceived to fulfil such a lacunae has chapters written by eminent specialists on mammalian and avian depredators of crops. Due to time constraints the chapters on bats and elephants are not authored by the experts, I contacted. With my 33 years of research on vertebrate pests, I felt confident to write these chapters in addition to the ones on rodents and general topics of vertebrate pest management. The book is an excellent example of co-operation of field biologists, conservationists and agricultural scientists to shed light on a contemporary problem of great economic, social, emotional, ethical and conservation importance.

A book of this magnitude and perspective is impossible to conceive and produce without inspiration, guidance, encouragement and patronage which I received in abundance from many directly and indirectly. First and foremost, I reverentially put on record my gratitude and affectionate regards to late Dr. Ishwar Prakash who was my Guru literally transforming me from an animal physiologist into an agricultural scientist specializing in rodent control and vertebrate pest management. Of the indirect sources of inspiration, I acknowledge the inherent wisdom of Dr. M.S. Swaminathan who as Director General of Indian Council of Agricultural Research (ICAR) recognised the role of rodents in Indian agriculture and initiated the process of rodent research under the umbrella of ICAR with this act a hither to scattered discipline of research, which was mostly oriented towards epidemiology got its due place in the agricultural research of the country. I am indebted to him. I am also thankful to Dr. Mangala Rai, Director General, Dr. O.P. Dubey, Assistant Director General (Oil seeds & pulses) and Dr. T.P. Rajendran (Plant Protection) of Indian Council of Agricultural Research, New Delhi for constantly guiding and patronising research on rodent control in the country in general and me in particular. I am highly indebted to the contributors to this book namely Abraham Verghese, Achiraman. S, Archunan. G, Butani. P.G., Chakravarthy. A.K., Chhangani. A.K., Dubey. O.P., Girish. A.C., Ikwal Imam, Iqbal Malik, Jayahari. K.M., Jayson. E.A., Kapadia. M.N., Mohnot. S.M., Parshad. R.K., Rajpurohit. L.S., Raveendra Babu. T., Shivayya. V., Srihari. K., Thyagaraj. N.E., Tripathi. R.S., Vasudeva Rao. V., Vipin Choudhary and Vyas. H.J. all of whom amidst hectic academic life wrote valuable chapters. I thank profusely each one of them.

On the home front I am grateful to Dr. M.N. Sheelavanthar, Vice-chancellor and Dr. T.K. Prabhakar Shetty, Director of Research, University of Agricultural Sciences, Bangalore for providing not only facilities for research and writing this book but also creating and maintaining an atmosphere of academic freedom and trust that is so much essential for the success of a scientist. I feel fortunate to be patronized by them. I also acknowledge the co-operation and help extended to me by my colleagues Mr. V. Shivayya, T. Raveendra Babu, A. Ravishankar, A.C. Shivaji, Y. Muniraju, Rajanna and Munivenkataramanappa. Ms. Archana S. Shet has painstakingly transferred my manuscript into this highly attractive book format with her sincere, dedicated efforts in computer setting. Her involvement was so thorough that she ended-up even as a technical assistant. I am grateful to her. Last but not least, I owe a lot, beyond words to my daughter, Sriranjini for her patience and interest in my work. The cover credit goes to her to a large extent. It is nice to have a youngster at home, unrelated to the subject giving inputs which largely enrich the contents of a book from a laymen's perception. One of the aims of this book is, to enlighten the public about the intricacies of vertebrate pest management apart from serving as a reference manual for details. To that extent, Sriranjini fine-tuned my presentation.

Bangalore **S. Sridhara**

CONTRIBUTORS

ABRAHAM VERGHESE
Fruit Entomology Laboratory
Division of Entomology & Nematology
Indian Inst Horticultural Research,
Hesseraghatta Lake – P.O.,
Bangalore-560 089
e-mail: avergis@iihr.res.in

ACHIRAMAN, S.
Department of Animal Science
Bharathidasan University
Tiruchirappalli-620 024
Tamil Nadu, India

ARCHUNAN, G.
Department of Animal Science
Bharathidasan University
Tiruchirappalli-620 024
Tamil Nadu, India
e-mail: garchu56@yahoo.co.in

BUTANI, P.G.
Department of Entomology
College of Agriculture
Junagadh Agri. University,
Junagadh – 362 001

CHAKRAVARTHY, A.K.
Department of Entomology
University of Agri. Sciences
GKVK Campus,
Bangalore – 560 065, Karnataka
Chakravarthyakshay@yahoo.com

CHHANGANI, A.K.
Indo-US Primate Project
Department of Zoology,
J.N.V. University
Jodhpur – 342 003 (India)

DUBEY, O.P.
Assistant Director General
(Plant Protection),
Indian Council of Agricultural Research
Krishi Bhavan, New Delhi-110 001

GIRISH, A.C.
Department of Entomology
University of Agricultural Sciences
GKVK Campus, Bangalore 560 065,
Karnataka

IKWAL IMAM
Department of Wildlife Sciences
Aligarh Muslim University,
Aligarh 202 002

IQBAL MALIK
Founder Director
VATAVARAN-NGO, 540,
Hawa Singh Marg,
Asian games village,
New Delhi- 110 049

JAYAHARI, K.M.
Division of Forest Ecology and
Biodiversity Conservation
Kerala Forest Research Institute,
Peechi,
Thrissur, Kerala-680 653

JAYSON, E.A.
Division of Forest Ecology and
Biodiversity Conservation
Kerala Forest Research Institute,
Peechi,
Thrissur, Kerala-680 653

KAPADIA, M.N.
Department of Entomology
College of Agriculture
Junagadh Agricultural University
Junagadh – 362 001

MOHNOT, S.M.
Indo-US Primate Project
Department of Zoology,
J.N.V. University
Jodhpur – 342 003 (India)

PARSHAD, R.K.
Department of Zoology & Fisheries
Punjab Agricultural University
Ludhiana 141004, India

RAJPUROHIT, L.S.
Indo-US Primate Project
Department of Zoology,
J.N.V. University
Jodhpur – 342 003 (India)

RAVEENDRA BABU, T.
AINP on Rodent Control
University of Agricultural Sciences
GKVK, Bangalore-560 065

SHAKUNTHALA SRIDHARA
AINP on Rodent Control
University of Agricultural Sciences
GKVK, Bangalore-560 065

SHIVAYYA, V.
AINP on Rodent Control
University of Agricultural Sciences
GKVK, Bangalore-560 065

SRIHARI, K.
Department of Entomology
University of Agricultural Sciences
GKVK Campus,
Bangalore – 560 065, Karnataka

THYAGARAJ, N.E.
Department of Entomology
University of Agricultural Sciences
G.K.V.K., Bangalore-560 065

TRIPATHI, R.S.
AINP on Rodent Control
Central Arid Zone Research Institute
Jodhpur-342 003

VASUDEVA RAO, V.
AINP on Agricultural Ornithology
Acharya N.G. Ranga Agricultural
University
Rajendranagar, Hyderabad-500 030

VIPIN CHOUDHARY
AINP on Rodent Control
Central Arid Zone Research Institute
Jodhpur-342 003

VYAS, H.J.
Department of Entomology
College of Agriculture
Junagadh Agricultural University
Junagadh – 362 001

CONTENTS

1

PRINCIPLES OF POPULATION ECOLOGY AND VERTEBRATE PEST MANAGEMENT

Shakunthala Sridhara

AINP on Rodent Control, University of Agricultural Sciences, G.K.V.K., Bangalore - 560 065

ABSTRACT

After defining population, the characteristics of population viz. density, natality, mortality and dispersal are described followed by a brief account of factors affecting population and theories of population regulation namely : Biotic school, Climatic school, Biotic and Climatic approach, Self regulation school and Evolutionary concepts. The different types of ecological associations — Mutualism, Predation and Parasitism are briefly mentioned. Measurement of pest populations including details of sampling is elaborated with emphasis on methods for rodent population estimation including, absolute estimates by capture-mark-release-recapture method, relative estimates inclusive of trap index, regression of daily captures on cumulative captures, change in ratio method, census baiting, tracking patches and tiles etc. Definition and explanation of life system concept is followed by methods of life table construction. A brief account of ecological basis of pest management forms the concluding part of the chapter.

INTRODUCTION

Population is defined as the fundamental unit of ecology consisting of a group of organisms of the same species occupying a specific area at a particular time. Biotic and abiotic factors of the environment as well as the genetic make-up of the species govern the survival and abundance of a population. These factors influence the population leading to fluctuations in density with definite upper and lower limits. In some species, the upper limit corresponds to the status of a species as a pest i.e. a level at which it causes economic losses. The quality and quantity of damage depends on the type of the pest and its density.

Growth of a population is influenced by the action of five factors namely physical environment, a food source including a host, space, population itself and other species of the organisms. All the five act by influencing natality, mortality and dispersal (Southwood & Way, 1970).

CHARACTERISTICS OF A POPULATION

Population characteristics are unique and do not belong to individuals of the population but is summation of individual characteristics. These characteristics are density, natality or birth rate, mortality or death rate, age distribution, dispersal and growth rate which distinguish one population from the other.

Density: Density of a population denotes population size in relation to a unit of space. In other words, the number of individuals of a species per unit area. Density can be crude density and ecological density. *Crude density* represents number of individuals per unit of total space. *Ecological density* denotes number of individuals per unit of habitat space i.e. the space actually occupied by the population.

Natality (*Birth rate*) refers to number of individuals that are added to the population by reproduction per unit of time. Natality is of two types – Absolute natality and ecological natality.

Absolute natality refers to maximum number of individuals that can be produced by an individual under optimal environment conditions. Here reproduction is assumed to be governed by only physiological factors and the role of ecological factors is not taken into consideration.

Ecological natality is increase in population due to specific environmental conditions. This is also called realized natality. It is not a constant but governed by environmental conditions, size and composition of population.

Mortality (*Death rate*) represents number of individuals that are lost due to death in a given population over a period of time. It is categorized into minimum mortality and ecological mortality. *Minimum mortality* refers to mortality under ideal or non limiting conditions of the environment and is constant for a population. *Ecological mortality* refers to mortality under specific environmental conditions. It is not a constant for a population but varies with changes in environment and population.

Survivorship: For a given population surviving members are more important than the dead ones. Thus survival rates are more important than mortality rates. Survival rates are expressed as *Survival Curves* which are categorized into: *Convex Curve* which is characteristic of those species which have high survivorship till the end of life span and mortality increases only due to old age. This situation is seen in animals with high degree of parental care. e.g.: humans, ungulates and honeybees. *Concave curve* is distinguished by high mortality rates during early life followed by a period of much lower but constant rates of mortality. E.g. parasites, grasshoppers, frogs, etc. *Constant curve* represents constant rate of survival independent of age. However in actuality there is no constant age-survival rate throughout the life of an organism. A modification of the constant curve is called *stair–step survivorship curve* where variations in survival occur at different stages of life history. This type of curve is seen in holometabolous insects like butterflies where the steep portions indicate egg, pupa and adults of short-life with high degree of mortality and the flatter segments represent larval and pupal stages of low mortality. Another modification of this curve is slightly sigmoid with high mortality when young which stabilizes during adult life. E.g. birds, rabbits, mice.

Dispersal: The rate at which individuals of a population move into or out of it is called *population dispersal* which may be in search of food, to avoid predators, prevent over-crowding, escape from unfavourable conditions, etc. Dispersal takes place in three ways. *Immigration* (one way inward movement of individuals), *Emigration* (one way outward movement from one place to another) and *Migration* i.e. mass movement of entire population, where some individuals of the population return to the area from which they moved. Such migrations occur during unfavourable conditions, from original unfavourable area to favourable places. These movements can be seasonal, prior to or after breeding, due to extreme climatic conditions or when crops are harvested over a wider area. Notable migrants are locusts, butterflies, aphids, coleoptera, hemiptera, some species of birds, elephants etc.

Dispersal bestows the following advantages:

(a) By supplementing natality and mortality, shapes population growth and density.

(b) Colonization of new or depopulated areas leading to density equilibrium.

(c) Genetic variability through interchange between populations.

(d) By increasing range of species, enhances genetic survival over evolutionary time.

Dispersion is the way in which individuals within a population are distributed at a particular time and space. The distribution of individuals in a population follows three broad patterns namely random distribution, uniform distribution and clumped distribution (the most common in nature).

Age distribution: The relative proportion of individuals belonging to different age groups of a population is called age distribution or age structure. It is an important attribute of population which determines current reproductive status and indicates future projections. Large proportion of young ones indicates a growing population, large proportion of old individuals represents a declining population and a stationery population has equal representation of young and old individuals.

Ecologically three age groups are recognized namely pre reproductive, reproductive and post reproductive. Most insects have a long pre reproductive stage and most vertebrates have a long reproductive period.

Patterns of population growth: Most of the time, natural populations are in a state of dynamic equilibrium wherein the changes in numbers are not easily detectable by mere observation. Or the populations may reach epidemic proportions and then crash. Conversely there may be a gradual decline due to habitat deterioration and other factors leading to threatened, endangered or extinct status. Factually rate of population growth is due to cumulative effects of natality, mortality and dispersal; all the three are influenced by several factors. Each factor is complex and is inter-related with other factors.

Two growth patterns are exhibited by animals-**logistic** and **eruptive**. Both resemble in having an initial slow phase of growth followed by a period of rapid geometric and exponential increase. Following this phase, the growth styles diverge. Eruptive growth patterns are characterized by sudden, catastrophic mortalities after the rapid increase phase while logistic growth curves level off gradually to a stable asymptote. There are many variations of the two patterns. In general insect and vertebrate pest populations in unstable environments exemplify eruptive growth pattern while logistic grow patterns characterize simple life histories in stable environments.

A given population in a particular environment will have a specific growth rate of individuals. This rate is determined by environmental factors and certain innate characters of the organism itself. The actual rate of increase (*r*) changes in response to changes in environment and changes within population with respect to age distribution and genetic composition. The innate capacity for increase (*rm*) can be determined in the laboratory by eliminating the unfavourable effects of environment.

Population growth is a result of interaction between **biotic potential** and **environmental resistance**. Biotic potential is the inherent capacity of an individual to reproduce and survive. Environmental resistance refers to the sum total of environmental limiting factors that prevent the biotic potential from being realized wholly.

FACTORS INFLUENCING PEST POPULATIONS

The rate of change in pest populations is determined by the fecundity, speed of development and survival among the members. The environmental factors that are favourable to fecundity, development and those which are not inimical to survival promote increase and those having a reverse effect result in declined populations. A factor responsible for growth of a population may result in negative growth of another population. Thus, it is necessary to consider the various factors that effect each pest population.

The environmental factors can be grouped into **abiotic** and **biotic** categories. The abiotic factors are physical and include temperature, light and moisture while biotic factors comprise food and other populations, primarily the natural enemies of the pest populations. The abiotic factors also affect pest population indirectly by modifying biotic factors.

ABIOTIC FACTORS like temperature, moisture, light and wind influence the development, growth, reproduction and fecundity of all animals including vertebrate pests. Extreme weather conditions directly kill pests.

Temperature: Both birds and mammals to which group most of the vertebrate pests belong are "homeotherms" i.e. capable of maintaining constant body temperatures by metabolic adaptations, behavioural changes and selecting optimal microhabitat. The upper lethal limits are usually 40-50°C and the lower varies widely and may also lie below 0°C. If the change in temperature is gradual homeotherms acclimate. If it is too cold and extends over a period of time some mammals 'hibernate' or enter winter sleep in temperate and Polar Regions. Similarly tropical poikilotherms (mostly invertebrates, fishes, amphibian and reptiles) 'aestivate' to tide over high temperature of tropical summers. Sudden seasonal fluctuations result in mortality.

Moisture: Optimum moisture is essential for metabolic reactions and to maintain the balance of salts in body fluids. Most vertebrates obtain their water requirements by way of food and its products, as well as by drinking. The loss of water from body is prevented by both metabolic and behavioural adaptations such as reduced urinary discharge, sweating and choosing microhabitats that can prevent water loss.

Light is an essential ecological factor, mostly by way of duration of day length or photoperiod. Most mammalian pests are nocturnal and bird pests are diurnal, thus photoperiod influences their behaviour considerably. A number of mammalian and bird species migrate seasonally for feeding and breeding purpose. Changes in photoperiod are a vital factor influencing the seasonal migration. Photoperiod also regulates the onset and termination of breeding activities in birds and mammals.

Air currents are of great value affecting bird flight influencing dispersal and migration.

Edaphic factors include the structure, texture and composition of soil, its physical and chemical characteristics. Many of the soil characteristics such as texture, moisture, drainage, chemical composition and topography affect the distribution and abundance of mammals living under soil such as rodents, hares, etc. Loamy soil is ideal for digging and burrowing. Several desert rodents have evolved into burrow diggers even in deserts.

BIOTIC FACTORS: Both *interspecific* and *intraspecific interactions* affect the growth, nutrition and reproduction of animals. The interdependence between pests of the same or other species or plants may exist throughout their life or may be casual and temporary. Such a relationship may be beneficial to both, harmful to both or beneficial or harmful to one and neutral for the other. Knowledge about the interaction of vertebrates for that matter any pest with other animals apart from its food source is essential to understand the ecological niche of the pest species and to develop appropriate management programmes. The three important interactions that are important for pest management are: *Interspecific competition, predation* and *parasitism.* These will be dealt in later.

POPULATION THEORIES

Populations-plant or animal are characterized by two principles: (a) The abundance of a species varies from place to place with favourable habitats having denser populations and unfavourable habitats supporting lesser number of pests. (b) Populations do not increase unlimitedly. In fact they are regulated and number of individuals in a population is prevented from reaching unlimited numbers by natural processes. Five theories are proposed to explain natural regulation of population.

1. Biotic School was first postulated by Howard & Fiske (1911) based on their research on parasites of two species of moth. They categorized agents of population control into two types namely facultative agents and catastrophic agents. Facultative agents exert a relatively severe restraint on increasing populations. Basically these are predators, parasites and diseases. The catastrophic agents act independent of population density and include unfavourable agents of the environment like storms, extreme temperature and other climatic factors. It was concluded that only facultative agents are capable of regulating excessive numbers of pest insects. Parasitism is the most effective in this process, disease acts rarely, that too when densities are very high, and starvation has no effect at all.

Based on a series of laboratory experiments on sheep blowfly, Nicholson (1933) ruled out the role of climate in population density. He strongly maintained that only biotic factors i.e. competition for food, habitat, competition between parasites and between predators can control population. Smith (1935) postulated that different species of animals tend to have average densities and the variations from this numbers are stable. He renamed facultative agents as density-dependent factors and catastrophic factors as density independent. He maintained that density-independent factors can never maintain the average density of a population and only under exceptional conditions climate acts as a density-dependent factor.

Based on his studies on bird populations, Lack (1954) supported Nicholson's theory. He excluded climatic factors in the regulation of population density of birds. Although unusual weather causes high mortalities, bird populations recover fast. Climate only helps in distribution of avian population and increasing their range. He asserts that food shortage, predation and diseases as the three density-dependent factors influencing bird populations. Of these food shortage is the most important factor in support of which he lists the following four reasons. (a) Very few adult birds die due to predation and disease. (b) Usually birds are more abundant in areas where there is greater food supply. (c) Since food acts as a limiting factor, birds have evolved species specific food preferences to avoid competition. (d) Bird migration is usually for food and takes place mostly in winter.

2. Climatic school: Based on the effect of low temperature on egg laying, development rates and large scale mortality of insects, Bodenheimer (1928) proposed that insect populations are primarily controlled by weather. Uvarov (1931) supported this view and proposed that all animal populations are unstable and unlikely to reach a stable equilibrium ever. Mostly climatic factors maintain insect densities.

Three hypotheses emerged from early work on climatic effects namely weather strongly influences population parameters, there is a correlation between insect outbreaks and weather factors and fluctuation in insect population is a major issue than the stability.

Andrewartha & Birch (1954) made the climatic theory very strong. They rejected the demarcation of abiotic and biotic factors as well as density-dependent and density independent factors in population regulation. According to them action of all environmental factors is density dependent and there is no existence of density–independent factors. They considered three means by which population is limited namely (a) shortage of food and shelter. (b) Inaccessibility of food and shelter due to limitations in dispersal and foraging. (c) Shortage of time when rate of increase (*r*) is positive. They considered the last as the most important and the first least important under natural conditions. Fluctuations in the rate of increase are due to weather, predator or any environmental factor.

3 Biotic and climatic approach: There also exists a comprehensive view that both biotic and abiotic factors operate to regulate density under different environmental conditions (Huffaker & Messenger, 1964). Under favourable environmental conditions density-dependent factors will determine the numbers. Such a condition is seen in the centre of specific range as well as in stable habitats of a species. e.g.: Birds. Density–independent factors become predominant in determining numbers in unfavourable and fluctuating environment. This situation is seen at the edge of a species distribution or in unstable, temporary habitats. e.g. insect populations described by Andrewartha & Birch (1954).

However, Milne (1958) emphasized combined action of density independent and imperfectly density-dependent factors in population regulation. In rare cases where such a combined action fails, unchecked increased numbers leading to mass suicide is prevented by competition between the individuals of the population. In contrast, decrease in numbers to zero is prevented mainly by density–independent factors.

Environmental factors regulating species numbers were categorized into three types viz. (a) density–independent factors which in turn include (i) physical factors mainly climate. Action of other species such as indiscriminate browsing, fouling, grazing, trampling of vegetation or occasional parasites or predators (ii) Imperfectly density–dependent factors include action of other species competing for the same resources (interspecific competition), parasites and predators. (iii) Density–dependent factors which include only competitors for food, shelter etc within the population i.e. intraspecific competition (Milne, 1958). Further density dependent factors were distinctly classified into perfect and imperfect categories and the density–dependent and density–independent factors were combined to explain population regulation. Because ecosystem was not taken into account as a whole this theory was not accepted.

4. Self regulation school: The theories of population regulation described so far considered only extrinsic factors namely climate, food, habitat, predators and disease and assumed that identical individuals comprised a population. A second school of thought maintains that individuals within a population differ from each other in terms of physiological and endocrine changes that occur due to changes in population density. Such genetic influences were termed intrinsic factors.

Based on his long term studies on population cycles of voles, Chitty (1960) postulated that populations are self regulating through changes in the physiology and reproduction of individuals resulting from changes in population density. He postulated that "All species are capable of regulating their own population densities without destroying the renewable resources of their own environment, or requiring enemies or bad weather to keep them from doing so". According to him, as the density increases, the interactions between members of a population increases. These increased interactions lead to overall deterioration in the quality of population and decline in reproductive rates. In course of time populations cease growing and density declines. This is followed by a reversal of changes-enhanced reproduction, increased numbers and the cycle becomes repetitive. Chitty (1960) postulated that increased densities lead to reduction of fecundity, increased susceptibility of offspring to environmental hazards leading to high mortalities, enlarged spleens, haemolytic anemia, reduction in the size of adrenals and thymus resulting in population decline. Wellington's (1960) work on the Western tent caterpillar substantiated general decline in the vitality of individuals as populations aged and progressive deterioration occurred in individuals in successive generations.

5. Evolutionary concept: Genetic properties of the organism in a population are also known to influence density. Pimentel (1961) explains such evolutionary changes through the genetic feedback mechanism. This mechanism operates through genetic changes in interacting populations brought about by density and selective pressure of such populations. In herbivore–plant interaction, animal density affects the selective pressure on plants influencing the genetic make up of the plant which in turn influences animal density. Thus the evolution and regulation of animal densities is the result of action and reactions of interacting populations of different animal species as well as population of animals and plants. A fine example for this scheme of population interactions is the case of myxomatosis virus, albeit a case of animals vs. animal populations. This South American virus when introduced into Australia to contain high densities of rabbit populations was exceedingly efficient in containing the pest species. Since its introduction in 1951 both the virus and rabbit are undergoing evolutionary changes. The virulent virus is mutating into milder forms killing fewer rabbits and takes longer time to kill. Meanwhile rabbits are also evolving resistance to virus, with the result the mortality of 98% in the aftermath of virus introduction has come down to 25% now. This turn of events demonstrates the role of evolution in abundance and also warns against continual introduction of new species into distant areas (Fenner & Ratcliffe, 1965)

ECOLOGICAL ASSOCIATIONS AT POPULATION LEVEL

Such associations are of three basic categories namely *intraspecific* (between individuals of the same species), *interspecific* (between individuals within populations of two different species) *and between different trophic levels* (Speight *et al.,* 1999). Such associations may be mutually beneficial (mutualism) or at the expense of one of them (predation, parasitism). Predation and parasitism have been studied more elaborately than other associations for their apparent role in pest management. Of course there are possibilities of manipulating many a competitive and mutualistic association for regulating pest populations.

Predation broadly includes all interactions wherein one species benefits at the expense of a second species by procuring nutrients directly from the second species. This definition includes pathogenesis and parasitism as well as the traditional consumption of one animal by the other (Dhaliwal & Arora, 2001).

As predation reduces prey numbers to some extent, this association has been manipulated over the centuries to manage pest populations. E.g. cat as a predator for commensal rodent control, mongoose to control rats in Hawaiian sugar cane fields etc.

Natural enemies: Not much work is done about the role of natural enemies in regulating vertebrate pest populations. The potential of predator control is obviously limited to smaller mammals notably rodents as they have well defined predators in the form of snakes, carnivorous birds predominantly owls and small mammals such as cats, jungle cats, jackals, mongoose, etc. To arrive at the concept of predatory control of rodent pests it is essential to quantify growth rates under a range of natural enemy levels. The ratio of pest: predator under field conditions assists in predicting trends in pest populations.

Inter specific competition: As a rule two closely related species do not occupy the same niche because it is impossible to co-exist and survive in the same habitat competing for similar food and shelter. If it is inevitable, for them to occur together for reasons unknown to us, there are three adaptations. They can exploit different resources or stay active at different times or they displace one another. Crop raiding is during daytime by Nilgai and deer but nocturnal in elephants. Granivorous birds are diurnal, rodent pests are nocturnal. Monkeys and squirrels, although diurnal and herbivorous, show some kind of dietary specialization. This ecological exclusion of a closely related species is known as competitive exclusion principle or Gause's hypothesis, after the Russian biologist (Gause, 1934) who postulated, "As a result of competition, two similar species scarcely even occupy similar niches, but displace each other in such a manner that each takes possession of certain peculiar kinds of food and modes of life in which it has an advantage over the other, its competitor". One established case of competitive exclusion amongst Indian rodents is the case of *Bandicota bengalensis*, a field rat gradually replacing *Rattus rattus*, the commensal rodent in Mumbai, Kolkata and even Bangalore (personal observation). Similar exclusion is achieved by *Rattus norvegicus* in United Kingdom with *Rattus rattus* being the replaced species. As complete competition cannot co-exist, there is great potential to use competitive exclusion to keep away a pest from its food source.

Mutualism is generally found in organisms having different ecological requirements. There are possibilities to reduce pest densities by using mutualists.

MEASUREMENT OF PEST POPULATIONS

It is essential to know population density to measure the intensity of a pest, to understand the influence of natural enemies on the population, to assess crop losses, monitoring pest populations and to adapt suitable techniques for pest management. They are also helpful in understanding factors affecting density and the functioning of life system of the pest. The pest populations are subjected to two types of studies:

(a) *Extensive studies* over large areas help in understanding distribution pattern of pests, to predict likely damages, to plan and initiate control measures and to correlate pest densities with climatic and edaphic factors. Such studies are carried out in a pre designated area once or at the most few times during a season.

(b) *Intensive studies* involve repeated studies in a given area. The main objective is to determine the contribution of various age intervals to the over all rate of change or dispersal of the species. Here the number of successive developmental stages is counted; life table and budget are prepared for determining key factor(s).

Reliable sound methods of population estimation are very vital to understand pest populations. This involves two important considerations namely the stage in the life history of a species at which density has to be measured and secondly the actual process of counting. Although it is desirable to count all the individuals of a population, logistically it is impossible to do so. This necessitates some method of sampling. A successful and satisfactory sampling has to take into account the nature, size, number and mode of sampling.

Nature of sampling: Although mechanism of sampling depends on the life history and habits of a pest, it is vital to decide the timing of sampling. It is advisable to do sampling when the population is at its peak. Regular and frequent sampling is necessary for intensive studies designed for the construction of life tables. The time of the day is also important for sampling and has to take into account whether the species is diurnal or nocturnal.

If the sampling is taken to determine timing of control measures, the timing should be such that it should give information in advance of an outbreak.

Since the estimation of population at any one point in the life history of a pest gives the information on the total effects of various factors on all the earlier stages in the generation, it is desirable that population levels be estimated at as many suitable points in the life cycle as desirable or practical.

Size of the sample should meet the following criteria (Morris, 1960):

- The size of the sample should be such that all units of the habitat have an equal chance of selection.
- The sampling unit should be easily delineated in the field.
- The sample size should balance variance and cost.

Cost includes man hours required to sample, count and time spent in moving from sample to sample. The sample size also depends on the spatial distribution of sampled organisms.

Number of samples: For practical purposes, an error of ten per cent is usually accepted which means at the chosen probability level the estimated mean will have the chance of being within ten percent of the true mean. In a Homogeneous habit the number of samples required is obtained by the formula.

$$N = \left\{ \frac{ts}{D\overline{X}} \right\}^2$$

Where,

N = No. of samples

t = it is a constant, depending on the number of samples and is obtained from the statistical tables for 'student t' (for more than 10 samples, t approximates to 2 at 5% level).

s = Standard deviation

D = Desired level of accuracy expressed as a decimal (normally 0.1)

$\overline{X}$ = Mean No. of individuals per sample.

Generally a fixed number of samples have to be taken at the time of sampling each population. However, when sampling is done in order to undertake control against a pest, sequential sampling is advised where number of samples taken is variable. Again at very high or very low densities very few samples are adequate which saves time and effort. However to adapt any of these methods, extensive knowledge about the distribution pattern, tolerable levels of densities and levels that are associated with crop damage is essential (Dhaliwal & Arora, 2001).

METHODS OF POPULATION ESTIMATION

Three broad categories of population estimation occur namely absolute estimates, relative estimates and population indices. The methods are described keeping rodents in mind as they are the only group amongst vertebrate pests that are conventional pests and also easily trappable, trapping being the basis of most of the methods of population estimation of rodents. Density of populations can be measured as absolute estimates, relative estimates and as population indices.

A. Absolute estimates: The total number of pests per unit area (e.g. hectare) represents absolute population. Density of the population is the numbers per unit of the habitat (hectare). Absolute population and population density are essential for preparation of life tables, to study population dynamics in fields and to calculate natality and mortality rates. The estimated population densities also indicate relationship between density of pest and its natural enemies. One of the popular methods of estimating absolute population level is:

Capture, mark, release and recapture: Here a sample of the population is captured, marked, released into their natural population and the sample is captured again. The population density is calculated based on the proportion of marked individuals in the total catch. The total population is estimated using the formula

$$P = \frac{Mn}{m}$$

where,

P = Total population

M= No. of animals marked, released

n = Total number of animals caught

m = No. of marked, recaptured animals

The limit within which the population lies (95% confidential limits) is calculated as below:

$$S.E. = \sqrt{\frac{M^2 n (n - m)}{m^3}}$$

Rodents can be marked by toe clipping or by cutting notches in the ears. Clipping gives a range of numbers from 1-10 on hind foot and from 20 to 90 on forefeet. The ears can be clipped to denote 1, 2, 4 and 7 in one ear and 10, 20, 40 and 70 in the other. By a combination of two notches it is possible to get numbers upto 121.

For successful estimation of animals by capture–mark–recapture method, the following criterion should be met:

- Marking should not have any impact on behaviour
- Marks should be weather proof and last till recaptures
- Marked individuals should be able to mingle with their population easily.
- Marking procedures should not in anyway affect the chances of recapture of marked individuals.
- Sampling must be at discrete intervals
- The population being sampled should be stable and not prone to sudden fluctuations.

B. Relative estimates: Here population is estimated in indiscriminate units, which allows comparison in time and space. Such estimates are useful to study activity patterns of species and to determine the constituents of a polymorphic population. Relative estimates are obtained by catch per unit time or effort and involve use of suitable traps.

(a) Trap index: By trapping animals (N) over a fixed number of nights (n) (a minimum of 3 days) trapping success can be measured as trap index using the formula,

$$\%\ \text{Trap success or Trap index} = \frac{\text{No. of animals caught (N)}}{\text{Total trap nights (n)}} \times 100$$

Generally 10% catch indicates high rodent population.

(b) Regression of daily captures on cumulative captures: From the decline in daily captures when trapping is carried over a period of 6-10 days, populations can be estimated. By plotting the numbers caught each day against the number previously caught a straight line is filled by linear regression using the method of least squares. The point where the line crosses the abscissa gives the population estimate.

(c) Change–in–Ratio method (CIR): Here the activity of the target animal is measured before and after trapping. It is termed change–in–ratio because some attribute of the population activity in case of rodents is changed consequent to removal of a part of it. Tracked tiles are kept for a night in the study area before and after intense trapping for 6-10 nights to measure activity. The rat population is calculated using the formula:

$$P_1 = \frac{T_1 \times n}{T_1 - T_2}$$

Where T_1 is the proportion of tiles scored positive before removal trapping.

T_2 is the proportion of positive scores after trapping n is the number of rats removed

P_1 is the original rat population

C. Population indices

(a) Census baiting: In this method, surplus amount of food is made available in big containers in the home range of pest populations for 8-10 days continuously. Baits are replenished every night. Daily intakes are plotted over the days. The consumption is initially low but gets steady after 4-5 days. The average removal for the last three days is taken for calculating rodent density, which is calculated by dividing the amount removed by the known average daily intake of food by the species under study. The method can be adapted only for indoor situations, harbouring a single species.

Alternately, food in open containers is exposed at regular intervals in the path/movement areas of rodents. Assuming rodents consume 10% of their body weight, the total number of rodents living in the given area can be calculated. Alternatively percent rodent activity is calculated by

$$\text{\% rodent activity (positive track patches or food intakes)} = \frac{\text{No. of positives patches or food intake points}}{\text{Total No. of patches laid/food point}} \times 100$$

(b) Tracking patches and tracking tiles: The activity index on tracking tiles can be used as a reliable population index. Approximately the same number of tracking tiles should be set for one or two nights every month and the number of positive tracks should be scored.

Similarly, setting of fixed number of traps each month in residential premises/storage facilities/shops will give estimates of catch/ unit time.

In a second method, patches in the size of 15x35x0.5cm are laid using fertilizer dust, talcum powder, gypsum, China clay, Fullers earth, etc. along the walls, beside the door, across the runways. Patches with foot prints are recorded as positive.

LIFE SYSTEM CONCEPT

The life system is a sub-system of the ecosystem comprising only such components which affect the population of a given species. The subsystem consists of primary and secondary ecological events. Birth, death, emigration and immigration, which directly determine the characteristics of a population, are called primary events. Secondary events are those which influence the magnitude, extent and frequency or duration of primary events such as quality and quantity of food available, weather, topography, etc. Primary and secondary ecological events can be additive or subtractive either reducing or increasing the number of pest species.

It is an established fact that in a given life system, some "key" ecological processes with age specific effect determine major trends compared to other processes. "Life tables" provide information on population density at different times during the life cycle and age–specific information on "key" ecological factors that influence mortalities of a population. A life table describes the number of deaths, survivors, rate of mortality and expectation of further life for successive age intervals of a population. They help in understanding changes in pest populations during various stages of development, throughout their life. It enables to determine the mathematical relationship between various environmental factors and aids to identify the key factors that contribute to changes in the size of populations.

Life tables are of two categorizes (Southwood, 1978):

(a) *Age-specific (horizontal) life table* is based on the fate of real cohort, the individuals of a population belonging to a single generation.

(b) *Time-specific (vertical) life table:* This is based on the fate of an imaginary cohort which is established by determining age structure in populations where there is considerable overlapping of generations. For time specific life tables age determination is a pre-requisite.

The difference in number of individuals of a species at different stages of development is due to mortality and dispersal. The developmental stage at which mortality is maximum is the critical age interval and the environment factor(s) that induce such mortalities are the key mortality factor(s). The generation survival is explained by the proportion of individuals in the first and last developmental stage. On the other hand, trend index represents the proportion of individuals at the same developmental stage in two successive generations.

DEMOGRAPHIC PARAMETERS RELEVANT FOR PEST MANAGEMENT AND CONSTRUCTION OF LIFE TABLES

(a) *CARRYING CAPACITY OF THE ENVIRONMENT (K)* is determined using the formula

$$K = \frac{b - d}{1}$$

where b = constant rate of reproduction

d = constant rate of death

(Christian & Fenchel, 1977)

(b) *INSTANTANEOUS RATE OF POPULATION INCREASE* is calculated by

$$r = \log \frac{(N_2)}{N_1}$$

Where, N_1 = Minimum number of rodents known to be alive in one week

Where, N_2 = Minimum number of rodents known to be alive 2 weeks later

(c) *MAXIMUM RATE OF POPULATION INCREASE (R)* is derived by the formula

$$R = \frac{(N - N_{t-1})}{N_{t-1}}$$

Or

$$N_t = N_{t-1} + R\,N_{t-1}$$

Where N = Population size per unit area

R = Individual rate of increase per unit of time

t = Time period (2)

N_{t-1} = Time period (1)

(Berryman, 1981)

(d) *MAXIMUM INDIVIDUAL RATE OF INCREASE* is calculated as

$$R = R_m - sn$$

Where R_m = Maximum individual rate of increase

S = Slope of the curve

N = Density

R = Realized per capita rate of increase

(Berryman, 1981)

(e) **Survival rates** are calculated as minimum rates per 14 days by determining the number of animals caught in week n that were known to be alive 2 weeks later to the total number caught in week.

(f) **Recruitment rate** is proportion of all animals caught in a particular trapping that are new captures.

(g) **Birth rate** includes immigration and is more often correctly called dilution rate and is given by

$$t = \frac{B}{1 \to 2} = \log \left\{\frac{r_{21} n_3}{n_2 r_{31}}\right\}^{t_2}$$

where, t_2 = the time interval between the 2nd & 3rd sampling

n_2 = the total no. of animals captured on the 2nd day

r_{21} = the no. of animals that had been captured on 2nd day that had been marked on the first

r_{31} = the no. of animals captured on the 3rd day that had been marked on the first

n_3 = the total no. of animals captured on the 3rd day

(h) **Death Rate** also called loss rate which is inclusive of Numbers actually dying and the numbers immigrating and is given by

$$t = O \rightarrow = -\log_e \frac{1}{\left\{\dfrac{1-r_{31}}{a_1 r_{32}}\right\}} t_1$$

where t_1 = the time interval between 1st and 2nd sampling

a_2 = the no. of newly marked animals released on 2nd day

a_1 = the no. of newly marked animals released on 1st day

r_{31} = the no. of animals captured on the 3rd day that had been marked on the first

r_{32} = the no. of animals captured on the 3rd day that had been marked on the second

(i) **Survivorship curves:** The nos. of living at a given age (l_x) are plotted against age (x)

Convex : low mortality throughout life

St. line : constant rate of mortality

Concave : high infant mortality

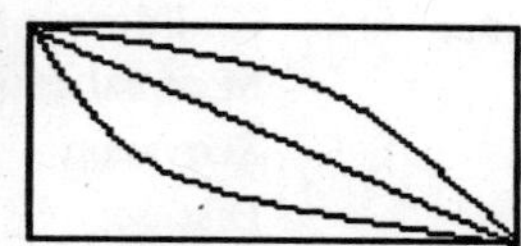

LIFE TABLES can be of two types: (a) Time specific life table and (b) Age-specific life table

(a) Time specific life table

X	l_x	d_x	O_x	e_x	L_x	T_x
1.						
2.						
3.						
4.						

Where x = age class

1_x = No. surviving at the beginning of the age interval out of 1000

d_x = No. dying in the age interval out of 1000

O_x = Mortality per 1000 at the beginning of the age intervals: $(1000. \frac{dx}{1_x})$

e_x = expectation of life

Two further columns facilitate the calculation of e_x

$$L_x = 1_x + 1_{(x-1)} \text{ and } T_x = L_x + L_{(x+1)} + \frac{1}{(x+2)} \text{ etc.,}$$

Then $e_x = \frac{T_x}{I_x}$ (Dempster, 1975)

(a) Age specific life table

To start with the maximum potential production of offspring for each generation i.e., the no. of females of reproductive age multiplied by the maximum number of litter per female is to be calculated. Next the no. of each successive stage in the life cycle is entered.

Age class	Due to Mortality	No. entering stage	No. dying	% dying	% initial no. dying
Adult	(Fecundity) →				
Litter	(....litter per female)				
	Maternal cannibalism				
	Starvation				
	Disease				
Post Weaned	Hunger				
	Predators				
	Intraspecies Aggression				
	Envl. Factors				
4 mo.	Competition				
	Food				
	Envt. Factors				
	Predation				
8 mo.	12mo., 12mo + etc. same as in 4 mo.				

(Dempster, 1975)

K-factor analysis (Varley & Gradwell, 1960) helps in recognizing:

(a) "Key-factor or factors" determining the population trend i.e. those factors which give the best prediction of nos. in future generators

(b) the effect of variations in natality and mortality on the population

(c) the recognition of any density–dependent relationship which exist

Analysis of k-factor from life tables: The maximum potential of off-spring for each generation is entered as the first figure in the table. Next the number of each successive stage in the life cycle is entered and the values are converted to logarithms. A series of "mortalities" can be calculated by subtracting each logarithm from the previous one, i.e.

$$K_1=k_g\ N_i\text{-log}\ N_1H$$

Example

	Number	log number	k-value
Maximum potential			
Natality	12	1.0792	0.0792
Litter born	10	1.0000	0.0458
Post weaned	9	0.9542	0.0511
Juvenile	8	0.9031	0.058
Adult (4 mo.)	7	0.8451	0.0669
8 mo.	6	0.7782	0.1761
12 mo.	4	0.6021	0.3011
12 mo.+	2	0.3010	
			0.7782=k

These are the k-values for each age class. Total mortality is given by the addition of successive k values i.e.

$$K = k_0+k_1+k_2+k_3\ \text{etc}$$

K value and mortality are related as follows

$$K = \frac{1}{\log_{10}(1-u)}$$

(Dempster, 1975)

COMPUTER SIMULATED MODELS OF RODENT POPULATIONS

Initially a simple, deterministic model using crop stage dependent matrices instead of projection matrix should be attempted. Projection preferably should be on a monthly basis to coincide with observed cropping pattern. Attempts should be made to relate population with damage and yield loss at harvest. The different parameters used in model building should be as per the following:

Parameters to be used in the model of a rat population in a cereal crops

Crop stage	Mean no. Tillers/ hill	Monthly fertility rate [a]		Monthly survival rate	
		age 0-2	class 2-12	age 0-2	class 2-12 A[c]
Land preparation	0.001			....( )[b]	...( )[b]
Transplanting	....	....	....		
Max. Tillering	....	....	....		
Booting, Heading	....	....	....		
Maturing, Harvest	....	....	....		
Post harvest	....	....	....		

[a]No. born/rat/month = No. embryos/female x % pregnancy x Sex ratio x 4/3 (3 wks gestation/ mo.)

[b]No. in parenthesis is weight factor (s) for each stage to adjust for proportion surviving rats remaining in the field

[c] Attack rate co-efficient

THE FERTILITY AND SURVIVAL RATE FROM THE BASIC PROJECTION MATRIX

Attack coefficient

Tillers cut = total tiller $(1\text{-}e^{a})^{N}$

Where N = Total number of rats (all ages)

e = base of natural logarithm

a = attack rate coefficient

Number of Rats

N(t) = M(t-1) x n (t-1)

Where n (t) = age-distribution vector at time t

N (t-1) = age distribution vector at time t-1

M (t-1) = the population–projection matrix at time (t-1)

Concept of *r*– and *k*– selection: In general, *r*–selection refers to selection favoring maximum growth in an uncrowded population and organisms adapting to such a condition are *r*-strategists (MacArthur & Wilson, 1967). In contrast *k*-selection is characteristic of survival in a crowded situation and the accompanying adaptations in a competitive atmosphere result in a *k* strategist (Horn, 1988). The following table distinguishes *r*- and *k*- strategy.

Characteristics of *r* - and *k* – strategy

Characteristics	Strategy	
	r-strategy	*k*-strategy
Fecundity	High	Low
Survivorship	Low	High
Competitive ability	Low	High
Relationship between generation Time (T) and habitat stability(S)	$T \geq S$	$T \ll S$

Agricultural crops particularly food crops with short growing seasons and regular disruptions are ideal for *r* selection. In fact many agricultural pest–insects and small mammals show "*r* strategy" namely short life span, greater ability to disperse and high rates of birth. During early conversion of land to agricultural fields, the ecological conditions are unfavourable, where only *r*– strategists can survive and are regulated by density independent factors, mainly environmental factors. After several generations the microhabitat changes, becomes more favourable and *k*- strategy is more suited. In contrast to herbivorous pests, predators and parasitoids take longer time to colonize. This differential capacity to invade an agricultural habitat frequently results in pest outbreaks during early phases of crop growth or after the application of pesticides as the later kill both the prey and predator directly as in insects or indirectly through secondary poisoning in vertebrates.

The differential population regulation of *r*- and *k*– selected species was depicted in a synoptic model by Southwood & Comins (1976). This model represents a theoretical population growth of an insect species beginning as a *r*– selected species in unstable environments, continuing as *k*– selected species in stable environments. The shapes of curves depend on the rate of growth of the population as influenced by the natural enemies of the species. Species with *r*–strategies, colonizing a new habitat outbreed the predation rates of natural enemies while *k*-related species in stable environment facing larger number of enemies invests more in defensive strategies against predation, surviving at least up to the stage of reproducing. This strategic adaptation of *r*- and *k*-strategy of pest species led Southwood (1978) to suggest that classical biological control can be effective only when pests have a life style intermediate between *r* and *k*.

CONCLUSIONS

An effective and sustainable pest management programme should be decided with the objective of modifying the life system of a pest so as to reduce its numbers below the economic threshold level. To achieve this several ecological aspects of the life history of the given species have to be taken into account. This include estimating density of pest population, assessment of crop losses, impact of natural enemies on pest population, monitoring the appearance of pest species and adaptation of suitable control methods. Depending on the density, aggregation and dispersal of population, suitable techniques of estimating populations and sampling methods have to be selected. To arrive at the economic threshold value to take up a control programme, information on relationship between the population of a given pest, time of its appearance, duration of damage to the crop at the susceptible stage and the consequent yield loss is a prerequisite. Construction of life tables help in determining the key mortality factors of a pest species, predicting the appearance of a pest and the most susceptible stage in the life of a pest when control measures will yield optimum reduction in pest numbers.

REFERENCES

ANDREWARTHA, H.G. & L.C. BIRCH. 1954. THE DISTRIBUTION AND ABUNDANCE OF ANIMALS. University of Chicago Press, Chicago, USA.

BERRYMAN, A.A. 1981. POPULATION SYSTEM: A GENERAL INTRODUCTION. Plenum press, New York & London. 282p.

BODENHEIMER, F.S. 1928. Welche Faktoren regulieren die individuenizahet einer Insektenalt in der Nature, *Biol. Zentrolbl.* 48: 714- 739.

BROOKS, J.E., E. AHMAD, I. HUSSAIN, S. MUNIR & A.A. KHAN. 1990. A TRAINING MANUAL ON VERTEBRATE PEST MANAGEMENT. Pakistan Agricultural Research Council, Islamabad, Pakistan 206 p.

CHITTY, D. 1960. Population process in the vole and their relevance to general theory. *Can J. Zool.* 38: 99-113.

CHRISTIAN, F.B & T.M. FENCHEL. 1977. THEORIES OF POPULATION IN BIOLOGICAL COMMUNITIES. Springer-Verlag, Berlin, Heidelburg & New York. 144p.

DEMPSTER, J.P. 1975. ANIMAL POPULATION ECOLOGY. Academic Press, London. 155p.

DHALIWAL, G.S. & R. ARORA. 2001. INTEGRATED PEST MANAGEMENT: CONCEPTS AND APPROACHES. Kalyani Publishers, Ludhiana, India. 427p.

FENNER, F & F.N. RATCLIFFE. 1965. MYXOMATOSIS. Cambridge University Press, Cambridge, UK.

GAUSE, G.F. 1934. THE STRUGGLE FOR EXISTENCE. McMillan Hafner Press, New York.

HORN, D.J. 1988. ECOLOGICAL APPROACH TO PEST MANAGEMENT. The GuilfordPress, New York, USA. 285 p.

HOWARD, L.O & W.F. FISKE. 1911. The importation to the United States of the parasites of gypsy moth and the brown-tail mouth. *U.S. Dept. Agri. Bur. Entomol. Bull.* 91, USA.

HUFFAKER, C.B & P.S. MESSENGER. 1964. The concept and significance of natural control. In: BIOLOGICAL CONTROL OF INSECT PESTS AND WEEDS. (Ed.): P. DeBach Chapman and Hall, London. pp. 74-177.

LACK, D. 1954. THE NATURAL REGULATION OF ANIMAL NUMBERS. Oxford University press, New York.

MacARTHUR, R.H & E.O. WILSON. 1967. THE THEORY OF ISLAND BIOGEOGRAPHY. Princeton University Press, Princeton, NJ, USA. 203p.

MILNE, A. 1958. Theories of natural control of insect populations *Cold Spring Harbour Symp. Quant. Biol.* 22: 253-271.

MORRIS, R.F. 1960. Sampling insect populations. *Ann. Rev. Ent.* 5:243-264.

NICHOLSON, A.J. 1933. The balance of animal populations. *J. Animal. Ecol.* 2: 132-178.

PIMENTEL, D. 1961. On a genetic feed-back mechanism regulating populations of herbivores, parasites and predators. *Am. Nat.* 95: 65-79.

SMITH, H.S. 1935. The role of biotic factors in the determination of population densities. *J. Ecol. Ent.* 28: 873-898.

SOUTHWOOD, T.R.E. 1978. ECOLOGICAL METHODS WITH PARTICULAR REFERENCE TO THE STUDY OF INSECT POPULATIONS. Chapman and Hall, London, U.K.

SOUTHWOOD, T.R.E & H.N. COMINS. 1976. A synoptic population model. *J. Anim. Ecol.* 45: 949-965

SOUTHWOOD, T.R.E & M.J. WAY. 1970. Ecological background to pest management. In: CONCEPTS OF PEST MANAGEMENT (Eds.): R.L.

Rabb & F.E. Guthrie. North Carolina State University Press, Raleigh, USA. pp. 6-29.

SPEIGHT, M.R., M.D. HUNTER & A.D. WATT. 1999. ECOLOGY OF INSECTS: CONCEPTS AND APPLICATIONS. Blackwell Science Ltd., Oxford, UK.

UVAROV, B.P. 1931. Insects and climate. *Trans. Ent. Soc. Lond.* 79: 1-247

VARLEY, G.C. & GRADWELL, G.R. 1960. Key factors in population studies. *J. Anim. Ecol.* 29: 399-401.

WELLINGTON, W.G. 1960. Qualitative change in natural populations during changes in abundance. *Can. J. Zool.* 38: 289-314.

2

GRANIVOROUS BIRD PESTS AND THEIR MANAGEMENT

*V. Vasudeva Rao and O.P. Dubey**

AINP on Agricultural Ornithology, Acharya N.G. Ranga Agricultural University, Rajendranagar, Hyderabad-500 030
**Assistant Director General (Plant Protection), Indian Council of Agricultural Research, Krishi Bhavan, New Delhi-110 001*

ABSTRACT

A variety of crops such as: cereals, millets, pulses, oilseeds, fruits and vegetable crops are depredated by birds during their vulnerable stages of growth mainly at dough and maturity stages. Even though, only 2% of the available bird species are depredatory in nature, the magnitude of damage caused by them is significant and needs immediate attention. In India, a total of 63 species of birds belonging to 19 families have been identified damaging several crops. The number of bird species that affected various crops was: cereals-52, pulses-14, oilseeds-15 and fruits-24. Amongst the 46 species of beneficial birds, which devoured insects and rodent pests, all fed on insects while six of them also consumed rodents. Twenty eight species of birds that inflicted damages to crops and fifteen of the beneficial species were omnivorous. The bird pests included common pea fowl, Sarus crane, Demoisella crane, Blue rock pigeon, Ring dove, Rose ringed parakeet, Slaty headed parakeet, Small green barbet, Rosy pastor, Common myna, Bank myna, House crow, Redvented bulbul, Jungle babbler, House sparrow, Baya weaver, Black throated weaverbird, White throated munia and Spotted munia. Composition of the species varied with season, cropping pattern, roosting sites and other climatic factors. Studies on the assessment of crop losses due to bird pests have given a varied picture because of different methods of damage estimates. Cereals damaged were paddy (0.6-22.9%), wheat (0.5-37%), maize (6-39%), pearl millet (0.5-26%) and sorghum (0.5-44%). Damage to pulses was negligible but oil seeds were affected seriously viz. 0.5 to 40% to sunflower, 1-36% to groundnut, 2-15% to safflower and soybean, and 3-30% to oil palm. Fruits depredated significantly were: mango (3-10%), pomegranate (2-20%), grape (2-27%), apple (5-13%), banana (14%), kinnow (3-16%) and papaya (6-50%). Effective management strategies evaluated and recommended are: wrapping maize cob with leaves, reflective ribbon, bioacoustics, pyro-techniques, screen crops, botanicals, habitat manipulation, netting and IBPM (Integrated Bird Pest Management) which are also farmer friendly.

INTRODUCTION

In India, there are about 1200 species of birds representing 20 orders. Of these, only about 25 species (2%) depredate crops. Even though only a few of them have so far become serious pests of crops, their population levels are so high so as to cause considerable economic losses to the major crops grown in India. There is a tendency among the granivorous birds to form large flocks for feeding and most of the major pest species have a well-developed foraging behaviour and a tendency to form large communal roosts. In the absence of natural wild food, these birds shift their feeding guild to the cultivated crops that are available to them in their habitat.

Ecological aspects of pestilence caused by birds to cereal crops have so far been studied only to a small extent by ornithologists in India. Whistler (1949), Dharamkumarsinhji (1956) and Ali (1971) mentioned house sparrow, baya, parakeet peafowl, etc. as serious bird pests. Sharma (1972) gave an account of damages caused to crops by the peafowl. Fitzwater & Prakash (1973) concluded that the house sparrow and the baya as most serious bird pests to crops. Doves, house sparrow, white throated munia and parakeet were the crop pests in the Indian desert (Rana, 1973).

HISTORICAL PERSPECTIVE

Although the beginning of ornithology in India was made in the mid-nineteenth century with the publication of the first book on Indian birds (Jerdon, 1863), research on applied aspects of birds including their relationship to agriculture was largely ignored until early twentieth century. Mason & Lefroy (1912) were probably the first to make a serious attempt in this direction by studying food of some birds in an intensively cultivated area at Pusa (Bihar) between 1907 and 1909. Later, D'Abreu (1920) described food of birds from central India and Hussain & Bhalla (1937) gave a brief description of food habits and importance of over 90 bird species affecting agriculture. The first bulletin advocating the role of birds in agriculture was published by Susainathan (1921).

Ali (1936, 1949 and 1971) stressed the need for undertaking serious studies in economic ornithology including the relationship of birds with agriculture from time to time. Nearly five decades ago, he wrote, "Economic ornithology is nowhere in the picture as yet, but it is a subject that should not escape the planners of our destiny in the greater and more prosperous India to be; and sooner a beginning can be made, the better for the cultivator" (Ali 1949). This beginning was made in the 1950s and 1960s when research on bird damage estimation and control was initiated selectively (Rao & Rao, 1953, Singh & Dungan, 1955; Varma *et al.,* 1960, Basu & Pramanik, 1964). The Government of India under the Indian Council of Agricultural Research (ICAR) launched the All India Coordinated Research Project on Agricultural Ornithology at six centers viz., ANGR Agricultural University, Hyderabad; Gujarat Agricultural University, Anand; Maharana Pratap University of Agriculture & Technology, Udaipur; Indian Agricultural Research Institute, New Delhi and Punjab Agricultural University, Ludhiana in 1983. Eventually, two more centers were added, one each at Kerala Agricultural University, Thrissur and Dr. YSP University of Horticulture & Forestry, Solan, during IX plan period. The last two decades of intense field studies in these centers has helped to identify key depredatory bird species and estimate the damage to important cereals, pulses, oil seeds and horticultural crops. A number of novel and field oriented technologies to minimize bird damage to the crops have been evolved. The other important thrust area was to identify the beneficial birds, which help in the control of insects and rodent pests and to develop methods to encourage them to derive maximum benefits from their activities.

Current status of granivorous birds in Agricultural eco system in India

The world's number one avian pest is Red-billed Quelea of southern Africa, while Roseringed Parakeet is number one avian pest of India (Ali & Ripley, 1983; Dhindsa *et al.,* 1993). In Punjab, the most intensively cultivated state in India, the rose ringed parakeet inflicts heavy damage to maize, sunflower, mustard, guava, almond, peach, *ber*, etc (Ramzan & Toor, 1972, 1973; Dhindsa *et al.,* 1992).

Table 1. Feeding guild of depredatory birds in different crops

Crops	No. of spp.	Feeding guild of birds					
		Carnivorous	Insectivorous	Frugivorous	Nectivorous	Granivorous	Omnivorous
PADDY	52	5	17	1	4	14	11
WHEAT	36	2	14	1	3	8	8
MAIZE	48	3	23	1	3	9	9
PEARL MILLET	39	2	20	1	2	9	5
SORGHUM	53	3	22	1	3	11	13
PULSES	14	-	4	-	-	7	3
OILSEED	15	-	10	2	-	1	2
FRUIT CROPS	24	2	13	5	1	-	3

Investigation on both basic and applied aspects of birds as harmful and beneficial components in agricultural ecosystems has generated valid information that has helped to evolve technologies for their management. A total of 63 species of birds belonging to 19 families have been identified damaging several crops (Table 1). The number of bird species that affected various crops was: cereals-52, pulses-14, oilseeds-15 and fruits-24. Amongst the 46 species of beneficial birds, which devoured insects and rodent pests, all fed on insects while six of them also consumed rodents. Twenty eight species of birds inflicted damages to crops and fifteen of the beneficial species were omnivorous. Omnivorous birds have a dual role in our agro-ecosystem.

Important depredatory birds of Agricultural crops (Table 2 & Plate 1)

Common Peafowl (*Pavo cristatus*): Male has blue neck and breast and long, trailing bronze, green upper tail feathers ending in blue green. Female duller, lower neck metallic green instead of blue. **Distribution :** Throughout the Indian union, upto 1800 m in the Himalayas. **Habits :** Fairly common in cultivated areas and in light wooded habitats and gardens. Male keeps a harem. Always excessively shy and alert. **Habitat:** Inhabits dense scrub and deciduous jungles, villages and cultivation. **Food :** Grains, vegetable shoots, insects, lizards, snakes. In recent times, high mortality of these birds was observed in agricultural habitats in different regions of the country due to indiscriminate use of pesticides and chemicals.

Table 2. Crop losses by Birds in different crops and their management

Crops	Stage of crop	Extent of Damage	Bird Depredators	Manage-ment Measures
CEREALS				
Paddy (*Oryza sativa)*	Nursery	0.6-5%	Baya, House Sparrow, White throated Munia, Spotted Munia, Black headed Munia, Black throated Weaver, streaked Weaver, Purple Moorhen, Rock pigeon, Ring Dove, Common Myna, House Crow	A,C,N,CTP
	Ripening	1.4-22.9%	House Sparrow, Baya, Black throated Weaver, Streaked Weaver, White throated Munia, Spotted Munia, Black headed Munia, Roseringed Parakeet, Large Grey Babbler, Common, Babbler, Jungle Babbler, Brahminy Duck, Lesser Whistling Teal, Gargany, Common Teal, Pintail, Purple Moorhen, Sarus Crane	A,C,D,E,RR
Wheat (*Triticum aestivum)*	Sowing and seedling stage	0.5-6%	House Crow, Jungle Crow, Rock Pigeon, Ring Dove, Peafowl, Grey Partridge, Ruff, Black tailed Godwit, Demoiselle Crane, Common crane, Sarus Crane, Black Ibis	A,C,D,E,B
	Ripening stage	0.6-37%	House Sparrow, Roseringed Parakeet, White throated Munia, Spotted Munia, Large Grey Babbler, Jungle Babbler, Indian Peafowl, Demoiselle Crane, Common Crane, Sarus Crane, Short toed Lark, Rufous tailed Finch Lark, Black headed Bunting, Redheaded Bunting, Ring dove	A,C,D,E
Maize (*Zea mays*)	Sowing and seedlings	10-20%	Rock Pigeon, Ring Dove, Spotted Dove, House Crow, Jungle Crow, Common Myna, Roseringed Parakeet, Indian peafowl, Grey partridge	A,B,C,D
	Ripening stage	6-39%	Roseringed Parakeet, House Crow, Large Indian Parakeet	C,D,E,F,SC
Pearl Millet (*Pennis-etum typhoideum*)	Sowing and seedlings	0.5-6%	Rock Pigeon, Ring Dove, Spotted Dove, Red Turtle Dove, Little Brown Dove, Baya, Black throated Weaver, Streaked Weaver, House Sparrow, Grey Partridge, Large Grey Babbler, Jungle Babbler	A,B,C,D
	Ripening	0.5-26%	House Sparrow, Baya, Black throated Weaver, Streaked Weaver, White throated Munia, Spotted Munia, Black	A,C,D, E,N,S

Crops	Stage of crop	Extent of Damage	Bird Depredators	Management Measures
			headed Munia, White backed Munia, Red Munia, Red headed Bunting, Black headed Bunting, Common Myna, Redvented Bulbul, White cheeked Bulbul, Large Grey Babbler, Common Babler, Jungle Babbler, White headed Babbler, Rosy Pastor, House Crow, Roseringed Parakeet, Ring Dove	
Sorghum (*Sorghum vulgare*)	Sowing and seedlings	0.5-8%	Rock Pigeon, Ring Dove, Spotted Dove, Little Brown Dove, Red Turtle Dove, Indian Peafowl, Grey Partridge, Large Grey Babbler, Jungle Babbler, Common Babbler	A,B,C
	Ripening	0.7-44%	Roseringed Parakeet, House Sparrow, Baya, Black throated Weaver, Streaked Weaver, White throated Munia, Spotted Munia, Black headed Munia, Rose Finch, Black headed Bunting, White backed Munia, Red headed Bunting, Ring dove, Rosy Pastor, Common Myna, Bank Myna, House Crow, Jungle crow, Blossom headed Parakeet	A,C,D,E,NS
PULSES				
Black gram (*Vigna mungo)*	Sowing and seedling	0.2-3%	Rock Pigeon, Ring Dove, House Crow, Grey Partridge.	A,C
	Ripening	0.5-2.5%	Roseringed Parakeet, House Crow, Rock Pigeon, Ring Dove, Grey Partridge, Pea Fowl	C,D
Green gram (*Phaseolus aureus)*	Sowing and seedling	0.3-3%	House Crow, Rock pigeon, Ring Dove, Grey Partridge	A,C
	Ripening	0.5-2%	Roseringed Parakeet, House Crow, Rock pigeon, Ring Dove, Grey Partridge, Pea fowl	A,C,D
Chickpea (*Cicer arietinum*)	Sowing	0.5-3%	House Crow, Rock Pigeon, Ring Dove, Grey Partridge, Pea fowl, Sarus Crane, Jungle Crow	A.C
	Ripening	0.5-2%	Rock pigeon, Ring Dove, Indian pea fowl, House Crow, Grey Partridge, Rose Ringed Parakeet	A,C,D
Pigeon Pea (*Cajanus cajan*)	Sowing and seedling	0.2-2%	Rock pigeon, Ring Dove, Indian pea fowl, House Crow, Jungle Crow, Grey Partridge	RR, A,C

Crops	Stage of crop	Extent of Damage	Bird Depredators	Management Measures
	Ripening	0.1-1%	Roseringed Parakeet, House Crow, Jungle Crow	A,C,D
Cow pea (*Vigna catjang)*	Sowing and seedling	0.5-3%	Rock pigeon, Ring Dove, House Crow, Grey Partridge, Indian pea fowl	A,C
	Ripening	10-4%	Roseringed Parakeet, Ring Dove, Rock Pigeon, Grey Partridge, Indian peafowl, House Crow, Small Green Barbet	A,C,PB
OILSEEDS				
Sunflower (*Helianthus annuus)*	Sowing & seedling	0.5-3%	House Crow	A,C,B
	Ripening	10-40%	Roseringed Parakeet	A,C,D,E,BS
Groundnut (*Arachis hypogaea)*	Sowing & seedling	0.5-36%	Rock pigeon, Ring Dove, House Crow, Roseringed Parakeet, Indian peafowl, Black Ibis	A,B,C
	Ripening	1-15%	House crow, Jungle Crow, Demoiselle Crane, Common Crane, Sarus Crane, Black Ibis, Indian Peafowl	A,C,D
Safflower (*Carthamus tinctorius)*	Sowing	2-15%	house Crow	A,C
	Ripening		Roseringed Parakeet	A,C,D,SC
Mustard (*Brassica nigra*)	Ripening		Roseringed Parakeet, Rock Pigeon, Large Grey Babbler, Ring Dove	A,C,D
Soyabean (*Glycine max*)	Sowing & seedling	2-15%	Roseringed Parakeet, Rock pigeon, Grey Partridge	A,C,D
	Ripening		Roseringed Parakeet, Indian Peafowl, Grey Partridge, Rock pigeon	A,C,D
Oil Palm (*Elaels quineensis*)	Ripening	3-30%	House Crow, Jungle Crow, Indian Myna	PB
FRUIT CROPS				
Mango (*Mangifera indica*)	Ripening	3-10%	Roseringed Parakeet	C,D,E,PN
Guava (*Psidium guajava)*	Ripening	3-5%	Roseringed Parakeet, Large Indian Parakeet, Red vented Bulbul, House Crow, Jungle Crow, Small Green Barbet	C,PN,E
Pomegranate (*Punica granatum)*	Ripening	2-20%	Roseringed Parakeet, Large Indian Parakeet, House Crow, Jungle Crow	C,D,E

Crops	Stage of crop	Extent of Damage	Bird Depredators	Manage-ment Measures
Grape (*Vitis vinifera*)	Ripening	2-27%	Purple Sunbird, Red vented Bulbul, Roseringed Parakeet, Koel, Rosy Pastor, Common Myna, Golden Oriole	C,D,E,PN
Date palm (*Phoenix dactylifera*)	Ripening	0.1-1%	Roseringed Parakeet, Bank Myna, Common Myna, Brahminy Myna, House Crow, White throated Munia	C,P,E,PB
Apple (*Malus sylvestris*)	Ripening	5-13%	Slaty headed Parakeet, Roseringed Parakeet, Large Indian Parakeet, Blossom headed parakeet, Himalayan Bulbul, Jungle Crow, Red billed Blue Magpie, Great Hill Barbet, White cheeked Bulbul, Streaked Laughing Thrush, Scaly bellied Green Woodpecker	A,C,E,PB,PN
Banana (*Musa paradisiaca*)	Ripening	19.4%	Small Green Barbet, House Crow, Jungle Crow, Common Myna	C,D,E,PB
Kinnow (*Citrus aurantium*)	Ripening	3-16%	Roseringed Parakeet, Jungle Crow, Indian Treepie, Black headed Oriole, Small green Barbet, Lesser Golden backed Woodpecker	A,C,E,PN
Pecan nut	Ripening	6-50%	Roseringed parakeet, Large Indian Parakeet	C,E,D
Papaya (*Carica papaya*)	Ripening		Koel, House crow, Jungle Crow, Red vented Bulbul, Red vented Bulbul, Brahminy Myna, Small Green Barbet	D,PB

A=scare crows, B = seed treatment, C = reflective ribbon, D = pyrotechniques, E = bioacoustics, F = wrapping, N = netting, CTP = covering bed with dry thorny plants, RR = removing weeds from field edge, NS = Neem spray, SC = screen crop, PB = physical barrier, GC = gunny bag covering, PN = Plastic Nets, BS = block sowing (21 acre land)

Source: *Survey reports of coordinated centres of AINP on Agricultural Ornithology.*

Sarus Crane *(Grus antigone)*: 165 cm: Huge, grey bird with long reddish legs with naked red head and upper neck, Sexes alike, but female slightly smaller and young birds are brownish grey, rusty brown on head. **Distribution:** Resident, subject to local movements, common in Gujarat and Rajasthan, eastwards to East Uttar Pradesh and South Madhya Pradesh and vagrant in North Andhra Pradesh. **Habits:** Normally seen in pairs or family parties or flocks. Well protected in central and western India where it is undisturbed, but habitat loss continues to be a great threat. **Habitat:** Open cultivation in well watered country, marshes, *Jheels*, village ponds, lakes and large rivers. **Food:** Grain, shoots and other vegetable matter, insects, reptiles, fish, frog, crustaceans, etc.

Demoiselle Crane *(Anthropoides virgo)*: **75 Cm: Overall** plumage grey; black head and neck; prominent white ear tuffs; long black feathers of lower neck fall over breast; brownish-grey secondaries are sickle shaped and drooping over tail. Large flocks in winter, often many thousands; Sexes alike. **Distribution:** Winter visitor and passage migrant. Commonest in Northwest India and eastern Rajasthan, Gujarat, Madhya Pradesh, North Karnataka and South Andhra Pradesh though sporadically Orissa. **Habits:** Highly gregarious and sometimes found with Common cranes. Mainly vegetarian; feeds in morning and late afternoon on winter crops and stubble fields, retires for the rest of the day on sand banks of rivers and edges of *jheels*. **Habitat:** winter crop fields, large rivers with sand banks and reservoirs, ponds, *jheel* edges. **Food:** Groundnut, wheat, paddy, gram; does extensive damage to winter crops. Besides grains, it also prefers insects and small reptiles.

Blue Rock Pigeon (*Columba livia*): 33 cm :Adult is greyish, with glistering metallic, purple green on neck, blackish terminal band on grey tail, and short, broad black bars across inner wing. Sexes alike. Flocks and colonies. **Distribution:** Resident throughout much of the subcontinent, except parts of the North-west and North-east. Both feral and wild populations occur. **Habits:** Live in colonies throughout the year and prefer cliffs, rocky hills and ruins. **Habitat:** Large numbers of birds inhabit villages, towns and cities and survive on the grains offered by people. Wild birds occupy cliff, gorges and ruins. **Food:** Feeds chiefly in cultivations; mainly on seeds; also eats green shoots.

Ring Dove (*Streptopelia decaoto*): 32 cm: A pale grey and brown pigeon distinguished by a prominent narrow black half ring or collar on the hind neck. The breast is lilac, which turns into ashy grey on abdomen and darker grey on remaining under parts; broad whitish tips to brown tail feathers, seen as a terminal band when fanned during landing. The bill is brownish, swollen at the base and the feet are dark pinkish red or magenta. **Distribution:** Resident, local migrant throughout much of the subcontinent, except extreme North East Himalaya, commonest in North-west, West and Central India. **Habits:** Locally abundant in the neighbourhood of human habitations, especially cultivated land. Small parties when not breeding, often associates with other doves; strong flyer, large gatherings, glean in cultivated country. **Habitat:** Open dry country with cultivation and groves of deciduous trees, open scrub, dry forest. **Food:** Paddy stubbles, newly sown millet fields, mature mustard, sunflower. Causes damage to the crop during post harvest stage on the threshing ground or in the godowns.

Plate - 1

Common Peafowl *Pavo cristatus* | Sarus Crane *Grus antigone* | Demoiselle Crane *Anthropoides vigro* | Blue Rock Pigeon *Columba livia* | Ring Dove *Streptopelia decaoto*

Roseringed Parakeet *Psittacula krameri* | Slatyheaded Parakeet *Psittacula himalayana* | Small Green Barbet *Megalaima viridies* | Rosy Pastor *Strunus roseus*

Common Myna *Acridotheres tristis* | Bank Myna *Acridotheres giginianus* | House Crow *Corvus splendens* | Redvented Bulbu *Pycnontus cafer*

Jungle Babbler *Turdoides striatus* | House Sparrow *Passer domesticus* | Baya weaver *Ploceus philippinus*

Blackthroated Weaverbird *Ploceus benghalensis* | Whitethroated Munia *Lonchura malabarica* | Spotted Munia *Lonchura punctulata*

Roseringed Parakeet (*Psittacula krameri*): 42 cm: Smaller than the Alexandrine Parakeet in size with bright red bill. Female lacks the chin stripe and collar and is all green. **Distribution:** Resident, almost throughout the country, except parts of the North-west, North-east and Himalayas. **Habits:** Very adaptable species often associated with human habitations and cultivation. Gregarious, large flocks of this species at roosting sites is a familiar sight in India, cause extensive damage to standing crops, orchards and garden fruit trees; also raids grain depots and markets. **Habitat:** Deciduous forests, open woodland, secondary growth, cultivation, gardens and vicinity of human habitation. **Food:** cereals, pulses, oilseeds, fruits, and vegetables.

Slaty headed Parakeet (*Psittacula himalayana*): 41 cm: A grass green parakeet with entire head dark bluish slaty, black chin, narrow neck, blue green hind neck collar, red shoulder patch, long pointed yellow-tipped tail, but female lacks red in hilly habitat (up to 2600 m), moves considerably, descending very low in winter. **Habits:** Very agile in flight, keeping in compact flocks, very destructive in hill orchards and upland maize fields. **Habitat:** Forest and well-wooded areas, especially near hill orchards and cultivation. **Food**: fruits, acorns, maize.

Small Green Barbet (*Megalaima viridis*): 23 cm: Brown and green barbet with prominent white cheek patch and white spotting on under parts, crown and nape dark brown with a blackish band running backward from eye. A broad white cheek stripe including the ear covers. Whitish chin and throat, bold white spotting on breast. Naked eye patch blackish. Sexes alike. **Distribution:** Resident and common in the evergreen and moist deciduous forests of the hills and plains of Southern India from Narmada southwards. Also frequents forest gardens and groves near human habitations and adjoining forests. **Habits:** Found singly or in small groups feeding on favoured-fruiting trees. Particularly fond of figs. **Habitat:** Broad leaved evergreen and moist deciduous wooded areas, gardens and grooves in town, villages and cultivation. **Food:** fruits, mainly wild figs, but eats cultivated fruits, especially coffee berries, banana and kinnows.

Rosy Pastor (*Sturnus roseus*): 24 cm: A rose pink bird with glossy black head, neck and upper breast, wings and tail; rest of plumage rose pink, brighter with the approach of spring emigration. A long, recumbent, pointed crest on crown and nape, sometimes erected. Sexes alike. **Distribution:** In winter all India, abundant in the North-western portions and Deccan, becomes less frequent further south. **Habits:** Gregarious, forms huge flocks at rich feeding sources. Follows locust swarms and helps to control this pest. Also causes extensive damage to cereal crops by feeding on grain. Attracted to flowering and fruiting trees; also seen following grazing cattle to catch the insects they disturb, and probes for insects in damp open lands; overall aggressive and extremely noisy bird, huge roosting colonies, resulting in deafening clamour before settling. **Habitat:** Open areas, cultivation, orchards, flowering trees amidst habitation. **Food:** grain, insects, flower nectar, fruits of Manila tamarind, Salvadora, Mulberry, Banyan and peepul figs.

Common Myna (*Acridotheres tristis*): 25 cm: One of the most familiar birds in villages, towns and cities; frequently seen scavenging on the streets and bazaars. Rich veinous-brown plumage, black head, neck, upper breast, yellow beak, legs and naked wattle around eyes distinctive. A large white patch in wing conspicuous in flight, blackish tail, with broad, white tips to all but central feathers; whitish abdomen. Sexes alike. **Distribution:** All India up to about 3500 m in the Himalaya. **Habits:** Usually in pairs or small parties. Commonly scavenges in built-up areas and flocks often follow grazing cattle or the plough or feed in cultivation. **Habitat:** Cultivation, villages, towns, cities and light forests. **Food:** omnivorous; fruits, cereals, nectar, insects, kitchen scraps, refuse.

Bank Myna (*Acridotheres ginginianus*): 23 cm: Similar to Indian Myna, but pale bluish grey instead of brown. Naked skin, round eyes brick red instead of yellow. Flocks in open country and human habitations. Sexes alike. **Distribution:** Locally common and patchily distributed, plains and hills, locally up to 800m. **Habits:** Similar to those of common Myna, but rather more gregarious, usually keeping in small flocks in breeding season and in large flocks in winter. Railway stations are particularly favoured. Feeds chiefly on the ground, also, regularly rides on backs of grazing animals seeking invertebrates. Haunts municipal refuse dumps. **Habitat:** Irrigated cultivation, sewage farms and damp grasslands, often with livestock and around villages and open scrub forests.

House Crow (*Corvus splendens*): 43 cm: Most common bird, both in rural and urban areas. Its body, beak, forehead, crown and throat are glossy black contrasting with dusky grey nape, neck, upper breast and upper back. **Distribution:** Widespread and common, reaching about 2500 m in the Himalayas. **Habit:** Gregarious when feeding and roosting. Most familiar and live in close association with man and obtain its livelihood from his works. Bold and cunning, omnivorous and an opportunistic feeder. Scavenger at rubbish dumps, seeks invertebrates and grain in cultivation near villages. They pull out freshly sown seeds of cereals, pulses and oilseeds causing considerable damage. **Habitat:**

Rural and urban habitats, cultivation, forest edges, range of habitats very wide. **Food:** Omnivorous, robs young birds from the nest, drives other birds from flowering trees.

Red vented Bulbul (*Pycnonotus cafer*): 20 cm :A distinctive bulbul, with crashed black head, dark sooty-brown plumage, pale edges of feathers on back and breast gives scaly appearance, white rump and red vent distinctive, dark tail tipped white. Sexes alike. **Distribution:** All India, up to about 1800 m in Himalaya. **Habits:** Bold, tame and quarrelsome. Keeps in pairs or in small loose flocks, according to the season. Arboreal, keeps to middle level of trees and bushes. **Habitat:** Light deciduous forests, secondary growth, gardens, roadside avenues and light scrub. **Food:** insects, fruits, grains of sorghum, pearl millet, flower nectar and kitchen waste.

Jungle Babbler (*Turdoides striatus*): 25 cm: A stocky, ash grey to grey brown babbler with stout yellowish bill. Throat and breast appear lightly mottled with grey or streaked with grey or brown and upper parts are lightly streaked; dark, rufous brown tail, sexes alike. **Distribution:** Throughout India, plains and hills upto 1200 m locally to 1800 m in Himalayas and to 1600 m in Peninsular hills. **Habits:** Gregarious, small parties, also referred to as "Seven sisters" in English as usually there are 5 to 7 birds in a flock. Spends much time on ground, rummaging amongst the fallen leaves and mulch for moths and other insects. Some times flies into branches above, especially when alarmed, appears ruffled and agitated and breaks into a squeaking chorus: Social birds; all come together if one member is in some problem. **Habitat:** Light forests, scrub, gardens, cultivation and vicinity of habitation. **Food:** Insects, wild figs, cereal grains, flower nectar.

House Sparrow (*Passer domesticus*): 15 cm : Wide spread and resident species, usually found in or near houses; male with grey crown, black chin, grey brown rump, chestnut wing having white shoulder patch, throat white, under parts greyish white; female dull grey brown above, lacks black chin and grey head, streaked darker, dull whitish brown below. **Distribution:** All India, to about 2000 m in the Himalaya. **Habits:** Small parties to large gatherings, mostly commensal on man, feeding and nesting in and around habitation, including most crowded locations, also feeds in cultivation; hundreds roost together. **Habitat:** Rural and urban cultivated fields. **Food:** Seeds, insects, and often omnivorous. Damage standing crop, particularly cereals at grain filling stage. Spilled over grains and weed seeds are picked up from ground. The young ones are largely fed on insects; most of them are pests of field crops.

Baya weaver (*Ploceus philippinus*): 15 cm: Female and male in non-breeding plumage look like female House Sparrow: dark-streaked fulvous brown above, plain whitish fulvous below. Stout conical bill, short square-cut tail. Breeding male has bright yellow breast, cream buff on under parts. Flocks about in open cultivation. **Distribution:** Throughout the Indian Union; Bangladesh; Pakistan; Sri Lanka; Myanmar. Resident and locally migratory. **Habits:** Flocks with considerable size glean paddy and other grain in harvested fields and damage ripening crops. **Habitat**: Roosts in large numbers in reedbeds bordering tanks, etc. Its seasonal, local movements are largely governed by paddy and cereal cultivation, which provide nesting material and food. Also eats insects. **Food**: Though omnivorous, they chiefly subsist on cereal grains and weed seeds. Crop field closer to the nesting colony or roosting place suffer more damage. Damages standing crops of bajra, sorghum, rice and other minor millets. It is also a pest in rice nursery.

Black throated Weaverbird (*Ploceus bengalensis*): 14 cm: Smaller than Baya. Male in the breeding plumage with glistering orange yellow cap, dark brown nape, chin and throat whitish, face and breast black, rest of under parts whitish, streaked with black on flanks. In winter lacks the golden cap, but has yellow super cilium, patches of yellow on face and neck, breast dark fulvous, mottled with black. Female similar to male in winter but with the yellow paler and breast less deeply mottled. Bill pale blue. **Distribution:** Resident, widely distributed mainly in Indus and Gangetic plains. **Habits**: Flocks with smaller size and colonies tend to be much smaller than those of Baya and Streaked Weaverbird. **Habitat:** Tall moist grassland, close to wetlands, reedy marshes and cultivation. **Food:** Mainly cultivated crops like cereals, millets, seeds and weed seeds. Young ones are fed with insect and animal matter.

White throated Munia (*Lonchura malabarica*): 15 cm: A dull little bird with a black pointed tail, black flight feathers, white rump and throat, whitish under parts, slaty/silver conical bill. Sexes alike. **Distribution:** Wide spread resident, except in North-west and North east subcontinent and Himalayas. **Habits**: Fairly common in small flocks in open waste country and areas of indigenous vegetation and also visit gardens. **Habitat:** Dry cultivation, grassland and thorn scrub. **Food:** Feeds chiefly on the ground, also on seedling grass and crops like bajra and small millets.

Spotted Munia (*Lonchura punctulata*): 15 cm : A typical munia, dull chocolate brown above with chestnut forehead, face, chin and throat, under parts white scalloped black except on the abdomen;

rounded brownish tail. Sexes alike. **Distribution:** Widely distributed throughout the Indian subcontinent. **Habits:** Large flocks hop about gleaning grass seeds etc; take refuse in near by bushes when disturbed. **Habitat:** Open forest, bushes, cultivation interspersed with grassland and gardens. **Food:** Grass seeds, cause considerable damage to crops like bajra and minor millets.

POPULATION DYNAMICS AND COMMUNITY STRUCTURE OF DEPREDATORY BIRDS (Based on annual progress reports of centres of AINP on Agricultural Ornithology)

Rose ringed Parakeet was the predominant depredatory species in Punjab. Its population fluctuated from month to month with a peak in April -June and low densities during July. Out of the twenty species recorded at a grain store in Punjab, House Sparrow, Weaverbirds, Ring Dove and Rock Pigeon were found to be most abundant, while Rosy Pastor (26%), House Sparrow (18%), Baya (8%) and Rose ringed Parakeet (8%) were the most abundant birds that depredated ripening pearl millet in Gujarat. The number of Rose ringed Parakeet roosts in Jamnagar city fluctuated in different months and their numbers were least in June (1860) and the highest in February (6775). In Andhra Pradesh, the size of the roost of Rose ringed parakeet varied from 500 to 20,000 and the most preferred habitat for roosting of Rose ringed parakeet was noticed in urban area followed by forest tracts and sacred grooves. Similarly, the populations of Common Myna, Bank Myna, Pied Myna and Jungle Myna in Punjab fluctuated seasonally depending on their breeding and social behaviour. In general, the flock size was greater in post-breeding period and winter.

The relative abundance of myna species in maize during sowing was: 29% common myna, 19% Bank myna and 16% pied myna in Punjab. Similarly, in wheat crop, important depredatory birds, viz., House Crow (44%), Ringed Dove (16%) and Common Myna (15%) were observed during sowing and sprouting stage. Black tailed Godwit (55%); Ruff (29%) and Rock Pigeon (4%) were the active depredatory birds in sown wheat fields in Gujarat. Relative abundance of the Rose ringed Parakeet in Gujarat in different crops was as follows: pearl millet (12%), sorghum (5%), maize (85%), wheat (6%) and date palm (51%). The bird community depredating pearl millet and sorghum in Punjab comprised 13 and 11 species, respectively. House Sparrow was the most abundant species followed by the Baya in both the crops. In Andhra Pradesh, the relative abundance of Rose ringed parakeet was: Sunflower (78%)> Maize (48%)> Sorghum (33%)> Pearl millet (9%)> Safflower (13%)>. Bird community of sorghum during winter in Gujarat comprised of 14 species. Baya (29.33%), Common Rose Finch (22.76%), Spotted Munia (23.13%) and Rosy Pastor were dominant among them. However, in case of groundnut crop in Gujarat Blue Rock Pigeon (38%) and Rose ringed Parakeet (36%) were the most abundant species from sowing to sprouting stage (Tables 1 & 2).

FOOD AND FEEDING HABITS OF DEPREDATORY BIRDS (Tables 1 & 2)

Ninety eight percent of the annual diet of Baya, Black throated Weaverbird and Streaked Weaverbird in Punjab constituted of plant matter. The major constituents were rice grains from stored straw or shed during crop harvesting, pearl millet (which was grown as a fodder crop) and two species of weeds. The relative proportion of insect matter (% by weight) was higher in the diet of the nestlings of Baya (46%), Black throated Weaver (74%) and Streaked Weaverbird (57%) in Punjab. However, in Gujarat, adult food of the Black throated Weaverbird comprised mainly of four cultivated cereals viz. rice (57.75%), pearl millet (30.22%), sorghum (5.49%), wheat (3.69%) as well as weed seeds (1.17%) of six different taxa (Fig. 1). Overall diet composition of sympatric weaverbirds (Common Baya and Black throated Weaverbird) and the House Sparrow was almost same in Gujarat. The diet of Rose ringed Parakeet in Punjab contained mainly cereals (45%) and tree seeds (38%). Among the cereals, sorghum (21%) was the most abundant. Guava ranked first among tree seeds (24%) followed by mulberry (10%). At Anand, however, cultivated crops comprised 26 percent to 100 percent of the diet of adult parakeets in different months. The annual diet composition consisted of cultivated crops like cereals (69.85%), oil seeds (1.35%) and pulses (1.52%) and seeds of wild trees (27.28%) (Fig. 1). The nestlings of Rose ringed Parakeet fed 57.19 percent (dry weight) cultivated crops and 43.05 per cent seeds of wild plants. Amongst the cultivated crops, wheat (29.85%), sunflower kernel (14.84%) and seeds of radish formed a major part. Amongst the seeds of wild trees, Manila Tamarind (33.76%) and pieces of bark (5.28%) were predominant. Dietary composition of Rock Pigeon in Punjab revealed that wheat, maize and paddy together constituted 66 per cent followed by pulses at 18 per cent and oil seeds at 5 per cent (Fig. 1). The experimental findings at different locations showed that the frequency of animal matter in the diet of adult Indian Myna was very high (60%) as compared to its dry weight (25%). On the other hand, the frequency of plant matter was only 40 per cent but it made up *ca.* 75 per cent dry weight of the diet (Fig. 1).

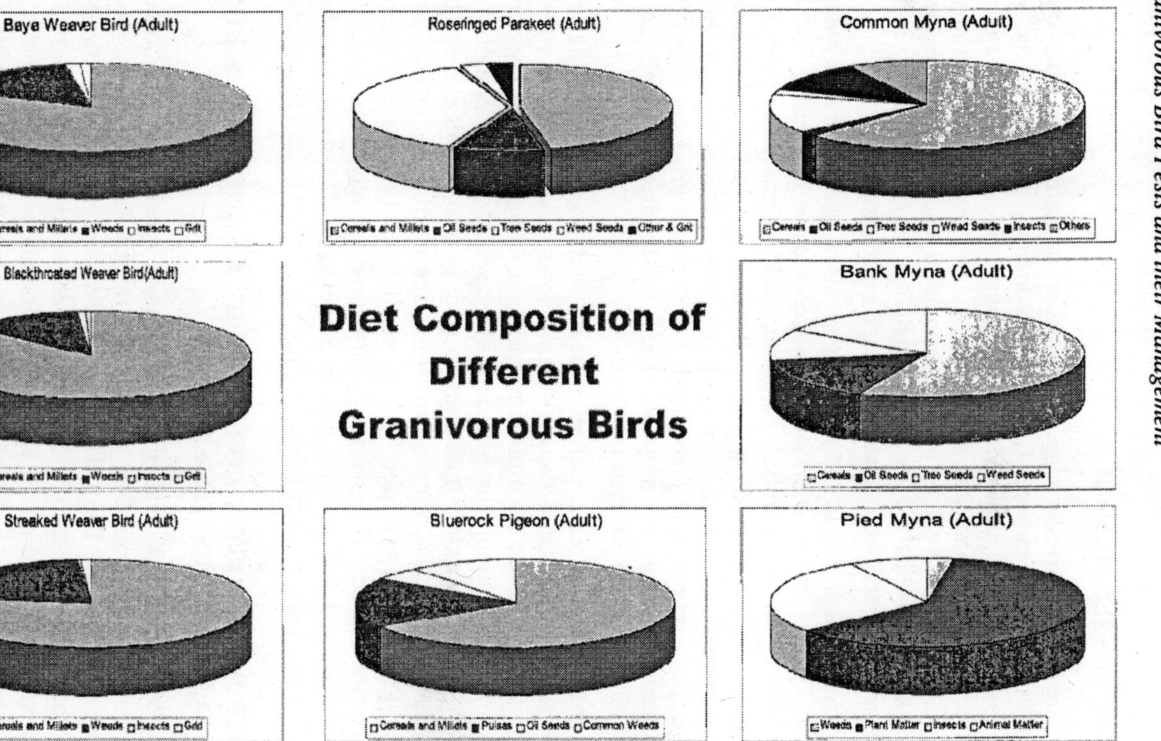
Baya Weaver Bird (Adult)
Cereals and Millets Weeds Insects Grit
Roseringed Parakeet (Adult)
Cereals and Millets Oil Seeds Tree Seeds Weed Seeds Other & Grit
Common Myna (Adult)
Cereals Oil Seeds Tree Seeds Weed Seeds Insects Others
Blackthroated Weaver Bird(Adult)
Cereals and Millets Weeds Insects Grit
Diet Composition of Different Granivorous Birds
Bank Myna (Adult)
Cereals Oil Seeds Tree Seeds Weed Seeds
Streaked Weaver Bird (Adult)
Cereals and Millets Weeds Insects Grit
Bluerock Pigeon (Adult)
Cereals and Millets Pulses Oil Seeds Common Weeds
Pied Myna (Adult)
Weeds Plant Matter Insects Animal Matter

FIGURE -1

CROP LOSSES (Table 2)

Due to intensive farming, many species of granivorous birds obtain food from the agricultural fields and its vicinity throughout the year. Moreover, a few species viz., Rose ringed parakeet, house sparrow, weavers, munias and doves manage to nest in close proximity to agricultural fields. So, the congenial condition available to these species in recent times has led to their population build-up causing serious concern amongst farmers and agricultural ornithologists.

Crop losses due to the depredatory birds can occur either due to single species as in sunflower (parakeet) or a community of bird complex as in pearl millet, sorghum, paddy, groundnut and apple. Negative impact of birds on agricultural crops vary from region to region, season to season, depending on varied factors like number of depredatory species and their density, area under crop, ecology of the area, concentration of migrants, their food habits as well as the physiological status of birds involved.

Almost all cereals, pulses, oilseeds and several vegetable crops were susceptible to bird damage during sowing, seedling and ripening stage. Birds attacked the fruits at maturing or ripening stage. The cereals were more vulnerable to bird attack especially at the dough stage. Damage to the crops of smaller grains such as pearl millet and sorghum was more serious as compared to large sized grains (e.g. maize). Small cereal grains were preferred by both smaller and larger birds, whereas, maize was depredated primarily by larger species such as parakeets and crows. Isolated fields were always prone to bird damage. Similarly, early or late maturing fields were highly susceptible to bird depredation.

CROP WISE LOSSES INFLICTED BY GRANIVOROUS BIRDS IN MAJOR CROPS (Table 2)

Pearl millet: A total of 23 depredatory species were recorded in pearl millet crop from five states. Rose ringed Parakeet, Rosy Pastor, House Sparrow and Baya were the predominant species that damaged the crop in North-western India. White cheeked Bulbul was reported for the first time to feed on the crop in large numbers in the arid and semi-arid zones. Estimation of bird damage varied highly in different states. It was highest in Gujarat (0.3 to 40%) followed by Andhra Pradesh (1.5 to 9%), Punjab (4-5%) and Delhi (60%) during kharif season. In Gujarat, bird damage to the summer crop was significantly less (0.2 to 2.1%), due to the absence of migratory birds especially the Rosy Pastor. Synchronization of crop cultivation also resulted in better yield compared to kharif season mainly due to prevention of damage by pests.

Wheat: Bird damage to wheat varied from 0.2 to 41% in different parts of the country. The damage was significantly high in Rajasthan compared to Gujarat and Punjab. About 13 species of birds damaged standing wheat. Although Rose ringed Parakeet, Ring Dove and Baya were the most common species damaging the crop in most of the fields, large flocks of Demoiselle Cranes, Common Cranes and Short-toed Larks, heavily damaged wheat fields in Bhal area of Gujarat. Ruff and Black tailed Godwit are reported for the first time to damage wheat at sowing and seedling stages in the coastal areas of Gujarat

Paddy: Although paddy is cultivated in several states of the country in vast areas, it was prone to heavy damage by birds only under congenial ecological conditions. Damage to paddy was highest in Punjab (0.1 to 6.5%), followed by Kerala (1.5 to 6%), Andhra Pradesh (1.5 to 3%) and Gujarat (0.1 to 1%). Thirty-nine species of birds feed on the grains of standing crop in Gujarat. Baya, House Sparrow and Rose ringed Parakeet were common and predominant depredatory birds in most parts of the state. Gargany Teal and Lesser Whistling Teal damaged paddy crop in Kole area of Kerala.

Sorghum: Bird damage to sorghum was highest in Rajasthan (2.3 to 48%) as compared to Gujarat (0.4 to 18.6%) and Andhra Pradesh (0.5 to 16.6%). This is one of the most preferred crops by the granivorous birds like Rose ringed Parakeet, Rosy Pastor and Bank Myna. A total of 26 species were recorded feeding on sorghum.

Maize: Bird damage to maize was relatively less in Gujarat (0.3 to 2.5%) compared to Punjab (3.3 to 7.5%), Andhra Pradesh (3 to 9.1%) and Rajasthan (0 to 20%). Ten species of birds were recorded to feed on maize, of which the Rose ringed Parakeet was the most important in all the states.

Sunflower: Rose ringed Parakeet and House Crow were the predominant depredators and caused 10 to 30% damage in Andhra Pradesh and 40 to 90% damage in Rajasthan. In Punjab the mean percentage of damage in different years of study ranged from 5.7 to 29%. Because of the bird problem, the crop could not be introduced in Gujarat state.

Oil palm: The damage was higher both in south Gujarat (10 to 27%) and west Godavari district (3.3 to 30%) of Andhra Pradesh. Common Crow, Jungle Crow, Rose ringed Parakeet and Common Myna were the depredatory birds of this crop.

Groundnut: During sowing to sprouting stage, 3 to 33% damage was caused by ten species of birds in Saurashtra region of Gujarat. The migratory Demoiselle Cranes caused damage up to 10% at the time of harvesting.

Safflower: Rose ringed Parakeet was the only species feeding mainly on the periphery of the crop fields. The damage level was negligible in Andhra Pradesh and Gujarat.

Apple: Apple orchards in Himachal Pradesh incurred 5 to 12.5% loss due to bird damage. Although 11 species of birds fed on apple, Slaty headed Parakeet and Rose ringed Parakeet were the most important.

Kin now: Damage to kin now fruits in Himachal Pradesh ranged from 3 to 16%. Seven depredatory species were recorded of which, Rose ringed Parakeet and House Crow caused maximum damage.

Date palm: Date palm fruits were almost free from bird damage (0.1 to 1%) in the Kutch region of Gujarat. Although eight species of birds damaged the fruits, the Rose ringed Parakeet, House Crow and Common Myna were the major species of concern.

Banana: Little Green Barbet, House Crow and Jungle Crow damaged Banana crop in Kerala.

Pomegranate: In Gujarat, bird damage was higher (3.6 to 20%) compared to Andhra Pradesh (3.6 to 5.5%). Rose ringed Parakeet was the common depredator in both the states. However in Andhra Pradesh the House Crow also damaged the fruits.

Guava: Damage was moderate in Andhra Pradesh (1 to 4.5%) and Punjab (2.5 to 5%), but in Rajasthan it was high (4.5 to 24.5%). Rose ringed Parakeet was the common bird that damaged the fruits in all the three states. House Crow also caused significant damage in Andhra Pradesh and Rajasthan.

Grape: Grape orchards in Andhra Pradesh suffered 2 to 27% bird damage due to seven species. Rose ringed Parakeet, Rosy Pastor, Koel and Sunbird caused significant damage.

Papaya: Koel, Brahminy Myna, Red vented bulbul, Jungle Babbler and Jungle crow damaged ripe fruits in Gujarat and Kerala.

Pecan Nut: Rose ringed Parakeet and Large Indian Parakeet caused severe damage to pecan nut (6 to 50%) in Himachal Pradesh.

MANAGEMENT OF PROBLEM BIRDS (Table 2)

Repellents: Since ages, man has developed various devices to scare or repel birds from cultivated crops. Several management practices such as frightening devices, traps, protective netting, shooting, cultivation of less susceptible crop varieties, modified planting and harvest schedules, alternative crops or foods, decoy crops, lethal baits, lethal sprays, repellent sprays, glues, predators, parasites, diseases and chemosterilants were advocated for bird pest management. However, the control technology should never aim at controlling the pest but should only decrease the population levels to below economic injury levels. Similarly, none of the management strategy will ever work with uniform effect for all the given situations. The management strategies should always be location specific and target oriented without causing any adverse effects on environment and always safe to human beings. The success of any management method will also depend on the economic feasibility and practical utility to the farming community. Some non-lethal devices include using of non-reflecting tapes, spotted balloons with painted eyes, glass balls etc. (Dhindsa *et al.,* 1993). However, they are effective only initially in repelling the birds but gradually the birds become habituated to these devices as they are harm-less and their effectiveness as repellent diminishes. Thus the natural response of the farmers has been to attempt to reduce the depredating population by lethal means i.e. killing them which appears to be more effective than the non-lethal ones.

Lethal methods: Dynamite, toxic baits and spray chemical repellents and shooting were employed to control the avian pests. However, killing birds to protect agricultural crop is now viewed critically as conservation has become a priority leading to new laws being enacted to protect bird species. In addition, research has revealed that agricultural losses to birds are often not as great as initially estimated (Dyer & Ward, 1979; Weatherhead *et al.,* 1982; Dolbeer, 1988) and that many bird pests also have attributes beneficial to man. Studies of bird damage to various agricultural crops have established that overall losses on a regional or national level are almost always less than one per cent of the total production (Dolbeer, 1988). Thus bird damage usually does not have a significant impact on the overall farm economy. Therefore, bird control programmes should be aimed at relatively few farms with high losses and not at entire farming area.

Dolbeer (1988) extensively reviewed the status and potential of lethal means of reducing bird damage in agriculture. He tried to focus on the fact that most of the field evaluation of lethal control techniques has put far more emphasis on numbers of bird killed than on how much damage was eliminated in relation to the cost of control.

Ecological approaches in bird pest management: Agricultural ecosystem is so dynamic and ever changing that it becomes imperative to evolve suitable bird pest management approaches from time to time to suit the changing pest situations. Sunflower introduced during 80s as oilseed crop in Punjab that matures in the month of May-June provides a feast for birds like parakeet. Before the introduction of sunflower, parakeets fed on seeds of native trees like mulberry and weed seeds of *Crotalaria medicaginea* despite the presence of mature wheat in the fields (Saini *et al.*, 1992). Earlier the population of Wood Pigeon was seen tremendously increasing and becoming a pest in Britain during 1970 as the area of oil-seed rape grown increased (Feare *et al.*, 1988). These types of difficulties posed by birds could be avoided by reversing the trends in farm management that had given rise to the problems, although this is unrealistic with respect to many aspects of food production. However, we should be aware of the changes that had lead to the damage as some alteration to cultivation practices might help to alleviate the bird problems without obstructing the affecting targets.

Denying access to preferred food, physically and chemically: Where birds show clear preferences for certain varieties of crop, growing of non-preferred cultivars might reduce damage (Feare *et al.*, 1988). The preferences of European Starling for certain cherry cultivars were reported to be based on colour so also the taste preferences of Red-winged Blackbird for corn cultivars was influenced by colour. Physically bird access to preferred food can be denied by wrapping the corn cobs by its leaves (Dhindsa *et al.*, 1993). Covering pineapple fruits by rosettes of spiky leaves and a thatch of dry grass (Srihari & Chakravarthy, 1998) or tying 6-8 dried (waste) bunches along a metal wire along the entire garland of bunches of oil palm also denies access to the damaging birds (Chakravarthy *et al.*, 1998). Chemical deterrence is conferred primarily by tannins (Feare *et al.*, 1988) or neem, *Mormodia foetida, Vernonia amydalina* and *Gliricidia sepium* extracts (Bright, 2000). Both physical and chemical deterrents become less effective when alternative foods are lacking (Feare *et al.*, 1988), suggesting that these methods may have only a limited role in the alleviation of bird damage.

Farming Practices: Bird damage rarely reverses trends in agricultural development that have stimulated the damage. However, partial reversals can contribute to protection. For example, European Starling damage to germinating cereals could be avoided by planting the seed deeper in areas where damage is anticipated (Dolbeer *et al.*, 1979), although that can have other disadvantages. In semi-arid parts of Africa by growing maize instead of growing small-seeded cereals, damage by Red-billed Quelea was reduced (Feare *et al.*, 1988). More subtle changes in the timing of events in the farming calendar can also reduce damage. In Scotland, Feare *et al.* (1988) found that early and late sown barley and oat fields suffered more damage than main season sowing, because crops that were sown out of phase were relatively rare so that rooks concentrated on them. A reduction of 8.5 per cent bird damage was reported in Karnataka when cucumber was grown on raised basins along with bottle gourd and creepers (Srihari & Chakravarthy, 1998). Growing legume catch crops with maize reduced bird damage to maize in Germany (Germeier, 1997). All of these modifications in management practices necessitate advance planning keeping bird damage in mind; unfortunately, bird damage generally appears low in farmer's priorities at the planning stage.

Habitat Management: Agricultural development has created ideal conditions for many pest birds but to consider altering agricultural habitats for the prevention of bird damage is unrealistic. However Feare *et al.* (1988) suggested removal of certain attractions or the provision of alternate food can be used with some success in damage prevention. Feare & Wadsworth (1981) found that cattle farms at which stock was fed indoors suffered less feed loss to European Starling than those whose cattle were fed outdoors. The total exclusion of birds from crop scenario is expensive, but exclusion can be economically justified when the crops themselves are of great value, for example, netting the cherry orchard (Feare *et al.*, 1988) or netting the cereal seed production areas (Bruggers & Jaeger, 1982). In some instances, it may be possible to reduce bird damage by denying access to perches and water or by removing roost sites.

The provision of alternative foods, in an attempt to draw birds away from the valued crops (Weatherhead *et al.*, 1982) necessitates detailed knowledge of the factors limiting bird population. The provision of food when natural foods are limiting might simply increase the population. However, provision of alternate resources can be of value when used in consideration with other techniques. The provision of undisturbed refuges to which birds can be scared from valuable crops enhances the efficacy of scares (Feare *et al.*, 1988). This kind of integrated approach is of particular value to meet situations where birds are legally protected.

Plant chemicals as "natural" repellents: Chemical repellents should be efficient, humane, environmentally safe and cost effective (Dolbeer, 1988). However, seldom such products are available. The substances used till now have drawbacks, such as narrow margin between repellent and lethal doses (Feare *et al.*, 1988) as most of the substances tested as repellents have been chosen from chemicals used as insecticides or fungicides. A more systematic approach is needed to ascertain the kind of aversive compounds, which birds may encounter naturally. For herbivores, one possibility is to examine plants producing toxins. The drawback is, most probably birds would have already evolved mechanisms to avoid such foods. The fact that many repellents work through conditioned aversion learning, led to addition of a colour to increase the repellence of such chemicals. This might allow birds to detect the presence of the chemical more quickly and easily, particularly at low concentrations. By analogy with batesian mimicry, it might cause birds to generalize their aversions to parts of the crop treated with the cue alone, thereby affording general protection with the use of a smaller quantity of the substance (Feare *et al.*, 1988).

ECO FRIENDLY MANAGEMENT METHODS FOR GRANIVOROUS BIRDS AND RECOMMENDATIONS DEVELOPED BY ALL INDIA NETWORK PROJECT ON AGRICULTURAL ORNITHOLOGY

As a part of All India Net work Project on Agricultural Ornithology, multi location trials for evaluation of different eco friendly management practices were carried out to determine their efficacy, feasibility and economical viability under different agro climatic zones. These methods were tested all over India, i.e., Himachal Pradesh in the north to Kerala in down South in varying cropping patterns stretched in different climatic regimes. As a result several eco friendly economical management practices have been evolved for minimizing losses caused by depredatory birds which are elucidated as under:

(1) Wrapping method on maize crop: Covering maize cobs by wrapping adjacent green leaves around them reduced parakeet and crow damage to a negligible level. Being hidden and camouflaged, the wrapped cobs escape detection by birds and so also damage. All cobs need not be covered. Since parakeet damage is restricted to peripheral rows, covering of 50 per cent cobs at random on outer three rows of the field is sufficient to effectively reduce bird damage (Plate 4). The method is very simple and effective, does not involve any material cost and is less laborious than scaring. There is no negative impact on the grain yield. Farmers can save approximately Rs. 1900 per ha. by adopting this technique (Plate 2).

(2) Reflective ribbon: Strips of 'Reflective ribbon', i.e. polyester film of 1.5 cm width, preferably 15 to 20 m long, when fixed parallel to the crop at 0.5 m height above the crop and at 5m intervals using bamboo poles and strings scare away the birds. For better reflection, the ribbon should be fixed north to south direction. During sunshine the

PLATE - 2

Integrated Bird Pest Management (IBPM)
— A new Approach to minimize bird damage

Netting-Complete Protection to the crops

PLATE - 3

Screen Crop — A Potential saving device from birds

Block Plantation of Sunflower minimize parakeet damage to 1% level

reflection of sunlight and humming noise produced by the wind scares the birds from the field. But the device is effective only for 15-20 days as birds get habituated to it gradually. Additionally ribbons do not scare birds under poor light condition and as also when the crop is grown in isolation. To protect the crop during vulnerable stages, the technique of bird scaring by ribbons is very effective and easily acceptable to the farmers. Birds like Rose ringed Parakeet, House Crow and mynas on crops like sunflower, maize, sorghum, pearl millet and orchards, Demoiselle Cranes in groundnut, depredatory birds in pomegranate, date palm and parakeets in apple and kin now orchards in Himachal Pradesh were scared by this device (Plate 2). Installation of these ribbons will cost the farmer Rs. 675/ha @ Rs. 200 for the Reflective tape, Rs. 375/- for the stakes and Rs. 100/- as labour charges, which is much lesser than the cost of employing two labourers @ Rs. 2000/- per ha (Plate 2).

(3) Bioacoustics: The acoustic equipment consists of one stereo tape recorder with 30 w amplifier, 2 speakers and one 12 v battery, with distress calls of birds pre-recorded on a tape. Broadcasting of such distress calls kept the depredatory birds away from sunflower fields, orchards and small acreage crops. However, the cost involved in this method is high limiting its use only for remunerative crops (Plate 2).

(4) Automatic mechanical bird scarcer or pyrotechnic method: It is a sound-producing device, works continuously for a whole day with one kg of calcium carbide and water. Areas of one-hectare can be covered with this method and crop losses by birds could be reduced effectively. Even though initial cost against purchase of the equipment is high community level adaptation will reduce the cost.

(5) Screen crop: Thick planting of sorghum fodder as well as of fodder maize significantly reduced parakeet damage to maize crop grown for grain production. Besides giving better yield, this practice also provided additional fodder and turned out to be highly economical (Plate 3).

(6) Seed treatment for protecting sprouting seeds: Thiram and Copper-oxychloride at 0.5% are very effective in reducing seedling losses to birds in maize, chickpea, soybean, sunflower and groundnut.

(7) Botanical repellents: Spraying of botanical formulations like Bio-bird Repellent (BBR) and Fortune Aza (Neem formulation) in the field reduced number of visiting birds and resulted in higher yield compared to control in Andhra Pradesh and Gujarat. Tobacco leaf decoction (10%) spray on sorghum at milky stage had similar effect. In addition to their effectiveness these botanicals are also economical as farmer can save upto Rs. 1225/- per ha by employing these repellents for bird pest management (Plate 2).

(8) Cultivation in large blocks: In Punjab bird damage to sunflower on block area showed lesser damage. A minimum block size of 21 acres is required to reduce parakeet damage to 1% level (Plate 3).

(9) IBPM (Integrated Bird Pest Management): Integrated bird pest management (IBPM) using reflective ribbons and botanicals in sorghum and wrapping along with ribbons in maize fields gave higher yields in Andhra Pradesh. Botanicals along with installation of reflective ribbons proved effective and controlled bird pests significantly leading to higher yields. The studies carried out with regard to economics involved in the IBPM clearly envisaged that near about Rs. 845/- and Rs. 1520/- could be saved in sorghum and maize, respectively as compared to normal method of employing bird scares for the above purposes (Plate 2).

(10) Habitat manipulation: Creating continuous disturbances to the nesting sites of the depredatory breeding birds in and around the cropped areas will force the birds to leave breeding grounds in crop fields and shift to another area. For parakeets, in addition to manual destruction of nests, closing the entrance of the nests proved effective in reducing their populations. Planting of some fruit bearing trees like Manila tamarind (*Pithecolobium dulce*), Flame of the forest (*Butea monosperma*), Mulberry (*Morus alba*) and Toothbrush Tree (*Salvadora persica*) in and around cropped areas attracts many granivorous birds during fruiting period and reduces the impact at vulnerable stage of the maize crop.

(11) Netting: It is one of the best methods and gives complete protection to the crops. The nylon net having a mesh size sufficiently small to prevent passage of the smallest granivorous bird is recommended for high value trials, breeder seed experiments and/or multiplication trials as it is very expensive for large acreage plots.

CONCLUSIONS

As about 55 per cent of the total land is cultivated, the agro ecosystem plays a vital role in the conservation of the avian biodiversity. Effective management of depredatory birds should be taken up by keeping in view the conservation aspects of birds while managing problem birds in the agriculture.

Ideally, a management strategy apart from considering the pest species concerned (its sensory abilities, social habits, population size and turn over), the crops being protected (capacity to compensate for damage, timing and pattern of cultivation, the need for immediate, complete or long term protection) and the way in which the repellent chemical is applied (addition of colour cues, timing and frequency of treatments and spot or strip-spraying) interact with one another. Hence, a holistic approach is needed, rather than individual application of a technology which will eliminate the use of broad-spectrum toxic substances, and would allow the usage of minimum amount of chemicals. The future thrust area thus should preferably be of multidisciplinary approach with an ultimate objective to reduce farmer-bird conflict and enhance the environmental conservation. The recommendations given by AINP on Agricultural Ornithology are certainly aimed in this direction. These methods however, have not yet been popularized/utilized to their fullest extent because of several constraints in extension activities. To develop a truly effective, environmentally sound approach for protecting crops from depredatory birds, use of both traditional and non-lethal techniques need to be employed in a balanced way. On the basis of these facts we believe that there must be intensive thrust on the following priorities:

- Distribution, population and habitat mapping of depredatory bird species for an effective quantification of damage to various agricultural crops in different agro-climatic regions.
- Identification, development and evaluation of eco-friendly bird management practices for crop protection.
- Exploitation of beneficial birds in suppression of insect/rodent pests in the crops.
- Application of modern Information Technology techniques such as use of Remote Sensing in combination with Geographical Information System (GIS) and Global Position System (GPS) to study bird habitat suitability and conservation strategies.
- Integrated Agricultural Landscape Management with special reference to bird–crop interdependency.

Special attention should also be given to bird pest forecasting and forewarning using weather-crop interdependency. Effect of agro-chemicals and their ill effects on avifauna and human health should be focused properly and priority should be given to the use of bio-pesticides and compounds of botanical origin as alternatives. Efforts must also be intensified to formulate suitable agricultural policies to reduce the bird-farmer conflicts and promote bio-diversity conservation.

REFERENCES

ALI, S. 1936. Economic ornithology in India. *Curr. Sci.* 4: 472-478.

ALI, S. 1949. Bird friends and foes of the cultivator. *Indian Farming* 10: 385-387.

ALI, S. 1971. Sunder Lal Hora memorial lecture: Ornithology in India: its past, present and future. *Proc. Indian Natl. Sci. Acad.* 37 B: 99-113.

ALI, S & S.D. RIPLEY. 1983. HANDBOOK OF THE BIRDS OF INDIA AND PAKISTAN. Compact edition, Oxford University Press, Bombay, India.

BASU, A.C & L.M. PRAMANAIK. 1964. Notes on the control of graminivorous birds causing seed loss of germination. *Sci. Cult.* 30: 297-298.

BRIGHT E.O. 2000. Plant products as seed protectants against weaverbirds. *Global. J Pure and Appl. Sci.* 6: 379- 383.

BRUGGERS, R. L & M.M. JAEGER. 1982. Bird pest and crop protection, Strategies for cereals of the semi arid and African tropics Sorghum in the eighties. *Proc. Intern. Symp. on Sorghum.* pp: 303- 312.

CHAKRAVARTHY A.K., K. SRIHARI., J. CHANDRAPRAKASH., P.P. REDDY., N.K.K. KUMAR & A.VERGHESE 1998. Management of bird pests in oil palm plantations in Karnataka. In: ADVANCES IN IPM FOR HORTICULTURAL CROPS (Eds.): Reddy, P.P & A. Kumar. Indian Institute of Horticultural Research, Bangalore, India. pp.

D'ABREU, E.A.D. 1920. Some insect prey of birds in Central Provinces. Report of the Third Ecological Meeting held at Pusa. Feb. 1919. Govt. Supdt. Press, Calcutta. Vol. 3. pp. 859-874.

DHARAMKUMARSINHJI 1956 BIRDS OF SAURASTRA (INDIA), Dilbahar, Bhavanagar, India.

DHINDSA, M.S., H.K. SAINI & H.S. TOOR. 1992. Wrapping leaves around cobs to protect ripening maize from rose-ringed parakeet. *Trop. Pest Manage.* 38: 98-102.

DHINDSA, M.S, H.K. SAINI., P.S. SANDHU & M.S. SAINI. 1993. *The World of Birds: A Souvenir on Ornithology.* Department of Zoology, Punjab Agricultural University Ludhiana, India.

DOLBEER, R.A. 1988. Current status and potential of lethal means of reducing bird damage in agriculture. *Proc. Intern. Ornithol. Cong.* 19: 474-483.

DOLBEER R.A., A.R.Jr. STICKLEY & P.D. WORONECKI. 1979. Starling *Sturnus vulgaris* damage to sprouting wheat in Tennessee and Kentucky, USA. *Protec. Ecol.* 1: 159-169.

DYER, M.I. & P. WARD 1977. Management of pest situations. In: GRAINIVOROUS BIRDS AND ECOSYSTEMS. (Eds.): Kendeigh, S.C & J. Pinowski. Cambridge University press, Cambridge, U.K 431p.

FEARE, C.J. & J.T. WADSWORTH. 1981. Starling damage in farm using the complete diet system of feeding dairy cows. *Anim. Prod.* 32: 179-185.

FEARE C J., P.W. GREIG-SMITH & I.R. INGLIS. 1988. Current status and potential of non-lethal means of reducing bird damage in agriculture. *Proc. Intern. Ornithol. Cong.* 19: 493-506..

FITZWATER & I. PRAKASH. 1973. HAND BOOK OF VERTEBRATE PEST CONTROL. Indian Council of Agricultural Research, New Delhi, India.

GERMEIER, C.U. 1997. Possibilities and limits of growing wintering catch crops and use of reduced soil cultivation methods in ecological maize production. In: *Proc 4th Scientific meeting on Ecological Agriculture.* Bonn, Germany.

HUSSAIN, M. F & H.R. BHALLA. 1937. Some birds of Lyallpur and their food. *J. Bombay nat. Hist. Soc.* 39: 831-842.

JERDON, T.C. 1863. THE BIRDS OF INDIA. Military Orphan Press, Calcutta, India

MASON, C.W & H.M. LEFROY. 1912. Food of birds in India. *Mem. Dept. Agric. Ent. Ser.* 3, 367 pp.

RAMZAN. M & H.S. TOOR. 1972. Studies on damage to guava fruit due to Rose ringed Parakeet *Psittacula krameri Scopoli* at Ludhiana, Punjab. *Punjab Hort. J.* 12: 144-145.

RAMZAN, M & H.S. TOOR 1973. Damage to maize crop by Roseringed Parakeet *Psittacula krameri Scopoli* in Punjab. *J. Bombay nat. Hist. Soc.* 70: 201-204.

RANA, B.D. 1973. Food and feeding habits of common Indian desert birds. *Indian Forester*. 99 : 219-223.

RAO, S.B.P & D.V.N. RAO. 1953. Bird damage to jowar. *Madras Agric. J.* 40: 466-467.

SAINI H K., M.S. DHINDSA & H.S. TOOR. 1992. Food of the Rose ringed Parakeet *Psittacula krameri:* A quantitative study. *J. Bombay nat. Hist. Soc.* 91: 96-102.

SHARMA, I.K. 1972, Ecological study of rest and roost behaviour of the peafowl. *Bird Watcher;* 7, 5-8.

SINGH, R. P & G.H. DUNGAN. 1955. How to catch crows. *Allahabad Fmr.* 29: 59-67.

SRIHARI, K & A.K. CHAKRAVARTHY. 1998. Ecological approaches for the management of vertebrate pests in Western ghat region of Karnataka. *Ind. J Ecol.* 25: 55-61.

SUSAINATHAN, P. 1921. *Bird friends and foes of farmers*. Madras Department of Agriculture, Bulletin No. 81, Madras, India.

VARMA, B.K., S.H. SHIYA & H.R. GAUBA. 1960. Control of birds by mechanical bird scarer. *Pl. Prot. Bull.* 12: 11-16.

WEATHERHEAD P.J., S. TINKER & H. GREENWOOD. 1982. Indirect assessment of avian damage to agriculture. *J. Appl. Ecol.* 19: 773-782.

WHISTLER, H. 1949. THE HAND BOOK OF INDIAN BIRDS. Oliver and Byod, London.

❑❑❑

3

FRUGIVOROUS BIRDS AND THEIR MANAGEMENT

Abraham Verghese

Fruit Entomology Laboratory, Division of Entomology and Nematology, Indian Institute of Horticultural Research, Hesseraghatta Lake – P.O., Bangalore-560 089
E-mail: <avergis@iihr.res.in>

ABSTRACT

The major frugivorous birds in India are: Rose ringed parakeet (*Psittacula krameri* S.), Slaty headed parakeet (*Psittacula himalayana*), Blossom headed parakeet, (*Psittacula cyanocephala* L.), House crow (*Corvus splendens* V.), Jungle crow (*Corvus macrorhynchos* W.), Indian myna (*Acridotheres tristis* L.) and Indian koel (*Eudynamis scolopaceae*). In addition, bulbuls (*Pycnonotus* sp.) and barbets (*Megalaima* sp.) also cause some negative impact on fruit yields occasionally. Slaty headed parakeets damaged fruits like apple, pear, peach and apricot in Himachal Pradesh, rose ringed parakeets fed on seeds, berries, fruit, blossoms and nectar and are serious pests on guava and peach. Blossom headed parakeet caused maximum damage to apple fruits (80%), Red Whiskered bulbul damaged ripening fruits in orchards, affected kitchen gardens, depredated on cultivated crops, guava, urban gardens and open shrub jungle and avoided heavy forest. Red vented bulbuls damaged apple, pear, peach and apricot in Himachal Pradesh and are a minor pest of sapota, guava and pomegranate. House crow is a major pest on grapes and peach. Jungle crow and house crow together caused 12% loss to sapota. Jungle crow also damaged guava, sapota and was a major pest on grapes and caused yield reduction in peach. Mynas of *Acridotheres* sp. were observed feeding on pomegranate fruits, immature and mature grape berries. Indian koel affects lightly wooded country-gardens, mango orchards and groves of trees in and around cultivation, towns and cities. Suggested management techniques to prevent bird damage to fruits are: scaring, netting, planting of alternate fruit crops and discouraging roosting by thinning heavily canopied trees or chasing them away by fire works or crackers.

INTRODUCTION

In India, there are about 1,200 species of birds of which less than 1% may be termed as pests on fruit crops. Generally birds become pests only if the damage done to fruits becomes economically significant. Among granivorous birds, pests are more evident than in frugivorous. For e.g. *Quelea* of Africa is worth sighting. From India's point of view, the birds of economic significance are: Roseringed parakeet (*Psittacula krameri* S.), slaty headed parakeet (*Psittacula himalayana*), blossom headed parakeet (*Psittacula cyanocephala* L.), house crow (*Corvus splendens* V.), jungle crow (*Corvus macrorhynchos* W.), Indian myna (*Acridotheres tristis* L.) and Indian koel (*Eudynamis scolopaceae*). In addition, bulbuls (*Pycnonotus* sp.) and barbets (*Megalaima* sp.) can cause some negative impact on fruit yields occasionally.

FRUGIVOROUS BIRDS OF ECONOMIC IMPORTANCE (Plate 1; Table 1)

Some description of the major frugivorous birds in Indian context is given below drawn mainly from Ali & Ripley (1969) and from my own observations:

I: PARAKEETS

(a) Slaty headed parakeet, *Psittacula himalayana*

Slaty headed parakeets (*Psittacula himalayana*) are found to damage fruits like apple, pear, peach and apricot in Himachal Pradesh.

(b) Rose ringed parakeet, *Psittacula krameri* (Scopoli)

Rose ringed parakeets feed on seeds, berries, fruit, blossoms and nectar and are serious pests on guava. They caused fruit loss in peach (*Prunus persica* B.) in Punjab.

(c) Blossom headed parakeet, *Psittacula cyanocephala* (L).

Blossom headed parakeet caused maximum damage to apple fruits (80%) at Nauni, Solan. The damaged fruits were subsequently fed by a number of insect pests and fungi adapted to decaying sweet resources. Such fruits become unfit for marketing. The cultivars suffered more bird damage.

Slaty headed parakeet
(*Psittacula himalayana*)

Blossom headed parakeet
Psittacula cyanocephala

Jungle crow,
Corvus macrorhycus

Common crow,
Corvus splendens

Red vented bulbul,
Pycnonotus cafer

Rose ringed parakeet
Psittacula krameri

Table 1. Frugivorous birds of India

Fruit	Scientific name	Frugivorous birds
Apple	*Pyrus* sp.	Mainly parakeets; occasionally crows and bulbuls
Ber	*Zizyphus* sp.	Mainly bulbul; occasionally barbets, crows and parakeets
Fig	*Ficus* sp.	Mainly parakeets, crows; occasionally bulbuls
Grapes	*Vitis* spp.	Mainly crows and mynas; occasionally orioles, koel, etc.
Guava	*Psidium guajava*	Mainly parakeets; occasionally bulbuls, barbets, etc.
Mango	*Mangifera indica*	Mainly parakeets and crows
Papaya	*Carica papaya*	Mainly crows
Pine apple	*Ananas comosus*	Mainly crows
Pomegranate	*Punica granatum*	Mainly parakeets and bulbuls
Sapota	*Achras sapota*	Mainly crows; occasionally bulbul
Temperate fruits (Pear, peach, etc.)	*Pyrus* sp. and *Prunus* sp.	Mainly crows and parakeets; occasionally bulbuls and sparrows

II: BULBULS

Red whiskered bulbul, *Pycnotus jocosus* (Gould)

This is the most common and widespread. Found in Madhya Pradesh, Maharashtra, Karnataka, Kerala, Western Tamil Nadu, western half of Andhra Pradesh and chiefly found in the Western Ghats and associated hills.

The bird frequently damages ripening fruits in orchards and affects kitchen gardens. It also depredates on cultivated crops, urban gardens and open shrub jungle. Red whiskered bulbuls are minor pests of guava.

Red vented bulbuls, *Pycnonotus cafer* (Linnaeus)

Found in Southern Indian peninsula, upto Godavari river in the east, Maharashtra in the west, in plains and hills following human settlements and cultivation. These are very similar to the Red whiskered bulbul. The birds are noisy, found usually in pairs, often in small flocks during non-breeding season and in large numbers during fruiting of *Ficus* trees or in lantana patches. Arboreal but occasionally come to ground to pick up insects. Flight is quick and strong though seldom sustained for any distance.

These birds feed on fruits and berries of *Loranthus*, *Lantana*, *Zizyphus*, *Ficus*, *Antalum*, *Salvadora* and others. They inhabit light scrub, gardens, secondary jungle and light deciduous forest. Although locally destructive to gardens, fruits, flowers or peas, they are efficient pollinating and seed-disseminating agents. They have been frequently observed bibbing toddy from pots hung for toddy collection on coconut palms. Red vented bulbuls are found to damage fruits like apple, pear, peach and apricot in Himachal Pradesh. Fruits become unfit for consumption because of accumulation of toxic residues. The species is a minor pest of sapota and guava. Seeds of pomegranate are extracted after damaging fruits.

III: CROWS

House crow (common crow), *Corvus splendens* (Vieillot)

The common crow is found throughout India except eastern Kerala and south of a line roughly from Himachal Pradesh to Kutch (including lowlands of Nepal, Sikkim, Bhutan and Assam). Crows are inseparable from human habitations, abundant to pest proportions in many urban and metropolitan areas and totally commensal of man. They have 'traditional' community roosts in ancient banyan trees, mangrove jungle, coconut and forest plantations and the like, where vast numbers, sometimes may be thousands, converge at sunset flying in a steady stream of straggling rabbles over considerable distances from all points of the compass. Roosts are commonly shared with parakeets and mynas contributing to the deafening clamour that precedes sleep. The birds disperse before sunrise, radiating in disorderly groups, each headed towards its respective foraging ground, which can include fruit orchards.

Their food is practically everything that can be eaten. In Bangalore, house crow is a major pest on grapes. Compensatory increase in number of grape bunches per vine offsets damage due to bird pests. Crows also feed on berries packed after harvest. House crow caused fruit loss in Peach (*Prunus persica* B) in Punjab. Jungle crow and House crow together caused a loss of 12% to sapota under Bangalore conditions.

Indian Jungle crow, *Corvus macrorhynchos culminates* (Sykes)

It is found in Gangetic plain, all over peninsular India southwards upto Kanyakumari. Affect well-wooded country, outskirts of forests, villages as well as towns and cities.

Less gregarious and sociable than house Crow, usually keeping singly or in pairs or small scattered parties and occasionally in large gatherings when food is abundant. It is a regular marauder of bird nests, and often becomes a serious menace to poultry keepers as chicken-lifter, especially so when rearing its own young. Serves as a useful indicator to sportsmen and vultures of the presence of tiger or panther kills hidden within forest. Has the usual evening and morning flights to and from communal roosts, which are often shared with house crows, mynas and other birds.

The bird is omnivorous and feeds on wild and orchard fruits. Jungle crow damaged guava fruits in the hill region of Karnataka. Birds gnaw even unripe fruits, wasting far more than they actually eat. In Bangalore, Jungle crow is a major pest on grapes. In home vineyards, harvested bunch are advised to be covered with muslin bags and nylon netting to prevent damage. Bird scaring with traditional methods is also adopted. Jungle crow caused fruit loss in peach (*Prunus persica* B) in Punjab. The bird was the dominant species feeding on oranges in Karnataka (Mudigere). The crows cause about 12% loss in sapota.

IV: MYNAS

Indian myna, *Acridotheres tristis tristis* (Linnaeus)

It is ubiquitous. One of the commonest and most familiar birds in the entire subcontinent. Very abundant, quite at home in Srinagar even in snow. Addicted to the neighbourhood of human habitations and cultivation, villages, towns and cities as well as outlying homesteads in desert or forest.

They are found along with crows and sparrows. Is a confirmed commensal of man, quick to colonize virgin areas in his wake, even far out in the desert. Sociable disposition, aggressive, self-possessed demeanour and omnivorous habits combine to make the myna a highly successful colonist. Except in the breeding season it keeps in family parties, usually of 5 or 6 birds, which join up into flocks sometimes of many hundred strong.

Feeds on trees as well as on ground saundering along or hopping jauntily sideways to dig for grubs and worms on inundated lawns and *maidans*. Grazing cattle and burnt grass fields or scrub jungle likewise afford it attractive feeding opportunities. Flocks follow the plough for the insects and grubs which turn up with the soil, the birds leap-frogging or flying ahead in their eagerness to keep up with the advancing bullocks.

Its food chiefly includes fruits, grains, insects and grubs but also everything that can be eaten. Often causes damage to orchard fruits and standing cereal crops, but on the other hand is highly beneficial as a destroyer of serious agricultural pests. Indian mynas damage both immature and mature grape berries. Mynas of *Acridotheres* spp. were observed feeding on pomegranate fruits.

Indian Koel- *Eudynamys scolopacea scolopacea* (Linnaeus)

They are nomadic and local migrant throughout the subcontinent excepting Assam. They are described as summer visitor in many localities in north and peninsular India, and as winter visitor in the south. Rare or absent in the more arid tracts of western Rajasthan and northern Gujarat, but spreading gradually in the wake of advancing colonization and its inseparable concomitant, the house crow. Occurs locally in the outer Himalayas from Punjab and Kashmir including Nepal, Sikkim and Bhutan. Occurrence and local abundance closely linked with its principal fosterers, the house and jungle crows. Affect lightly wooded country, gardens, mango orchards and groves of trees in and around cultivation, towns and cities.

REVIEW OF FRUIT LOSS AND MANAGEMENT OF FRUGIVOROUS BIRDS

In Himachal Pradesh, it was observed that 20% of the fruits in apple, pea, peach and apricot were damaged by birds like Red billed blue magpie (*Cissa erythrorhyncha*), Red vented bulbul (*Pycnonotus cafer*), White cheeked bulbul (*Pycnonotus leucogenys*) and Slaty headed parakeets (*Psittacula himalayana*).

The depredators could be managed by spraying landarin 1.5 kg. *a.i.* ha. but the fruits become unfit for consumption due to accumulation of toxic residues. Much of the damage to orchards is because of the conversion or encroachment into the natural habitats of wild animals (Howe, 1990; Prakash *et al.*, 1971).

Roseringed parakeet (*Psittacula krameri* S.) is reported as a serious pest on guava. It also feeds on berries, fruit, blossoms and nectar (Bhatnagar *et al.*, 1982). In Hyderabad, guava fruits were consumed by birds throughout the year (Shivanarayan, 1982). In the hill region of Karnataka, small green barbet (*Megalaima viridis*) and jungle crow (*Corvus macrorhynchos* W.) were found damaging guava fruits, where birds on an average incurred loss of 14% unripe, 24% partially ripe and 33% ripened fruits (Chakravarthy, 1993). Birds gnaw even unripe fruits, wasting far more than they actually eat (Ali & Ripley, 1969). The loss caused was positively correlated to the age of fruit (Chakravarthy, 1993, Verghese & Tandon, 1993). In Ludhiana, guava suffered 20-30% fruit losses due to birds (Toor, 1982; Simwat & Sidhu, 1973; Ramzan & Toor, 1972). Red vented bulbul (*Pycnonotus cafer*) and Red whisked bulbul (*Pycnonotus jocosus*) were minor pests (Toor, 1982).

In guava orchards of Mudigere (Karnataka), the loss caused by parakeets was estimated to be 23.5%, warranting protection (Chakravarthy, 1993). Shooting, trapping, scaring, destruction of bird roosts and nests and encouragement of natural enemies like owls helped to reduce the pests (Singh & Kumar, 1982). Covering vulnerable portions of canopy during fruit-bearing period with thatch saved 20% fruits in Mudigere and Ckikmagalur in Karnataka (Chakravarthy, 1993).

Bank mynas (*Acridotheres ginginianus* L.) and Indian mynas (*Acridotheres tristis* L.) damage both immature and mature grape (*Vitin vinifera*) berries. In Ludhiana, Bank myna caused fruit loss of 19% and damage was severe during May-June (Toor, 1982), when alternate food sources were meagre. In Punjab, more damage (54.60%) occurred in grape vines trained with the Head System than in Bower System (35%). Birds preferred 'beauty' seedless variety of grapes having purple berries (Sandhu & Dhindsa, 1995).

In Bangalore, Jungle crow, (*Corvus macrorhynchos* W.) and House crow (*Corvus splendens* V.) are the major pests on grapes. The barbet (*M. viridis*) also incurs minor losses (Prasad & Verghese, 1985). Fruit loss due to birds in Bangalore ranged about 30-36% in varieties Arka Hans, Arka Shyam, Arka Kanchan and Bangalore Blue (Verghese, 1993). Compensatory increase in number of grape bunches per vine offsets damage due to bird pests (Verghese & Tandon, 1993). Birds also fed on berries packed after harvest in home vine yards. Bunches should be covered with muslin bags to prevent damage. The vineyards are covered with nylon netting to deter bird pests. Bird scaring with traditional methods are also adapted (Jindal, 1990).

Roseringed parakeets, house crow and jungle crow together caused 21.20% fruit loss in peach (*Prunus persica* B.) in Punjab (Chahal *et al.*, 1973; Toor & Ramzan, 1974; Toor, 1982; Sandhu & Dhindsa, 1995). In Punjab, Flordasun peach cultivar attracted flocks of House sparrow (*Passer domesticus* L.) during second week of January as new flush begin growing. As soon as the greenish-white flower buds appear, sparrows feed on flower buds to the extent that no seeds are produced. The average yield of Flordasun cultivar is 75 kg/tree. But the yield from bird affected trees was only 0-2 kg (Mann, 1986).

Narang & Chandel (1995) at Solan, Himachal Pradesh assessed the extent of damage caused by seven species of birds to apple (*Malus domestica* L.) orchards. The damage was maximum during 0600-0900 h., mostly confined to the upper and middle canopy of the tree. Maximum damage to fruits (80%) was caused by Blossom headed parakeet (*Psittacula cyanocephala* L.). The damaged fruits were subsequently fed by a number of insect pests and fungi adapted to decaying sweet resources. Such fruits became unfit for marketing. The early maturing compared to late maturing cultivars suffered more bird damage.

In hill region of Karnataka, oranges (*Citrus sinensis*) are grown in coffee (*Coffea arabica* and *Coffea robusta*) estates. Shade trees planted or naturally present in estates served as nesting and roosting sites for more than 90 bird species. Among the birds of Karnataka, jungle crow was the dominant species feeding on oranges. An individual required 12+ 1.4 min. (n= 18) to empty the contents of a single fruit. On an average, a crow destroyed eight fruits/day and in 15 days in an estate, crows destroyed 2000 fruits. Golden fronted Chloropsis (*Chloropsis cochinchinensis*) siphoned out juice by inserting beak into the rind while the fruit remained intact on the tree. Such fruits rot subsequently (Chakravarthy, 1993).

To sapota (*Achras sapota* L.), jungle crow and house crow caused a loss of 12% under Bangalore conditions. The crows are major pests while Red vented bulbul is a minor pest. In Mudigere, Chikmagalur (Karnataka), crow (*Corvus* spp.) along with other vertebrates caused 18% loss of fruits (Nirula *et al.*, 1954).

In pomegranate (*Punica granatum*), the main bird pests encountered were Small green barbet (*Megalaima viridis*), Coppersmith (*Megalaima haemacephala*) and mynas of *Acridotheres* spp. The damage is mainly by feeding on fruits (Chakravarthy, 1993).

Birds also inflict damage to papaya (*Carica papaya* L.). Bird damage commences when the fruit starts developing yellow colour. Two green barbets consumed on an average, pulp of a medium sized papaya in 12 min. (n=8) in Dakshina Kannada. Covering fruits with thatch, gunny bag or a festoon of thorny sticks, nylon nets and scaring facilitated management of vertebrate pests including birds (Chakravarthy, 1993).

In Punjab, the main bird pest recorded on mango (*Mangifera indica* L.) was Roseringed parakeet, which fed both on immature and mature fruits (Toor, 1982).

In pineapple (*Ananas comosus* L.) the loss caused by vertebrate pests mainly Jungle crows, squirrel and rat in coastal and hill regions of Karnataka amounted to 22% and 12.5%, respectively. In a field trial, 70% fruits in a control plot were destroyed by vertebrate pests compared to no damage on fruits covered with either leaf or dry thatch (Chakravarthy, 1993).

MANAGEMENT

The best means to manage bird's pestilence is to overlook it! However, if the damage exceeds the threshold of tolerance a farmer may have to resort to control strategies. Some of the successfully practiced and suggested management tactics are as follows:

1. Scaring: This is the most economical form of management. Scaring by the use of predatory decoys such as stuffed falcon is common. But, birds get habituated to this and the control may become less and less efficient as the days go by. The same would happen when reflective tapes are used or calls of predators are broadcast. I have found that manual scaring using slings and rattling an empty tin acts as the best repellent. A single labourer can manage easily a two-acre grape orchard. I have worked out the cost to scare away birds which amounts to Rs 700 from fruit ripening to harvest, which is the vulnerable stage for bird attack. Though it is believed that birds are active in the morning, vigil and watch should be from sunrise to sunset.

2. Netting: There are two types of netting. Placing fishnets all around the orchard as in grapes can prevent entry of birds into the orchards, In case of fig each tree is covered with net after full maturity. However, it is also possible to trap some of these birds using mist-nets and later releasing them away from the site of pestilence. But before this, the permission of the local conservator of forest or wildlife authorities should be obtained.

3. Planting of alternate hosts: I would put this category under habitat manipulation, wherein, every orchard should devote the border fence area to plant some wild fruit and even sunflower seeds to divert attention of the birds from the main crops.

4. Discouraging roosting: It is relevant to mention that the major frugivorous birds like crows, mynas and parakeets are community roosters and therefore, when leaving or arriving at a roost can indulge in group feeding. Discouraging birds from gathering near fruit ecosystem and from roosting by thinning heavily canopied trees. By chasing them away frequently at late dusk after they roost using fire works or crackers it is possible to prevent fruit damage on small scale.

CONCLUSIONS

The above strategies are non-destructive and chemical free and need to be advocated. Birds should be appreciated more for their ecological role and aesthetics. Pestilence up to a level is tolerable.

REFERENCES

ALI, S & S.D. RIPLEY. 1969. HANDBOOK OF THE BIRDS OF INDIA AND PAKISTAN. Vol. 3. Oxford University Press, Bombay, India.

BHATNAGAR, R. K., K.P. SRINIVASA., AGNIHOTRI., M.G.JOTWANI & V.T.GAJBHIYE. 1982. Efficacy of bird repellents in sorghum at maturity stage of crop. In: MANAGEMENT OF PROBLEM BIRDS IN AVIATION AND AGRICULTURE. (Eds): R.A. Agarwal & R.K. Bhatnagar. Indian Agricultural Research Institute, New Delhi, India. pp. 224-226.

CHAHAL, H. S., G.S. SIMWAT & H.S. BRAR. 1973. Bird pests of crops and their control. *Pesticides* 50(9): 715-719.

CHAKRAVARTHY, A.K.1993. *Vertebrate Pest Management*. Final Report of ICAR ad-hoc project, University of Agricultural Sciences, Bangalore, India. 98 p.

HOWE, H.F. 1990. Seed dispersal by birds and mammals; Implications for seedling demography In: REPRODUCTIVE ECOLOGY OF TROPICAL FOREST PLANTS (Eds.): K.S. Bawa & M. Hadley, MAB Series. pp 191-218.

JINDAL, P.C. 1990. Grape. In: FRUITS: TROPICAL AND SUB-TROPICAL. (Eds.): T.K Bose & S.K.Mitra. Naya Prakash, Calcutta, India. pp. 186-225.

MANN, G.S. 1986. House sparrows (*Passer domesticus* Linn.) causing severe damage to peach buds in residential areas at Ludhiana, Punjab. *Tropical Pest Manage.* 32 (1): 43.

NARANG, M.L. & J.S. CHANDEL. 1995. Preliminary studies on bird damage to apple orchards in Himachal Pradesh. Abstract, *Environment and Birds*, Secretariat, Box 45, Srinagar Garhwal, U. P, India pp: 54-55.

NIRULA, K.K., J. ANTONY & K.P.V. MENON. 1954. The flying fox, *Pteropus edwardsii*- a pest of coconut palm in India. *The Indian Coconut Journal* 12: 71-76.

PRAKASH, I., R.K. GUPTA., A.P. JAIN., B.D. RANA & B.K. DUTTA. 1971. Ecological evaluation of rodent populations in the desert biome of Rajasthan. *Mammalia* 35: 384-423.

PRASAD, V. G & A. VERGHESE. 1985. Birds as pests of horticultural crops. *Bull. Entomol.* 26(1): 94-96.

RAMZAN, M & H.S. TOOR. 1972. Studies on damage to guava fruit due to rose-ringed parakeet, *Psittacula krameri* (Scopoli) in the Punjab. *J. Bombay nat. Hist. Soc.* 70(1): 201-204.

SANDHU, P.S & M.S. DHINDSA.1995. Orchard birds- a review. In: ENVIRONMENT AND BIRDS. Abstracts, Secretariat Box 45, Srinagar, Garhwal, U. P, India. 57 p.

SHIVANARAYAN, N. 1982. The Roseringed Parakeet, the problem bird in agriculture and its management. In: MANAGEMENT OF PROBLEM BIRDS IN AVIATION AND AGRICULTURE. (Eds.): R.A. Agarwal & R.K. Bhatnagar, Indian Agricultural Research Institute, New Delhi, India. pp. 151-158.

SIMWAT, G.S. & A.S. SIDHU.1973. Note on the feeding habits of Roseringed Parakeet, *Psittacula krameri. Ind. J. agri. Sci.* 43(6): 607-609.

SINGH, A. K & P. KUMAR.1982. Damage to maize and guava crops by the Parakeets. In: MANAGEMENT OF PROBLEM BIRDS IN AVIATION AND AGRICULTURE. (Eds.): R.A. Agarwal & R.K. Bhatnagar., Indian Agricultural Research Institute, New Delhi, India. pp. 164-167.

TOOR, H.S. 1982. Problem birds and their management in Punjab. In: MANAGEMENT OF PROBLEM BIRDS IN AVIATION AND AGRICULTURE. (Eds.): R.A. Agarwal & R.K. Bhatnagar, Indian Agricultural Research Institute, New Delhi, India. pp. 132-142.

TOOR, H. S & M. RAMZAN. 1974. A study on bird pests of grapes. *Punjab Horticulture Journal* 14: 46-48.

VERGHESE, A. 1993. Report submitted to Indian Institute of Horticultural Research, Hessaraghatta, Bangalore, India.

VERGHESE, A & P. L. TANDON, 1993. Enhanced number of grape bunches per vine offsets damage due to bird pests. *Newsletter for Birdwatchers* 33(4): 70-71.

❑❑❑

4

BIRD PEST MANAGEMENT – IS THERE A SATISFACTORY SOLUTION?

Shakunthala Sridhara

AINP on Rodent Control, University of Agricultural Sciences, G.K.V.K., Bangalore – 560 065

ABSTRACT

The paper gives an overview of non-lethal and lethal methods of managing pest birds worldwide. It deals with various approaches such as: removal methods (trapping and netting), frightening devices (visual and auditory scares) and biological methods (encouraging predators and cultivating trap crops) under non-lethal approaches. The extensive data on lethal methods of bird pest management furnishes information on chemicals used as : avicides, contact poisons, repellents, chemosterilants and miscellaneous chemicals. The avicides described include starlicide, DRC-1339 and CPT. Contact poisons used are endrin, fenthion and malathion. The successful avian repellents are classified as primary and secondary repellents with methyl anthranilate, sucrose and thiram falling under the former and methiocarb, 4-aminopyridine, anthraquinone and copper oxychloride coming under the latter group. There is only one chemosterilant namely Ornitrol. The other chemicals used to manage with less than satisfactory results are malathion, captan, trimethocarb, copper oxalate, capaicin, neem products, strychnine and alphachlor-alose. In the final analysis, no single method is adequate to prevent bird damage to agricultural crops. The approach has to be an integrated one based on the biology, ecology and behaviour of the pest as well as its relationship to the crop being damaged.

INTRODUCTION

The problem of birds in Indian agriculture is negligible. Only few species of granivorous and frugivorous birds cause damage which is not alarming. The granivorous birds are the baya weaver *Ploceus philippinus*, munia of *Lanchura* sp., the Roseringed Parakeet, (*Psitticula krameri*), Blue-rock pigeon (*Columba livia*), Common peafowl (*Pavo cristatus*), Demoiselle Crane (*Antrhopoides virgo*), Ring dove (*Streptopelia decaoto*), Rosy pastor (*Sturnus roseus*), Common myna (*Acridotheres tristis*), Red-vented bulbul, (*Pycnonotus cafer*) and Jungle babbler (*Turdoides striatus*). The frugivorous are Slaty headed parakeet (*Psittacula himalayana*), Blossom headed parakeet (*Psittacula cyanocephala*), House crow (*Corvus splendens*), Jungle crow (*Corvus macrorhynchos*), Indian myna (*Acridotheres tristis*), Indian koel (*Eudynamis scolopaceae*), bulbuls (*Pycnonotus* sp.) and barbets (*Megalaima* sp.).

The magnitude of damage and the density of pest bird population in India is no where near the millions of quaelea in Africa and North-America. Most bird species of our agro ecosystem are in fact beneficial by depending on many insect and rodent pests of agriculture. It is only in pockets of food grain and fruit cultivation that birds become a major problem. In the absence of acceptable, non-lethal management techniques suited to each problem situation, the problem appears unmanageable. It is in this context that this review paper describes and evaluates the different techniques of bird pest management that is in vogue worldwide. The aim is to create awareness about these approaches so that a person with the responsibility of deciding the *modus operandi* of bird management in India has an array of successful techniques at his disposal which he/she can suitably modify to suit local problems and environment.

I NON CHEMICAL BIRD MANAGEMENT (Table 1)

(i) TRAPPING

Trapping of small sized bird pests can be successful at locations where a static population is causing damage or where other techniques cannot be used. There are a variety of bird traps especially for sparrows, starlings and pigeons. Two types of entrances are used: Funnel type and Bob type. In funnel type it is easy for the birds to get in and difficult to get out. The Bob type, has vertical rods called bobs which are easily pushed inside as the bird enters the trap but pushing them out is difficult. Pre baiting is a requisite for both types of traps.

The Australian crow trap (MAC) has been suitably modified to capture parakeets and sparrows in the agricultural fields of Pakistan (Brooks *et al.*, 1990) and is called PAROTRAP. It consists of six pieces along with one entrance board with a longitudinal slot to allow the entry of the birds. With the help of bolt and nuts the trap can be assembled or dismantled within 10-15 minutes. It should be provided with 2-4 decoy birds of both sexes, a poultry drinker, enough feed consisting of fresh fruits and millet seeds. Food and water should be replenished. Traps should be placed at the edge of ripening crops, favourite perching and roosting sites for effective captures.

Table 1. Non-chemical approaches to Bird Pest Management

MAJOR APPROACH	SUB-DIVISIONS WITH EXAMPLES
I. REMOVAL METHODS	1. Trapping e.g. Funnel type, bob type, MAC 2. Netting e.g. Mist net, fish net, plastic nets 3. Shooting
II FRIGHTENING DEVICES	1. Visual scares e.g. Balloons, kites, scare crows, mylar tape, spot lights, flash light, strobe light, laser-bean. 2. Auditory scares (a) Physical: Bangers, propane exploders, pyrotechnics, shell crackers, Rockets, Acetylene exploders. (b) Biological: Distress calls, Alarm calls.
III BIOLOGICAL METHODS	(i) Installing nest boxes for predatory birds. (ii) Trap crops

(ii) NETTING

Netting is useful for protecting fruit crops such as cherries and grapes from birds. It is cost effective in protecting high value crops. Netting, if used carefully lasts five years. Properly installed and maintained, nets have been shown to block the entry of pest birds to cereal crops in several African countries. They can be laid directly or can be supported with tree branches slightly above the crop. The protection becomes more effective by supporting nets over crops on portable or permanent metal frameworks or poles. However, the cost of such supported nets limits its use to high value crops, research trials and small areas of seed multiplication crops. The different types of nets used are nylon, fish-net, ultraviolet–stabilized plastic nets and meshes of varying size (2.0-2.5 cm^2 to 4.0 x 6.0 cm) (Bruggers & Ruelle, 1982).

Nets should be installed before the damage starts. Trapped birds should be regularly removed; otherwise captured birds attract other birds.

Nets are non-toxic, noiseless, can be reused and are potentially 100% effective (Stuckey, 1973). The high cost of netting is the major constraint in using it in developing countries to protect crops. But in Africa it was found that the cost of netting was less than hiring people to scare birds (Bruggers & Ruelle, 1982).

(iii) SHOOTING

Shooting is effective more as a dispersal technique than as an attempt to reduce pest numbers. Though the number of pests killed is small, shooting can supplement and reinforce other methods of mitigating damage mainly auditory scares (Table 1). Shooting was effective against ducks, pigeons and pukeko in New Zealand (Coleman & Spurr, 2001).

(iv) FRIGHTENING DEVICES

Birds can be frightened by using one or more visual or auditory scaring devices. These tools are most effective when used for short periods of time because the birds get habituated to the devices after a few harmless exposures.

(a) Visual scares include Helium filled balloons with or without eyespots, Hawk or owl kites, Scare crows, Spot lights, Flashing light, Strobe lights, Mylar tape and laser. Of these, only Scare crows and Mylar tapes are used in India.

Scare crow is the oldest device used to discourage birds from depredating on crops. With proper knowledge of feeding and behaviour of birds, they can be made quite effective. Scare crows in India are made of a variety of materials–old clothing, grain sacks filled with straw, painted mud pots hung on a pole, white plastic flag atop a stick or pole, etc. When cloth and plastic sheets swing in the wind repellency increases. It is important to change the position of scare crows every 3 days, a practice not employed by farmers in India.

Mylar tape is an elastic, transparent, synthetic polypropylene metallic shining tape with a silver metal layer on one side and red resin on the other side. By reflecting sunlight it flashes regularly. On being stretched between poles it pulsates making a noise which ranges from hissing to loud humming and sometimes like a thunder. The tapes are usually 10-15 meters long and 15mm in width. About 20-30 strips are recommended to be fixed in an acre with the help of bamboo poles, one foot above the crop level. The tapes should be 5-8 meters apart.

Lasers have been reported to be potential in repelling birds. Earlier experiments have shown them to be effective in disbursing gulls, double crested cormorants. However, a later study gave mixed results. European starlings were not repelled but rock doves, mallards and geese exhibited marked avoidance of the beam, under experimental conditions (Blackwell *et al.*, 2002).

(b) Auditory Repellents

Bangers make a big noise ("bang") are extensively used as frightening devices. Eg. Cracker shell fired from a shotgun explodes loudly twice after travelling 100-150m, once when the shell is fired and again when it explodes. Modifications include starter pistols, rocket bombs, screamer shells, whistle bombs and rope of fire crackers.

Propane exploders are devices in which propane is ignited in a firing chamber by a spark. This is a very commonly used device, producing multiple loud explosions. Shot gun used intermittently reinforces the effect.

Shell crackers are shells fired from a 12-gauge shotgun, which explodes 65 to 75m away. But the major drawbacks is the possibility of dry vegetation getting burnt.

Rockets need a launching rod. They have a much greater range than shell crackers. The hissing sound as it travels through the air frightens the birds.

Acetylene exploder was effective in deterring parakeets from damaging guava in Punjab (Simwat, 1975).

Automatic mechanical bird scares or pyrotechnique method is ideal for fields of 4-5 acres, especially in orchards. It should be operated only at the time of bird visit to fields and steps to prevent habitation are essential (Raheja, 1992).

Distress calls are emitted when a bird is captured by a predator which shocks the predator into dropping the prey. Factors that influence the effectiveness of distress calls are: its repellency, the rate of habituation and the effect of regional dialect. A distress call has multiple functions. It can attract and warn conspecifics, can startle the predator to loosen its grip or attract another predator. Parakeets were frightened and repelled on hearing recorded distress calls both in the laboratory and field conditions in Punjab (Simwat, 1975). Recorded alarm and distress calls, when played in the habitat of pest birds are very effective and are commercially available. Amplified calls are more successful. The broadcast equipment of distress calls is very important for successful dispersal and repellence of birds. So also accuracy, signal strength and clarity. Mobility of broadcast units further enhances the effectiveness of calls. Electronically produced sounds such as AV-alarm or other sounds also scare birds but they are not as effective as amplified distress and alarm calls of birds.

Alarm calls are softer than distress calls and are emitted to communicate the presence of predators to conspecifics. Alarm calls are not investigated as thoroughly as distress calls. They are emitted to alert conspecifics when the bird is in danger and may result either in fleeing or in crouching down and hiding. They are considered to be superior to distress call in dispersing or repelling birds (Schmidt & Johnson, 1983). They are commercially available.

Habituation: Although frightening devices produce immediate effects, birds get quickly habituated and the devices become useless. Habituation is affected by the type, duration and intensity of the stimulus and the animal's hunger. They are most effective when used in short spells such as during germination of seed. The impact can be enhanced by combining with other stimuli such as light, by random timing pattern in using and habitat management (Bushel & Giban, 1968).

II BIRD MANAGEMENT USING CHEMICALS

Bird management involves use of a limited number of chemicals. Because of high level of public displeasure against and disapproval of using toxicants for preventing bird damage, the choice of chemicals that are effective and non lethal narrows down to very few. Basically chemicals used to prevent bird damage can be categorized under lethal chemicals, repellents, chemosterilants and miscellaneous (Table 2).

(i) LETHAL CONTROL can be by ingestion or absorption.

(a) Avicides

(1) Ingestive poisons

Starlicide is the only registered chemical to kill pest birds. Starlicide, 3-chloro-p-toulidine hydrochloride is a slow acting but highly toxic avicide to starlings, blackbirds and crows and is used in grain baits around livestock and poultry in USA. It is slow acting and painless to birds. Time to death ranges 3-50 hours after ingestion. Toxic symptoms include restlessness, inactivity, slightly increased and difficult breathing. There are no convulsions before dying. Death is due to uraemia i.e. uric acid accumulation in blood leading to deterioration of blood circulation in the liver and kidney. The compound is completely metabolized within 2-5 hours before death and the metabolites are excreted which are non-toxic. Thus there is no secondary poisoning and environmental pollution.

Table 2. Bird pest management using chemicals

TYPE OF CHEMICAL	EXAMPLES & MODE OF ACTION
AVICIDES	Starlicide- Death due to accumulation of uric acid, painless.
	DRC-1339-Believed to cause irreversible necrosis of kidney.
	CPT-Lowers surface tension of water and enhances wetting leading to heat loss and death.
CONTACT POISONS	Endrin (chlorinated hydrocarbon, insecticide)
	Fenthion (organophosphate chemical, insecticide)
	Malathion (insecticide)
REPELLENTS	Primary repellents
	Methylanthranilate-Irritant
	Sucrose-Birds lack sucrase-hence aversive
	Thiram (fungicide)-depresses CNS
	(ii) Secondary repellents
	Methiocarb-(insecticidal origin) inhibits AChase
	4-aminopyridine –causes post ingestional distress
	Anthraquinone - causes post ingestional distress
	Copper oxychloride (fungicide)-irritation of gut
CHEMOSTE-RILANTS	Ornitrol inhibits egg production
OTHER CHEMICALS	Malathion (insecticide) } aversive repellents
	Captan (insecticide) } aversive repellents
	Trimethocarb-repellent
	Copper oxalate – repellent
	Capsaicin – not satisfactory
	Neem products – potential antifeedents
	Commercial products ingrained with bitter chemical compounds, naphthalene etc.
	Strychnine-lethal
	Alphachloralose-narcotic

Starlicide is used at 1% in grain or poultry feed baits. The treated bait is mixed with untreated bait in the ratio of 1: 9 and broadcast. It is always advisable to use fresh baits of starlicide as storage affects the potency of the chemical. Prebaiting is necessary, preferably using high quality poultry pellets that resemble toxic bait in colour, size and texture. Baiting should be continued for 6 continuous weeks to achieve 99% kill (Brooks & Hussain, 1990).

Cautions: Starlicide is poisonous to chicken, turkey, ducks and other birds. All precautions should be taken not to expose the poisoned baits to poultry, livestock and non-target wildlife.

DRC-1339 (3-chloro-4-methyl-analine-hydrochloride) is emerging as a successful avicide. It is highly toxic to most pest birds but low to moderately toxic to most predatory birds and almost all mammals. Its primary action is yet to be understood although it is believed to cause irreversible necrosis of the kidney. The affected bird is unable to excrete uric acid. Death occurs 1-3 days from uraemia after most of the poison has been metabolised and excreted. Thus there is no threat of secondary poisoning and environment toxicity. It is successfully used to control blackbirds and starlings at feedlots, problem gull species in and around their nesting site.

LD_{50} value of DRC–1339 has been determined for 55 bird species, acute toxicity for six species and dietary tests on 12 species which include blackbirds, corvids, rock doves, gulls and starlings in the laboratory (Eisemann *et al.*, 2003) and in some field conditions (Linz *et al.*, 2002).

CPT (3-chloro-4-methylbenzenamine, CAS 95-74-9) or PA 14 is a potential toxicant for aerial application to control blackbird roosts in America (Mott *et al.*, 1987). The avicide is formulated as CPT concentrate (15.6%) with the composition of 35.9% propylene glycol, 46.9% isopropyl alcohol, 1.6% linseed oil and 15.6% CPT. This formulation is mixed with florescent particle marker and water (10% v/v.). When applied at the rate of 4.3 to 11 lbs/a, it killed 99% of the red-winged blackbirds in the laboratory (Lefebure *et al.*, 1987) and 3% of blackbird and starling population at the roost site (Heisterberger *et al.*, 1990). The toxicant was demonstrated to be safe and effective in reducing large populations of starlings and blackbirds (Lefebure & Seubert, 1970). The chemical lowers the surface tension of water and enhances wetting. Birds die when the plumage is sufficiently wetted with PA-14 during or before the cold weather due to excessive loss of heat. In addition to low temperature there should be about 10-12mm rain during or after the treatment. The chemical is toxic to fish and to plants to some extent.

(iv) Endrin (Avicide, organochlorine insecticide)

1,2,3,4,10-hexachloro-6,7-epoxy-1,4,4a,5,6,7,8,8a-octahydro-exo-1,4-exo-5,8-dimethanonaphthalene $C_{12}H_8Cl_{16}0$

Endrin is recommended to be used at 9.4%. Pearl millet containing 0.3% endrin was found lethal to house crow (Sidhu & Simwat, 1973) and house sparrow (Narang & Bindra, 1971). Endrin treated baits when kept on roof top, were avoided by parakeets but readily consumed by doves and mynas which later succumbed (Simwat, 1975). Acutely toxic to both invertebrates and vertebrates.

(v) Fenthion $C_{10}H_{15}O_3PS_2$

O,O-dimethyl*O*-[3-methyl-4-(methylthio)phenyl] phosphorothioate. Fenthion is recommended at 11% to repel birds. Moderately toxic. Cholinesterase inhibitor. Potential ground water contaminant.

(vi) Malathion

Dusting parakeet nests with 5% malathion dust was suggested to control the pest bird (Simwat, 1975).

(2) Contact poisons

Birds usually perch briefly at a convenient, safe, peripheral point, survey the surroundings and then fly to a nest, perch or food source. Control with contact poisons makes use of this behaviour of birds. Suitable toxic wick perches are developed to act as temporary perches while using contact poisons. Essentially these perches are hollow metal tubes with a permeable wick which comes in contact with the feet of perching bird and facilitates transfer of toxic solution into the body of the bird through skin. Death follows a few hours later.

Two insecticides namely endrin and fenthion are used as contact toxicants to poison birds. Both are readily absorbed through the skin. The wick inside the hollow perch is soaked in either of the solution. Endrin is a chlorinated hydrocarbon and fenthion is an organophosphate chemical. Fenthion is comparatively less hazardous to humans and non-target animals in addition to being less persistent in the environment.

(ii) REPELLENTS

Avian repellents are of two types- primary and secondary. Chemicals that irritate the peripheral senses and hence reflexively avoided are called primary repellents. Primary repellents are usually derived from natural products mostly human food and flavour ingredients. These are less potent than secondary repellents. Secondary repellents are those that induce gastrointestinal illness upon ingestion. Birds learn to associate the illness with the paired sensory cues and avoid such feeds. These are highly effective bird repellents and are mostly derived from synthetic agricultural pesticides (Table 2). They are generally regulated because of their toxicity and unfavourable effects on the fauna, flora and environment.

(a) **Primary repellents** include Methyl anthranilate, Copper oxychloride, Sucrose and Thiram

Methyl Anthranilate (MA) (O-aminobenzoic acid methyl esther) is a naturally occurring compound in 'corcurd' grapes. Methyl anthranilate is extensively used to impart a 'fruity' flavour to candy, gum, soft drinks etc, but it is an irritant to birds acting through the trigeminal nerve. Birds treated with Methyl anthranilate consistently reject food and water which is a reflex response, not learnt. As the irritation caused is not strong, birds tend to return to the tasted food and thus damage cannot be prevented (Curtis *et al.*, 1994). Field evaluations of Methyl anthranilate as a repellent are very few. Blueberry treated with MA did not deter depredation by birds but aerial application of Methyl anthranilate on corn and sunflower reduced blackbird damage to these crops (Askham, 2000). The chemical effectively reduced grazing by geese at golf courses, parks, winter wheat and dispersed nuisance birds from roosts, airports, land fills and temporary water bodies (Cummings *et al.*, 1995). Recent experiments suggest the possibility of using Methyl anthranilate as a secondary repellent. The idea is to encapsulate the Methyl anthranilate so that the chemical is ingested without any irritation but once inside the gut, irritation by the chemical produces emetic reaction leading to learned avoidance response (Sayre & Clark, 1999).

Sucrose has been found to reduce bird depredations in fruits. Sucrose solutions sprayed repelled red winged blackbirds, American robins and European starlings (Socci *et al.*, 1997). The repellency is due to lack of sucrose hydrolysing enzyme in birds.

Thiram (Tetram-ethylthiaram disulphide fungicide). The chemical depresses the activity of central nervous system. It has low oral toxicity. When used as a dressing at 0.5%, thiram reduced crow damage to maize seedlings significantly. Thiram treatment was as effective as covering the seed bed with netting (Sandhu *et al.*, 1987).

(b) Secondary repellents

(i) Methiocarb (Mesurol) 3,5-Dimethyl-4-(methylthio)phenyl methylcarbamate:

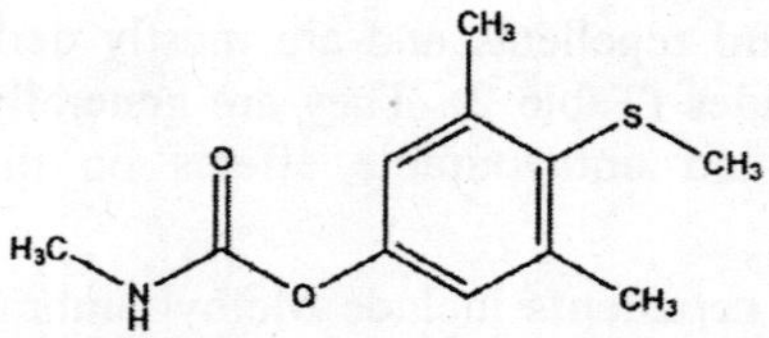

Methiocarb has its origin in a carbam insecticide. It acts by inhibiting acetylcholinesterase activity at synapses in the nervous system. However this action is reversible. Poisoned birds exhibit distress symptoms such as retching, vomiting and temporary paralysis. Symptoms may appear within 10 minutes after ingestion and paralysis may occur within 30 minutes. Treated food is readily accepted. Feeding stops when the bird starts sensing physiological effects. It is a secondary repellent. Birds associate post ingestional distress with the ingested food and learn to avoid it. Often the avoidance is location specific. Methiocarb is extremely toxic.

In North-America it is extensively used to protect newly seeded and sprouted crops, ripening grain, soft fruits such as cherries, grapes and blueberries. The crops protected were rice in USA (Mott *et al.*, 1976), West Africa (Ruelle & Bruggers, 1979) and Philippines (Garrison & Libay, 1982), maize in India (Sandhu *et al.*, 1987) and sprouting wheat in Bangladesh (Brooks *et al.*, 1982).

Toxic to amphibians, crustaceans, echinoderms, fish, insects, molluscs, phytoplankton and zooplankton. The registration of methiocarb in USA lapsed in 1989 as the registrant declined to furnish additional data as specified by the U.S. Environmental Protection Agency.

4-Aminopyridine (Avitrol) Pyridine-4-amine $C_5H_6N_2$

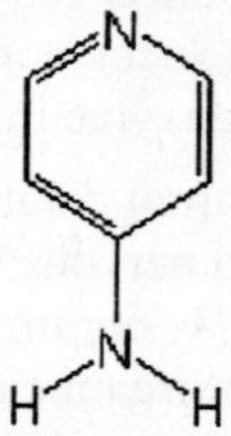

Avitrol or 4-aminopyridine is a restricted use pesticide and highly toxic chemical used in bird management. It is used as a frightening repellent to scare social birds that feed in flocks.

It is available in several bait formulations and used as a chemical frightening agent. Such baits contain a small number of treated grains or pellets mixed with more quantity of untreated grains or pellets. Birds after eating the treated portion of the bait behave erratically; give warning cries which frightens the rest of the flock forcing them to leave the treated area. Usually birds which consume treated particles die. If the dosage is sub-lethal, then the chemical induces bait shyness. After one exposure, the bait-shy birds do not visit the treated fields again necessitating prebaiting for several days with untreated pellets. Persistence of bait shyness can be tackled with changing location of baits and with additional prebaiting.

The chemical is highly toxic to invertebrates, birds and mammals with LD_{50} values generally less than 10mg/kg. Toxicity ranged 2.5 to 5mg/kg. for doves, parakeets, sparrows and munias of Bangladesh. Avitrol is used at 3% concentrations in the treated grain, usually cracked corn. Treated grains are mixed with untreated grains in the proportion of 1:29 and broadcast in the field at the rate of 1kg/ha (Sultana *et al.*, 1986).

Broadcasting aminopyridine treated sunflower and corn reduced blackbird damage to sunflower, field corn, sweet corn, peanut and pecan in USA (De Grazio *et al.*, 1972; Knittle *et al.*, 1988). Spraying maize heads at milky stage with 2% or 3% 4-aminopyridine solution just before harvest reduced parakeet damage substantially in Bangladesh. Although the chemical is expensive, the treatment may not be costly as the amount required is very small (Sultana *et al.*, 1986).

Anthraquinone is a secondary repellent causing post-ingestional distress. However it does not affect the nervous system as methiocarb and does not immobilize the birds. There is no aversion to treated food until it is consumed. It is a registered chemical for bird management in Europe. Recently it has been registered in the name of **flight control** as a repellent for birds at turf, grass and roosting site of nuisance birds. It also repelled birds from corn and rice (Blackwell *et al.*, 2002). However its potency was found reduced 6-7 days after application indicating that additional applications are necessary at regular intervals (Dolbeer *et al.*, 1998).

The chemical has low toxicity to birds and mammals and is harmless to insects. There is no hazard to non-target animals.

Copper Oxychloride: Copper oxy-chloride is a fungicide which was found to reduce consumption of treated food (Avery & Decker, 1991). Although the mode of action is not clear, copper ingestion is thought to lead to irritation of the gut lining and thus act as a secondary repellent.

It is readily available and inexpensive in Bangladesh where it is used as a fungicide in potato cultivation. In parts of the country farmers have been traditionally using it to protect germinating wheat from birds. It was found to be an effective repellent at 1-2% giving a cost: benefit ratio ranging 1:30 to 1:60 (Brooks *et al.*, 1982). Although copper oxychloride is comparatively less repellent than methiocarb, its availability and lower cost make it a recommended chemical to repel birds from crop fields of Bangladesh.

Other Repellents

Malathion dust, **captan** and **thiram** when coated on groundnut kernels at 1.5% induced aversion towards the treated seeds both in the laboratory and field conditions (Simwat, 1975).

Trimethacarb, a chemical related to methiocarb was found to be an effective repellent but was not registered.

Copper oxalate at 0.5 to 1.0% concentrations protected maize and wheat seedlings from bird damage (Table 2).

Although capsaicin, the active principle in hot capsicum peppers is a doubtful deterrent to birds, various capsaicin based products are being sold to repel birds. In fact it is well documented that birds are relatively insensitive to capsaicin (Mason & Clark, 1995).

Derivatives of the neem plant, *Azardirachta indica* are reported to have bird repellent properties (Shivanarayan & Rao, 1988) which was attributed to the compound azadirachtin present in them.

There are several commercial products in USA, UK and Australia which are effective in repelling birds. These are incorporated with substances that have offensive properties to humans. **Ropel** in spray or granular form repel geese, ducks and wood peckers. It contains the active ingredient, denatonium sachharide and Thimal. **By Birdie** marketed in granular forms contains 100% naphthalene and deters starlings, pigeons, sparrows and other birds from structures. An aluminium ammonium sulphate based product called **Duck off** deterred ducks and other species from golf courses and other areas. The compound protected ripening cereals for several weeks from granivorous birds in Africa (Bruggers, 1979). Bitter compounds are aversive at least to some bird species so it is likely that research in this line may lead to formulation of safe, effective bird repellents.

(iii) CHEMOSTERILANTS

The lone registered avian chemosterilants is Ornitrol® (azacosteral, 20,25-diazacholesterol dinydrochloride). It has been found highly effective for pigeons. Used at 1% in treated maize, it caused temporary sterility in pigeons, the effect lasting six months and thus requiring two baits a year (Wolford & Elder, 1967). Sterility is due to failure of egg implantation. The treated baits have to be scattered for 10 days in fields/areas frequented by pigeons. As the compound is water soluble, baiting should be avoided during rainy days. The compound is not used any longer as the cost of using in a year round breeding pigeon is prohibitive.

(iv) MISCELLANEOUS CHEMICALS

(a) Strychnine (Avicide, Botanical Rodenticide)

Strychnine-10-one

$C_{21}H_{22}N_2O_2$

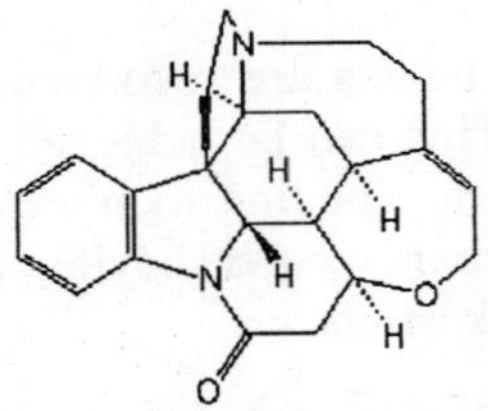

Baits containing 0.4% strychnine alkaloid were effective against pigeons in laboratory studies (Schaffer & Eschen, 1986).

(b) Narcotics

Alpha Chloralose

$C_8H_{11}Cl_3O_6$

1,2-*O*-[(1*R*)-2,2,2-trichloroethylidene]-α-D-glucofuranose

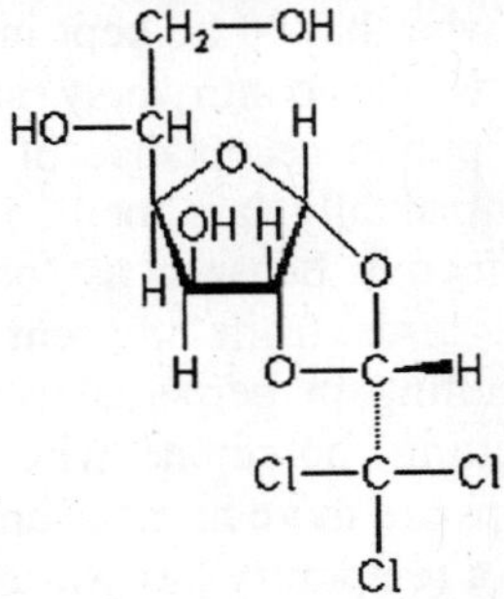

Alphachloralose ($C_8H_{11}Cl_3O_6$) is a chloral derivative of glucose which depresses the cortical centers of the brain but medulla is not affected. Alphachloralose was used to control corvids during world war

two in France. Later, it was also utilized to capture several species of birds for research as well as to reduce population of pest species. It is marketed as a rodenticide in West Germany and England and has been registered as an avicide in England, France, New Zealand and Australia. Between 1988 and 1990, alphachloralose was used to immobilise and remove Canada geese, mallards, ducks and coots from commercial and residential sites in USA. The optimum dose used was 20-30mg/kg (Waronecki *et al.*, 1990).

Studies carried out in Punjab revealed that parakeets, common myna, bank myna, ring dove and large grey babbler readily ate maize treated with 2% alphachloralose initially but soon they flew away from the treated area (Simwat, 1975).

(v) BIOLOGICAL CONTROL

Kestrels or sparrow hawks are recommended to control starling, blackbird and sparrows. This can be achieved by installing nest boxes and perches. As kestrels do not line their nest, nest boxes should be lined with wood chips or coarse wood shaving. It may take a year or two to attract a nesting pair of kestrels.

CONCLUSIONS

Once damage levels reach economic threshold, management of pest population of birds becomes essential. There are several methods to achieve this such as exclusion by way of netting, physical deterrents, acoustic, ultrasonic and scare devices, population reduction by trapping, shooting, poisoning, irritants and taste deterrents. However, no single method is adequate to get the desired results. When used in combination, a successful management procedure can be evolved using a series of methods. But what has to be kept in mind in evolving such management techniques is the controversy surrounding the use of avicides and the lack of public acceptance of such toxicants and the need to decide an environmentally safe method. Poisoning birds is not only cruel but very ineffective because as long as the area remains attractive and accessible, birds from adjacent fields will move into newly vacant niches, resulting in perpetual conflict which cannot be resolved with killing. Netting, porcupine wire, frightening devices in combination with trap crops are more humane and environment friendly. Killing birds can only be a temporary but wasteful solution to problem birds in India as elsewhere.

REFERENCES

ASKHAM, L.R. 2000. *Proc. Vertebr. Pest Conf.* 19: 22-25.

BLACKWELL, B, F., G.E. BERNHARDT & R.A. DOLBEER. 2002. Lasers as non lethal avian repellents. *J. Wildl. Manag.* 66 (1): 250-258.

BROOKS, J.E & I. HUSSAIN. 1990. Chemicals for bird control. In: VERTEBRATE PEST MANAGEMENT-TRAINING MANUAL. (Eds.): J.E. Brooks, E. Ahmed, I. Hussain, S. Munir & A.A. Khan. Pakistan Agricultural Research Council, Islamabad, Pakistan. pp. 193-195.

BROOKS, J.E, P. SULTANA., Md. M. RAHMAN., P. BRINK & Md.A. SIDDIQUE. 1982. *Control of pest bird damage to sprouting wheat.* Bangladesh Agricultural Research Institute, Vertebrate Pest Section, Dacca, Bangladesh. Technical report No. 15. 8pp. Mimeo.

BROOKS, J.E., E. AHMAD, I. HUSSAIN, S. MUNIR & A.A. KHAN. 1990. A TRAINING MANUAL ON VERTEBRATE PEST MANAGEMENT. Pakistan Agricultural Research Council, Islamabad, Pakistan 206 p.

BRUGGERS, R.L & P. RUELLE. 1982. Efficacy of nets and fibres for protecting crops from grain-eating birds in Africa. *Crop Prot.* 1(1): 55-65.

BRUGGERS, R.L. 1979. Vertebrate Pest control and management materials, ASTM, STP 680. American Society for Testing and Materials, Philadelphia, USA. pp. 188-197.

BUSHEL, R.G & J. GIBAN. 1968. In: THE PROBLEMS OF BIRDS AS PESTS (Eds.) R.K. Murton & E.N. Wright. Academic Press, New York. pp. 17-18.

COLEMAN, J.D & E.B. SPURR. 2001, Farmer perceptions of bird damage and control in arable crops. *New Zealand plant Prot.* 54:184-187.

CUMMINGS, J.K., P.A. POSCHOP, J.E. DAVIS, Jr. & H.W. KRUPA. 1995. *J. Wildl. Manage.* 59: 47-50.

CURTIS, P.D., I.A. MERWIN., M.P. PRITTS & D.V. PETERSON. 1994. Chemical repellents and plastic netting for reducing bird damage to sweet cherries and grapes. *Hort. Science* 29(10): 1151-1155.

DeGRAZIO, J.W., J.F. BESSER., T.J. DeCINO, J.L. GARRISON & E.W. SCHAFER, Jr. 1972. Protecting ripening corn from blackbirds by broadcasting 4-aminopyridine baits. *J. Wildl. Mgmt.* 36(4): 1316-1320.

DOLBEER, R.A., T.W. SEAMANS., B.F. BLACKWELL & J.L. BELANT. 1998. Anthraquinone formulation (Flight control^{TM}) shows promise as avian feeding repellent. *J. Wildl. Mgt.* 62(4): 1558-1564.

EISEMANN, J.D., P.A. PIPAS & J.L. CUMMINGS. 2003. Acute and chronic toxicity of compound DRC-1339(3-chloro-4-methyl-analine-hydrochloride) to birds. In: MANAGEMENT OF NORTH AMERICAN BLACK-BIRDS. *Proc. of Special Symp. of the Wildl. Soc. 9th Ann. Conf.* (Eds.) G.M. Linz, Bismark, N. Dakota, USA. pp: 49-63.

ESCHEN, M.L. & E.W. SCHAFER, Jr. 1986. Registered bird damage controls chemical. 1985. Denver Wildlife Research Centre. Bird section, Colorado, USA, Research Report No. 356.

GARRISON, M.V & J.L. LIBAY. 1982. Potential of methiocarb seed treatment for protection of sprouting rice from Philippine bird pests, Lonchura spp. *Phil. Agri.* 65: 363-366.

HEISTERBERG, J.F., J.L. CUMMINGS, G.M. LINZ., C.E. KNITTLE., T.W. SEAMANS & P.P. WORONECKI. 1990. Field trail of a CPT-Avicide aerial spray. *Proc. 14th Vertebr. Pest Conf.* (Eds.) L.R. Davis & R.E. Marsh. Univ. Calif. Davis, USA. pp. 350-356.

KNITTLE, C.E., J.L. CUMMINGS., G.M. LINZ & J.F. BESSER. 1988. An evaluation of modified 4-aminopyridine baits for protecting sunflower from blackbird damage. *Proc. Vertebr. Pest. Conf.* (Eds.) A.C. Crab & R.E. Marsh, University of California, Davis, USA. pp. 248-253.

LEFEBVRE, P.W & J.L. SEUBERT. 1970. Surfactants as blackbird stressing agents. *Proc. Vertebr. Pest Control. Conf.* West Sacramento, California, USA. 4: 156-161.

LEFEBVRE, P.W., R.E. MATTESON., D.G. DECKER & C.O. NELMS. 1987. *Toxicity to blackbirds of candidate DRC-1347 formulations*. Bird Damage Research Report. No. 396. Denver Wildlife Research Center, Colorado, USA. 6pp.

LINZ, G.M., G.A. KNOTSEN., H.J. HOMAN & W.J. BLEIER. 2003. Baiting blackbird (Icteridae) in stubble grain fields during spring migration in South Dakota. *Crop. Protec.* 22: 261-264.

MASON, J.R. & L. CLARK. 1995. Evaluation of methyl anthranilate and activated charcoal as snow goose grazing deterrents. *Crop Prot.* 14: 467-469.

MASON, J.R & L. CLARK. 1996. Grazing repellency of methyl anthranilate to snow geese and enhanced by a visual cue. *Crop Prot.* 15: 97-100.

MOTT, D.F., J.L. GUARINO., E.W.SCHAFER, Jr., & D.J.CUNNINGHAM. 1976. Methiocarb for preventing blackbird damage to sprouting rice. *Proc. 7th Vertebr. Pest. Conf.* 7: 22-25.

MOTT, D.F., G.M. LINZ & E.W. SCHAFER, Jr. 1987. *An evolution of CPT-avicide aerial sprays for control of roosting blackbirds and starling*. Denver Wildlife Research Center, Colorado, USA. 46pp.

NARANG, D.D & O.S. BINDRA. 1971. Studies on the poison baiting of house sparrow, *Passer domesticus* L. *Pl. Prot. Bull.* 23(3).

RAHEJA, A.K. 1992. (Ed.) *Research highlights of agricultural ornithology.* All India Co-ordinated research project on agricultural ornithology-Technical bulletin No.1. Andhra Pradesh Agricultural University, Hyderabad, India.

RUELLE, P & R.L. BRUGGERS. 1979. Evaluating bird protection to mechanically sown rice seed treated with methiocarb at Nijanga, Senegal, West Africa. In: VERTEBRATE PEST CONTROL AND MANAGE-

MENT MATERIALS. (Ed.) J.R. Beck. ASTM STP 680. American society for Testing & Materials, Philadelphia. pp. 211-216.

SANDHU, P.S., M.S. DHINDSA & H.S. TOOR. 1987. Evaluation of methiocarb and thiram as seed treatments for protecting sprouting maize from birds in Punjab (India). *Tropical pest Manage.* 33 (4): 370-372.

SAYRE, R & L. CLARK. 1999. Effect of primary and secondary repellents on European starlings: An initial assessment. *J. Wildl. Mgt.* 65(3): 461-469.

SCHMIDT, R.H & R.J. JOHNSON. 1983. Bird dispersal recordings: an overview In: VERTEBRATE PEST CONTROL AND MANAGEMENT MATERIALS: Fourth symposium ASTM STP 817. (Ed.) D.E. Kaukeinen. American society for testing and materials, Philadelphia, USA. Pp. 43-65.

SEHAFER, E.W. Jr. & M.L. ESCHTIEN. 1986. Laboratory efficacy studies with strychnine baits in pigeons. *Proc. 12th Vertbr. Pest Conf.* (Ed.) T.P. Salmon, University of California, Davis, California, USA. PP. 275-280.

SHIVANARAYAN, N & M.A. RAO. 1988. *Pavo.* 26: 49-52.

SIDHU, A.S & G.S. SIMWAT. 1973. Studies on the control of house crow (*Corvus splendens* Viellot) through poison baiting. *J. Res. Punjab agric. Univ.,* Ludhiana. 10(2): 216-219.

SIMWAT, G.S. 1975. Control of rose ringed parakeet in Punjab. *Indian J. agri. Sci.* 45(6): 269-272.

SOCCI, A.N., M. P. PRITTS & M.J. KELLY. 1997. Potential use of sucrose as a feeding deterrent for frugivorous birds. *Hort. Technology.* 7: 250-253.

STOUT, J.F & E.R. SCHWAB. 1979. *Long term dispersal of sea gulls from U.S. Air Force Bases.* Final report. Andrews University, Michigan.

STUCKY, J.J. 1973. Use of plastic netting. *Proc. 6th Bird Control Sem.* Bowling Green, Ohio, USA. (Eds.): H.N. Cones, J. Jackson & W.B. Jackson. pp. 195-197.

SULTANA, P., J.E. BROOKS & R.L. BRUGGERS. 1986. Repellency and toxicity of bird control chemicals to pest birds in Bangladesh. *Tropical Pest Manage.* 32(3): 246-248.

WOFFORD, J.E & W.H. ELDER 1967. Field trials of the chemo-sterilant SC-12937, in feral pigeon control. *J. Wildl. Manage.* 31(3): 507-515.

WORONECKI, P.P., R.A. DOLBEER & T.W. SEAMANS 1990. Use of Alpha-chloralose to remove waterfowl from nuisance and damage situations. *Proc. 14th Vertebr. Pest Conf.* (Eds.). L.R. Davis & R.E. Marsh, Univ. California, Davis, USA. Pp. 343-349.

5

BATS IN AGRICULTURE

Shakunthala Sridhara

AINP on Rodent Control, University of Agricultural Sciences, G.K.V.K., Bangalore-560 065

ABSTRACT

India has 12 species of fruit bats of which only three species, the short-nosed fruit bat, *Cynopterus sphinx,* the fulvous fruit bat, *Rousettus leschenaulti* and the Indian flying fox, *Pteropus giganteus* are common throughout the country. The three fruit eating bats raid guava, mango, sapota, custard apple, black berry, fig, coconut, areca nut, orange, plantain, grapes, tamarind, etc. Quantified information is available only for grapes, oranges and areca nut. The suggested methods of management of fruit eating bats are: use of scaring devices, beating drums, lighted torches, playing recorded alarm calls, use of repellents, erecting nylon and mist nets, block plantation, planting trap trees and covering grape bunches with dry foliage and thorns. However, bat damage is seasonal and far less when compared to losses caused by other vertebrates like birds, squirrels, monkeys and mongoose. Moreover only ripe fruits which are beyond the stage of harvest are consumed and factually do not affect marketable yields. Scaring them could be very effective coupled with netting for short term solutions. Planting trap trees seems a long term solution but needs evaluation.

INTRODUCTION

Bats commonly called flying foxes seem to be the most misunderstood among animals. They are traditionally associated with witchcraft, haunted houses and cemeteries. For centuries they have been associated with evil or the mysterious unknown. The blood sucking habits of bats of South and Central America have been exaggerated. Some people believe that bats pass 'stool' through their mouths. Still others think that they are birds.

Bats live on trees in forests. They also live in man made structures such as abandoned houses, forts, sparingly used office blocks, temples, monuments, mosques, tombs, dilapidated buildings, ruins, disused roads, railway bridges and caves.

Bats play several important roles in maintaining the ecosystem. Insectivorous bats are voracious depredators of insects including pests of agriculture and mosquito. Frugivorous bats play a decisive role in pollination and dispersal of seeds mainly in forests and are crucial in regeneration of degraded forests and maintenance of established forest canopy. Bats also act as indicators of environmental degradation. Their role in causing yield losses in horticulture, nuisance disfiguring and discolouration of structures and disease transmission is sporadic and almost negligible compared to the benefits derived by insectivory, pollination, seed dispersal and regeneration of forests.

DISTRIBUTION, NUMBERS, SIZE AND BIOLOGY

There are nearly 1000 species of bats distributed all over the world. This is nearly one quarter of all mammal species. They live in all conditions except extreme deserts, polar and mountain regions. All bats belong to the order chiroptera meaning hand winged. A bat's wing in reality is just an extended hand, with long fingers connected by a thin sturdy membrane. This membrane extends from greatly elongated forearms and fingers to the hind limb and body. Their bodies are furred and eyes are small. The world's smallest bat is called bumblebee bat and weighs less than a 10 paise coin. Some are the size of a house mouse when wings are spread whereas in the largest fruit bats wings extend four feet across. Most bats live in colonies of hundreds to thousands. The largest colony lived in North America with some 30 million numbers around 1960. Today only around 30, 000 of them left. Once bats colonize a place it is difficult to get rid of them. Most bats show a great degree of place fidelity.

Bats have no social organization. There is no close parent–infant or any other form of social life. Although frugivorous bats leave en mass in the same direction at dusk to forage, they eat alone and return alone to the roosting site where they again squabble for individual places. *Rousette* species are fruit eating and roost in thousands in caves, tunnels etc. In contrast smaller frugivorous and insectivorous bats live in small colonies. Amongst insectivores it is common to find colonies of two species in the same cave or tunnel with bats of each species occupying different portions and avoiding intermixing. In most European species males and females are known to live in separate roosts after the young are born. There are limited number of reports on males and females living separately in different roosts or occupying different portions on the same tree. Data on segregation of sexes during non–breeding season, separate roosts for gravid females as well as separate nurseries for females with young is yet to be firmly established. There are preliminary reports on the segregation of sexes in the black beard tomb bat, *Taphozous melanopogon* (Vanitharani, 2001) and the short nosed fruit bat, *Cynopterus sphinx* (Sandhu & Gopala Krishna, 1984).

Scent plays an important role in the establishment and maintenance of territories and colonies. Propatagial sacs are used to store and display odours collected from genital and gular glands. Social communication and threat behaviour are facilitated by high degree of vocalizations. These fascinating animals 'see' with their ears, hang upside down to sleep by day and can catch insects while flying even in the darkest of nights.

Some bats are capable of detecting changes in barometric pressure associated with changes in weather. They use this information to decide whether to go out to hunt insects or not. The pressure changes are detected by Vitali organ, a receptor in the middle ear.

The fruit bats are reported to mate during October–December and the young are born between March and April coinciding with or just before flowering and fruiting season. Insectivorous bats on the other hand have two breeding seasons-at the beginning of summer (March-May) and majority during the second half of the year. Both breeding seasons have evolved so as to assure abundant insect supply for the young.

There are no records on gestation period. Generally one young is born which clings to the mother quite for some time. In some species young are left in the roost, with mothers visiting them to feed while in some other species all the young are left behind and feeding is communal, with the first approaching young getting fed irrespective of parentage. However *Rousettes* and flying foxes carry their young till the later can fly independently which is a few weeks or a month after birth. Depending on the species bats live upto 18-22 years.

CLASSIFICATION AND SCIENTIFIC STATUS

CLASS : MAMMALIA

ORDER : CHIROPTERA

SUB ORDERS: (1) MEGACHIROPTERA- Frugivorous bats

(2) MICROCHIROPTERA- Insectivorous bats

Depending on the diet, bats are classified into two sub groups namely *Megachiroptera*, the fruit bats and *Microchiroptera*, the insectivores. The insectivore bats are major predators of night flying insects. A single bat may catch up to 3000-4000 mosquito like insects in a night.

Microchiroptera (Insectivorous bats): There are 32 species of microchiropterans in India. Of these, only four species occur in South India. They are:

(1) *Pipistrellus mimus* and *Pipistrus dormeri* commonly called pigmy bat. Weight is 3g. Found in old buildings and cinema theaters. If cave dwelling they are the first to leave and last to return to cave a little before sunrise. Flight is erratic resembling that of a butterfly.

(2) *Hipposiderus speoris,* Leaf nosed bat. Body weight is 11g. This is a very common species and the one we often come across in the night sky. These never cluster. They live in man made constructions, excavations and caves.

(3) *Rhinopoma hardwickei*, Mouse tailed bat, weighs 13-18 g with a wing span of one feet. They live in natural caves, caverns and crevices. They are noisy and diurnal, feed on insect pests of crops and can co-exist with other insectivorous bats.

(4) *Taphozous melanopogon* is called Tomb bat. weighs 24-28 g and the wing span is about 2 feet. Roosts in hills, temple towers and in the forest area. Develop beard during breeding season.

Insectivorous bats by virtue of their voracious feeding and generalized diet of almost all insects are the most important predators of nocturnal and crepuscular insects. As they hunt at different levels i.e. from above ground to tree tops, their diet is naturally varied.

In addition to the above, there occurs a single species of carnivorous bat, *Megaderma lyra* commonly called Indian false vampire bat. It has big head, huge ears, peculiar nose leaf and large eyes. It is adapted to man built structures, lives in temples and unoccupied buildings, feeding on insects, geckos, lizards, frogs, rats, mice, other small bats and small birds. The bat is beneficial to agriculture.

The most frightening mode of feeding seen in animal kingdom is perhaps sanguivory or blood drinking by vampire bats. The vampire bats are native to the Americas – from Mexico to Brazil, Chile and Argentina. They make a small cut and suck the trickling blood. Vampire bats generally feed on livestock and very rarely on human beings. These bats are the most sociable chiropterans showing the extreme of social behaviour namely reciprocity of blood sharing.

Megachiroptera (Frugivorous bats): Fruit bats are restricted to tropical and subtropical regions. They rarely inhabit the Himalayan forests or the evergreen rain forests. They preferably live in tropical, semi- evergreen, moist deciduous and dry deciduous forests of southern India. Introduction of fruit bearing trees into deserts has opened up new habitats for them. The distribution and movement of fruit bats, their changing feeding grounds and the type of the diet is largely influenced by the seasonal flowering and fruiting trees. Trees also provide roosting sites with the exception of *Rousette* which is a cave dweller. Many colonies of bats exhibit great affinity for their roosting site. While *Cynopterus sphinx* selects palms, *Pteropus giganteus* prefers banyan, fig, and tamarind trees. Fruit bats are important agents of pollination and seed dispersal. Very rarely they become serious pests of orchards.

The old world fruit bats belonging to the family pteropodidae are fructivorous, feeding on fruits, berries, leaves and bark, often nectar and pollen and by accident a few larvae that are found on leaves and fruit. After picking fruits, bats travel to a roosting site where the fruit is crushed, its juices swallowed and the pips along with seeds are spat out. Smaller seeds along with soft parts are swallowed with seeds being voided later. Occasionally some species reach pest status.

Although there are 170 species of fruit bats distributed throughout the tropical areas of Africa, Asia and Indo Australia, there are only eight genera occurring in the Indian sub continent. India has twelve species of fruit bats of which only three species namely the short-nosed fruit bat, *Cynopterus sphinx* (45g), the fulvous fruit bat, *Rousettus leschenaulti* (90g) and the Indian flying fox, *Pteropus giganteus* (900g) are common throughout the country. The remaining are rare and restricted in their distribution to the undisturbed mountains of south and north-east India, Andaman and Nicobar islands. Two of them, the long tongued *Eonycteris spelaea* and *Macroglossus sobrinus* are specialized nectar feeders. The other fruit bats although visit flowers for nectar and pollen, are primarily frugivorous. Following is a brief account of the major frugivorous bats of India, their habits, distribution, biology and behaviour, based on Bates & Harrison (1997).

1. *Rousettus leschenaulti* (Desmarest, 1820), Fulvous fruit bat

SCIENTIFIC STATUS

Class	: Mammalia
Order	: Chiroptera
Suborder	: Megachiroptera

Family : Pteropodidae

Genus : *Rousettus*

Species : *Leschenaulti*

The length of forearm is 75-86 mm with shorter hind feet. The muzzle is relatively short and slender. The crown, back, flanks and throat are fulvous brown and the belly is grayish.

Distribution: *R. leschenaulti* is widely found throughout India. It is widely reported from Jammu and Kashmir, Himachal Pradesh, Rajasthan, Gujarat, Maharashtra, Kerala, Orissa, Bihar, West Bengal, Sikkim, Arunachal Pradesh, Meghalaya, Tripura, parts of Karnataka, Tamil Nadu, Andhra Pradesh and Uttar Pradesh.

Biology and behaviour: *R. leschenulti* occurs in a wide range of habitat extending from coastal areas to high altitude mountain ranges of Himalayas as well as tropical, humid forests of Western Ghats and the arid areas of Rajasthan. It is colonial, the numbers ranging from two to three thousands. A huge colony of 15000 individuals is reported from the ruins of a temple in Tamil Nadu (Vanitharani, 2000). They are diurnal and roost in caves, tunnels, deserted buildings and large leafed trees. Hanging upside down they are actively engaged in noisy wrangling, mating or squabbling. When disturbed the entire colony flies out and repeatedly swarms in circles, finally settling on rocks or trees.

Females breed twice a year- generally in March-April and July-October giving birth to a single young. Gestation lasts around 15-18 weeks and lactation 35-40 days. Females reach sexual maturity in five months and males in 15 months.

They generally feed on fruits and flowers. They may share same fruit trees with other bat species such as *Pteropus giganteus* and *Cynopterus sphinx.*

Agricultural damage: In Rajasthan they are reported to feed on guava (Advani, 1982)

2. *Pteropus giganteus* (Brunnich, 1782), Indian Flying Fox

SCIENTIFIC STATUS

Class : Mammalia

Order : Chiroptera

Suborder : Megachiroptera

Family : Pteropodidae

Genus : *Pteropus*

Species : *giganteus*

This is a large bat with its forearms extending 152-183 mm. The crown of the head is rich chestnut and the rest of the body is light yellowish tan to deep chestnut brown. The colour becomes a shade darker around the eyes and mouth. The long snout covered with hairs. The ears are black, hairless, tall and pointed.

Distribution: *P. giganteus* has a very wide distribution throughout the country.

Biology and behaviour: *P. giganteus* is colonial and lives in large diurnal roosts on huge trees. Colony size ranges from hundreds to thousands and varies seasonally with low numbers in summers, which increases during rainy season and remains steady during winter. They live close to human habitats and are found both in cities and villages. Semal (*Bombax malabaricus*), sheesam (*Dalbergia sisso*), peepal *(Ficus bengalensis*), mango (*Magnifera indica*) and tamarind (*Tamarindus indica*) trees are preferred for colonizing.

Individuals groom themselves several times a day and there is no social grooming amongst adults. Social ranks are present amongst males in colony. Thermoregulation during summer is achieved by flapping wings and salivating on bodies. They sleep wrapping the wings around the body and hang head downwards using one or both feet to cling to the branches.

Mating takes place from July to October. Pregnancy extends 140-150 days. Foetus is single. Young are delivered in March and are carried by mothers for a few weeks after birth.

Pest status: Around twenty minutes after sunset, the bats leave the roost and fly one by one, following each other covering long distance ranging from 2-20km in search of food. The major fruits consumed are: mango (*Magnifera indica*), jack fruit (*Artocarpus integra*), soursop

(*Annona maricata*), guavas (*Psidium guajava*), banana (*Musca paradisiaca*), papaya (*Carica papaya*), Indian blackberry (*Syzygium cumini*), peepal (*Ficus bengalensis*), custard apple (*Malus domestica*), palm (*Borassus flabeliter*), date (*Phoenix silvestris*), mahua (*Basica latifolia*), wood apple (*Feronia* sp.) and fruits and flowers of many forest trees.

3. *Pteropus vampyrus* (Linnaeus, 1758), Large Flying Fox

SCIENTIFIC STATUS

Class	: Mammalia
Order	: Chiroptera
Suborder	: Megachiroptera
Family	: Pteropodidae
Genus	: *Pteropus*
Species	: *vampyrus*

This bat is similar to *P. giganteus* but bigger, with the fore-arm extending 195-209 mm. The ventral side is darker than that of flying fox, a dark russet brown on the throat and blackish brown on the lower chest and belly.

Distribution: *P. vamprus* has a restricted distribution and is found only in Andaman and Nicobar Islands.

Biology and behaviour: Not much is known. It is group living, roosting in colonies of hundreds of individuals. After dusk, they leave the roost, and fly in long files but not in flocks.

4. *Pteropus faunulus* (Miller, 1902), the Nicobar Flying Fox

SCIENTIFIC STATUS

Class	: Mammalia
Order	: Chiroptera
Suborder	: Megachiroptera
Family	: Pteropodidae
Genus	: *Pteropus*
Species	: *faunulus*

This is smaller than *P. giganteus* and *P.m. melanotus*. The length of forearm is about 113mm. The hind feet are small. The pelage on the crown of the head is grey-brown speckled with white. The dorsal pelage is dark brownish black intermixed with some white hairs. It becomes paler on the rump. The ventral side is uniform grey-chestnut brown.

Distribution: *P. faunulus* is endemic to Nicobar Islands.

Not much is known about its biology and behaviour.

Conservation status: It is included in list 1 "Threatened species" in the 1996 IUCN Red list of Threatened animals. It is vulnerable due to its limited distribution and loss of habitat.

5. *Pteropus satyrus*, the Island Flying Fox

SCIENTIFIC STATUS

Class	: Mammalia
Order	: Chiroptera
Suborder	: Megachiroptera
Family	: Pteropodidae
Genus	: *Pteropus*
Species	: *satyrus*

This is smaller than *P. giganteus* with the fore arm length of 141mm. Face is dark brown, crown is fawn and the mantle deep russet brown. Belly is fawn to dark brown.

Distribution restricted to Andaman Islands.

Habits: Not much is known about the species in India. It is reported to frequent disturbed forests. Also reported in coconut grooves and orchards where several hundred individuals are known to roost together.

Biology: In Philippines breeding is reported to occur in April-May. Lactation extends upto three months and sexual maturity occurs after more than a year.

Pest status: It is known to feed on mango and chiku (*Pouteria sapota*).

6. *Pteropus melanotus*, Blyth's Flying Fox

It is a medium sized fruit bat with arms extending upto 153mm. Head is brown, mantle golden tawny or reddish and the ventral side is dark brown. The back and rump are blackish brown sprinkled with grey hairs.

SCIENTIFIC STATUS

Class	: Mammalia
Order	: Chiroptera
Suborder	: Megachiroptera
Family	: Pteropodidae
Genus	: *Pteropus*
Species	: *melanotus*

Distribution: Found only in Andaman and Nicobar Islands.

Behaviour and biology: Reported to live in huge colonies with several thousand individuals in mangroves of Nicobar Islands. They were known to be fearless and not prone to disturbance. It is reported to breed during February-March. Usually a single young is born. Females reach sexual maturity in 6 months and males in 18 months.

Pest status: Mickleburg *et al.,* (1993) found that more than 20 species of fruit plants are consumed by *P. melanotus* which included mango, papaya, (*Carica papaya*) and guava (*Psidium guajava*) in Christmas islands near Java.

7. *Cynopterus sphinx,* the short nosed fruit bat

It is a medium sized frugivorous bat with forearm measuring 70mm. In males, the chin, anterior part of shoulders, sides of the chest, belly and thighs are orange tinted. The forehead and the nape of neck are russet brown in colour while the posterior side of the back is greyish brown. Females have tawny brown collar, grey brown rump and pale grey back.

SCIENTIFIC STATUS

Class	: Mammalia
Order	: Chiroptera
Suborder	: Megachiroptera
Family	: Pteropodidae
Genus	: *Cynopterus*
Species	: *sphinx*

Distribution: *C. sphinx* is very commonly found throughout the country except in Gujarat, Punjab and Himachal Pradesh.

Behaviour and biology: *C. sphinx* is most abundant in agricultural and forested areas, and city parks mostly living at lower altitudes. It lives in small colonies of 3-4 individuals. The bats roost on the underside of palm leaves, crevices in banyan, peepal, palm and coconut trees feeding on the flowers and fruits of kitul palms and Ashoka trees. The species constructs shelter or stem tents usually by a single male taking 30-50 days. The fronds of the palm are chewed to construct simple tents. They also construct tents by closely interweaving leaves and twigs of vines which cover buildings. This usually occurs when there are no palms. In Tamil Nadu nest construction is reported to correspond to biannual breeding season. Females along with her pups and a breeding male roost inside these (Balasingh *et al.*, 1993). Males and females are reported to live separately in Uttar Pradesh except during breeding when groups of 5-10 males and 10-15 females are formed.

Breeding season varies geographically. In Maharashtra they are fertile during September–October and February–March while July–August seems to be breeding period in West Bengal. In central India breeding occurs from October to March and again in July. In Karnataka breeding extends from December to August. Gestation lasts 115-125 days. Usually there is a single foetus. Females mature in about six months and conceive by seventh month.

Feeding: *C. sphinx* leaves the roosting site well before sunset and is active throughout the night. The bat's acute sense of olfaction aids in detecting fruits. There is a definite seasonal variation in feeding depending on availability of fruits. In Rajasthan it feeds on guavas in winter, mango, date palm and the seeds of *Phoenix sylvestris* in summer, custard apple during monsoon and fruits of *chicku* in post monsoon season. It also consumes bananas and *litchis* (Advani, 1982). It is a pest of guava, pour sops, mangoes, fruits and flowers of plantain. Due to availability of plenty of wild fruits the damage to orchards is negligible. These bats are considered as possible vectors of Japanese encephalitis (Bhat, 1994).

Although *C. sphinx* feeds on fruits in groups, flowers are visited by single individuals. When it feeds with megachiroptera on the same tree, there is stratification with *P. sphinx* confining its feeding to lower parts of the fruit tree. Earlier it was assumed that fruits are sucked during flight but recent studies indicate that fruits are plucked, carried to feeding perches on trees or to a verandah of a house where the soft parts are chewed, juice swallowed, seeds, and chewed fibres are dropped. Thus it aids in pollination and seed dispersal.

Behaviour: These bats are gregarious, roosting in unisex groups of 8-9 individuals. The sexes live together for a short period during breeding season. After mating, males return to male groups (Sandhu & Gopalakrishna, 1984).

8. Cynopterus brachgotis, lesser dog faced fruit bat

It is smaller than *C. sphinx* with a fore arm length of 63mm.

SCIENTIFIC STATUS

Class	: Mammalia
Order	: Chiroptera
Suborder	: Megachiroptera
Family	: Pteropodidae
Genus	: *Cynopterus*
Species	: *brachgotis*

Distribution: It is found in Karnataka and Tamil Nadu and may possibly occur in Andhra Pradesh.

Behaviour and biology: Mostly inhabits mountains ranging 2500-3000 feet. It roosts in talipot palms and kitul palms where it hangs from seed clusters. It also lives in forests, orchards or open country.

Feeding: In Sri Lanka it is reported to feed on two species of guava, eating huge quantities of fruits, sucking the juices and scattering the residue near human dwellings.

Females become pregnant at about 6-8 months while it takes one year for the males to become mature. After a gestation period of 3-4 months, the females give birth to a single young one. Lactation period lasts 6-8 weeks.

9. *Magaerops niphanae*

The species is a small fruit bat, similar to *Cynopterus brachgotis* in size and appearance. The pelage is uniformly greyish brown on the back. Ventral side is grey and the wings are uniformly brown.

SCIENTIFIC STATUS

Class	: Mammalia
Order	: Chiroptera
Suborder	: Megachiroptera
Family	: Pteropodidae
Genus	: *Megaerops*
Species	: *niphanae*

Distribution: It is found in West Bengal and Arunachal Pradesh.

Behaviour and biology: Not much is known about the species.

10. *Latidens salimali,* Salim Ali's fruit bat

It is a medium sized bat, similar to *Cynopterus sphinx* in size sans the external tail. Pelage is soft, dense, brown to block on the dorsal surface, wings are dark and ventrum has pale buffy hairs.

SCIENTIFIC STATUS

Class	: Mammalia
Order	: Chiroptera
Suborder	: Megachiroptera
Family	: Pteropodidae
Genus	: *Latidens*
Species	: *salimali*

Distribution: Reported only from Tamil Nadu.

One of the rarest bats of the world, the Salim Ali's fruit bat, *Latidens salimali* is endemic to Tamil Nadu. It was discovered 50 years ago at high mountains (Hutton, 1949) and rediscovered in the same habitat in 1994 (Bates *et al.,* 1994), 2003 (Singaravelan & Marimuthu, 2003) and 2004 (Vanitharani *et al.*, 2004). It is now assumed that the extent of Salim Ali's fruit bat is about 1040km^2 and the area of occupancy is about 10.05km^2. Parturition is reported to occur as early as February. The bat seems to feed mainly on the seeds of *Ficus racemosa,* which is known to bear fruit much of the year. In addition, fruits of *Elaeocarpus tuberculatus*, *Dichapetalum gelonioides* and *Elaeocarpus serratus* also seem to be consumed (Vanitharani *et al.*, 2004).

Nothing is known about reproduction except about the male genitals of two bats captured (Bates *et al.,* 1994).

Of the ten species of frugivorous bats only three species, *P. giganteus, C. sphinx and R. leschenaulti* are in conflict with farmers as they occasionally raid guava, mango, sapota, custard apple, blackberry, fig, coconut, plantain, tamarind etc.

THE ECONOMIC IMPORTANCE OF BATS

Bats are economically important primarily because of their feeding habit and secondly as vectors and transmitters of diseases. There are contrasting claims about the impact of insectivorous bats which are recorded to consume insects one to one and a half times of their body weight. On the other hand the role of frugivorous bats in pollination and seed dispersal is an acknowledged fact. Folivory, resorted to by frugivorous bats in times of fruit shortage has not been regarded as damaging or destructive so far. Compared to the positive role of frugivorous bats in pollination, seed dispersal, regeneration of forests, their pestiferous activities, which are extremely rare and sporadic, are economically negligible. Gaum, the bat excreta has been found highly nutritious to plants and crops. Following is the critical appraisal of the role bats play in the ecosystems.

(1) BATS AND DISEASES

Several bacterial, viral micotic, protozoan, rickettsial and viral diseases, a few helminthic and parasitic arthropods infest and infect bats. Although there is no evidence of rabies transmission from bats to man, rabies remains the greatest threat to humans of all the pathogens that bats carry and transmit. At least fifty arboviruses, half of them capable of producing diseases in human beings are present in bats. Histoplasmosis in all the bats of the world, trypanosomiasis in African bats and toxoplasmosis in European bats are well known infections of bats. In spite of lack of reports, bats have all the potential of carrying and harbouring bacteria pathogenic to human beings. However none of the pathogens harboured by bats have any great epidemiological importance. In fact compared to rodents and small carnivores, the danger of disease transmission by and through bats is not serious and does not warrant regular control measures of the transmitter/vector.

However, some recent viral discoveries (McKenzie *et al.,* 2001) seem to raise some concern. Hendra virus was formerly called equine morbillivirus. It is thought to be transmitted by bats to horses and then from horses to humans. The Australian paralysis tick may be the missing link in this transmission chain. The Australian bat lyssavirus is related to rabies virus. Porcine stillbirths and foetal malformations in pigs is associated with Menangle virus while Nipah virus is reported to cause pneumonia in pigs and encephalitis in humans (Bannerjee *et al.,* 1984). Tioma virus has been recently isolated from bats but its disease profile is still not clear. The latest threat is the detection of leptospiral antibodies in flying foxes of Australia (Smythe *et al.,* 2002)

Bats carry ectoparasites, particularly the bat bug, which closely resembles the bedbug. These may attack man. Mites, ticks and fleas are also listed among other parasites and possibly create minor health problem at times. Bat guano attracts flies, cockroaches and other coprophagous insects. Whenever bats are controlled by exclusion or some other method, consideration must also be given to the control of ectoparasites and other insects which may be left behind.

(2) INSECTIVOROUS BATS

Nearly 75% of the known 1000 species of bats in the world are insectivorous. They consume a large variety of insects. A fair amount of these insect prey include agricultural pests like moths, beetles, corn borers, June bugs, cucumber bugs and also mosquitoes. A few studies claim substantial reduction in insect populations due to the voracious feedings rates of bats. For instance, the Mexican free tailed bat, *Tadarida brasiliensis* is known to consume insects more than half of its body weight every night. As the species is colonial living in millions, the consumption of insects at this rate will amount to ten tons per million bats each night. A similar high density destruction of insects of around 7500 kg/night is reported for a cave dwelling colony of insectivorous bats in Borneo. Similarly the little brown bat, *Myotis lucifugus* is known to consume insects equivalent to its body weight a fair amount of which are mosquitoes thus contributing to vector control (Molur *et al.,* 2002). In spite such highlights on insectivory the claims on possible reduction of insect pests of agriculture are not supported by long term research data.

The same situation of lack of data about predation on crop pests exists in India in spite of the country being agro based spending huge amounts on insecticides. This situation is slowly leading to environmental and human health hazards apart from affecting the survival of fauna and flora of the country. There are only two studies on the diet of insectivorous and carnivorous bats although studies on the foraging habits and prey composition of insectivorous bats has lot of economic potential. Earlier studies (Sinha, 1986, 1999) revealed that the carnivorous bat, *Megaderma lyra* consumes rats and mice, which again has a beneficial impact on agriculture as rodents are known to devour 10-30% of agricultural produce both in the farm and in storage. The species is known to predate on small vertebrates like fishes, amphibians, lizards, small mammals, birds and large insects like locusts, grasshoppers, beetles etc (Rajan & Marimuthu, 1999). Two studies reported the species predating on Kelaart's *Pipistrella pipistrella ceylonicus* (Brosset, 1962; Srinivasalu, 2001). The small mouse tailed rat, *Rhinopoma hardwickei* is reported to consume insects belonging to Orthoptera, Dictyoptera, Lepidoptera, Hymenoptera, Coleopteran and Diptera throughout the year while Isoptera was not consumed during winter. With such an array of insect diversity in its food, the bat is considered to play an important role in regulating the density of crop depredating insects. The large rat-tailed bat, *Rhinopoma microphyllum kinneani* and also the lesser tailed bat, *R. hardwickei* had winged termites in its stomach (Sinha & Advani, 1976; Sinha, 2001).

(3) BATS AS POLLINATORS

Frugivorous bats are renowned dispersers of pollen and seed in palaeo-and neotropics. A detailed study on pollinating behaviour of bats in the cloud forests of Venezuela revealed that 5.4% of the 33 monocotyledons of them were pollinated by bats (Seres & Ramirez, 1995). Not only two bat species, *Leptonycteris lurasoae* and *Glossophaya longirostris* were critical pollinators of two species of cacti in Netherlands but there existed a mutualism between the two (Petit, 1995, 1996). In a diptocarp forest in Malaysia 13.7% trees were partially dependent on bats for pollination and seed dispersal (Hodgkison *et al.*, 2003). Two interesting but diverse types of pollination have been observed wherein the inflorescence of the tree and bat's foraging behaviour have evolved in such a way as to be mutualistic. The flowers of *Helicteres barnensis* lack suitable structures to enable a heavy pollinator to cling or perch. Nectar being available deep inside the flower, only glossophagine bats with their sophisticated ability to manoeuvre and hover are capable of exploiting the flowers for nectar and bring about pollination (Helversen *et al.,* 2002). In contrast the inflorescence of the palm, *Calyptrogyne ghiesbreghtiana* enables perching while bats forage on it. The nectarivorous bats consumed flower tissue while hovering. With comparatively inferior fruit setting in inflorescences that received nectarivores, perching bats were considered to be superior in pollen transfer in this case (Tschapka, 2003)

There are instances where pollination and fruit dispersal of the same tree are effected by two different species of bats. The flowers of *Dissochroma viridiflorum*, an epiphyte endemic to Brazil open during night and are visited by the nectar feeding *Glossophaga soricina* whereas the fruits are consumed by frugivorous *Carollia perspicillata* and *Sturnira lilum* (Sazima *et al.*, 2003)

Limited studies are available on nectar feeding behaviour facilitating pollination by megachiroptera in India (Taglianti, 1979).

(4) BATS AS INDICATORS OF ENVIRONMENTAL DISTURBANCE

To take effective and fast steps to determine the conservation status of an ecosystem, degree of ecological disturbance is a quick prerequisite. Such evaluations should be not only relatively inexpensive but accurate and fast. Bats with their numerical abundance, diversity, wide distribution in all trophic levels especially in neotropical forests seem to serve as good indicators of deterioration of environment. Bat species diversity, number of rare bat species and the bat diversity index are positively correlated with vegetation scores. While relative abundance of the most abundant species is negatively correlated with vegetation scores, relatively higher density of phyllostomine species is considered to reflect low levels of environmental disturbance. In contrast, the presence of *Desmodus rotundus* and *Carollia perspicillata* indicated disturbed habitat (Medellin *et al.*, 2000). Although a single indicator group is insufficient to make decisions in conservation programmes, bat population may be a good initial step in assessing the conservation value of an area especially rain fed regions.

(5) BATS AND ENVIRONMENTAL DETERIORATION

When bats start roosting in a residential building which is in regular use, there is always the possibility of some nuisance mainly the disfiguring and deterioration of the structure. Secondly the vocal and physical disturbance they create while roosting, departing and returning to roosting sites is a nuisance. Though this is a common occurrence in old and historical buildings, not much attention has been paid to them. Often the bat excreta cause irrepairable losses to the building and artefacts therein. Paine (1993) records the damage caused to precious paintings displayed in an old church in England. Similar disfiguring of walls and corridors of the Armed Forces Medical College at Pune by the urine and faeces of bats was reported recently (Pradhan & Mahabal, 2001). In this case they were found roosting in the narrow gap between a huge signboard and the wall to which it was affixed. By using iron angles to fix the signboard to the wall, the gap was widened between the two making it unsuitable for roosting.

(6) FRUGIVOROUS BATS

Fruit bats play two vital roles in terms of human economy namely regeneration of forests and predation on cultivated fruits. Their role in regeneration of forests is far more vital and sustained than their occasional damage to orchards which in fact is generally inconsequential in terms of economic losses.

Seed dispersal and pollination: Bats play a significant role in tropical and other forests pollinating and dispersing seeds of a variety of trees which are economically important as they are the sources of commercial products such as dyes, tannin, medicine, fibre, fuel, timber etc. This has been illustrated by studies on the impact of short-tailed fruit bat on the forest ecology in Costa Rica (Fleming, 1988). Although bats eat up thousands of fruits per night individually, taking into consideration only 1000 seeds/per bat/night it was calculated that 146 million seeds were dispersed annually by a single bat colony. Even if only 10% of these germinate, the numbers would be 1,46,000 seedlings which is way above human plantings in terms of economy and success. Further, bat dispersed plants are among the first and most abundant plants to invade man-made and natural forest clearings.

In the forest savannas of West Africa, the survival rate of fallen seeds is very low due to high density of seed eating insect populations. To overcome this 80-95% of the trees and shrubs of this area have evolved to produce fruits of abundant, nutritious pulp making them very attractive to frugivores, a large proportion of which (95%) are bats. The seeds from the faeces and bolus (spat out pulp along with seeds after infesting the juices) germinated faster and better than fallen seeds (de Figureiredo, 1999). Similar results were obtained later in Malaysia (Racey, 2002).

Fruit bats play a crucial role in the regrowth of disturbed and deformed forest areas in the neotropics. But the opposite trend i.e. dispersal of old growth and canopy species is also attributed to fruit bats in India (Mistry, 1995). This difference in distribution of seeds of different types of trees is attributed to the bigger size of Indian fruit bats which range from 50g to more than one kilogram. Their larger size enables them to feed on a variety of flowers and fruits ranging from the smallest flowers of eucalyptus to large sized fruits of mangoes and guavas. Thus seed dispersal encompasses from smallest to biggest fruits and trees of old growth forests. The three common Indian fruit bats, *Pteropus giganteus, Cynopterus sphinx* and *Rousettus leschenaulti* are reported to pollinate and disperse seeds of more than 114 species of trees and shrubs of economic, ecological, medicinal and religious importance. Natural populations of several wild fruits including bananas, guavas, and mangoes are maintained by bats. Fig tree, one of the most important key stone species of India is the favourite abode of bats and its seeds are dispersed far and wide. Recent studies (Vanitharani, 2005) indicate that the Salim Ali's fruit bat, *Latidens salimali* pollinates some important tree species and is a prime seed disperser of seeds aiding in the restoration of tall fruiting trees of Agasthiyar malai range situated in the Southern parts of Western Ghats. *C. sphinx* is reported to disperse 36 plant species compared to 39 by Bonnet macaque, 23 by civets, 22 by jackal and 12 by spotted deer (Balasubramanian & Bole, 1993). The considerable foraging and migrating flights coupled with the capacity to retain viable seeds for long periods of more than ten hours in the gut of fruit eating bats indicates that seeds of small size can be dispersed over hundreds of kilometres. Thus fruit bats play a major role in the gene flow between isolated plant populations (Shilton *et al.,* 1999).

There are several adaptations to frugivory in bats. One of them is the bimodal reproductive activity in two species of fruit bats, *Artibeus toltecus* and *Sturnia ludovici.* Their breeding activity is so evolved that lactation coincides with seasonal peak in fruit availability (Pinerstein, 1984). Another adaptation is consumption of huge amounts of fruits by palaeotropical fruit bats belonging to pteropodidae upto 2.5 times their body weight in a single night. This is two times more reported for neo-tropical phyllostomacid fruit bats. The high intake in the former is attributed to the need to obtain adequate protein from the otherwise carbohydrate rich fruit source alone. The phyllostomalid fruit bats are thought to compensate their protein deficiency by insect supplements in their diet. This difference in diet affects the relative capacity to select fruits, evaluation of fruit nutrition, regulation of fruit consumption and seed dispersal (Thomas, 1984).

Based on the available reports, farmer's complaints about damage to orchards and fruit crops, all twelve species of fruit bats in India were included under the vermin category (schedule V) of the Wildlife Protection Act, 1972. However, systematic studies and scientific publications on the degree, nature and other details of damage to fruits by bats are extremely poor and limited to few fruits in south India. Grapes were damaged substantially by *C. sphinx* in Karnataka, Andhra Pradesh and Tamil Nadu. In Karnataka damage was positively correlated with fruit maturity and was higher when there was an open space adjacent to vineyard (Verghese, 1998). In Andhra Pradesh, *C. sphinx* fed on grapes from dusk to dawn causing damage anywhere between 10-100% (Srinivasulu & Srinivasulu, 2001). The calculated yield loss was 1182 kg amounting to Rs.10,683/ha in vineyards of Tamil Nadu. Grapes were also attacked by *R. leschenaulti* (Elangovan & Marimuthu, 2001). In Karnataka *P. giganteus* damaged 18% of areca nut (*Arecha catechu*) in coastal areas and 12.5 to 22.3% of sapota in hilly regions. Guava was damaged at 18% by *P. giganteus* and *C. sphinx* (Chakravarthy & Girish, 2003).

From the studies of foraging behaviour of *C. sphinx* two interesting behaviours emerged. Foraging behavior was negatively correlated with moonlight with feeding declining during the waning phases of moon and vice versa. Cloud cover enhanced foraging while lighting and thundering diminished it. During lunar eclipse bat visits to grape orchards increased threefold. The observed suppression of foraging activity of bats during bright moonlight is an adaptation to avoid nocturnal predators (Elangovan & Marimuthu, 2001). The second behaviour of interest was the possibility of weaning bats learning about foraging sites from their mothers while accompanying her on her flights to food sources. (Gopukumar *et al.*, 2003)

MANAGEMENT OF FRUGIVOROUS BATS

The major thrust in managing bat problem in fruit orchards is to adapt non-lethal approaches. These include using scaring devices like beating drums, flashing torches, broadcasting alarm calls, partially covering vulnerable portions of the canopy, use of olfaction based repellents, erecting nylon and mist nets, block plantation and planting trap tree i.e. edible fruits in and around orchards to divert bats feeding on commercial fruits (Hall & Richards, 1987; Mickleburgh *et al.,* 1993; Verghese, 1998; Srinivasulu & Srinivasulu, 2001; Chakravarthy & Girish, 2003; Marimuthu, 2004). However, none of them is effective when used alone.

There are two ways to manage tree roosting bats using nets. If the fruit involved is costly type and the orchard is small, then entire fruiting tree can be covered with fine mesh fishing net. In larger orchards with not too expensive fruit crops, bats are managed by capturing using mist nests. Nets measuring 10 M long, 2 M wide, having 6.0 mm mesh should be strung at heights of 5-15 M between trees from ropes. The nets should be located near roosting sites and fruit trees placing them in flight pathways. Captured bats should be killed humanely before they are removed from nets by placing a chloroform soaked cloth over their noses and mouths.

Netting programs should be repeated only at 3-4 year intervals. The goal of netting is to manage population to reduce loss of fruit rather than to exterminate this interesting species. Additionally timely harvest of fruit to prevent damage should be incorporated whenever control measures are planned.

Although netting is effective, non-destructive, its exorbitant cost makes it impractical. Sonic auditory scares have been found economical and less time consuming in Israel (Kornie *et al.,* 1999) but need to be evaluated in India. Covering grape bunches with dry springs of foliage but leaving bat damaged bunches on the vine itself (Verghese, 1998) and use of fire crackers (Srinivasulu & Srinivasulu, 2001) seem to be economical, ecofriendly for small orchards. Block plantation is planting trees in smaller block or dividing orchard into smaller plots makes it easy to cover fruit bearing trees and vines with sprigs, foliage, thatch and/or nylon nets. These measures minimize damage (Chakravarthy & Girish, 2003). Growing trees like Singapore cherry, *Muntingia calabura* etc in and around orchards can divert bats away from the commercial fruit crops to these trees thereby reducing damage (Chakravarthy & Girish, 2003; Marimuthu, 2004). But the time taken from planting to fruiting of these trees is considerable and hence this approach is feasible only in large scale fruit growing areas and when fruit cultivation is taken up either permanently or for long durations which is unlikely in Indian agriculture where market trends and vagaries of monsoon as well as the economic status of the farmer decides cultivation practices.

Preliminary studies of Marimuthu and his colleagues (Marimuthu, 2001) have shown that far more damage is caused by 16 species of birds, squirrels and mongoose to grapes in Tamil Nadu than by *R. leschenaulti, C. sphinx* and *P. giganteus* combined which forage vine yards only from November to January. This could be true to situations of bat depredation elsewhere.

In case of bats roosting in caves, old houses and temples etc several non-lethal methods like spreading Naphthalene or Paradichlorobenzene inside the caves or dusting the roosting area with 50% DDT is also effective. Fumigants like calcium cyanide and aluminium phosphide can also be used. Sometimes illuminating bat roosting site with floodlights causes them to migrate to other areas. The best method however is to seal all the entry points after the bats have left the roosting sites.

(7) FOLIVORY

Leaf eating is reported for about 14 species of old world megachiroptera and four species of new world microchiroptera. Leaves of 44 species of plants belonging to 23 families are eaten by bats. Generally leaves with high protein and low fat are selected. The phenomenon of folivory is attributed to deriving protein supplement from leaves, a process which is especially beneficial to pregnant and lactating females. Once considered a rare phenomenon, folivory may be quite common amongst species that generally feed on fruits with low protein content (Kunz & Diaz, 1995). This is a second adaptation towards protein supplementing; the first was excess ingestion of fruits in old world bats to obtain body requirements of protein compared to insectivorous bats which derive proteins from insects.

DECLINING BAT POPULATIONS AND BAT CONSERVATION

Bats are declining all over the world. As in the case of declining populations of natural ecosystems the reasons are basically related to developmental activities namely expanding human settlements and increasing agricultural land resulting in habitat loss as well as due to predators like cats, hawks, snakes etc. Tourism, hunting for meat and medicine, assumed pest status of bats are additional causes. Although there are no systematic studies on population densities, the trend in their declining numbers has been observed by many bat specialists. The frugivore colonies in India were in thousands historically but now have dwindled to an average of 500 or less. Similarly the age of colony has also been reduced, which is an indicator of roost disturbance. Around 70% of the roosts are under some kind of threat while 40% are affected by human activities such as tree cutting, cave destruction etc (Mistry, 1995).

Insect eating bats are prone to secondary poisoning due to large scale use of pesticides and automobile pollutions. Severe population decline of Brazilian free-tailed bat, *Tadarida brasiliensis* is attributed to DDT (Clark, 2001). Urban bats are vulnerable to toxic effects of lead poisoning. The tissues of dead urban bats had high concentrations of lead in their organs and tissues compared to non-urban bats in Australia. The lead emissions from automobiles deposited on the fur of bats seems to have entered the gut while grooming.

CONCLUSIONS

So far the frugivorous bats seem least to qualify for pest status. Their attack on cultivated fruits is sporadic and rare. Few instances of fruit depredation occur randomly. Most of the time fruit bats subsist on natural sources of food. It appears that the lack of substantial data on the economic impact of fruit bats is not due to lack of research but possibly due to absence of regular, sustained depredation on fruits. If and when a situation of pestilence and damage of considerable severity occurs, non-lethal methods like illumination, auditory scares, selective netting and covering of fruits/bunches with sprigs, leaves, thatches, thorny twigs etc to deter bats from fruits offer adequate protection. The removal, trapping, killing of bats should be the last step, to be employed in cases of extreme damage and when all other attempts fail to prevent/reduce bat damage. They have to be resorted only once in 3-4 years. Timely harvest is one major method to prevent bat damage because bats consume only ripe fruits. Because of their declining numbers and restricted distribution, fruit bats of India should be brought under schedule 1 of the Wildlife Protection (WP) Act giving them the status of vulnerable or endangered species with provision of licensing to kill only in cases of severe damage as is done for Nilgai in Haryana, Punjab and Rajasthan.

REFERENCES

ADVANI, R 1982. Feeding, foraging and roosting behaviour of the fruit eating bats and damage to fruit crops in Rajasthan and Gujarat, *Saeugeteir. Mitteil.* 30(1): 46-48

ADVANI, R. 1992. Seasonal fluctuation in the diet composition of *Rhinopoma hardwickei* in Rajasthan desert. *Proc. Indian. Acad. Sic.* (Animal Sic.). 91(6): 563-568.

BALASINGH, J., S. SUTHAKAR-ISSAC & R. SUBBARAJ. 1993. Tent roosting by the frugivorous bat, *Cynopterus sphinx* in Southern India. *Cur. Sci.* 65(5):418

BALASUBRAMANIAN, P & P.V. BOLE. 1993. Seed dispersal by mammals at point Calimore Wildlife Sanctuary, Tamil Nadu. *J. Bom. nat. Hist. Soc.* 90(1): 33-44

BANERJEE, K., M.A. ILKAL & P.K. DESHMIKH. 1984. Susceptibility of *Cynopterus sphinx* (frugivorous bat) and *Suncus murinus* (house shrew) to Japanese encephalitis virus. *Indian J. of Med. Res.* 79(1): 8-12.

BATES, P.J.J., D.L .HARRISON, N. THOMAS & M. MUNI. 1994. The Indian fruit bat *Latidens salimalii* Thonglongya, 1972 (Chiroptera: Pteropidae) rediscovered in Southern India. *Bonn Zool. Beitr.* 45:89-98.

BATES, P.J.J & D.L. HARRISON. 1997. BATS OF THE INDIAN SUBCONTINENT. Harrison Zoological Museum, Kent, UK. XIV+268 p.

BHAT, H.R 1994. Observation on the food and feeding behaviour of *Cynopterus sphinx* Vahl. (Chiroptera, pteropodidae) at Pune, India. *Mammalia* 58: 363-370.

BROSSET, A. 1962. The bats of central and Western India. Part I. *J. Bombay nat. Hist. Soc.* 59: 583-624.

CHAKRAVARTHY, A.K & A.C. GIRISH. 2003. Crop protection and conservation of frugivorous bats in orchards of hill and coastal regions of Karnataka. *Zoo's print Journal.* 18(8):1169-1171.

CLARK, D. R. Jr. 2001. DDT and the decline of free-tailed bats (*Tadarida brasiliensis*) at Carlsband cavern, New Mexico. *Archieves Envt. Contam. Toxicol.* 40(4): 537-543.

De-FIGUEIREDO, R.A. 1999. A comparison of the quality of dispersion of *Ficus eximia* Schott (Moraceae) by birds and bats in South-eastern Brazil. *Leandra.* 14:37-42.

ELANGOVAN, V & G. MARIMUTHU 2001.Effect of moonlight on the foraging behaviour of a megachiroptera bat, *Cynopterus sphinx. J.Zool.* 253(3): 347-350.

FLEMMING, T.H. 1988. THE SHORT TAILED FRUIT BAT: A STUDY IN PLANT-ANIMAL INTERACTIONS. The University of Chicago Press, Chicago. pp. 177-179

GOPUKUMAR, N., P.T. NATHAN., P.S. DOSS., A. PRAKASH. K. EMMANUEL, J. BALASING., G. MARIMUTHU & T.H. KUNZ. 2003. Early ontogeny of foraging behaviour in the short-nosed fruit bat, *Cynopterus sphinx* (Megachiroptera): preliminary results. *Mammalia.* 67(1): 139-145.

HALL, L & G. RICHARDS. 1987. FLYING FOXES, FRUIT AND BLOSSOM BATS OF AUSTRALIA. University of New South Wales Press, Australia, VIII+135p.

HELVERSON.O.VON & C.C. VOIGT. 2002. Glossophagine bat pollination in *Helicteres baruensis* (Sterculiaceae). *Ecotropica.* 8(1): 23-30

HODGKISON, R., S.T. BALDING., A. ZUBALD & T.H. KUNZ. 2003. Fruit bats (Chiroptera : pteropodidae) as seed dispersers and pollinators in a low land Malaysian rain forest. *Biotropica.* 35:491-502

HUTTON, A.F.I. 1949.*J. Bombay Nat. Hist. Soc.* 48: 454-460

IBID. 1999. Contribution to the knowledge of bats (Mammalia : Chiroptera) of North East Hills, India. *Records of the Zool. Survey of India*, Paper 174: 1-52.

IBID. 2001. Excellent association between three species of tropical bats at Sri Kallapiran Tirukovil Tower Sriraikundam, Tamil Nadu, *BAT NET* 2(1).

IBID. 2001. Rat-tailed bats feeding on winged termite in rainy season. *Bat Net* 2(1): 1

IBID. 2004. A non commercial fruit attracts bats and saves commercial fruits-a boon to orchardists. BAT NET 5(1).

IBID. 2005. Special report about *Latidens salimali* of Kalakad Mundanthurai Tiger Reserve. BAT NET-CCINSA Newsletter. 6(1):7-9.

KORNIE, C., I.KHAKI & ARAD. 1999. The Egyptian fruit bat *Rousettus aegyptiacus,* a pest in Israel? An analysis of the bat's diet and implication for its control. *Biol. Conserv.* 88:301-306

KUNZ, T.H & C.A. DIAZ. 1995. Folivory in fruit eating bats with new evidence from *Artibeus jamaicensis* (Chiroptera : Phyllostomidae). *Biotropica.* 27(1): 106-120.

MacKANZIE, J.S., K.B. CHUA., P.W. DANIELS., B.T. EATON., H.E. FIELD, R.A. HALL., K. HALPIN., C.A. JOHANSEN., P.D. KIRKLAND & S.K. LAM., P. McMINN., D.J. NISBET., R. PARU., A.T. PYKE., S.A. RITCHIE., P. SIBA., P.W. SMITH., G.A. SMITH., A.F. KANDEN HURK., L.F. WANG & D.T. WILLIAMS. 2001. Emerging viral diseases of Southeast Asia and western pacific. *Proc. International Conf. on emerging infectious diseases.* Atlanta, Georgia, USA. pp. 497-504.

MARIMUTHU, G. 2001. BAT NET 2(9)

MEDELLIN, R.A., M. EQUIHUA & M.A. AMIN. 2000. Bat diversity and abundance as indicators of disturbance in neotropical rain forests. *Conserv. Biol.* 14(6): 1666-1675. Special section: Habitat disturbance and tropical rain forest mammals.

MICKLEBURGH, S., A. HUTSON & P.A. RACEY. 1993. *Action plan for old world fruit bats.* Gland: IUCN, 252p

MISTRY, S. 1995. The bats of India. *Bats.* 13(2):11-15.

MOLUR, S., G. MARIMUTHU., C.SRINIVASALU., SHAROUKH MISTRY, A.M. HUTSON, P.J.J. BATES., S. WALKER., K. PADMA PRIYA & A.R. BINO PRIYA. 2002. *Status of South asian chiroptera-conservation assessment and management: C.A.M.P workshop report,* 2002. Zoo Outreach Organization, Coimbatore, India. 141p

PAINE, S 1993. The effect of bat excreta on wall paintings. *Conservator* 17:3-10.

PETIT, S. 1995. The pollinators of two species of columnar cacti on Curacao, *Netherlands antilles. Biotropica.* 27(4): 538-541

PETIT, S. 1996. The bat-cactus mutualism of Curacao. *Cactus and Succulent J.* 68(3): 150-152.

PINERSTEIN, E. 1984. In: Desert Abstracts International. B-Sciences & Engineering. 44:11, 3252, 146p.

PRADHAN, M.S & A. MAHABAL 2001. Observations on the roosting sites of free-tailed bat (*Tadarida aegyptiaca*) in Pune. BAT NET 2(1).

RACEY, P.A. 2002. Germination rate of seeds passing through the guts of bats is significantly greater than that of seeds taken from ripe fruits and from faecus of frugivorous birds. *Proc. of Twelth International Bat Research Conference*, Kuala Lampur, Malaysia.

RAJAN, K.E & G. MARIMUTHU. 1999. Localization of prey by the Indian vampire bat, *Megaderma lyra. Mammalia.* 63(2): 149-158.

SANDHU, S & A. GOPALAKRISHNA. 1984. Some observations on the breeding biology of *Cynopterus sphinx* in central India. *Curr. Sci.* 53(22): 1189-1192.

SAZIMA, M., BUZATO & I. SAZIMA. 2003. *Dyssochroma viridiflorum* (Solananaceae): a reproductively bat dependent epiphyte from the Atlantic rainforest in Brazil. *Annals of Botany.* 92(5): 725-730.

SERES, A & N. RAMIREZ. 1995. Floral biology and pollination of some monocotyledons of a Venezuelan cloud forest. *Annul. Miss. Bot. Garden.* 82(1): 61-81.

SHILTON, L.A., J.D. ALTRINGHAM., S.G. COMPTON & WHITTAKER. 1999. Old world fruit bats can be long distance seed dispersers through extended retention of viable seeds in the gut. *Proc. Royal. Soc. London. series B. Biol. Sci.,* 266(2416): 219-223.

SINGARAVELAN, N & G. MARIMUTHU. 2003. Mist net capture of the rarest fruit bat, *Latidens salimali. Curr. Sci.* 84:24-26.

SINHA, Y.P & R. ADVANI. 1976. Notes on food and reproduction of some Rajasthan bats. *Geobios.* 3:37-40.

SINHA, Y.P. 1986. Bats of Bihar: Taxonomy and field ecology. *Rec. Zoological Survey of India. Misc. Publ.* 77: 1-60.

SMYTHE, L.D., H.E. FIELD, L.J. BARNETT, C.S. SMITH, M.F. DOHNT, M.L. SYMONDS, M.R. MOORE & P.F. ROLFE. 2002. Leptospiral antibodies in flying foxes in Australia. *J. of Wildlife Dis.* 38(1): 182-186.

SRINIVASALU, C. 2001. Kelaart's *Pipistrella pipistrella ceylonicus* (Kelaart, 1852) in the diet of greater false vampire, *Megaderma lyra,* Geotra, 1810. *BAT-NET.* 2(1):5.

SRINIVASULU, B. & C. SRINIVASULU. 2001. Magnitude of depredation on grapes by short nosed fruit bats, *Cynopterus sphinx* Vahl, 1997 in Secunderabad. *Curr. Sci.* 80(1): 14-15

TAGGILANTI, A.V. 1979. THE WORLD OF MAMMALS. Berkshire House, Italy. pp. 202-206.

THOMAS, D.W. 1984. Fruit intake and energy budgets of frugivorous bats. *Physiol. Zool.* 57(4): 457-467

TSCHAPKA, M. 2003. Pollination of the under storey palm *Calyptrogyma ghiesbreghtiana* by hovering and perching bats. *Boil. J. Linean Soc.* 80:2, 281-288.

VANITHARANI, J. 2000. Some behavioural observations on *Rousettus leschenaulti* at Cheranmahadevi, Tamil Nadu. *BAT NET.* 1(2)

VANITHARANI, J., M. PEARCH., L. JEYAPRABA & R. ANNAMALAI. 2004. A review of the distribution and status of *Latidens salimali*

(Chiroptera: pteropidae) with new records from the Western Ghats, India. *Lutra* 47(1): 21-32.

VERGHESE, A 1998. Non destructive control of the bat, *Cynopterus sphinx* vahl (chiroptera : pteropodidae) in grapes (*Vitis vinifera Linnaeus*) in India. *Intl. J. Pest. Manag.* 44: 81-85.

❑❑❑

6

RODENT PEST MANAGEMENT IN CULTIVATED CROPS OF INDIA

Shakunthala Sridhara and O.P. Dubey*

AINP on Rodent Control, University of Agricultural Sciences, G.K.V.K., Bangalore, India-560 065
** Assistant Director General (PP) Indian Council of Agricultural Research New Delhi, 110 011, INDIA*

ABSTRACT

Of all the vertebrate pests, rodents are by far the most successful causing considerable damage to most cultivated crops throughout the world and undoubtedly qualify for the status of conventional pests. The major factors that have contributed to their unparalleled success are their '*r*' strategy, invasive capacity, agricultural practices such as irrigation, monoculture, continuous cropping, abnormal rainfall, drought, crop residues in fields, improper storage both in the fields and inside storage structures. About twenty species affect Indian agriculture namely *Funambulus pennanti, F. palmarum, F. tristriatus, Hystrix indica, Tatera indica, Meriones hurrianae, Gerbillus gleadowii, Rattus rattus, R.r. brunneusculus, R. nitidus, R.r. wroughtoni, R. norvegicus, Millardia meltada, Golunda ellioti, Mus musculus, M. platythrix, M. booduga, Nesokia indica, B. bengalensis* and *B. indica*. The damage varied from crop to crop and was specific to stages of crop growth. Pearlmillet, maize, jowar, pigeon pea, green gram, soybean, *zizyphus*, coconut and oil palm were highly susceptible at germinating/seedling stage forcing the farmer to sow several times in some areas. Around 5-10% damage was seen in cereals like rice, wheat, finger millet, vegetables like tomato, cauliflower, carrot, cucumber, musk melon, bottle gourd, chillies, fruits such as water melon, summer squash, sapota, pomegranate, spices like cardamom and the oil palm. Severe damage was seen in apple, pecan, coconut, cocoa and oil palm. Three species of rodents have invaded new habitats successfully viz. two wild species, *B. bengalensis* and *T. indica* have successfully adapted to commensal life in some cities. *B. bengalensis* has, in addition colonised the irrigated areas of arid Rajasthan. In contrast, the xeric, *M. hurrianae* is successfully living in the Chamundi hills near Mysore.

The biotic and abiotic factors of population regulation that affect natality, mortality and movement of rodents in the light of their control is briefly discussed. The strategies of rodent control is detailed under four steps namely preventing access, discouraging infestation, reducing density and finally Integrated Pest Management. Tillage, bund reduction, agro forestry and barriers are reviewed under preventing access while aspects of synchronized planting, clean cultivation and resistant varieties are dealt under discouraging infestation. The four methods of density reduction evaluated are physical control (hunting and killing, trapping, Trap Barrier System), chemical control (using acute rodenticides, chronic rodenticides and fumigants), biological control (predators and pathogens) and fertility control (steroids, immunocontraceptive vaccines and predatory odours). The validity of integrated pest management which is based on understanding the bio-ecology of pest species, the action threshold and population dynamics is highlighted. The three steps of IPM namely prevention of infestation, non lethal control or use of weak chemicals in the first stage followed by pesticide application later is elaborated. The frequent and repeated failures of rodent control is attributed to the high capacity of rodents to adapt to varied environments, resistance to anticoagulants, development of bait shyness towards zinc phosphide, neophobia on one hand and to poor understanding of rodent control and hence poor response of farmers to rodent management on the other hand. Additionally factors contributing to sub-optimal success of rodent control are improper execution of control measures, lack of proper guidance, dilution of technology by the time it reaches the end user i.e. the farmer and blind adaption of temperate zone models of rodent control which is not suitable for tropics. The importance and benefits of rodent control using multiple media channels, consideration of socio-religious attitude of farmers while planning a control programme and the need to constitute a national task force to plan and execute a five year rodent control programme throughout the country with major emphasis on educating and training the farmer on the feasibility and benefits of community based rodent control are discussed.

INTRODUCTION

Rodents are serious pests of agriculture particularly in tropical Asia, Africa and South America. In Asia alone they are reported to reduce yield of food grains to the extent of 10-20% in Indonesia, 2-5% in Malaysia (Singleton *et al.*, 1999), 6-7% in Thailand (Boonsong *et al.*, 1999), 5-15% in India and Bangladesh (Islam *et al.*, 1993; Sridhara, 1999; Parshad, 1999). Being vectors or transmitters of diseases rodents play a very significant role in the transmission of several zoonotic diseases such as resurgent plague, Leptospirosis, rat bite fever, salmonellosis, murine typhus, etc. Structural damage and production losses due to loss of man hours in crop fields and factories as a result of diseases, power/machine/computer breakdowns and failure of communication system consequent to cable damage by rodents are enormous though not estimated precisely.

Agriculture provides livelihood to 65-70% of the 100 billion plus population of India. Although there is surplus production, to achieve the target of increasing yields to feed the ever increasing populace not only increase in agricultural production is essential but to save what has been grown is equally important. Insect and disease control has been standardised for almost all crops grown in India but rodent control in spite of considerable damage still remains an unconquered problem. Factors responsible for such a scenario are: lack of adequate data on crop-pest relationships, biology and reproduction strategies of pest rodents, reliable damage estimates, non availability of crop and area specific strategy for rodent control as the country is ecologically highly diverse to have a uniform control programme, lack of awareness in farmers about the technology to be adapted as well as the absence of community approach to rodent control. This chapter outlines the major steps for effective rodent pest management centring around five major steps namely preventing access, discouraging infestation, killing rodents that have become pests, biological control and fertility control followed by strategies for integrated rodent management. A brief account on rodents in agro-ecosystem and rodent damage to agriculture in India will precede discussion on management.

AGROECOSYSTEMS AND RODENT PESTS

a. The *r* and *k* strategy of pests: Many agricultural pests have evolved a complex life history pattern fitting with the concept of *r* strategy. Agricultural ecosystems can be broadly considered as 'ecological islands' because pest infested crop fields are interspersed with non-crop habitats which are favourable for survival of dispersing species during lean periods. Cultivated crops especially annuals with a short growth period and changing crop cultivation area are ideal for *r* strategists like rodents. Similarly the new lands brought under cultivation by clearing forest, scrub and marshy areas are also colonized by rodents. While the former is comparatively less stable habitat, the later is not only highly unstable but harsh with extreme physical conditions. Thus the colonizing rodents are subject to heavy mortality due to limitations of physical environment and exhibit *r* strategy of producing many offspring and having a short life span. Most rodent outbreaks are associated with such disturbance resulting in severe losses in marginal lands. Once crop growing is stabilized, the microclimate is modified, animal communities are established and predation and competition for resources become limiting factors. The '*k*' strategy of investing more energy in improving the competitive ability of pregnancy and defending against predation is suitable for such situations. Being highly adaptable rodents under these conditions exhibit even '*k*' strategy (Price & Waldbauer, 1994).

b. Agricultural practices affecting rodent population: Rodent problems in agriculture become severe due to invasion, irrigation, monoculture, post harvest crop residues etc. Invasion is a serious problem of dry land agriculture. In tropics rodents usually breed during wet seasons. As dry seasons progress food supply and shelter begin decreasing. Considerable population dies due to starvation and predation. The survivors seek resources elsewhere. If irrigated crop fields are nearby rodents migrate there and colonise causing substantial damage.

Irrigation and replacing the existing crops with high yielding varieties with short growing periods also affect the natural breeding cycle of rodents. In diversified agriculture where different crops mature at different times early maturing varieties are damaged more. Rodents migrate from harvested areas to standing, maturing crops to access food and shelter.

In rain fed agriculture where only one crop is raised, a large number of rodents die during the long dry seasons. This seasonal mortality maintains chronic damage levels at reasonably low levels. During abnormal rainfall food availability and cover increase as also the duration of their availability resulting in declined mortalities. This may trigger an outbreak of population in the next season as reproduction occurs in a larger population.

Attempts to increase crop yields in areas already under cultivation such as intense irrigation, increased fertilizer input and raising several crops a year aggravate rodent problem further. Such measures facilitating cropping all round the year in areas which had rain fed crops earlier bring changes in rodent biology and behaviour leading to breeding of rodents throughout the year increasing rodent numbers and damage. Bunds and canals in rice and other irrigated crops provide additional shelters for fossorial species.

Small scale seasonally irrigated crops such as summer crops grown in traditional dry land areas surrounded by uncultivated land are highly susceptible to rodent damage. The post monsoon breeding in a wide variety of natural and wild habitats results in high rodent populations which face a resource depleted habitat as dry season progresses. These rodents are attracted to patchily irrigated crop fields.

Similarly soon after harvest rodents start seeking newer sources of food and shelter which they find in unprotected grain in the field or openly stored or poorly stored agriculture produce in houses and godowns.

The above, more or less universal pattern of rodent infestation and damage in tropical countries emphasises the need for developing and incorporating rodent control as an integral part of agricultural practices. Such a planning requires data on the ecology of pest species involved, appropriate techniques to reduce rodent damage and training of local resource persons who can supervise long term rodent control programmes.

RODENT SPECIES INFESTING CROPS IN INDIA

To design an effective rodent control strategy it is essential to have data on species infesting, species responsible and species likely to cause damage, value of crop to be protected, probability of damage, value of potential damage, possible method to be employed, cost of control method selected, size of the area to be controlled and side effects of the baiting programme.

India, by virtue of its diverse ecological and agricultural conditions supports 128 species of rodents belonging to 46 genera. Of these nearly 20 are pests of agriculture both in the field and storage occupying diverse ecological niches (Table 1). The common commensal rodents occurring throughout the country are *Rattus rattus* and *Mus musculus*. Of late *Bandicota bengalensis* is also becoming a commensal. These species are serious pests in warehouses, godowns, poultry and livestock facilities. There are three species of arboreal squirrels, *Funambulus pennanti, F. palmarum* and *F. tristriatus* and two species of *Rattus, R.r. wroughtoni* and *R.r. blanfordi* all of which have wonderfully adapted to food available on tree tops of coconuts, cocoa, cashew and shrubs like cardamom and coffee mostly in the South-western, Western coastal areas. The coconut plantations of Lakshadweep islands are infested by *R. rattus* while the Andaman Islands have *R. r. andamanensis* and three more sub-species of *Rattus* infesting coconut and oil palm plantations. The arid areas of Rajasthan and North West Gujarat have the xeric species, *Meriones hurrianae* and the mesic species *Tatera indica* and *Millardia meltada* as major pests with *Gerbillus gleadowii* and *Golunda elioti* occasionally becoming serious pests. North-east India has, apart from *B. bengalensis*, *B. indica, Mus musculus, M. booduga*, its own repertoire of indigenous rodent pests viz. *R.r. nitidus, R. sikkimensis, R.r. brunneusculus* and *Dremomys lokriah*. While *Nesokia indica* is restricted to Punjab, Haryana, Uttar Pradesh and parts of West Bengal, *Hystrix indica* is ubiquitous occurring at the edges of forests and wilderness throughout India. The state of Jammu & Kashmir, apart from the usual complex of *B. bengalensis*, *T. indica, M. musculus* and *Rattus* species is distinct with several species of marmots, hamsters, voles and indigenous squirrels whose pest status is yet to be established. In the vast majority of crop fields of India the rodent species of high impact are *B. bengalensis, T. indica, M. meltada, M. platythrix, M. booduga* and *M. musculus*. Of these irrigated fields are predominantly infested by *B. bengalensis, M. booduga* and in northern parts by *N. indica*.

Table 1. Rodent pests in Indian agriculture

Species	Distribution	Pest status
1. Five stripped Northern palm squirrel, *Funambulus pennanti* (Wroughton 1905)	South of Sikkim to northern districts of Karnataka	Damage to fruits and vegetables (Prakash *et al*., 92; Parshad & Malhi '94)
2. Southern palm squirrel, *F. palmarum* (Linnaeus 1766)	Entire South India	Common pest of *chiku*, pomegranate (Sridhara, '99), coconut, cocoa, coffee, areca nut, cashew nut & cardamom (Bhat, '92; Chakravarthy, '93)
3. Western Ghat squirrel, *F. tristriatus* (Water house 1792)	Limited area of Western and south Western India from Mumbai to Travancore, West coast & Western ghats	Major pest on Cocoa, cashew nut and areca nut (Bhat, '92).
4. The Indian crested porcupine, *Hystrix indica* (Kerr 1792.	Throughout India upto 2750M	Feeds on tubers, bulbs, tree barks-damage severe at forest edges (Agarwal & Chakravarthy, 92; Sharma, 94' Girish '05)

Species	Distribution	Pest status
5. The Indian gerbil, *Tatera indica* (Hardwicke 1807)	Throughout India.	Major pest of rain fed agriculture, consuming almost all crops at every stage (Jain, 92; Sridhara, 99).
6. The desert gerbil, *Meriones hurrianae* (Jerdon)	Restricted to North west Gujarat, Rajasthan desert, parts of Punjab and Haryana	Serious pest of cereals, vegetables and fodder (Prakash, '81).
7. The hairy-footed gerbil, *Gerbillus gleadowii* (Murray 1886)	Rajasthan desert and parts of Gujarat.	Occasionally becomes a serious pest of crops (Tripathi *et al.*, '92).
8. The House rat, *Rattus rattus* (Linnaeus 1758)	Throughout India as a commensal; as a field pest in plantation crops of S. India and crop fields of N.E. India.	Huge losses in godowns, poultry and serious damage in livestock facilities, cocoa, coconut and field crops in N.E. India (Bhat, 92 Sridhara & Krishnamoorthy '79; Parshad, '99).
9. Sikkim or Hodgson rat, *R.r.brunneusculus* (Hodgson 1845)	Restricted to the hill states of N.E. India.	Considerable loss to paddy, maize and vegetables. Population outbreak correlated with bamboo flowering (Chauhan & Saxena, '92; Pathak & Kumar, '01).
10. The Himalayan rat, *R. nitidus* (Hodgson 1845)	North East India & Kumaon region in Uttar Pradesh.	Damages paddy, maize and pineapple (Singh *et al.*, '94).
11. The Wroughton's rat, *R.r.wroughtoni* (Hinton 1919)	Kerala, Karnataka, Andhra Pradesh, parts of Maharashtra.	Serious pest of coconut, cocoa and oil palm (Bhat *et al.*, '90).
12. The Norway rat, *R.norvegicus* (Berkenhout 1769)	Occurs in only port cities of Mumbai & Kolkata.	Pest of warehouses (Jain *et al.*, 1993).
13. The soft furred field rat, *Millardia meltada* (Gray 1837)	Throughout India except North-East mountains.	Serious pest of cereals, pulses and oil seeds. Also damages natural grass-

Species	Distribution	Pest status
		lands and fodder crops in Rajasthan (Rana, '92)
14. The Indian bush rat, *Gollunda ellioti* (Gray 1837)	North-West region of India.	Minor pest of agriculture in Punjab & Rajasthan (Saini & Parshad, '93; Prakash *et al.*, '95)
15. The House mouse, *Mus musculus* (Linnaeus 1758)	Throughout the world.	Nuisance in houses, damage in storage and a pest of sugarcane, groundnut (Rao & Balasubramanyam, '92).
16. The Brown spiny mouse, *M. platythrix* (Bennet 1832)	Throughout India except the North-East, Jammu & Kashmir.	Pest of paddy, ragi, wheat, oil seeds, pulses and vegetables (Rao & Balasubramanyam, '92; Sridhara, '99)
17. The Indian field mouse, *M. booduga* (Gray 1837)	Throughout India	Pest in paddy, vegetables and groundnut (Rao & Balasubramanyam, '92; Sridhara, '99)
18. The short-tailed mole rat, *Nesokia indica* (Gray 1830)	North-Western and Northern India.	Pest on lawn grass, cereals, groundnut and vegetable crops (Ramesh, 92).
19. The lesser bandicoot rat, *Bandicota bengalensis* (Gray 1835)	Throughout India including the semi arid Rajasthan.	Very serious pest of cereals, pulses, sugarcane, oil seeds, almost all vegetables, selectively commensal in godowns. Burrowing activity affects fruit trees like apples etc. (Chakraborthy, '92; Sridhara, '99; Sridhara & Tripathi, 05)
20. The larger bandicoot rat, *B. indica* (Bechstein 1800)	From the south of Rajasthan to the southern tip of India, eastwards to W. Bengal and North East India.	Damage to paddy, wheat, maize and vegetables. Also to aquaculture (Chakraborthy, '92; Girish *et al.*, '05).

Following is the pest status of Indian rodents (Plate 1; Table 1)

*1. **Funambulus pennanti*** **(90g)**, the five stripped squirrel is medium sized, occurring from south of Sikkim to northern tip of Karnataka, from Baluchistan in the West to West Bengal in East. It is a serious pest in pomegranate, sapota (*Chiku*), grapes, guava and *zizyphus*.

*2. **Funambulus palmarum*** **(90g),** the southern palm squirrel is a medium sized squirrel whose biology, habits, habitat and economic importance are similar to *F. pennanti* except for a minor morphological difference. Major pest on *chiku*, pomegranate, coconut, cocoa, areca nut, cashew nut, cardamom and coffee.

*3. **Funambulus tristriatus*** **(125g),** the Western Ghat squirrel. Its distribution is limited to Western and South Western India extending from Mumbai to Travancore, mainly occurring in the coastal and Western ghat area. The squirrel is omnivorous preferring fruits and produce of plantation crops. Causes heavy damage to cocoa.

*4. **Hystrix indica*** **(11-18kg),** the Indian porcupine is the largest Indian rodent occurring throughout the country from 2750 meter altitude of Himalayas to the coast of Kanyakumari. It causes considerable loss to maize, sugarcane, potato, sweet potato, ripe melon, onion, carrot etc. apart from girdling and debarking forest trees, coconut, rubber etc.

*5. **Tatera indica*** **(70 to 150g)**, the Indian gerbil is ubiquitous occurring throughout the country. Being highly flexible in food habits, it is destructive to all kinds of cultivated crops at all stages of growth. It switches over readily to wild vegetation or becomes insectivorous and carnivorous eating smaller rodents and ground nesting birds. The species is a serious pest of all crops in arid zones and on millet, jowar, maize, pulses, groundnut and vegetables in other states. It always lives in association with other major rodent pests of an area.

*6. **Meriones hurrianae*** (40-60g), the desert gerbil. Its distribution is limited to North-West arid areas of the country. Lives in almost all habitats of Thar Desert and is a common pest of bajra, wheat, vegetable crops, grass and other fodder vegetation.

*7. **Gerbillus gleadowii,*** the hairy footed gerbil is found only in arid parts of Rajasthan and Gujarat, generally inhabiting sand dunes and uncultivated patches. Not much is known about its biology and behaviour. Occasionally it becomes a serious pest of crops.

*8. **Rattus rattus*** **(100-150g),** the house rat is the most abundant, widely distributed commensal rodent of India with at least 14 subspecies (Biswas & Tiwari, 1969). It inhabits a wide variety of man made structures, poultry, livestock facilities, selected crop fields, plantation crops, open country, forests and also hills. It is a serious pest on coconut, cocoa in South India and field crops in North-East India.

9. R. rattus brunneusculus, the Sikkim or Hodgson rat is found only in the North-Eastern states of India mostly in paddy fields and also in hilly slopes, forests and in tribal settlements. Considerable damage is inflected to paddy, maize and vegetables. Population outbreaks are cyclic, correlated with bamboo flowering.

***10. Rattus nitidus* (100-175g),** the Himalayan rat is found only in North-East India and Kumaon hills of Uttar Pradesh. It damages paddy, pineapple and maize.

11. R. rattus wroughtoni, the Wroughton's rat has restricted distribution covering south India and Maharashtra. It is found abundantly in the semi evergreen forests, scrub jungles and plantations crops of Karnataka, throughout Kerala where it inhabits houses, coconut and cocoa plantations. The species is a pest on coconut, cocoa, cotton seeds and oil palm seedlings.

***12. Rattus norvegicus* (250-350g),** the Norway rat is basically a temperate zone commensal and has invaded the port cities of Mumbai and Kolkata and is a pest in urban godowns.

***13. Millardia meltada* (40-60g),** the soft furred field rat is distributed throughout the country in dry habitats except North-Eastern parts of the country. Metads are serious pests of *kharif* and *rabi* crops. In Rajasthan they inflict damage to natural grassland and fodder crops also.

14. Gollunda ellioti, the Indian bush rat has a restricted distribution in North-West region of the country. It is a minor pest of agriculture in Punjab and Rajasthan.

***15. Mus musculus* (15-20g),** the house mouse is commensal and universal in distribution. Reported in groundnut and sugarcane fields.

***16. Mus platythrix* (20-30 g),** the brown spiny mouse occurs throughout India except the North-East, Jammu & Kashmir. It is a pest of paddy, ragi, wheat, oil seeds, pulses and vegetables.

***17. Mus booduga* (10-12g),** the Indian field mouse is the smallest Indian rodent and is distributed throughout the country more so in irrigated crops and is reported to be a pest in radish, groundnut, sugarcane and rice.

***18. Nesokia indica* (200g),** the short tailed mole rat restricted to Punjab, Rajasthan, Haryana, Uttar Pradesh, Himachal Pradesh and Delhi. Prefers irrigated crop fields. Pest on lawn grass, barley, wheat, groundnut, sugarcane, mustard and vegetables.

***19. Bandicota bengalensis* (200-300g),** the lesser bandicoot rat, occurs throughout India including parts of Rajasthan that are under irrigation. It is a very serious pest of agriculture in India causing damage to cereals, pulses, vegetables, plantation crops and fruit trees. The last mentioned are damaged by their digging behaviour.

20. Bandicota indica **(500g to 3 kg),** the larger bandicoot rat is distributed from the south of Rajasthan to the tip of Kanyakumari in South India. It lives in close proximity to human habitat but never inside the houses or structures but always making burrows close to houses, in gardens, near compounds, stables, poultry and outhouses. It is an occasional pest of vegetables and cereals, but a serious pest on aquaculture in West Bengal.

RODENT DAMAGE TO AGRICULTURAL CROPS IN INDIA

The extent of rodent damage to agricultural crops in India is varied and extensive. Almost all cultivated crops are vulnerable to rodent depredation at some stage or the other of crop growth and maturity. The various rodent species are adapted to feed on tuberous vegetables below the ground, the small sized cereals, the medium sized oil seeds, the larger fruits and vegetables of various kinds, seeds of cocoa, coconut, cashew nut, fruits and nuts of oil palm. The various modes of living namely fossorial, nest living, and arboreal are all adaptations to the seasonal availability of cultivated and wild foods. They switch over from vegetarian to animal food mostly insects during periods of non availability of crop produce. Though not studied in detail, their foraging strategy seems to have highly evolved wherein pattern of search facilitate not only consumption of easily available nutritious food but also evasion of predators. Rodent damage to various crops in India and the species responsible are presented in Tables 2 and 2a and Plate 2. Some crops are extremely susceptible at seedling stage especially pearl millet, maize, jowar, pigeon pea, green gram, soybean, sunflower, groundnut, *ber,* coconut and oil palm. The damage is so serious in some cases that repeated sowing has to be carried out. More than 10% damage was seen in tomato, cauliflower, carrot, cucumber, musk melon, bottle gourd, chillies, water melon, summer squash, sapota, pomegranate, cardamom and oil palm apart from cereals. Severe damage was seen in apple, pecan, coconut, cocoa and oil palm (Sridhara & Tripathi, 2005). The following is the brief account of rodent damage to different cultivated crops in India.

FOOD CROPS

Wheat: Rodent damage is at seedling and during ripening. It ranged from 2.7% in Gujarat (Rana *et al.*, 1994) to 21.3% in Rajasthan

(Jain *et al.*, 1993). *B. bengalensis, T. indica, M. musculus* were the major pests. *M. hurrianae* in Rajasthan, *R. rattus* and *M. musculus* in Himachal Pradesh also depredated on wheat (Malhi & Parshad, 1989; Jain *et al.*, 1993; Rana *et al.*, 1994; Sridhara, 1999; Butani & Vyas, 2005).

Rice: Damage is at seedling, milky and grain maturation stages and ranged from 1.1% in Punjab (Anon, 1991) to 79.7% in hill regions of Karnataka (Chakravarthy *et al.*, 1992). The countrywide damage to paddy ranged from 3.28% to 24.4%. The major pests were *B. bengalensis, M.meltada, R.rattus* and *M.booduga* (Rana *et al.*, 1994; Sridhara, 1992; Parshad, 1999). There was some geographic variation in species damaging with *M. musculus, R. nitidus* in Meghalaya and Mizoram (Pathak & Kumar, 2001), *B. indica* in Himachal Pradesh (Sridhara & Tripathi, 2005) and *R. sikkimensis* in Assam (Anon, 2004a).

Table 2. Rodent damage and species infesting food crops in India (Sridhara & Tripathi, 2005)

CROP/CATEGORY	SEEDLING STAGE		PRE HARVEST STAGE	
	Dam. %	Species infest*	Dam. %	Species infest
WHEAT (*Triticum aestivum*)	-	-	2.7-20.9	Mh, Ti, Mm
RICE (*Oryza saliva*)	8.22	Bb, Mb, Mp	1.1-17.5 1-79.7 (Hilly reg. Karnataka)	Bb, Rm, Rr, Rn, Mm, Mb, Mp, Bb, Mm, Mb, Rr
PEARL MILLET (*Pennisetum glaucum*)	Almost 100% (Resown, Rajasthan)	Ti, Mh	3.0-12.0	Bb, Ti, Mm
FINGER MILLET, (*Eleusine coracana*)	-	-	6.0-12.0	Bb, Mm, Ti
MAIZE (*Zea mays*)	10.7 (Avr.) 50-80 (sometimes)	Bb, Ti, Mb	1.9-24.0	Rn, Bb, Ti, Mm, Rr
JOWAR (*Sorghum vulgare*)	4.0-7.0	Bb, Mm	4.0-10.0	Bb, Ti, Mm
BENGAL GRAM (*Cicer arietinm*)	-	-	2.5-25.0	Mm, Bb, Ti
PIGEON PEA (*Cajanus cajan*)	50.0-100%	Bb, Mm, Mp	2.0-7.0	Bb, Ti, Mm, Mp
COW PEA (*Vigna catiang*)	-	-	4.0-18.0	Bb, Ti, Mm

GREEN GRAM (*Phaseolus aureus*)	10.0 (Vegetative growth)	Bb, Mm, Ti	5.0-6.0	Bb, Mm, Ti
SOYBEAN (*Glycine max*)	27-45 (Green pods)	Bb, Mm, Rr	0.6-3.0	Mm, Ti
GROUNDNUT (*Arachis hypogea*)	30-40 (green pods)	Bb, Ti	0.6-19.0 85% (outbreak)	Ti, Rm, Bb, Mb, Mm
SUNFLOWER (*Helianthus anus*)	70	Bb, Mm, Ti	-	-

Table 2a. Rodent damage and species infesting Horticultural and Plantation crops in India (Sridhara & Tripathi, 2005)

I.HORTICULTURAL CROPS	Dam. %	Species
(i) VEGETABLES		
TOMATO (*Lycopersicum esculentum*)	2.6-37.3	Bb, Ti, Mm, Mp Ti, Mh, Mm (Arid zone)
KNOL-KHOL (*Brassica oleracea* var *gongylodes*)	2.0-6.0	Bb, Ti
POTATO (*Solanum tuberosum*)	5.0 (seedling) 3.5-9.0	Bb, Mm, Ti Bb, Mp,Ti
PEA (*Pisum sativum*)	1.8-8.0	Bb, Rm, Mms, Mb, Rr, Ge, Mh, Ti (Arid zones)
CABBAGE (*Brassica oleracea var capitata*)	5.5	Mm, Mb, Rr, Ge, Bb
BRINJAL (*Solanum melongena*)	2.0-6.0	Bb
CAULIFLOWER (*Brassica oleracia*)	2.3-13.9	Bb, Mms, Mb, Mm, Rr, Ti
CARROT (*Daucus carota*)	4.0-11.0	Bb, Mm, Ti, Mp
CUCUMBER (*Cucumis sativa*)	4.8-19.9	Bb, Ti, Mm
MUSK MELON (*Cucumis melo*)	5.3-11.8	Bb, Ti, Rm
FRENCH BEAN (*Phaseolus vulgaris*)	4.0-7.0	Bb, Ti, Mm, Mp
BOTTLE GOURD (*Lagenaria siceraria*)	4.0-14.6	Bb, Ti, Mm, Mh, Ti (Arid zone)
BEETROOT (*Beta vulgaris*)	4.0	Bb, Ti, Mp

RIDGE GOURD (*Lufa acutangula*)	Upto 10%	Bb, Mm
SWEET POTATO (*Ipomea batatus*)	3.0-9.0	Bb, Mm, Ti
SPONGE GOURD (*Lufa cylindrica*)	9.8	Bb, Ti
CHILLI (*Capsicum annum*)	0.6-11.7	Bb, Ti, Rr, Mp, Mb
LUCERN (*Medicago sativa*)	5.0	Bb, Ti, Mm
(ii) FRUITS		
WATER MELON (*Citrullus vulgaris*)	9.9-19.8	Bb, Rm
SUMMER SQUASH (*Cucurbita smoschta*)	1.4-18.4	Bb, Ti
APPLE (*Pyrus malus*)	17.0-40.0	Bb, Rr, Mm, Ge, Mb
PEACH (*Prunus persica/ P.platycarpa*)	2.0-7.0	Bb, Mms, Mb, Rr, Ge
BER (*Zizyphus jujuba*)	8.0-80.0 (Nursery)	Ti, Fp, Mm, Mh (Arid zones)
PECAN (*Carya illinoensis*)	1.6-17.4 (Root damage)	Bb, Mms, Mb, Rr, Ge
SAPOTA (*Acrus sapota*)	5.0-10.0	Fpm, Bb, Ti
PINEAPPLE (*Ananas comosus*)	2.6-44.4	Rn, Bb, Rm, Rr
POMEGRANATE (*Punicia granatum*)	6.1-12.0	Fp, Fpm, Bb
II. PLANTATION CROPS		
COCONUT (*Cocos nucifera*)	6.8-8.0 (nursery)	Bb
	4.5-55.0 (Tender nuts)	Fpm, Rr, Rrw, Rra
	7.0-15.0 (nuts)	Sv, Sc, Rr
COCOA (*Theobroma cocoa*)	10.28-60.0	Rrw, Ft, Fpm
CARDAMOM (*Elettaria cardamomum*)	1.4-20.0	Fpm, Bb

OIL PALM (*Elaeis guineensis*)	10.0-45.0 (Seedling) 50.0-57.3	Rrw, Bb, Rra Rra
CASHEW (*Anacardium occidentale*)	5.7 nuts/day	Rrr, Rb, Bb, Fpm, Hi
RUBBER (*Hevea brasiliensis*)	Root damage (Nursery) Debarking	Bb, Ti, Mm, Bi

**Ti-Tatera indica, Mm-Millardia meltada, Bb-Bandicota bengalensis; Mh-Meriones hurrianae; Mp-Mus platythrix; Rr-Rattus rattus; Ge-Gollunda ellioti; Mb-Mus booduga; Mms-Mus musculus; Rn-Rattus nitidus; Fp-Funambulus pennanti; Fpm-Funambulus palmarum; Rrw-R.r.wroughtoni; Rra-R.r.adamanensis; Sv-Sciurus vulgaris; Sc-Sciurus carolinensis; Ft-Funambulus tristriatus; Hi-Hystrix indica; Bi-Bandicota indica; Rrr-R..r. rufescens; Rb- R. blanfordi.*

Pearl millet: The damage at seedling stage was 100% caused by *G. gleadowii.* At milky and grain stage considerable damage was inflicted by *T. indica* and *M. hurrianae* in Rajasthan and by *B. bengalensis, T. indica* and *M. meltada* in Gujarat (Tripathi *et al.*, 1992).

Finger millet: *B. bengalensis, M. meltada* and *T. indica* damaged 6-12% of the crop at milky and grain formation stage in Karnataka (Sridhara, 1999).

Maize: Damage at seedling stage ranged from 10.7% in Punjab (Anon, 1995) to 50-80% in Karnataka (Sridhara, 1999). Damage at cob formation and cob maturation ranged from 1.9% in Gujarat (Butani & Vyas, 2005) to 24% in North east hill states (Pathak & Kumar, 2001). The seedling stage was damaged by *B. bengalensis, T. indica* and *M. meltada* and cob formation stage was damaged by *B. bengalensis, T. indica, R. nitidus, M. musculus* and *R. rattus* (Singh *et al.,* 1994; Sridhara, 1999; Pathak & Kumar, 2001).

Redgram, Sunflower, Jowar and Maize: Damage is to sown seeds and seedlings and ranged 50-100% for red gram, 70% for sunflower, 4-7% to jowar and 50-80% to maize. The species damaging were *B. bengalensis, R.nitidus, M. meltada, M. platythrix* and *T. indica* (Sridhara & Tripathi, 2005).

Groundnut: Damage was 30-40% to seedlings, 6-9% at pod maturation and 4-9% to mature pods. The species inflicting damage were *B. bengalensis, M. booduga, M. meltada* and *T. indica* (Sridhara & Tripathi, 2005).

Legumes: Damage to whole plant or the shoots and pods of lentil, *arhar, moong,* soybean and Bengal gram are well recorded. *B. bengalensis, M. meltada* and *R. rattus* damaged soybean and red gram (Sridhara & Tripathi, 2005).

Vegetables: Almost all vegetables are attacked by rodents at seedling and mature stages. The species involved were *M. musculus, B. bengalensis, M. booduga, M. meltada, M. platythrix* and *R. rattus* (Sridhara & Tripathi, 2005). Extent of damage was 0.6 to 19.9%.

Fruits: Damage is of three types viz to the seeds, seedling in the nursery by rats and squirrels causing death or stunted growth of saplings. It ranged from 1.6 to 44.4%. Secondly, young trees are damaged by the burrowing activity of fossorial species such as *B.bengalensis, N. indica* and sometimes *T. indica* and *H. indica* and lastly to the fruits by the arboreal species like *R. rattus, F. pennanti* and *F. palmarum.* The fruits susceptible to rodent attack are apple, pecan, peach, pineapple, *ber*, pomegranate and strawberries (Sridhara & Tripathi, 2005).

Plantation crops: Almost all plantation crops grown in India viz. coconut, cocoa, cardamom, oil palm, areca nut, cashew and rubber are attacked by rodents. Except oil palm, all others are grown in the hilly tracks of Western Ghats in South India. The damage and control strategy is different for each of them and has been worked out exhaustively for coconut, cocoa and cardamom (Bhat, 1992; Srihari & Chakravarthy, 1992).

a. Coconut: 6.8-8.0% seedlings are damaged by *B. bengalensis, B. indica* and *H. indica;* Tender coconuts are damaged by *R. r.wroughtonii* and *R. rattus* in South India and by *Sciurus vulgaris* and *S. carolinensis* in North East India. In the islands, it is *R. rattus* in Lakshadweep and *R. rattus andamanensis* in Andaman. Porcupines damaged by debarking and burrowing (Guruprasad & Srihari, 1983; Advani, 1984, 1985; Sarkar, 1986).

b. Cocoa is heavily damaged by *R.r. wroughtonii, F. tristriatus* and *F. palmarum.* The damage, ranging 47-60% is by way of consumption of seeds and the mucilage around it (Bhat, 1992).

c. Cardamom capsules are damaged by rodents notably *B. bengalensis* during night and five species of squirrels but mostly by *F. palmarum* during the day. The damage is to the mature capsules and is by seed consumption. It ranged 8.7 to 12.6% (Srihari & Chakravarthy, 1992).

d. Areca nut was damaged at 20% by squirrels in North-East India. Although damage is reported elsewhere at fruiting stage, published estimates are nil. The author has come across farmers complaining about squirrel damage to inflorescence especially during summer (Sridhara, personal observation).

Hystrix indica *Bandicota indica* *Tatera indica*

Meriones hurrianae *Millardia meltada*

Mus booduga *Bandicota bengalensis*

Mus musculus *Rattus rattus*

Gerbillus gleadowii *Rattus rattus wroughtonii*

Plate 1. Some common rodent pests of agriculture in India

Tomato Coconut Groundnut

Paddy French bean

in Cabbage Marigold

in Ragi Ragi + Redgram + cowpea

Plate 2. Rodent damage and burrows in selected cultivated crops

e. Oil Palm: Rodent damage begins at the seedling stage (10 to 45%) when the succulent, tender stem base is eaten. In mature palms (50 to 57.3%) rodents feed on female inflorescence, the tender and ripe nuts. The sweet mesocarp of tender nuts as well as hard nuts was devoured. But the most preferred portion is the fruit in mature palms. Infestation is mostly by *Rattus* sp. Notably *R.r. andamanensis* and squirrels (Subiah, 1983; Bhat, 1992).

f. Cashew nut: Cotyledons of germinating cashew are damaged by *R. r. rufescens, F. palmarum, H. indica and B. bengalensis*. Nuts are consumed by *B. bengalensis, R. blanfordi* and *H. indica* (Basheer & Jayraj, 1964)

g. Rubber: There are no estimated figures of damage but several species viz. *R. meltada, B. bengalensis, B. indica* and *H. indica* are known to damage rubber in the nursery and young plants. Bandicoots damage the root system by their digging and porcupines debark at the base in order to eat the bark (Nehru & Jayarathnam, 1985).

COMMERCIAL CROPS

Sugarcane: Damage occurs by feeding on buds on seed cane, young shoot, apical growing part, and at inter-nodes. By far the greatest damage occurs by feeding on internodes leading to adverse changes in sugar content and lodging if the injury level is very high. Secondary damage is by way of injured cane becoming susceptible to insect and microbial attack. Damage becomes intense around 150 days, reaching a peak during cane maturity. An average 10-30% yield loss is common due to rodents. *B. bengalensis, T. indica and M. meltada* are the species inflicting damage (Srivastava, 1992).

Forestry: Young saplings are reported to be damaged by debarking and slicing activities of desert rodents in Rajasthan. *M. hurrianae*, *T. indica* and *R. meltada* debarked *Acacia tortilis* and *Parkinsonia aculeata* while slicing of roots of *A. tostilis*, *A. nilotica* and *Prosopis juliflora* below the ground and consequent death is due to *N. indica* (Jain & Tripathi, 2000).

CHANGING HABITAT AND SPECIES REPLACEMENT

Changing and establishment of new crop patterns is accompanied by changes in topography and the availability of food. Not only rodents adapt to changing environments but also invade new habitats. Sometimes the migrants replace predominant native species. Certain species may acquire entirely new areas for habitation. The invasion of newer habitats is well illustrated by *B. bengalensis*. Primarily a fossorial species infesting wetlands, the species has exhibited remarkable adaptability by successfully establishing as a commensal in Calcutta (Seal & Bannerji, 1966), Bombay (Deoras, 1966) and Karnataka (Sridhara, Personal observation) and has also succeeded in invading the irrigated crops in Rajasthan (Prakash, 1978). Several intrinsic characters of lesser bandicoots such as very high rates of reproduction all through the year, burrowing activities which confer better protection against predators and higher degree of aggressive behaviour are considered to have contributed to its invasion of newer habitats. When Bombay city expanded by encroaching agricultural grasslands, the species readily adapted from its wild habitat to a life inside the houses. Similarly it arrived from Punjab to the newly irrigated fields of Rajasthan through the Sri Ganga canal where it quickly adapted to irrigated crop fields of the hither to xeric environment. *T. indica* also successfully invaded houses and shops in the desert city of Bikaner over the past several decades. The most surprising adaptation is that of *M. hurrianae* to the hilly environs of Chamundi hills in Mysore. Similarly *R.rattus* has evolved into a complete arboreal existence on coconut palms in Lakshadweep islands (Shah & Subiah, 1978).

POPULATION DYNAMICS

The primary and basic objective of a rodent control programme is to suppress rodent numbers to a very low level for a long time with no undue harm to man, other animals and the environment. Understanding some facts of population dynamics of rodents facilitates their better management and hence damage reduction.

Rodent control is applied ecology where the principles of population growth, limitations and regulation are exploited to maintain minimal numbers of a pest population. Normally the population fluctuates seasonally with alternating periods of high and low density, active and inactive periods, reproductive and non breeding periods as well as periodical dispersals. Several environmental factors of the habitat like food, shelter, predators, disease and competition limit the size of rodent populations. Since these factors change with seasons, differ from place to place and time to time, rodent control becomes a highly specialised strategy for each cropping system. As it involves alteration of all the factors that provide food, shelter and protection from predators, a combination of cultural, mechanical and physical methods yields better results.

The environmental factors that affect changes in birth, mortality and movement of rats include habitat, predation and competition. These aspects act in a density dependent manner in regulating populations.

Habitat includes all aspects of physical environment including food. Destruction or disturbance of habitat significantly affects the population. Poisoning which is considered as a special type of pollution, is actually a different type of habitat manipulation. It temporarily reduces the carrying capacity of the environment by making the food toxic and some times by reducing the supply of oxygen when fumigants are used.

***Mortality factors*:** Predation and disease are prime causes of mortality in rodent populations. Man using poisons, traps or gases is the main predator of rats and mice. Predation in nature is caused by bacteria, owls, cats, foxes, ferrets, weasels, mongoose and snakes. However, these rarely alter prey population as in the failure of introduced mongoose (Davis *et al.*, 1976). Trapping is a form of predation. Intensive trapping can give good reduction of pest rodent populations.

Rodent parasites can be regarded in a similar manner; they can cause debility in rodents but severe mortality is rare. The plague organism, *Yersinia pestis*, is often highly fatal in rats but other pathogens, such as *Rickettsia mooseri* or *Leptospira icterohemorrhagiae*, produce few symptoms and are rarely fatal. Some types of *Salmonella* organisms may cause severe epizootics among rodent populations but the effect is generally short-lived and populations recover rapidly. This ineffectiveness coupled with potential danger to non-target species, including man are the reasons for not considering biological control by pathogens unless there is a rodent or species specific pathogen.

Competition within the species regulates rodent populations by increasing mortality and inhibiting reproduction especially in crowded conditions. Often both act together bringing down the carrying capacity of the area (Christian, 1950).

Inter specific competition also results in species replacement. When Norway rats were eliminated, roof rats increased and vice versa (Ecke, 1954). In New Zealand *R. norvegicus* has replaced the local *R. exulans.* In Mumbai and Kolkata, *B. bengalensis* has displaced *R. rattus* (Deoras *et al.*, 1975).

***Population growth and rodent control*:** Under conditions of optimum climate, surplus food, and abundant shelter, rodent populations tend to breed throughout the year. However, free living populations of rodents do not increase indefinitely beyond certain levels even if food and shelter are abundant. When the upper threshold of density is reached, the self-limiting factors cause a reduction in births or increase mortality, cause emigrations or a combination of these processes takes place. A rodent population that has been reduced in size recovers slowly at first and then at an increasingly faster rate. As it approaches the carrying capacity of the environment, the primary limitations being food, water and shelter, growth slows down and the population tends to level off. Two inherent characteristics of rodent populations facilitate fast restoration of density to its original levels if there is a decline in population. These are the great adaptability of rats to various crop conditions and their exceptionally high fertility rate. A control operation to be successful must reduce numbers to a very low level; otherwise, rats reproduce so rapidly that many times they exceed their precontrol density levels. It is as important to destroy the food sources and shelter of rodents as rodenticide application is. In the absence of such measures, poisoning has to be carried out relentlessly.

Table 3. Strategies for rodent pest management suitable for Indian agriculture

Sl. No.	STRATEGY	STEPS/TECHNOLOGY TO BE ADAPTED
1.	Preventing access to crop fields	(i) Tillage
		(ii) Bund reduction
		(iii) Agro forestry
		(iv) Barriers
2.	Discouraging infestation	(i) Synchronized planting
		(ii) Clean cultivation
		(iii) Resistant varieties
3.	Density reduction	(i) Physical control
		(a) Hunting and killing
		(b) Trapping
		(c) Trap barrier system
		(ii) Chemical control
		(a) Acute rodenticides
		(b) Chronic rodenticides
		(c) Fumigants
		(d) Timing of control
		Prophylactic control
		Symptomatic control

Sl. No.	STRATEGY	STEPS/TECHNOLOGY TO BE ADAPTED
		(iii) Biological control (a) Predators (b) Pathogens and diseases (iv) Fertility control (a) Steroids (b) Immunocontraception vaccines (c) Predator odours
4.	Integrated pest management of rodents	(i) Understanding pest species (ii) Action threshold (iii) Population dynamics (iv) Management (a) Prevention of infestation (b) Non lethal or weak chemical use (c) Pesticide application

STRATEGIES FOR RODENT PEST MANAGEMENT

Measures to control rodent pests are varied, diverse, traditional, historical and conventional (Table 3). They can be physical, mechanical, chemical, biological and ecological. In spite of all these attempts and enormous effort and expenditure in developing effective rodenticides, rodent damage continues to be unabated throughout tropical agriculture. The problem is very severe with heavy rodent infestation in almost all types of cultivated crops. Subsistence agriculture by marginal farmers who are mostly uneducated compounds the problem. The rodent population is not only dense but there are several species with varying biological traits and behavioural ecology infesting the same crop simultaneously or at different stages of its growth in a clear case of resource partitioning both temporally and spatially. Managing multiple species in varied crops and motivating farmers to adapt crop and area specific management techniques in such situations calls for evaluation of different methods of control and evolving an integrated approach. The literature is abundant with instances of success, partial success and failure of several approaches to manage these pests. In this chapter these various methods are categorised into four major steps namely preventing access to crop fields, discouraging infestation, density reduction when they reach pest status and preventing reinfestation and population build up. Finally an integrated rodent pest management suitable to Indian agriculture and how to make rodent pest management more effective will be dealt with.

1. Preventing access to crop fields: Theoretically preventing access to vulnerable crops is the most effective way of reducing rodent damage. However denying access to an attractive habitat with good sources of food and shelter, though difficult, is possible in commensal rodent control as the habitat includes structural enclosures. But in an open, resource rich environment like crop fields keeping rodents away is almost impossible or has limited possibilities.

Most of the suggestions to deter rodents from crop fields are physical in nature with one or two being chemically based. The basic principle is to make the terrain less hospitable and less safe to live. Measures include destruction of burrows, rendering fields unfit for fresh burrows, reducing vegetation cover so that exposure to predators is increased, erecting barriers and using chemical repellents.

(a) *Tillage:* Several cultural and traditional practices such as crop rotation and deep tillage inadvertently reduce rodent density. Deep tillage carried out by farmers while rotating crops destroys burrows of rodents and exposes them to aerial predators. Although data is inadequate for tillage effects on burrow destruction there is substantial indirect evidence to support this view. When vacant lands around wheat fields were ploughed the burrow density declined by 49% in Haryana (Ramesh & Katiyar, 1985). In Punjab, the predominant cropping system of rice-wheat rotation with interspersed sugarcane cultivation has evolved into an ideal habitat for rodents which have exploited the ready and continuous supply of food and shelter in this cropping system. After harvest of rice, the fields are subjected to mulching with rice straw and minimum tillage followed by sowing of wheat. The nearly undisturbed fields not only retain the native population of rodents but the availability of continuous food facilitates uninterrupted life cycle of these crop pests. Rodent infestation and damage are higher in such fields compared to deep tilled fields in conventional agriculture. Further the long duration cane fields cultivated in adjacent fields with their thick canopy and almost undisturbed land conditions act as temporary shelter whenever land preparations disrupt the rodents living in rice-wheat fields (Parshad, 1989).

(b) *Reduction in bund size* : Habitat can be rendered less favourable for digging burrows by keeping the bunds low and narrow (Sharma & Rao, 1989). However, this could be true if crops are cultivated over vast areas which render it difficult for rodents to travel long distances with their fixed home ranges. But in smaller, individual farms of tropical countries, higher level and bigger bunds are not a limitation as rodents are adept at colonising adjacent uncultivated areas and entering crop fields only for feeding.

(c) *Agro forestry*: A very interesting approach of discouraging rodents from infesting crops, in this case cassava, using agro forestry has been found successful in Africa. Planting of the bush, *Tephrosia vogelii* in and around the crop offered a high level of protection by discouraging burrow digging inside the crop fields (Anon, 2001-02).

(d) *Electric barriers:* Sub lethal electrical barriers were found effective in protecting rice in experimental plots in Philippines (Reidinger *et al.,* 1985) and America (Fitzwater, 1972). There are no published reports about the feasibility of electric fences to protect crops from rodents in India except a lone study (Sreenivasalu *et al.,* 1971). However, for farmers of developing countries this method is not only expensive, but difficult to install and maintain. Moreover rats can easily burrow under the fence or learn to cross the fence when there is no power.

(d) *Metal barriers* have been found successful in preventing rodent damage in coconut (Guruprasad & Srihari, 1978). A smooth metal sheet of 0.2mm thickness, measuring 18cm, fixed 2-3 meters above the ground around the trunk of coconut trees was effective (Anon, 2004b).

(e) *Repellents*: Farmers throughout the world have been using few traditional methods to repel rodents using odour and visual cues. These include placing screw pine leaves along the edges of paddy fields (Subiah, 1978), flagging off palm leaves or polythene pieces atop wooden poles of about one meter height (Neelanarayanan *et al.*, 1995a) or some kind of plant material which makes a rattling noise (Sharma, 1994). Preliminarily investigations in Tanzania have indicated that maize coated with thiram and cinnamamide prior to sowing reduced rodent damage considerably. Field trials of the repellency of these chemicals are under investigation (Ngowo *et al.*, 2003). Several laboratory studies in temperate countries have established that oleoresin of capsicum, Thiram and cinnamamide reduced feeding and damage by porcupines (Wagner & Nottle, 2000), house mouse (Nottle & Barnett, 2000) and Norway rats (Cracker *et al.*, 1993). A lone field trial established the repellency of capsaicin in poultry when capsaicin treated poultry feed was rejected in field trials (Jensen *et al.*, 2003).

Repellent odours derived from predator urine, faeces, gland secretions and compounds isolated from these sources suppressed feeding behaviour in rodents. Using this principle predator odours are used as repellents to protect plants from herbivores. Under predatory pressure, prey species change their activity rhythm to minimise risks, decrease locomotor activity, change their activity level and spatial distribution (Sullivan *et al.*, 1988).

2. Discouraging Infestation

By eliminating or reducing sources of food rodents can be prevented from infesting a field. Two measures which contribute to this effect are synchronizing cropping and clean cultivation. Raising rodent resistant varieties has remained theoretical so far.

(a) *Synchronized planting* specially of cereals adversely affects rodent infestation as it minimises the duration of availability of food and shelter and increases the duration of shallow period. It also distributes damage and infestation more evenly. Devoid of food and shelter, shallow periods limit the growth of rodent populations. Contrarily, areas with crops planted at different times are conducive to sever rodent infestation and damage. Crops planted early or late provide food and shelter when the surrounding areas have a low carrying capacity. In synchronised cultivation rats were found shifting their activity centres from harvested fields to yet to be harvested patches and also to piles of crop straw left in the field after harvest (Singh *et al.*, 1983; Pasahan & Singal, 1994; Jacob *et al.*, 2003). Rodents in the interior of crop fields are known to shrink their home range to avoid predation (Brown *et al.*, 2001).

Shifting of infestation to the interior of crop periods during harvest has several management implications. Primarily these aggregates cause heavy damage in yet to be harvested crops. Secondly this shift renders rat control by farmers less successful as they concentrate on poisoning burrows in refuge habitat, channel banks and bunds. But at the same time targeting the concentrated populations in the yet to be harvested patches is economical and effective.

(b) *Clean cultivation* prevents rodent infestation. Weeds provide shelter, breeding place and food during shallow period. Consequently their removal adversely affects rodent population. Elimination of shelter by weed control makes rodents vulnerable to predation resulting in decreased rodent population (Drost & Moody, 1982; Sablok & Pasahan, 1985).

(c) *Resistant varieties*: Attempts to raise rodent resistant varieties are almost non-existent although several studies have indicated definite preference towards some varieties of rice and sugarcane (Singh *et al.*, 1983; Parshad & Nijjar, 1995). However, the characteristics of rice that attracted rodents were not identified. Rodents being extremely sensitive to odour, taste, flavour, taste as well as nutritional composition of food in addition to grain size and texture exhibited definite preference in a laboratory choice situation (Sridhara, 1982). Only detailed studies can reveal preference for specific characteristics of certain rice varieties. However, in sugarcane some characters have been identified that are attractive to rodents such as thin barrel, soft rind, low fibre and logging habits. Damage is less to varieties with a thick barrel, hard rind, more fibre and non-lodging character (Prakash & Avasthy, 1980).

3. Density reduction measures

This is mostly a lethal approach with death of as many numbers of pest rodents as possible involving both traditional and conventional methods. These measures provide immediate results and are most popular but unless steps are taken to suppress surviving populations, the benefit derived is a short term one. The major methods of achieving satisfactory mortalities are physical (physical killing, mostly trapping) and chemical (rodenticide). The other two approaches are biological control and fertility control. Biological control has proven to be successful in some instances but population reduction by inhibiting reproduction is yet to achieve the status of field application.

During population reduction it is advisable to kill a large number of rats at a rate before rats learn to avoid the food or develop resistance. The response of population depends on poisoning regime. A single application for a short time reduces population, which then increases. A persistent poisoning programme may keep the population level low as long as poisoning regime is maintained (Drummond *et al.*, 1977).

(a) *Physical control*

It is mainly by hunting either by humans or dogs or by trapping. Both are oldest methods of rodent control worldwide. These are labour intensive and are not cost-effective over large areas of cultivated crops. Their utility is restricted to situations where using rodenticides poses a threat to environment and health, and as a supplement to chemical control. Only in small areas such as fields of breeding experiments for varieties, nurseries etc they are viable.

(i) *Hunting and killing* is an age old, traditional method of reducing/eradicating rodents. The two popular methods of hunting are flooding and smoking. When the burrows are flooded during irrigation, the immature, too old and sick get drowned but the healthy and young try to escape the water by coming out of burrows. Most farmers kill such rats either by sticks or allow dogs to hunt them. Several tribes in South India, Bihar and North-East India are habitual rat catchers, depending on rats for food during harvest season. They either dig or smoke the burrows forcing the rats to come out which are retrieved for consumption by using nets, catapults or by hand (Whitaker, 1970; Jain *et al.*, 1983; Ahmad, 1992).

In Lakshadweep islands the villagers have converted rat hunt into an annual ritual with the entire community participating and reportedly killing 3000-4000 rats every year. In this hunt, one group of people climb the coconut tree and shake the tree vigorously so that rats fall down which are killed by a second group of people waiting on the ground (Shah & Subiah, 1978).

(ii) *Trapping* rodents is a historical method of rodent control all over the world and even today reflects the ingenuity of farmers. The method is quite successful if used properly in small areas. Trapping is environmentally safe, facilities identification of species, serves as an index of density and gives great satisfaction to the user. Rodent traps basically belong to kill/ snap traps and live traps (Plate 3). The snap traps are of several types, mostly fabricated by farmers except the commercial varieties. These include Tanjore bow trap, a low cast bamboo trap used in Tamil Nadu (Neelanarayanan *et al.*, 1995b), wooden snap trap fabricated from timber splinters (Srihari & Chakravarthy, 1992) and the *urang* (Prakash & Mathur, 1987). The most popular is the commercially available break-back, spring-loaded snap trap with a serrated wooden or metal base.

The different live traps in vogue in India are the primitive pit fall or plot trap (Neelanarayanan *et al.*, 1994a), Sherman trap (Rana, 1982), the age old, single floor, box type wooden trap (Pasahan & Sabhlok, 1981) and multi catch wonder traps of different types (Prakash & Mathur, 1987).

Controlling field rodents by trapping is very expensive and should be resorted to only if the area involved is small. To be successful, trapping should be carried out meticulously. Some of the suggestions for effective trapping include the use of adequate numbers of suitable type, preferably using the ones with both the ends open and always in perfect working condition. Large, complicated types are unsuitable for field rat control. Bait selection is also very important. Peanut butter seems to be a universal favourite. *Vada*, a fried black gram preparation was found very attractive in South India (Sridhara, 1999). Similarly *nan* and *chapathi* are favoured baits in Pakistan (Hussain & Ahmad, 1999). Bait should be placed on the treadle with a small amount sprinkled around the entrance. Nesting material kept inside also attracts rodents. Fresh signs of activities such as runways, pathways, droppings, entrance of burrows are ideal spots for keeping traps. Pre baiting for 2-3 days is very important. After each capture the traps should be washed to get rid of the pheromone smell. Although trapping is a method of limited use in field rodent control, significant reduction of rodent population was achieved by trapping in deep water rice fields in Bangladesh (Islam & Karim, 1995) and in the vegetable fields of Himachal Pradesh (Sheikher & Jain, 1992).

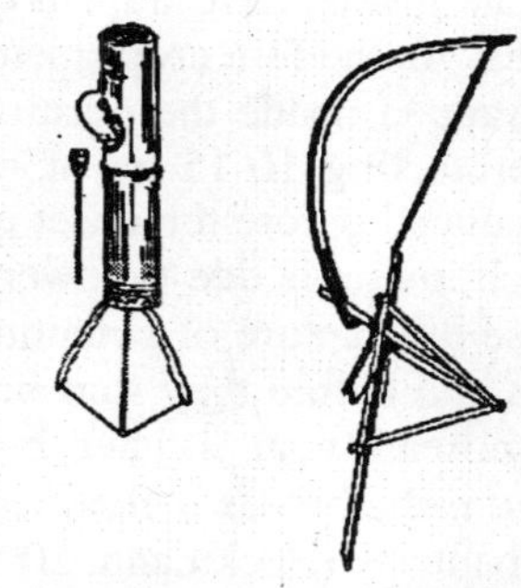

a. Urang b. Tanjore bow trap

c. Snap trap

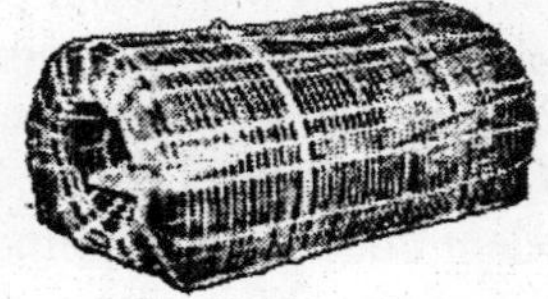

d. live trap

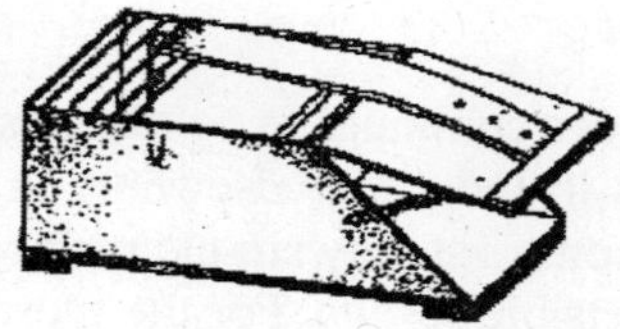

e. Wooden trap

Plate 3. Some popular traps in use in India

A spectacular breakthrough in rodent control is the trap-barrier system (TBS), a combination of trapping and barriers to reduce damage to rice in South-East Asia. The technology was developed by Singleton *et al.* (1999) and is based on attracting rats to a small, central area of lure crop which has been planted in advance (usually three weeks) of the main crop in a 20-50 meter square with a barrier. The barrier is nothing but a thick plastic sheet below and above the ground with strategically made holes to allow rats inside. Live traps are kept at regular intervals inside the enclosed area. As the lure crop ripens, rats are attracted to the food source and get trapped inside the plastic barrier. A single barrier has the potential of protecting 10-15 ha of rice fields around it. The rodent population is reduced before the onset of breeding. The less than impressive catch, which in fact is due to lower population levels should not dishearten. Because the capture of potential prolific breeders before the main crop matures and before they start breeding profusely reduces population levels significantly at a later but vulnerable stage. The removal of one rat before she breeds is reported to be equal to killing 32 rats when the crop is maturing (Beckmann, 2003).

TBS has been proved successful in Indonesia and Vietnam, and now is being extended to Laos, Cambodia and Philippines. Some of the unfavourable aspects of the methods are initial high cost of erecting the barrier, re-using of plastic by farmers in spite of it not being fit and labour intensive procedure as it involves daily checking and disposal of trapped rats. Despite this, the advantages are overwhelming and the method is environment friendly without the use of or recourse to chemicals. It is being popularised and funded by the government in Indonesia and is the basis of a new national policy on rodent control. A first round analysis of cost-benefit ratio and adaptability of TBS in Cambodia has convinced the farmers that it is an improvement in rodent control (Russell *et al.*, 2003). The TBS to be successful, apart from being a community rodent control programme, should be supplemented by other community actions like clean cultivation, synchrony of cropping and targeted rat control campaigns.

(b) *Chemical control*

Using rodenticides continues to be the most widely used method to manage rodent pests in India and elsewhere in the world. The obvious advantages include comparatively simple methodology, economy, effectiveness and quick, visible results. For the mostly uneducated farmer of tropical countries an approach of this kind is ideal and fits into his thinking and schedule of agricultural practices. Of the seven rodenticides registered in India (Insecticides Act 1976), only four are marketed, and these are widely used. The four include an acute rodenticide, zinc phosphide, a fumigant, aluminium phosphide and two anticoagulant based chronic rodenticides, coumatetralyl and bromadiolone.

Zinc phosphide is the most commonly used and popular rodenticide in India, widely used by farmers across the country for all the crops except sugarcane where the area range is too big, and plantation crops such as coconut which need bait placement on tree tops. It is quite effective against all species of Indian rodents with LD_{50} ranging from 25-40 mg/kg and is used at 2% in the bait. Baits are recommended based on local studies as there is great variability in the food preferences of Indian rodents in different habitats. Generally it is rice or millet mixed with 5% roasted groundnut seeds and 2-5% locally used vegetable oil (Sridhara, 1982).

Zinc phosphide is quite safe if used according to specified instructions. As it is detoxified rapidly in baits and in the gut of animals, and is not stored in any part of the animal there is no threat of secondary poisoning and environmental toxicity. It usually kills 60-70% of rodents. The Indian Council of Agricultural Research recommends its use in cereal crops, pulses, oil seeds, vegetables and boundaries of sugarcane. Development of bait shyness consequent to ingestion of sub lethal doses of rodenticide which may extend for periods ranging 6-170 days and lack of specific antidote are the limitation in its use (Idris & Prakash, 1992).

Aluminium phosphide tablets available at 0.4 to 0.5g and are used at one to several tablets per burrow depending on the size. The chemical is more effective in moist soils. It is quite popular because of the small size, easy handling and very rapid death of rodents inside burrows and godowns. However, because of its extreme toxicity, fire hazard and lack of antidotes, it is banned in many parts of the country and is registered for restricted use by qualified and licensed people only.

Coumatetralyl, the first generation anticoagulant is recommended to be used as 0.075% tracking powder for commensal rodent control and 0.0375% in cereal bait for outdoor use. Studies in Punjab have shown it to be as potent as second generation anticoagulants against *B. bengalensis* in both laboratory and crop fields infested by the species (Parshad & Malhi, 1995). Low toxicity, chronic action and availability of Vitamin-K, the antidote render the anticoagulant a practical rodenticide.

Bromadiolone is the only second generation anticoagulant marketed in India. It is available as ready to use wax blocks since 1988 and also as loose powder since past several years. The former is ideal for residential rodent control and the later for field rodent control. At

0.005% in cereal bait, the rodenticide is recommended for crop fields having moderate level of infestation. Since there is no bait shyness, a second application is recommended 15 days after the initial poisoning to kill the surviving populations (Rana & Tripathi, 1999). The limiting factor in using anticoagulants is genetic resistance after a long period of use. Resistance to bromadiolone has been reported in Europe and UK (Quy *et al.*, 1992; Myllmaki, 1995). This may not happen in India in the near future because rodent control has not yet attained the status of a regular agricultural practice as insect control schedules are. It is still mostly symptomatic control. In future, if pockets of resistance are detected, it is advisable to use other rodenticides alternatively with anticoagulant.

Timing of rodent control is very important with climate and seasons affecting agricultural practices. Parshad (1999) has admirably summarised the prophylactic or preventive control during lean periods and symptomatic treatments during crop growth.

Prophylactic control: In India the months of May-June and November-December are the fallow periods between the summer (*kharif*) and winter (*rabi*) crops respectively. Rodents at this time are confined to reservoir habitats such as bunds, dykes, borders of water channels, fallow, uncultivated lands or inside long duration crops such as sugarcane, orchards and plantations. A prophylactic control in these reservoirs during fallow periods breaks the natural breeding cycle of rodents and prevents population explosion during the cropping season. During such lean periods in the absence of adequate food resources, rodents are easily attracted to bait and mortalities are high. The months of May-June are also economical and more effective as the population is minimal and non-breeding during the summer season of scarce resources (Sridhara, 1999). The only drawback is to persuade the farmers to take up community based rodent control when there are no crops in the field. At this time farmers are relatively free of crop raising activities and a well planned, motivational campaign would give optimum results.

Symptomatic control is the strategy to kill rodents after observing signs of activity such as their movements, faecal pellets, fresh burrows and actual damage to crops. This type of control is not of much use in field crops because by the time damage becomes visible, much yield has been lost. However long term studies on crops susceptible to damage have shown that there are vulnerable stages during the growth of the crop when rodent damage is predictable. Carrying out control operations prior to the susceptible stage would prevent rodent damage substantially. These stages for various Indian crops are: 30-60 days after transplantation of rice (Rao & Singh, 1983; Bhaskaran *et al.*, 1995); 80-100 days after sowing of peanut (Parshad *et al.*, 1987); the late tillering stage of wheat (Rao, 1992), July-August and October-November for sugarcane (Parshad *et al.*, 1986; Ahmad & Parshad, 1991). These stages do not have any plant parts worth feeding but soon will produce seeds providing easily accessible food for considerable time. During this period rodents are less likely to consume the poison baits. In case where seedling damage is significant rodent control in the uncultivated fields surrounding crop fields is desirable during the stages of land preparation itself. For instance in view of reports of heavy damage to seedlings of jowar, maize, sunflower, red gram and groundnut, ranging 40-100% often forcing the farmer to resow the fields more than once (Sridhara, 1999), such a preventive control reduces damage at the vulnerable stage of germination.

Practical consideration during rodenticide use

Many practical problems are encountered while controlling rodents using toxicants in agricultural crops. Some of the important problems are:

a. ***Toxicity considerations:*** While using rodenticides all precautions must be taken to protect humans, livestock and the environment. Toxic rodenticides are extremely poisonous to all vertebrates and some result in secondary poisoning. The factors to be considered while using rodenticides are: possible residue effect, hazard to farmers poisoning, effect on beneficial species and wildlife, accumulation in the soil and contamination of water bodies.

b. ***Pest species:*** For effective control of pests it is essential to consider all aspects of pest biology, ecology and behaviour. This requires detailed knowledge of the bionomics of the pest. Further biological control which increases the effectiveness of chemical control also requires detailed knowledge of the pest species.

c. ***Baits and Baiting:*** An ideal bait should be safe, economic, convenient, easily available, acceptable to the rodents and weather resistant. Hence during monsoons it is advisable to use wax blocks of rodenticides or loose bait in plastic covers. Loose cereal baits cost less and can be conveniently prepared from locally available grains and poison concentrates. The following factors affect the success of baiting:

(i) *Choosing the bait*: It is advisable to conduct a bait choice test in the habitat of rodents to identify the most attractive bait. If bait preferences has already been established preference test may not be necessary. Baits similar to what rats and mice are eating are often good choices. Perishable baits can be used only when acute rodenticides are used, because they decompose fast and are ideal only for overnight exposure. Similarly, cereal baits are ideal to mix anticoagulants where baits are exposed for long duration. In situations where water is limited, moist grains, watery vegetables like tomatoes, corn, carrots, potatoes and lettuce, fruits like melons, apples, bananas and water baits are accepted better than dry baits. Baits can be prepared by mixing fresh ground cereals with groundnut and oil. However, baits made of meat and fish pose a hazard to pets like dogs, cats and other carnivores.

(ii) *Attractants and additives*: In spite of repeated claims of attractants enhancing bait acceptance, so far none have been found practical and satisfactory. On the other hand certain additives are known to make baits more acceptable. These include 5% sugar, 3-10% vegetable oil, mineral oils and non rancid animal fat. The oils not only increase acceptance but also bind the rodenticide to the cereal grain. Although various flavours and aromas are blended with rat bait, none have proved to be attractive. Mould and insect inhibitors maintain bait quality but reduce bait acceptance.

Coloured dyes or pigments have become a popular ingredient of rodent baits primarily as a warning signal and thus reduce hazards of accidental poisoning. They are also helpful in indicating thorough mixing. Brightly coloured baits repel birds. But it is important that colours used should not repel rodents. The proven safe colours are fast green, aqua green, methylene blue and mono astral green.

(iii) *Prebaiting* for 3-4 days is carried out while poisoning with acute rodenticides to overcome the neophobic behaviour of rodents. Prebaits contain all the ingredients of poison bait except the poison. The uneaten, left over prebait should be removed prior to poison baiting. Sometimes one day gap between pre and poison bait enhances acceptance of toxic baits. Prebaiting has several benefits. The amount of food consumed indicates the density of pest population, helps in determining the quantity of poison bait, prevents waste of bait material, ensures maximum acceptance of poison bait and eliminates chances of underbaiting. Sometimes prebait is not accepted satisfactorily. It could be due to lower density of rodents or there could be other factors like taste of bait or location of bait placement which can be corrected by changing bait material and location of bait placement respectively.

(iv) *Bait forms and bait placement*: Baits can be best placed near the runways of rodents. It is advisable to place them hidden as rodents prefer to eat in concealed locations. Large quantities of bait should be used because underbaiting specially while using acute rodenticides results in development of bait/poison shyness which in turn reduces the success of control programme.

For field rodent control, the most practical and popular method is burrow baiting. Wax blocks or loose bait formulations are introduced into 'live' burrows and closed. All opened burrows are closed with mud balls or small stones on the evening of first day. Those found open next day are considered 'live' burrows. By this only the target species is poisoned and additionally burrow count serves as an index of population density. While baiting burrows, long handled wooden or plastic spoons are ideal to transfer baits to the interiors of the burrow. Alternatively dry cereal baits of anticoagulants or acute poisons can be transferred into suitable, thin plastic covers and placed inside burrows, sewage systems and places which are unsuitable for placing bait boxes or stations. These are dragged by rodents into their burrows or carried off to secluded places and consumed. If the surrounding uncultivated land is wild and inaccessible and if the sugarcane and plantation crops are too dense, then broadcasting of wax blocks of second generation anticoagulants is the ideal alternative.

(v) *Sociological consideration:* Understanding of farmer's attitude and behaviour is important as it influences the success or failure of rodent control program. When planning rodent control campaign it is essential to take into account the culture of the people, their agricultural practices, local economics and other factors that may have a bearing on acceptance of new approaches.

Simply killing large number of rats in short term campaign will not resolve rodent problem. The control programme apart from reducing pest population should also be designed to lower the carrying capacity of the environment so that there is permanent reduction of damage and hazards. Intense killing is justified in emergencies and after harvest. But such temporary killings should be followed by plans for permanent control.

(c) *Biological Control*

Basically biological control means suppression of pest populations by increasing mortality, reducing natality and/or facilitating emigration by using predatory pressure, introducing pathogens, by fertility control and manipulating habitat. Of these, habitat management has already been dealt with and fertility control will be evaluated after reviewing predators and pathogens. Biological control is based on detailed knowledge of the ecology of the species and predator-prey interactions.

Biological control can be an effective measure for bringing down the population level of vertebrate pests, below economic injury level (Stern *et al.*, 1959).

(i) Predation: It is vital to select an appropriate, prey specific predator which is quite a daunting task if not impossible. Introduced predators may become ecological nuisances as in the case of Indian mongoose introduced to control *R. rattus* is Hawaiian sugarcane. Once the rat population was suppressed, the predator devoured the native indigenous prey to extinction (Davis *et al.*, 1976). It is thus more practical and safe to use indigenous reptilian, avian and mammalian predators. A second factor unfavourable to predatory control of pests is the low population of predators in nature compared to enormous population of pest animals like rodents. Further predators feed only on surplus population of prey rather than reducing their density.

Amongst the reptilian predators in India, lizards and snakes are important. Two species of lizards, *Varanus bengalensis* and *V. griseus* are known to predate on diurnal rodents such as *M. hurrianae, G. ellioti* and *F. pennanti* (Minton, 1966). Pythons were used to control rodents in some Australian islands, granaries in Singapore and river banks in Bankong (Whitaker & Bhaskar, 1978). Some of the Indian snakes known to prey upon rodents are: *Ptyas mucosus* (rat snake), *Erix johni* and *E. conicus* (sand snakes), *Banaarus caeruleus* and *B.sindanum* (the kraits), *Naja naja* (cobra) and *Echis carinata* (viper). However, in view of the low feeding rate of snakes and their slower metabolism, their predatory pressure is no match to the fast breeding rates of rodents.

After detailed studies on hawk and owl predation on rodent population, Craghead & Craghead (1956) concluded that to meet their heavy dietary requirement, this category of birds feed voraciously and thus qualify for further studies as biological control agents of rodent pests. Barn owl, *Tyto alba* (Neelanarayanan *et al.*, 1994b) and spotted owlet (Jain *et al.*, 1982) predated heavily on field rodents. It may be beneficial to some extent to use artificial perches and bird houses to facilitate perching and nesting of rodent eating raptors. The problem arises when the rodent population declines after harvest forcing the owls and other raptors to migrate elsewhere.

(ii) Pathogens and diseases: There is considerable potential in using micro parasites like virus, bacteria and protozoans and macro-parasites such as helminths and arthropods in controlling rodents. The classic case of introducing myxomatosis virus to control rabbit population is Australia (Fenner & Ratcliff, 1965) is an eternal impetus to search for similar pathogens for other rodent pests. Although

Salmonella was found effective against rats in Europe (Singleton & Redhead, 1990), it was ineffective in India (Bindra & Mann, 1975).

A recent breakthrough in biological control is the potential of the protozoan, *Sarcocystis singaporensis* to kill rodents. The spores secreted inside the intestine of the python host are excreted in the faeces which enter the rat's body through infested water or by ingestion of invertebrates that have eaten the snake dung. The protozoan multiplies in the blood of rats, initially forming cysts, making the rat sluggish and slightly infertile. However heavy infestation causes pneumonia and subsequent mortality. A team of Thai and German scientists have succeeded in harvesting spores from the faeces of specially farmed snakes. The spores are purified, their concentrate inoculated into palatable pellets of 1 gm made of cereal flour, coconut and fish extract. The parasite was found potent against *R. nitidus, R. rattus, R. argentiventer* and *Bandicota* sp. It has been found effective in different environments and reduced crop damage significantly. A Thai firm has already registered the parasite pellets as a rodenticide. At present the current estimate of cost: benefit ratio is approximately 1:3 (Herwati & Sudarmaji, 2003).

The nematode, *Capillaria hepatica* caused density dependent mortality during the outbreaks of wild house mice, *M. musculus* in Australia (Barker *et al.*, 1991). Mathematical models evaluating *C. hepatica* as a potential agent for the control of mouse populations suggest that the nematode is potentially effective when released at least one year before the mouse outbreak occurs. The model concluded that a parasite affecting only fecundity is not likely to succeed. Another limitation is the inability of macro parasites to spread sufficiently rapidly when there are sudden outbreak of vertebrate populations (McCallun, 1993). Two Indian studies have shown the nematode, *Hepatojarakas bandicoti* (Parshad & Sood, 1976) and *Trypamosoma evansi* as potential rodent control parasites (Biswas *et al.*, 2001) but unfortunately the research was not continued.

(d) *Fertility control*

Fertility control is defined as any technique that reduces the production of young and therefore induces a reduction in either fertility or fecundity (Bomford, 1990). Chemosterilants cause permanent or temporary sterility in either sex, reduce the number of offspring or alter the fertility of off spring produced (Marsh & Howard, 1977). Females can be rendered sterile by suppressing secretion of anterior pituitary, preventing follicular development and maturation, blocking the passage of ova in the oviduct, preventing fertilisation and implantation and interfering with gestation. In males it is by suppression of gonadotrophin secretion, inhibition of spermatogenesis and interference during transport and storage of sperm. An additional mechanism could be by blocking production and release of sexual pheromones and thus interfering with reproductive behaviour.

Fertility control, instead of increasing mortality reduces birth rate to achieve reduction in population levels, making it more humane and morally acceptable than the conventional lethal methods. It has great acceptance where killing is socially and legally unacceptable. The other advantages of chemosterilants include safety to non-target animals and environment, no threat of secondary poisoning, mostly species-specificity. As they are active for short or long terms, they are a flexible tool for population management (Kirkpatric & Turner, 1985).

Some of the antifertility compounds are (Bomford, 1990; Kirkpatric, 2002):

Alpha-chlorohydrin is an approved antifertility compound in USA and is marketed as Epiblock. Although it caused dose dependent sterility which could be permanent or temporary, it failed as a convincing reproductive inhibitor. It is no longer marketed.

Bromocriptine acts as an enzyme inhibitor of prolactin and was effective in kangaroos.

Diethylstilbestrol is a synthetic oestrogen and acts as an antifertility agent in dogs, foxes, many rodents and deer. Delivered as oral bait or applied on nesting material, it is particularly potential to reduce population growth of small rodents having very short breeding period.

Mestranol is an orally active oestrogen. Though effective, it has little value in rodent control due to induced bait shyness.

Quinoestrol is synthetic-oestrogen steroid inhibiting implantation but not suitable for rodents.

BDH 10131 is the most promising steroid for rat control but needs further experimental evidence.

Immunocontraception vaccines are the latest agents for inducing infertility in livestock, problem animals and humans. The approach is based on using the animal's immune system to produce antibodies against gamete proteins, reproductive hormones and other proteins essential for reproduction. These antibodies interfere with the normal physiological activity of the reproductive system (Talwar & Gaur, 1987). The induced antibodies interfere with reproduction without the need for constant or repetitive treatment with chemicals which is the major advantage of vaccines (Turner & Kirkpatric, 1991).

These vaccines can potentially block gonadotrophin releasing hormone (GnRH), prevent sperm penetration of Zona Pellucida (ZP), prevent implantation and development of fertilized egg by producing antibodies to the hormone involved and affect sperm proteins. At present immunocontraception vaccines are successfully used to manage problem ungulates but this technology can also be applied to rodents, bird pests, coyotes and foxes if suitable delivery systems are developed (Miller *et al.*, 1998).

A most promising evaluation of immunocontraceptive virus for rodent control is taking place in Australia. A multidisciplinary team has demonstrated that genetically engineered mouse virus, such as mouse cytomegalovirus (MCMV) and electromelia virus expressing mouse zona pellucida is capable of inducing long-term infertility in both experimental and field trials (Hinds *et al.*, 2003). However to be a mass scale strategy, information is needed about the mode of transmission of natural virus, how its transmission is affected by genetic modification, whether it can create immune response in a population that is already infected naturally, the proportion of population to be infected for satisfactory results and effect on social and sexual behaviour of mice. This apart there are social and ethical aspects of releasing a genetically modified virus.

Predatory odours and reproductive inhibition: Predatory odorus have been shown to affect reproduction in several rodent species. The estrous cycle of *R. norvegicus* and bank roles was extended by the anal sac secretions of mink and by the odour of weasels respectively. Cat urine and urine of mice from overcrowded situations gave similar results (Feotistova *et al.*, 2003). Laboratory studies reveal that reproductive success of rats as measured by litter size, number of pups, number of placental scars and number of corpora lutea was reduced when they were housed close to the predator, *Lynx lynx* (Naidenko *et al.*, 2003). Extending laboratory results to field rodent control is definitely a distant possibility.

POPULATION OUTBREAKS

Rodent control strategies planned for an area and a crop are adequate under normal situations and if meticulously carried out in an integrated manner when population fluctuations are within the predictable limits. But it is likely that unpredictable environmental factors may lead to sudden rodent outbreaks. Under normal circumstances both floods and drought result in initial decrease of rodent pest population. This is followed by outbreaks once the climatic conditions become normal. For instance, flash floods in 1996 reduced rodents in Andhra Pradesh drastically followed by an outbreak of rodents in 1997-98. A dry spell in the Cauvery delta of Pondichery was followed by good monsoon resulted in rodent outbreak in 1994. Similarly the cyclic flowering of bamboo in North-East India is always followed by plagues of rodents. Such an eventuality calls for a contingent plan to eradicate the pests rapidly and maximally using highly toxic rodenticides like zinc phosphide and sodium floroacetate.

POPULAITON MODELS

(a) Under normal situations

Although the most successful of all control methods, chemical control is not universally adapted as a farm management tool by farmers even in India. The subsistence level farmer postpones investing in rodent control till damage becomes visibly alarming. It is only at this stage he is convinced about the economic benefits of controlling rodents. However, by this time enough damage has already been inflicted to have any real benefit by such belated control which he is unaware of. In a situation like this, if there is a possibility of predicting damages based on pest ecology in the cultivated crop it becomes easier to convince the uneducated farmer about preventive rodent control. Population models are useful tools for such predictions and also to predict the expected outcome of various control measures (Stenseth, 1984). Models link bits and pieces of ecological information to provide insights into growth of pest population and connected damage potential. Such predictions about population trends and damage occurrence are highly accurate and precise and are of great help in planning timely rodent control. A model developed for the multimammate rat, *Mastomys natalensis* of Africa (Stenseth *et al.*, 2001) revealed that control measures if applied when rodent densities are very high will not have persistent effects even if very high mortalities are achieved. On the contrary if control measures are applied to reduce the survival of rodents during low densities, long term effects can be anticipated. Further reducing more than 75% reproduction not only prevents outbreaks of *Mastomys* but will keep densities low over a long period.

(b) During population outbreaks

The cyclical outbreaks, if they occur, are easy to manage. They require constant surveillance and certain parameters notably the reproductive status of females and their proportion in the population are good indicators of such eruptions. Timely intervention by IPM of rodents with more emphasis on mortality will prevent massive damages and economic losses. In contrast irregular eruptions of rodent population occur unexpectedly taking the farmer and government agencies by surprise and unprepared. Several studies have been undertaken to develop productive models for such outbreaks. However, the supposed benefits of such a model are debatable. The stoichasticity in the population dynamics of rodents makes predictions unreliable to a large extent. Yet another dilemma is taking action based on model generated cost-benefit ratios of controlling rodents. If the cost is justified in terms of higher benefits reaped, controlling pest rodents is justified. Otherwise it is pragmatic to accept the losses occurring. It was suggested that the outbreak of Australian mice should choose the former option while the unexpected eruptions of multimammate rat in Tanzania are better left unattended rather than incurring higher expenses of controlling rodents to save the agricultural produce (Leirs, 2003).

INTEGRATED RODENT PEST MANAGEMENT

Integrated Pest Management (IPM) is the co-ordinated use of pest and environmental information to design and implement pest control methods that are economically, environmentally and socially sound. IPM promotes prevention over remediation and advocates integration of at least two or more strategies to achieve long term solutions.

The IPM strategy is based on few basic principles of pest-crop inter-relationship. The two preliminary requisites are the database on the pest species and threshold levels of damage to initiate control. The actual management of pest involves multiple strategies in a three-tiered management. The first step is prevention of infestation mostly by non-toxic methods. The second step is to use non lethal but if necessary weak chemical or other methods to keep away the pests failing which appropriate pesticide application is resorted to in the third level.

(i) Preliminary database for IPM

(a) Understanding the pest species: More than one rodent pest species affects crops in India. Broadly speaking four major rodent pest complexes dominate the crop fields in the country with agro-climatic conditions and the resultant varied crops grown being the major influencing factors. The arid and semi-arid regions of Western Rajasthan, semi-arid zone of Western Gujarat and Western Punjab to some extent have the species complex of *T. indica, M. hurrianae,*

G.gleadowii and *G. ellioti* while the north-east hilly regions have a complex of *B. bengalensis, R. nitidus, R. rattus, M. booduga,* and *M. meltada.* The south Western, tropical evergreen and moist forests on the Western ghats and coastal areas are dominated by squirrels of the species *Funambulus*, few species of *Rattus, B.bengalensis, B. indica* and *H. indica.* Rest of the country has various combinations of *B. bengalensis, M. meltada, T. indica, M. booduga* and *M. platythrix.* The species exhibit geographical variation in density, breeding and life history. Consequently management strategies have to be locally planned.

(b) Action threshold: It is not necessary to kill all the pests in a given environment. The term action threshold represents the level of pest presence that causes enough damage to make rodent control beneficial costwise. This threshold is linked to some critical population level or appearance of damage level on a particular number of plants or the area being protected. It varies from pest to pest. In case of rodents it is practical to link action threshold to critical levels of damage as population levels are difficult to arrive at because in addition to those dwelling inside the crop fields, there are daily foragers from adjacent cultivated and uncultivated fields to the affected area.

(c) Population dynamics and control: Leirs (2003) recognises three basic patterns of population cycles in rodent pests of agriculture. First, relatively stable or irregularly fluctuating. e.g. *R. tiomanicus* in the oil palm plantations of Malaysia. The second type exhibits seasonal fluctuations combined with internal differences as in the African multimammate rat, *Mastomys natalensis* in maize fields. Indian rodents fit into this model. In the third category irregular peak years alternate with periods of almost no population as in *M. domesticus* in Australia. True cyclic populations are rare in agricultural crops. This is attributed to non-occurrence of feedback mechanisms in the tropic interactions of agro ecosystem (Stenseth, 1999). Absence or rarity of plant–herbivore interactions, predator-prey interactions, and social regulation are the feed back mechanisms lacking in agricultural situation. This in turn is attributed to instability of habitat due to annual or seasonal cropping, absence of predators and lack of stable social structures within the population.

Species specific population dynamics is a result of differences in the life history and the interactions with the environment, Most of the Indian rodents are seasonal breeders with breeding coinciding with monsoon rains, an adaptation to availability of food post rainy season. The maturing adults produce young ones repeatedly and rapidly within a short period of few months before the onset of winter followed by summer. Although it is likely that those born in the earlier stages of breeding cycle may litter young ones in the same year, those born at later stages wait until the next breeding season to reproduce.

Rodents with a stable pattern of population fluctuation are density dependent. In seasonally fluctuating species, density-independent factors are responsible for the patterns of population. The control strategy for such populations is to lower the long term equilibrium around which population size fluctuates.

In species that show population outbreaks, rodent management has to be different. Here the focus should be on preventing population build up. If not, to initiate control in the early stages of populations growth. It is also possible that some species of cyclic populations also have relatively low average densities, and yet exhibit irregular eruptions. The 18 and 21 year old eruptions of rodents in the bamboo forests of North-East hilly region of India which spreads to crop fields subsequently fits into this category. Management of pests in this situation is by constant surveillance and control measures when the population shows signs of increasing abnormally. It may be worthwhile here to recall the key factors of survival in the rat population of Tanzania (Anon, 2001-2002). Here factors favouring survival were important during non-breeding season whereas features favouring reproduction were critical during the breeding season. This implies killing rodents during their reproductively quiescent period is most effective not only to effect high mortalities but to keep population at low levels subsequently. Similar suggestion was made for rodent management for Karnataka earlier (Sridhara, 1999).

(ii) IPM strategies

(a) Preventing infestation is the key to management in IPM. This has to be based on a combination of several cultural and mechanical methods. The basic objective is to make the habitat less attractive to the pest.

Reducing habitat by keeping bund height and width discourages fossorial species. So also flooding and smoking of burrows. Fixing metal sheets on coconut acts a barrier.

Crop synchrony also discourages rodent damage. If planting and irrigation are not synchronised over vast areas, the differently maturing crops at adjacent fields facilitate movement and migration to yet to be harvested fields.

Crop hygiene influences infestation and survival of rodents to a great extent. It is very important to keep crop fields and field edges free of weeds. Active weed control is an important component of IPM for rodent control. For tree nesting rodents, regular cleaning of nests discourages rodent infestation.

The two promising approaches of alternate, non-chemical methods are reducing fecundity and reducing resource availability. Habitat manipulation reduced rodent damage considerably in Macadamia orchards of Australia (White *et al.*, 2003). The large, temporally, stable adjacent non-crop habitat with dense vegetation which provided harbourage to rodent population was converted into highly modified grasslands to reduce the resources for rodent pests inhabiting Macadamia orchards. The exercise resulted in declined rodent population and reduced damage. The approach was cost effective. This preliminary study was highly successful but the authors advice caution as to possible damage to environment and adverse effect on biodiversity of the habitat.

Large monocultures discourage rodents. Damage is comparatively less in these fields compared to small holdings that are interspersed with different crops, fallow and uncultivated lands.

Use of predators is also a component of prevention of rodent damage. As agricultural lands are mostly devoid of terrestrial predators with the exception of snakes and lizards due to constant disturbance in land, avian predators seem to have some potential. Predation by birds can be facilitated by erecting perches and nest boxes. Of course snakes, in spite of their low metabolic rate and hence low requirement of prey, do play a role in keeping natural populations of rodents low and therefore there should not be any killing of non-poisonous snakes which invariably happen to be rat eaters.

At present the biological control of rodents using pathogens is not only highly controversial but lacks specific pathogens of commercial viability.

(b) Non lethal/weak chemical approaches: The next step in IPM is use of 'soft pesticides' such as repellents, growth inhibitors etc. These approaches are successful with insects but there are no methods of repelling rats away from crop fields with only preliminary studies on seed repellents. Products that interfere with growth and metabolism are non-existent.

Fertility control is one of the best methods to manage species which breed rapidly, have both high fecundity and high rates of mortality such as rodents. It is also more humane than lethal control. However so far attempts to find a suitable chemical to induce sterility has not been successful. Only recently promising results both in the laboratory and enclosures were obtained with immunocontraception strategy for the Australian mouse, *M. domesticus* (Chambers *et al.*, 1999).

Leirs (2003) discusses whether mortality or fecundity is the limiting factor in population growth. For instance annual growth rates in an African and South American rodent were affected more by mortality factors in contrast to population with eruptions where fertility control is a better tool to manage.

Trap crops are a neglected area of research. Based on the foraging strategies of rodents wherein food is selected based on the amount of energy spent and predator risk taken in relation to nutritive value of the foraged food, providing alternate foods to protect crops is considered a good strategy. This has been tested positively in orchards (Sullivan, 1979), forestry (Sullivan & Sullivan, 1988) and is a regular practice in sugar beet cultivation (Pelz, 1989). The alternate food is provided on the edges of field till the seeds sprout and are beyond the stage of rodent damage. This strategy has not been evaluated much in spite of its high potential.

***(c) Lethal approach using rodenticides*:** Rodents being intelligent, adaptable cannot be contained fully with all the above approaches. Killing using chemicals is the ultimate but essential step. Care should be taken to use appropriate rodenticides in a sequence so that adverse effects if any are taken care off. For instance rodent control using zinc phosphide should be followed by bromadiolone baiting so that bait shy rodents do not get a chance to reject the later. Similarly if bromadiolone is used initially the subsequent operations should use zinc phosphide to eliminate anticoagulant resistant residue population.

Crop growth and rodent damage are intrinsically inter-related. Damage to sown seeds and at seedling stage is seen in maize, jowar, sunflower, ground nut and pulses, grain formation stage of cereals and pulses, while losses are inflicted at the maturity/fruit formation of fruits and vegetables. Apart from killing during non-breeding seasons, these stages prior to rodent vulnerability are suitable for controlling rodents.

The discussion on IPM of rodents will remain incomplete without recalling its successful applications in sugarcane fields in Australia.

The cane field rat, *Rattus sordidus* and the climbing rat, *Melomys burtoni*, the native grassland rodent species of Australia cause sporadic but significant damage to sugarcane at several stages of crop cycle. An integrated management replaced the earlier reactive approach to rodent control. The components of IPM included improvements in short and long term management of harbourage areas, in-crop weed control, monitoring of rodent population and poison baiting (Smith *et al.*, 2003). Harbourage management was by weed control. In pockets that were inaccessible for physical weed control, harbourage was reduced by revegetation with trees that shaded grasses and by grazing. Revegetation along waterways reduced erosion, stabilised banks, improved water quality and provided wildlife corridors. This combined approach reduced the potential large-scale rat plagues and reduced damage to sugarcane considerably.

WHY RODENT CONTROL CAMPAIGNS OFTEN FAIL AND HOW TO MAKE IT SUCCESSFUL

Indian research on rodent pest management has contributed significantly towards preventing damage to various standing crops, in storage and in residential premises which in turn is based on three decades of intense, government backed research on the biology of pest species, rodenticide and non-chemical approach to the problem. Still the problem of rodent damage is not only unabated but there are sudden unexpected outbreaks here and there. The government of India has a number of training programme to impart education on rodent control at different levels of department of agriculture both at state and central level before it is passed on to the farmer. These include routine and advanced training for senior government officers, extension specialists, agricultural development and plant protection officers, village level training and field demonstrations, radio and television programmes by both government and private channels.

There are two major sets of reasons for rodent control not yielding the desired/projected levels of success. On one hand rodent biology, behaviour and adaptability contributes to such failures. At the other end socio-economic conditions and farmer's attitude affects adversely the adaption of rodent control technology. Rodents are *r* strategists with great potential for reproduction and adaptability to environmental changes including control measures. To be effective the various methods of reducing rodent numbers such as rodenticides/predators/ parasites/ trapping/habitat management should act not only complimentarily but faster and eliminate more number of rodents than they reproduce. Actually the rate of reproduction drastically increases as compensation to partial reduction of population. The capacity of rodents not only to detect and colonise newer habitats but getting adapted to newer environments quickly and successfully enables them to spread to expanding agriculture. Their genetic resistance to anticoagulant rodenticides, neophobia and bait shyness towards acute rodenticides are additional factors contributing to less than satisfactory levels of rodent control.

The major reasons for improper and unsatisfactory levels of adaption of rodent control technology by farmers in India is attributed to general neglect, lack of awareness of economic losses, small land holdings which make organisation and execution of large scale rodent control operations difficult, low or no education and poverty, earlier failures which were mostly due to improper execution of rodent control and the resigned attitude of farmers that rodent problem is minor and unmanageable (Parshad, 1999).

As regards the outcome of training of personnel in the departments of agriculture and development, most of them are basically ignorant of rodent control. The technology learnt during the training is forgotten once they return to their parent department as other more important tasks are entrusted to them.

Yet another reason is attributed to so called "Trickle down" process. It means that the content and essence of technology dwindles, getting diluted progressively from its origin i.e. the scientist to the ultimate receiver, the farmer.

Most of the rodent control programmes in tropics are adaptations of Western urban models. They are unsuitable and ineffective for the tropical agricultural fields because of the density and varied distribution of tropical rodents in several drastically different habitats. The other reasons for the failure of such models are the ability of tropical species to shift and rapidly adjust to new habitats and their prolific breeding. The sanitation and habitat management based programmes of temperate zones being intense, and costly cannot be sustained by individual farmers and governments of developing countries of the tropics. In fact one of the major reasons for lack of adaption of rodent control in these countries is the absence of visual or perceived benefits compared to the cost of losses and/or control operations.

Once the reasons for non adaption of rodent control or failure/ partial success are detected, it is time to design, a technology with additions, improvements and deletion of mistakes. First and foremost rodent control as a routine sanitary and pest problem has to be ingrained in people. If the farmer is convinced that rodent control is possible, and it is cost–effective, half the problem is solved because it has been universally proved that all technologies, however difficult they are, are readily and quickly adapted by the farmer throughout the world, if he is convinced that such a technology increases agricultural yield. Thus educating the farmer about the possibility of success, cost effectiveness and community approach to rodent control is the foundation of successful rodent control campaign. Apart from making the farmer realise the profits gained, it is important to understand the indigenous knowledge and socio religious attitude of the local farmers. Such initiatives will definitely make rodent control more acceptable and successful. Motivating farmers should be intense and through special extension education programmes and multimedia campaigns as is done for pulse polio programme. Involvement of celebrities for message giving on the eve of campaign should be seriously considered. Ultimately it is time to constitute a special task force to design a countrywide rodent management programme and execute it all over the country annually as a five year plan at the end of which farmers would definitely become not only aware of but proficient in carrying out the control programme on their own.

CONCLUSIONS

A quick overview of rodent pest management reveals that the basic approach is killing mostly using rodenticides. Trapping, flooding, hunting and fumigating may be used to varying degrees. Induced mortalities using rodenticides have solved the rodent damage to agriculture to a large extent when applied taking into consideration the biology of the pest species and the features of agro-ecosystem. Fertility control remains theoretical even now. Recent investigations on immuno-contraception vaccines have been found promising in controlling mice pests in Australia. Habitat management has been neglected pathetically. There is an urgent need to evolve area and crop specific IPM strategies for rodent control. It is very important to convince the farmer that rodent control is feasible economically and yields sustainable reduction in rodent populations and prevents damage considerably.

REFERENCES

ADVANI, R. 1984. Ecology, biology and control of black rat, *Rattus rattus* in Minicoy Island. *J. Plant Crops*. 12: 11-16.

ADVANI, R. 1985. Rodent pest management in the coconut plantations of India and its islands. *Indian Coconut J*. 15(10): 3-9.

AGARWAL, V.C & CHAKRABORTY, S. 1992. The Indian crested porcupine, *Hystrix indica* (Kerr). In: RODENTS IN INDIAN AGRICULTURE. (Eds.). I. Prakash & P.K. Ghosh. Scientific Publishers, Jodhpur, India. pp. 25-30.

AHMAD, E. 1992. The rat eaters of Bihar. *India Today* (October). 15:21.

AHMAD, N & V.R. PARSHAD. 1991. Evaluation of rodenticidal baits in fields of sugarcane (*Saccharum afficinanum*). *Indian J. agri. Sci.* 61: 281-284.

AHMAD, N. & V.R. PARSHAD 1985. Rodent pest problems in sugarcane and their remedies. *Prog. Farm.* 22: 17-9.

ANON, 1991. Report of the second quinquennial review team 1985-89: All India Coordinated Research Project on Rodent control, Indian Council of Agricultural Research, New Delhi, India. 45p.

ANON. 1995. Control of rats and mice. In: PACKAGE OF PRACTICES FOR CROPS OF PUNJAB. *Rabi* 1995-96. Punjab Agricultural University, Ludhiana, India.

ANON 2001-2002. Second annual report of the project: protecting staple crops in eastern Africa: Integrated approaches for ecologically based field rodent pest management from http://www.dpil.dk/stapleart/www.lb.htm. 84p.

ANON. 2004a. *Annual Progress Report of AICRP on Rodent Control*, Assam Agricultural University, Jorhat, India.

ANON. 2004b. *Annual progress report. All India Networking Project on Rodents* (AINP). University of Agricultural Sciences, Bangalore, India.

BALDWIN, P.H., C.W. SCHARTZ & E.R. SCHWARTZ. 1952. Life history and economic status of the mongoose in Hawaii. *J. Mammal.* 33: 355-356.

BARKER, S.C., G.R. SINGLETON & D.M. SPRATT. 1991. Can the nematode, *Hepatica capillaria* regulate abundance of wild house mice? Results of enclosure experiments in Southern Australia. *Parasitology* 103(3): 439-449.

BASHEER, M & S. JAYRAJ. 1964. Cashew nut pests. In: ENTOMOLOGY IN INDIA. Entomological Society of India, New Delhi, pp. 261-266.

BECKMANN, R. 2003. Tougher times for raiders of the rice field. http:// www. cse.csiro.au/research/program4/community ecology/rodents/ ACIAR rodent article.

BHASKARAN, J., R. KANAKASABAI & NEELANARAYANAN, P. 1995. Evaluation of two rodenticides in paddy fields during Samba and Thaladi seasons. *Indian J. exp. Boil.* 33: 113-121.

BHAT, S.K. 1992. Plantation corps In: RODENTS IN INDIAN AGRICULTURE (Eds.) I. Prakash & P.K. Ghosh. Scientific Publishers, Jodhpur, India. pp. 271-278.

BHAT, S.K., P.S.P.V. VIDYASAGAR & A. SUJATHA 1990. Record of *Rattus rattus* Wroughtoni as a pest of oil palm seedlings. *Trop. Pest Manage.* 36: 75-76.

BINDRA, O.S & G.S. MANN 1975. Ineffectiveness of murine typhus fever bacterium against field rats and mice at Ludhiana. *Indian J. Plant. Prot.* 3: 98-99.

BISWAS, B & TIWARI, K.K. 1969. Taxonomy and distribution of Indian rodents In: *Proc. Indian Rod. Symp.* Calcutta, India. (Ed.) K.L. Haris. pp. 9-45.

BISWAS, D., A. CHOUDHARY & K.K. MISRA. 2001. Histopathology of *Trypanosoma (Trypanozoan) evansi* infestion in bandicoot rat. I. visceral organs. *Exp. Parasitol.* 99(3): 148-159.

BOMFORD, M. 1990. *A role for fertility control in wildlife management.* Bulletin No. 7. Bureau of Rural Resources, Australian Government Publishing service, Canberra, Australia. 50pp.

BOONSONG, P., S. HONGNARK., K. SUASA-ARD, Y.K. HOPRASER., P. PROMKERD., G. HAMARIT., P. NOOKAM & T. JAKEL. 1999. Rodent management in Thailand. In: ECOLOGICALLY BASED MANAGEMENT OF RODENT PESTS. (Eds.). G.R. Singleton., L.A. Hinds., H. Leris & Z. Zhang. AICAR Monograph No. 59. Australian Centre for International Agricultural Research, Canberra, Australia. Pp. 338-357.

BROWN, P.R., G.R. SINGLETON & SUDARMAJI. 2001. Habitat use and movement of the rice field rat, *Rattus argentiventer* in West Java, Indonesia, *Mammalia*. 2: 151-166.

BUTANI, P.G. & H.J. VYAS. 2005. Rodent management in groundnut in Gujarat. In: VERTEBRATE PESTS IN AGRICULTURE-THE INDIAN SCENARIO. (Ed.) Shakunthala Sridhara. Scientific Publishers, Jodhpur, India. (this volume).

CHAKRABORTY, S. 1992. The larger bandicoot rat, *Bandicota indica* (Bechstein) In: RODENTS IN INDIAN AGRICULTURE. (Eds.) I. Prakash & P.K. Ghosh. Scientific Publishers, Jodhpur, India. pp. 193-210

CHAKRAVARTHY, A.K. 1993. *Vertebrate pest management in cardamom and other crops in Malnad region:* Final project report. Indian Council of Agricultural Research, New Delhi, India. 54p.

CHAKRAVARTHY, A.K., SHADAKSHARI, Y.G & E. GANGAPPA. 1992. Rodent damage to rice germplasm in Mudigere, Karnataka. *Rod. Newsl.* 16:6.

CHAMBERS, L.K., M.A. LAWSON & L.A. HINDS. 1999. Biological control of rodents. The case for fertility control using immunocontraception In: ECOLOGICALLY BASED MANAGEMENT OF RODENT PESTS. (Eds.). G.R. Singleton., L.A. Hinds., H. Leris & Z. Zhang. AICAR Monograph No. 59. Australian Centre for International Agricultural Research, Canberra, Australia. pp. 215-242.

CHAUHAN, N.P.S & SAXENA, R.N. 1992. Hodgson's rat, *Rattus rattus brunneusculus* (Hodgson) In: RODENTS IN INDIAN AGRICULTURE. (Eds.) Prakash, I & P.K. Ghosh. Scientific Publishers, Jodhpur, India. pp. 95-114.

CHRISTIAN, J.J. 1950. The adreno pituitary system and population cycles in mammals. *J. Mammal.* 31: 247-256.

CRACKER, D.R., C.B. SCANLON & S.M. PERRY. 1993. Repellency and choice: feeding responses of wild rats (*Rattus norvegicus*) to cinnamic acid derivatives. *Appl. Anim. Behav.* 38: 61-66.

CRAGHEAD, F. & J. CRAGHEAD. 1956. HAWKS, OWLS AND WILDLIFE. Stack Pole, Pennsylvania, USA.

DAVIS, D.E., K. MYERS & J.B. HAY. 1976. Biological control among vertebrates. In: THEORY AND PRACTICE OF BIOLOGICAL CONTROL. (Eds.) C.B. Huffaker & P.S. Messenger. Academic Press, New York and London.

DEORAS, P.J. 1966. Significance of probable change of rat population in Bombay. In. *Proc. Indian Rodent Symp.* (Ed.) K.L. Haris. Calcutta, India. pp: 58-68.

DEORAS, P.J., M. MITTAL. & M.S. PRADHAN. 1975. Notes on a collection of rats from Goregaon- Malad in Bombay. *J. Bombay nat. Hist. Soc.* 72: 534-537.

DROST, D.C & K. MOODY. 1982. Rat damage in weed control experiments in rainfed transplanted rice. *Trop. Pest Mgmt.* 28: 295-296.

DRUMMOND, D.C., E.J. TAYLOR & M. BANP. 1977. Urban rat control: further experiment studies at Folkestone. *Envir. Health.* 85: 265-267.

ECKE, D. 1954. An invasion of Norway rats in Southwest Georgia. *J. Mammal.* 35:521-525.

FENNER, F.N & R. RATCLIFF. 1965. MYXOMATOSIS. Cambridge University Press, England.

FEOKTISTOVA, N. Yu., S.V. NAIDENTO., A.E. VOZNESSENSKAYA., G.J. KRIVOMAZON., L. CLARK & V.V. VOZNESSENSKAYA 2003. The influence of predator odours and over crowded mouse odours on regulation of oestrous cycles in house mice (*Mus musculus*) In: RATS, MICE AND PEOPLE: RODENT BIOLOGY AND MANAGEMENT (Eds.): Singleton G.R., L.A. Hinds., C.J. Krebs & D.M. Sprat. Australian Centre for International Agricultural Research, Canberra, Australia. pp. 173-175.

FITZWATER, W.D. 1972. Barrier fencing in wildlife management. *Proc. 8th Vert. Pest Conf.* Sacramento, California, USA. pp. 88-92.

GIRISH, A.C. 2005. Feeding behaviour of porcupines in cultivated ecosystems of Western Ghats of Karnataka, South India. Ph.D thesis submitted to Kuvempu University, Shimoga, Karnataka.

GURUPRASAD, B.K & K. SRIHARI. 1978. Rat damage and its control in coconut groove. *Rod. Newsl.* 2: 1-2.

HERAWATI, N.A & SUDARMAJI. 2003. Helminths of rice field rat, *Rattus argentiventer*. In: RATS, MICE AND PEOPLE: RODENT BIOLOGY AND MANAGEMENT (Eds.) G.R. Singleton., L.A. Hinds., C.J. Krebs & D.M. Sprat. Australian Centre for International Agricultural Research, Canberra, Australia. pp. 55-56.

HINDS, L.A., C.M. HARDY., M.A. LAWSON & G.R. SINGLETON. 2003. Development of fertility control for pest animal management. In: RATS, MICE AND PEOPLE: RODENT BIOLOGY AND MANAGEMENT

(Eds.) G.R. Singleton., L.A. Hinds., C.J. Krebs & D.M. Sprat. Australian Centre for International Agricultural Research, Canberra, Australia. pp. 31-36.

HUSSAIN, I & E. AHMAD. 1999. Traditional and non-chemical methods of rodent control In: A TRAINING MANUAL ON VERTEBRATE PEST MANAGEMENT (Eds.) Brooks, J.E., E. Ahmad., I. Hussain, S. Munir & A. Aziz khan. Pakistan Agricultural Research Council, Islamabad, Pakistan. Pp. 63-65.

IDRIS, M & I. PRAKASH. 1992. Shyness behaviour In: RODENTS IN INDIAN AGRICULTURE (Eds.) Prakash, I & P.K. Ghosh, Scientific Publishers, Jodhpur, India. Pp. 433-444.

ISLAM, Z & A.N.M.R. KARIM. 1995. Rat control by trapping in deepwater rice yield in Bangladesh. *Intl. J. Pest Mgmt.* 39: 250-254.

ISLAM, Z., R.G. MORTON & B.P. JUPP. 1993. The effects of rat damage on deepwater rice yield in Bangladesh. *Intern. J. Pest Manage.*39: 250-254.

JACOB, J., D. NOLTE., R. HARTONO., J. SUBAGJA & SUDARMAJI 2003. Pre and post harvest movements of female rice field rats in West Javanese fields In: RATS, MICE AND PEOPLE: RODENT BIOLOGY AND MANAGEMENT (Eds.) G.R. Singleton., L.A. Hinds., C.J. Krebs & D.M. Sprat. Australian Centre for International Agricultural Research, Canberra, Australia. Pp. 277-280.

JAIN, A.P. 1992. The Indian gerbil, *Tatera indica* Hardwicke. In: RODENTS IN INDIAN AGRICULTURE. (Eds.) Prakash, I & P.K. Ghosh, Scientific Publishers, Jodhpur, India. Pp. 31-47.

JAIN, A.P., R. ADVANI & I. PRAKASH. 1982. Winter foods of spotted owlet, *Athena brama* in Rajasthan. *J. Bombay nat. Hist. Soc.* 80: 415-416.

JAIN, A.P & R.S. TRIPATHI. 2000. An integrated approach to rodent pest management. In: IPM SYSTEM IN AGRICULTURE. Vol. 7 (Eds.). R.K.Upadhyay, K.G. Mukherjee & O.P. Dubey. Aditya Books Pvt. Ltd., New Delhi, India. pp. 275-307.

JAIN, A.P., R.S. TRIPATHI & B.D. RANA. 1983. Rodents-cheap source of proteins for tribals of NEH region. *Rod. Newsl: NEH special issue.* 18: 16-18.

JAIN, A.P. R.S. TRIPATHI. & B.D. RANA. 1993. *Rodent management: The state of art*. Technical bulletin No. 1. Indian Council of Agricultural Research, New Delhi. 38p.

JENSEN, P.G., P.D. CURTIS., J.A. DUNN., R.E. AUSTIC & M.E. RICHMOND. 2003. Field evaluation of capsaicin as a rodent aversion agent for poultry feed. *Pest Manage. Sci.* 59: 1007-1015.

KIRKPATRIC, R. 2002. *Wildlife fertility control*. Technical Review. 02-2. The Wildlife Society, Bethesda. USA. 30pp.

KIRKPATRIC, J.F & J.W. TURNER. 1985. Chemical fertility control and wildlife management. *Biosciences.* 35: 485-491.

LEIRS, H. 2003. Management of rodents in crops- the pied piper and his orchestra. In: RATS, MICE AND PEOPLE: RODENT BIOLOGY AND MANAGEMENT (Eds.) G.R. Singleton., L.A. Hinds., C.J. Krebs & D.M. Sprat. Australian Centre for International Agricultural Research, Canberra, Australia. pp. 183-190.

MALHI, C.S. & V.R. PARSHAD. 1989. Preharvest rodent damage to wheat in Punjab, India. *FAQ Quart. Newl. Asia Pacific Plant Prot. Comm.* 32:25-7.

MARSH, R.E & W.E. HOWARD. 1977. Testing rodent chemosterilants. *EPPO Bulletin.* 7: 485-493.

McCALLUN, H.I. 1993. Evaluation of nematode (*Capillaria hepatica* Bancraft, 1893) as a control agent for populations of house mice (*Mus musculus domesticus,* Schwartz and Schwartz, 1943). Revue scientifique–et–tech-office-international-des-epizootics. 12(1): 83-93.

MILLER, L.A., B.E. JOHNS & D.J. ELIAS. 1998. Immunocontraception as a wildlife management tool: some perspectives. *Wildlife Soc. Bull.* 26: 237-243.

MINTON, S.A. 1966. A contribution to the herpetology of West Pakistan. *Bull. Amer. Mus. nat. Hist.,* 134: 27-184.

MYLLMAKI, A. 1995. Anticoagulant resistance in Europe: appraisal of the data from the 1992 EPPO questionnaire. *Pestic. Sci.* 43: 69-72.

NAIDENKO, S.V., S.V. NAIDENKO., L. CLARKE & V.V. VOZNESSE-NSKAYA. 2003. Predator presence affects the reproductive success of prey in outdoor conditions. In: RATS, MICE AND PEOPLE: RODENT BIOLOGY AND MANAGEMENT. (Eds.) Singleton G.R., L.A. Hinds., C.J. Krebs & D.M. Sprat Australian Centre for International Agricultural Research, Canberra, Australia. pp: 148-150

NEELANARAYANAN, P., R. NAGARAJAN & R. KANAKASABAI. 1994a. Pitfall traps for the control of field rodents in cotton field. *Rod. Newsl.* 18: 7.

NEELANARAYANAN, P., R. NAGARAJAN & R. KANAKASABAI. 1994b. Vertebrate pests as prey and their composition in the diet of common barn owl, *Tyto alba. Rod. Newsl.* 18: 5-6.

NEELANARAYANAN, P., R. NAGARJUN & R. KANAKASABAI. 1995a. Traditional rodent control methods in Cauvery delta. *Rod. Newsl.* 19: 9-11.

NEELANARAYANAN, P., R. NAGARJUN & R. KANAKASABAI. 1995b. Tanjore bow trap: an ecofriendly tool for rodent management. *Rod. Newsl.* 19: 8-11.

NGOWO, V., J. LODAL., L.S. MULUNGU., R.H. MAKUNDI., A.W. MASS-AWE & H. LEIRS. 2003. Evaluation of thiram and cinnamamide as potential repellents against maize seed depredation by the multi mammate rat, *Mastomys natalensis* in Tanzania. In: RATS, MICE AND PEOPLE: RODENT BIOLOGY AND MANAGEMENT (Eds.) G.R. Singleton.,

L.A. Hinds., C.J. Krebs & D.M. Sprat. Australian Centre for International Agricultural Research, Canberra, Australia. pp. 260-261.

NOLTE, D.L & J.P. BARNETT. 2000. A repellent to reduce mouse damage to long leaf pine seed. *Biodet. Bidegr.* 45: 169-174.

PARSHAD, V.R. 1989. Ecological and agricultural considerations in the management of rodent pests. *Proc. 5th Internatl.* Theriological Congress, Rome, Italy (Abstract).

IBID. 1997. Ecology of rodent populations and development of strategies for their management in intensive agricultural system. *Internatl. Conf. Ecol. Agric.* Chandigarh, India. pp. 159-160.

IBID. 1999. Rodent control in India. *Integrated pest management reviews.* 4: 97-126.

PARSHAD, V.R & C.S. MALHI. 1994. Rodents In: HANDBOOK OF FIELD PROBLEMS OF VEGETABLE CROPS OF PUNJAB (Eds.) G.S. Gill & I.S. Mangat. Punjab Agricultural University, Ludhiana, India. pp. 217-224.

PARSHAD, V.R & C.S. MALHI. 1995. Comparative efficacy of two methods of delivering an anticoagulant rodenticide to three species of South Asian rodents. *Int. Biodet. Biodgr.* 36: 89-102.

PARSHAD, V.R & C. NIJJAR. 1995. Effects of rice variety and germination on the feeding preferences of the lesser bandicoot rat, *Bandicota bengalensis. Rod. Newsl.* 19: 6-7.

PARSHAD, V.R & M.L. SOOD. 1976. Prevalence and pathology of *Hepatojarakus bandicoti,* Sood et Parshad, 1973 (Trichostrongylidae: Nematoda) in natural infections of rodents. *Z. Parasitenk* 49: 139-44.

PARSHAD, V.R., N. AHMAD & G. CHOPRA. 1986. Effects of ecological factors on the performance of rodenticides in sugarcane. *Proc. Indian Acad. Sci. (Anim. Sci.,)* 95: 729-737.

PARSHAD, V.R., C.S. MALHI., N. AHMAD & B. GUPTA. 1987. Rodent damage and control in peanut in India. *Peanut Sci.* 14: 4-6.

PASAHAN, S.C & R.K. SINGAL. 1994. Burrowing patterns of lesser bandicoot rat (*Bandicota bengalensis*) in wheat (*Triticum aestivum*) crop in sandy loam soil. *Indian J. agri. Sci.* 64(10): 717-722.

PASAHAN, S.C & V.P. SABHLOK. 1981. Efficacy of wooden cage traps in field rodent control. *Rod. Newsl.* 5: 4-5.

IBID. 1993. Information on rodent infestation in some vegetable crops. *Rod. Newl.* 17:2.

PATHAK, K.A. & D. KUMAR. 2001. Rodent pests in north-eastern hill regions and their management. *Indian farming*. 45: 26-30.

PELZ, H.J. 1989. Ecological aspects of damage to sugar beet seeds by *Apodemus sylvaticus.* In: MAMMALS AS PESTS. (Eds.) R.J. Putman. Chapman and Hall, London. Pp: 34-48.

PRAKASH, I. 1978. Impact of changing land use pattern on the rodent communities in Indus valley. *Proc. Symp. on land and water management in the Indus basin.* Vol. 2. Ludhiana. Pp. 381-385.

IBID. 1981. *ECOLOGY OF THE INDIAN GERBIL, Meriones hurrianae.* Central Arid Zone Research Institute. Monograph No. 10. Jodhpur. 87p.

PRAKASH, O & P.N. AVASTHY. 1980. Rodent damage to sugarcane. *Rod. Newsl.* 4: 1-2.

PRAKASH, I., N. KASHYAP & M. MATHUR. 1992. The five stripped northern palm squirrel, *Funambulus pennanti* (Wroughton) In: RODENTS IN INDIAN AGRICULTURE. (Eds.): Prakash, I & P.K. Ghosh. Scientific Publishers, Jodhpur, India. pp. 17-24.

PRAKASH, I & R.P. MATHUR. 1987. MANAGEMENT OF RODENT PESTS. Indian Council of Agricultural Research, New Delhi. 133p.

PRAKASH, I., P. SINGH & A. SARAVANAN. 1995. Small mammals of the Abu Hill, Aravali ranges in Rajasthan, India: A comprehensive and ecological study. *Zoologie.* 5: 55-64.

PRICE, P.W & G.P. WALDBAUER. 1994. Ecological aspects of pest management. In: INTRODUCTION TO INSECT PEST MANAGEMENT (Eds.). R.L. Metcalf & W.H. Luckmann. John Wiley & Sons, New York. pp. 35-72.

QUY, R.J., D.S. SHEPHERD & I.R. INGLIS. 1992. Bait avoidance and effectiveness of anticoagulant rodenticides against warfarin and difenacoum resistant population of Norway rats, *Rattus norvegicus. Crop Prot.* 11:14-20.

RAMESH, P. 1992. The short tailed mole-rat, *Nesokia indica* Gray. In: RODENTS IN INDIAN AGRICULTURE. Vol. 1 (Eds.): Prakash. I & P.K. Ghosh. Scientific Publishers, Jodhpur. pp. 167-171.

RAMESH, P & R.N.KATIYAR. 1985. Effect of ploughing on the population of short-tailed bandicoot rat, *Nesokia indica. Rod. Newsl.* 9: 3-4.

RANA, B.D. 1982. Relative efficacy of two small mammal traps. *Acta oecologia.* 3: 149-153.

IBID. 1992. The soft furred field rat, *R. meltada pallidior* Ryley. In: RODENTS IN INDIAN AGRICULTURE (Eds.) I. Prakash & P.K. Ghosh, Scientific Publishers, Jodhpur, India. pp. 119-46.

RANA, B.D. & R.S. TRIPATHI. 1999. *RECENT ADVANCES IN COORDINATED RESEARCH ON RODENT CONTROL.* Central Arid Zone Research Institute, Jodhpur, India.

RANA, B.D., A.P. JAIN. & R.S. TRIPATHI. 1994. *FIFTEEN YEARS OF CO-ORDINATED RESEARCH ON RODENT CONTROL.* Central Arid Zone Research Institute, Jodhpur, India.

RAO, A.M.K.M. 1992. Integrated rodent management. In: RODENTS IN INDIAN AGRICULTURE (Eds.) I. Prakash & P.K. Ghosh. Scientific Publishers, Jodhpur, India. pp. 651-667.

RAO, A.M.K.M & BALASUBRAMANYAM, M. 1992. The Mice, *Mus* spp. In: RODENTS IN INDIAN AGRICULTURE (Eds.) I. Prakash & P.K. Ghosh. Scientific Publishers, Jodhpur, India. pp. 147-64.

RAO, A.M.K.M & C.E. SINGH. 1983. Notes on the distribution of rodent damage in Tella Hamsa rice around Hyderabad. *Rod. Newsl.* 7:9

REIDINGER, R.F.Jr., J.L. JIBAY & A.L. KOLZ. 1985. Field trial of an electric barrier for protecting rice fields from rat damage. *Phil. Agri.* 68: 168-179.

RUSSEL, I., E. SOTHEARY., U. SOKCO., A. FROST., J. POWELL & L. LEUNG. 2003. Community costs and benefits of rodent control in Cambodia: a first-round analysis of adaptive management In: RATS, MICE AND PEOPLE: RODENT BIOLOGY AND MANAGEMENT (Eds.) G.R. Singleton., L.A. Hinds., C.J. Krebs & D.M. Sprat. Australian Centre for International Agricultural Research, Canberra, Australia. pp. 395-398.

SABHLOK, V.P & S.C. PASAHAN. 1985. Migration of field rodents under stress. I Removal of wild vegetation, *Rod. Newsl.* 9:9.

SAINI, M.S. & V.R. PARSHAD. 1993. Field Evaluation of alphachlorohydrin against the Indian mole rat: studies on toxic and antifertility effects. *Ann. appl. Biol.* 122: 153-160.

SARKAR, B.B. 1986. Control of coconut pests in North-Eastern Hill region. *Indian Coconut J.* (16): 3-8.

SEAL, S.C & R.N. BANNERJI. 1966. Changing patterns of rodent population in Calcutta and Howrah In: *Proc. Indian Rodent Symposium.* (Ed.). K.L.Harris, Calcutta, India. pp. 69-83.

SHAH, D.R & K.S. SUBIAH. 1978. Warfarin conquers Lakshadweep, notes on rodent control in Lakshwadweep islands with warfarin wax block. *Pestology.* 2: 36-42.

SHARMA, D. 1994. The damage by Indian crested porcupine. *Workshop on wildlife damage Problems and control* (Abstract), Dehradun, India. Pp. 1.

SHARMA, P.K & A.M.K.M. RAO. 1989. Effect of bund dimensions on rodent infestation in rice. *IRRI Newsl.* 14: 40.

SHEIKHER, C & S.D. JAIN. 1992. Efficacy of live trapping as a rodent control technique in vegetable crops. *Trop. Pest. Mgmt.* 38: 103-105.

SINGH, J., G.S. DHALIWAL & M.N. SHAHI. 1983. Reaction of rice varieties to mole rat under different dates of planting and ages of seedling. *Indian. J. Ecol.* 10: 87-89.

SINGH, Y.P., D. KUMAR & S.K. GANGWAR. 1994. Status paper on rodents in North-Eastern hill region and their management. *Rod. Newl.* 18: 3-11.

SINGLETON, G & T. REDHEAD. 1990. Future prospects for biological control of rodents using micro and macro parasites. In: *Rodents and rice: Report and proceedings of the expert panel on rice and rodent control.* (Ed.) G.R. Quick. International Rice Research Institute, Manila, Philippines. Pp. 75-82.

SINGLETON, G.R., SUDARMIJI., T.Q. TAN., JUMANTA & N.Q. HUNG. 1999. Physical control of rats in developing countries. In: ECOLOGICALLY BASED MANAGEMENT OF RODENT PESTS. (Eds.). G.R. Singleton., L.A. Hinds., H. Leris & Z. Zhang. AICAR Monograph No. 59. Australian Centre for International Agricultural Research, Canberra, Australia. Pp. 178-198.

SMITH, M., D. BRENDAN. A. BRODIE., W. HUNT & L. STAPLES. 2003. Overcoming rat infestation in Australian sugarcane crops using an integrated approach that includes new rodenticide technology In: In: RATS, MICE AND PEOPLE: RODENT BIOLOGY AND MANAGEMENT (Eds.) G.R. Singleton, L.A. Hinds, C.J. Krebs & D.M. Sprat. Australian Centre for International Agricultural Research, Canberra, Australia. pp. 238-241.

SRIDHARA, S. 1982. Food preference studies in rodents and their importance in rodent control. *Plant Prot. Bull. XXV-XXX:* 49-54.

IBID. 1992. Rice In: RODENTS IN INDIAN AGRICULTURE. (Eds.): I. Prakash & P.K. Ghosh. Scientific Publishers, Jodhpur. Pp. 211-230.

IBID. 1999. Quarter century of rodent research on rodent control (1973-98) in the University of Agricultural Sciences, Bangalore-highlights and recommendations. University of Agricultural Sciences, Bangalore, India. 126p.

SRIDHARA, S & R.V. KRISHNAMOORTHY. 1979. Grain losses by spillage and spoilage under laboratory conditions. *Prot. Ecol.* 1: 103-108.

SRIDHARA, S & R.S.TRIPATHI. 2005. DISTRIBUTION OF RODENTS IN INDIAN AGRICULTURE. Technical Bulletin No. 13. Central Arid Zone Research Institute, Jodhpur, India.136p.

SRIHARI, K. & CHAKRAVARTHY, A.K. 1992. Cardamom In: RODENTS IN INDIAN AGRICULTURE (Eds.) I. Prakash & P.K. Ghosh. Scientific Publishers, Jodhpur, India. pp. 289-308.

SRINIVASULU, N., B. VELAYUDHAN & A. SUBRAMANIAM. 1971. Observations on the use of electric fence for protecting rice crop from rat damage. *Madras agri. J.* 58: 233-238.

SRIVASTAVA, D.C. 1992. Sugarcane In: RODENTS IN INDIAN AGRICULTURE (Eds.) Ishwar Prakash & P.K. Ghosh. Scientific Publishers, Jodhpur, India. pp. 231-248.

STENSETH, N.C. 1984. Why mathematical models in evolutionary ecology? TRENDS IN ECOLOGICAL RESEARCH OF THE 1980S. (Eds.) J. Cooley & F.B. Golley. Plenum Press, New York. pp. 239-267.

STENSETH, N.C. 1999. Population cycles in voles and lemmings: density dependence and phase dependence in a stochastic world. *Oikos.* 87: 427-460.

STENSETH, N.C., H. LEIRS. S. MERCELIS & P. MWANJABE. 2001. Comparing strategies for controlling an African pest rodent: an empirically based theoretical study. *J. Appl. Ecol.* 38: 1020-1031.

STERN, V.M., R.F. SMITH., R.V. BOSCH & K.S. HAGEN. 1959. The integrated control concept. *Hilgardia* 29: 81-101.

SUBIAH, K.S. 1978. Some prevalent methods for eradicating rats adapted in a Tamil Nadu village. *Rod. Newsl.* 2: 3-4.

IBID. 1983. Rodent problem in oil palm plantation in Hut Bay, Little Andaman and suggested control measures. *Indian J. Farm Chemicals.* 1: 32-40.

SUBIAH, K.S & R.P. MATHUR. 1992. Andamans with special reference to oil palm plantations In: RODENTS IN INDIAN AGRICUTLURE. (Eds.) Scientific Publishers, Jodhpur, India. Pp. 343-356.

SULLIVAN, T.P. 1979. The use of alternative foods to reduce conifer seed predation by the deer mouse *(Peromyscus maniculatus) J. Appl. Ecol.* 16: 475-495.

SULLIVAN, T & D. SULLIVAN. 1988. Influence of alternate foods on vole population and damage in apple orchards. *Wildl. Soc. Bull.* 16: 170-175.

SULLIVAN, T.P., D.S. SULLIVAN., D.R. CRUMP., H. WEISER & E.A. DIXON. 1988. Predator odours and their potential role in managing pest rodents and rabbits. *Proc. Vertbr. Pest. Conf.* 13: 145-150.

TALWAR, G.P & A. GAUR. 1987. Recent developments in immunocontraception. *Amer. J. Obs. Gynec.* 157: 1075-1078.

TRIPATHI, R.S., A.P. JAIN., N. KASHYAP. B.D. RANA & I. PRAKASH. 1992. North-Western desert. In: RODENTS IN INDIAN AGRICULTURE (Eds.): I. Prakash & P.K. Ghosh. Scientific Publishers, Jodhpur, India. pp. 357-395.

TURNER, J.W & T.F. KIRKPATRIC. 1991. New developments in feral horse contraception and their potential application to wildlife. *Wildl. Soc. Bull.* 19: 350-359.

WAGNER, K.K & D.L. NOLTE. 2000. Evaluation of Hot sauce® as a repellent for forest mammals. *Wildl. Soc. Bull.* 28(1): 76-83.

WHITAKER, R. 1979. Biocontrol of rodents by direct capture. *Rod. Newsl.* 3: 6.

WHITAKER, R & S. BHASKAR. 1978. Rodents in the Lakshasweep islands. *Rod. Newsl.* 2:1.

WHITE, J.G., J. WILSON & K. HORSKINS. 2003. Habitat manipulating and rodent damage in Australian Macadamia orchards. In: RATS, MICE AND PEOPLE: RODENT BIOLOGY AND MANAGEMENT (Eds.) G.R. Singleton., L.A. Hinds., C.J. Krebs & D.M. Sprat. Australian Centre for International Agricultural Research, Canberra, Australia. pp. 213-216.

7

RODENT PEST MANAGEMENT IN POULTRY AND OTHER ANIMAL FACILITIES

Shakunthala Sridhara

AINP on Rodent Control, University of Agricultural Sciences, G.K.V.K., Bangalore- 560 065

ABSTRACT

Rodent damage and control is fairly well studied and attempts to remedy the problem have been well investigated in poultry but totally neglected in live stock and swine facilities in India. *Rattus rattus*, *Mus musculus* are the major pests inside poultry throughout the country. However, *Tatera indica*, *Rattus meltada* and *Meriones hurrianae* are found in poultry facilities of Rajasthan and North India. *Bandicota indica* is a minor pest in most of the country. Damage is not quantified as such but is severe to poultry feed and structure, and by way of transmission risk of salmonellas. Anticoagulants are the recommended rodenticide. Zinc phosphide can be used provided prebaiting is done and bait stations are used. In live stock facilities and piggery although no reports are available, observations indicate damage to animal feed by *R. rattus*. The potential of disease transmission is always there. As a precaution, trapping and bait stations are advised.

INTRODUCTION

Rodent damage in standing crops and storage facilities in India as elsewhere has been a subject of intense study resulting in standardized measures to reduce their numbers to minimum with varying degree of success. The indoor situations like godowns and warehouses provide ideal living conditions by way of uninterrupted food and water supply, shelter, absence of predators and most of the time an undisturbed habitat. Such optimal conditions have resulted in spectacularly high density of commensal rodents in grain storage facilities as revealed by the landmark study of Spillet (1968) in Calcutta. Poultry and livestock also offer an excellent habitat for rodents. In fact they are more hospitable with open availability of plenty of nutritious food, water, a secure shelter that facilitates infestation of arboreal, fossorial and nest dwelling species of rats and mice in the ceilings, below the ground and ground/walls respectively. Another factor that is contributing to high rodent infestation in poultries, dairy and piggery is the structure of the facility itself. With most of the poultry farms situated in rural and semi-urban environment in India, managed by poor as well as rich farmers, the structure of the poultry houses receives very little attention from a poor or small farmer. With muddy or loose floors, thatched roofs, no or weak foundation, lacking rodent preventing girdle around the structure, entry into poultry sheds is almost a 'walk in' for rodents. However, the big, renowned poultry manufacturers take all preventive and reductional measures to keep rodent infestation low or nil inside poultry facilities.

Additionally the agricultural fields around the poultry ensures easy migration of field rodents into poultry houses. The rodent problem in poultry is much more serious than godowns and crop fields because the damage is not just for the feed but also to the structure of the poultry house, to poultry equipment, to electric insulations and to poultry birds all around, i.e. in neighbouring sheds and to humans as rodents transmit infectious diseases by their regular movements from infected to noninfected poultry houses. In spite of significant damage and disease risk, farmers and poultry scientists have paid little attention to rodent problem in poultry. Actually the problem is accepted as an occupational hazard as poisoning of rodents to reduce their population has the inherent risk of affecting birds themselves unless the pest population reaches alarming levels. Thus, there is an urgent need to create awareness about rodent problem and its management in poultry and livestock facilities.

INFESTATION, DAMAGE AND MANAGEMENT IN POULTRY

RODENT SPECIES (Table 1)

Poultry houses across India are intensely infested by the roof rat, *Rattus rattus* followed by the house mouse, *Mus musculus.* The morphology, behaviour, life history and biology of these species are detailed else where in this book i.e. chapters on commensal rodents and VPM in storage (Sridhara *et al.*, 2005; Sridhara & Raveendra Babu, 2005). In parts of Karnataka, particularly in poorly constructed poultry houses close to human habitat, *Bandicota indica* is the sole species infesting (Fig. 1; Table 1). Other wild species have also some regional distribution such as *B. bengalensis* in Andhra Pradesh, Uttar Pradesh, Punjab, *Tatera indica* in Punjab, Rajasthan and Uttar Pradesh, *Millardia meltada* in Andhra Pradesh, *Rattus cutchicus* in Gujarat and *Meriones hurrianae* in Rajasthan. The insectivore, *Suncus murinus* and the carnivore, *Herpestes edwardsi* are reported in the poultry facilities of Rajasthan which would have come in search of rodent, insect and other prey fauna (Table 1).

Table 1. Mammals found in poultry farms in India

SPECIES	STATE	REFERENCE
*R. rattus**	AP, MP, Haryana, Maharashtra, Punjab, Rajasthan, S. India	Siddiqi' 74, Ahmed *et al* '84, Christopher *et al* '84, Sridhara & Krishnamurthy' 92
*M. musculus***	Haryana, Punjab, Rajasthan, AP	Bharadwaj & Prakash '83, Christopher *et al* '84
*B. bengalensis**	AP, UP, Punjab	Christopher *et al* '84, Sood & Malhi '80
*T. indica****	Punjab, Rajasthan, UP	Bharadwaj & Prakash' 83, Sood & Malhi' 80
*B. indica***	AP, Karnataka	Christopher *et al*' 84, Krishnamurthy' 90
*R. meltada**	AP, Rajasthan	Christopher *et al* ' 84
*M. hurrianae***	Rajasthan	Saxena, 1999
Suncus murinus	Rajasthan	Bharadwaj & Khan '83
Herpestes edwardsi	Rajasthan	Bharadwaj & Khan '83

* Major pest AP-Andhra Pradesh
** Minor pest MP-Madhya Pradesh
*** Seasonal pest UP-Uttar Pradesh

The house rat is an agile climber and lives in the space between girders and beams and in the thatched roof of poorer constructions. Mice may live in the lofts or in nests amidst feed bags, in electric boxes, in holes inside walls etc. The wild rats and gerbils dig burrows either inside the poultry itself or along the outer periphery of the poultry shed.

When poultry units are infested by a mixed population of rats and mice, rats may exclude mice. Once the rats are eliminated by control measures, the mice flourish.

Two aspects of rodent biology namely their feeding and breeding has great relevance to their success and management. Although cereals are the most preferred food, rodents can survive on a variety of food like garbage, insects, meat, fruit, vegetables and even manure. House mice are sporadic feeders, visiting the food site 20-30 times a night. At each visit they nibble at the food and eat small amounts. On the other hand rats and gerbils obtain their daily quota from one or two locations. Generally rats require water even if it is in small amounts. In contrast mice can survive long periods without water, getting their water requirements from the food they consume.

Rodents breed prolifically more so in poultry facilities. Thus it becomes essential to control them meticulously before they reach high numbers causing severe damage. Generally a female rodent litters 4-8 times a year with an average litter size of 8-12. The pups reach sexual maturity in about 12 weeks. After mating, pregnancy lasts about three weeks, so also weaning. The average life span is 5-12 months for both rats and mice.

Rodents breed throughout the year inside poultry units with one or two peaks which varies from state to state depending on climatic conditions. The predominant species *R. rattus* exhibited an average 40% pregnancy rate with annual productivity of 70 young/female/rat (Prasad, 1999; Sridhara, 1999). Consequently heavy populations exist in poultry houses which are calculated to be 196 rodents/2205 sq.ft. and 292 rodents/sq.ft floor area in Punjab and 784 rodents/shed (4000 sq.ft.) in Karnataka. A positive correlation is reported to exist between density of rodent population and age of the structure with numbers reaching maximum 3-4 years after the poultry house is established (Prasad, 1999).

ECONOMIC LOSSES

The very activities that have made rodents highly successful and adaptable render them equally highly undesirable pests of poultry. These include activities such as gnawing, nibbling, burrowing, defecation, urination, elaborate exploratory behaviour which not only deteriorate environment of poultry farm but also physically affect poultry and egg production, cause feed loss and potentially transmit diseases. It is difficult to arrive at cumulative losses caused by rodents and as such very few estimates are available. The damage has variously been calculated as Rs. 94/1000 birds and Rs. 15,000 while raising 20,000 broiler chicken in Punjab, Rs. 8744 in a medium sized poultry farm in Madhya Pradesh and Rs. 0.63/100 eggs/day in Rajasthan (Prasad, 1999).

Gnawing activity causes structural damage to wooden doors, windows and electrical wiring. Not only power shut downs resulting from rodents gnawing electrical wires and installation affects the productivity of poultry in general and broilers in particular but also increases cost of maintenance due to repair and reinstallation of wires. Structural damage to floor and foundation also occurs due to burrowing activity. The damage to equipment is due to gnawing of wire mesh of rearing cages, feed trays and containers, egg trays and gunny bags containing chicken feed.

Feeding activity results primarily in feed loss as rodents inside poultry completely depend on the poultry feed for their nutritional requirements. Quantified feed losses caused by rodents has been elaborately discussed by Chopra (1992) and is reported to be 0.05 to 0.14kg in sheds, 0.18 to 6.38kg in feed stores, 50g destruction per day by a single rat by way of consumption and spillage and upto 20kg/day in heavily infested poultry houses. A latter study recorded a loss of 8.6g/rat/day which comes to 2 to 50 kg poultry feed a day and contamination of poultry feed by 0.4 to 3.3%. Feed contamination results in indirect losses as inferior quality feed affects the consumption and growth of chicks ultimately adversely influencing egg and meat production.

Normally damage to eggs is negligible at 0.5% which may reach upto 10% under poor storage conditions. Although chicks below the age of 30 days are attacked, as a rule rodents prefer the grain based poultry feed to live chicks and eggs and hence damage is exceptional not a rule to the later. However the fast running and jumping movements of *R. rattus* scare the hens affecting their food intake, growth and egg yield. In case of broilers such startling movements of house rats are reported to have resulted in heart attack and death of chicks as the broilers are fed hormone rich feed and are susceptible to excitation, increased heart rate, palpitation leading to cardiac arrest (Sridhara, 1992-93, Personal observation).

DISEASE TRANSMISSION

Rodents are implicated in the spread of about 35 diseases from poultry to humans, animals and poultry. Notable amongst these are bacterial diseases like fowl typhoid, Avian paratyphoid, salmonellas, pseudo tuberculosis, fowl cholera, erysipelas, viral diseases like New castle disease, Laryngotracheitis, ornithosis, protozoan infections like toxoplasmosis and coccidiosis (Malhi & Prasad, 1993). To the list, the BIRD FLU, now spreading across far eastern countries of Asia menacingly threatening human health can be added. The disease is so virulent and feared that millions of birds in China, Taiwan, Thailand, Hongkong are being killed and incinerated. They are also hosts to several ectoparasites mostly ticks which are transported to unaffected sheds. However there are no published records of the extent of occurrence of these diseases and the damage caused. Although salmonella infection is low in wild rodent populations, several virulent forms of salmonella have evolved causing concern in the food industry as the infection is transmitted to man through bacteria contaminated eggs and meat. Weils disease affects the farm labourers and Newcastle diseases (*Ranikhet*) regularly kills large flocks of chicks every year. Other diseases are likely to be transmitted from contaminated sheds to uncontaminated sheds especially when rodents migrate from poultry farms where chicks have been sold and are empty at the end of production cycles. Spread of diseases is through droppings, feet, fur, urine, saliva and blood. In most cases rats directly transmit diseases by picking up the pathogens from infected birds as they eat and drink from the same containers as the birds.

RODENT CONTROL IN POULTRY FACILITIES

Five steps form the basics of effective rodent control in poultry houses namely:

1. Rodent proof construction
2. Sanitation
3. Inspection of infestation
4. Population reduction
5. Maintenance baiting and trapping

Rodent proof construction and sanitation are preventive methods of rodent pest management in poultry. If infestation is suspected, a thorough survey is necessary to be followed by measures to reduce rodent numbers for which the most effective and practical approach is by using rodenticides supplemented by trapping. Evaluation of success of control followed by implementing preventive measures will ensure rodent free poultry facilities for considerable length of time.

Rodent Proofing: Rodents can be permanently excluded from poultry houses by "building them out". This can be achieved by eliminating all openings through which rodents enter and exit. The details of preventing rodent entry into and establishment inside the poultry sheds based on Parshad (1999) and Loven (2003) are as follows:

(a) Fixing tightly fitted, perforated metal sheet drain covers on drain holes and keeping sewer and gutters properly covered.

(b) Since rats climb water and drain pipes, electric cables, installing guards on outside pipes and by sealing the space around them and cables will prevent rodent entry.

(c) Because low plinth structures facilitate rodent entry, plinth constructed 90 cm above the ground level with a 30cm collar will solve this problem.

(d) Open and damaged windows and ventilators allow rodents inside which can be prevented by fixing a permanent screen of wire mesh with less than 6mm holes and removing ventilators which heips rats to climb.

(e) Doors should fit properly in the sills. Further a thin galvanized iron sheet should be fixed on the lower border of doors upto 24 cm. Door sweeps and door closers are also helpful.

(f) Digging of burrows can be stalled by using concrete and mortar in foundation and also by maintaining a weed free, gravely perimeter of 60cm all around the poultry house.

(g) Seals and cracks in the wall and other structures should be sealed with concrete. Fixing metal sheets on joints will prevent nesting and hiding of rodents.

(h) Roofs without girders made of concrete slab makes it difficult for rodents to construct nests.

In spite of all the above measures of rodent proofing, it is difficult to eliminate the established rodent populations. This can be achieved by meticulous and systematic population control by chemical and physical methods. Such a population control should precede a thorough inspection of poultry houses for the nature and degree of rodent infestation.

Sanitation in a newly established poultry house is comparatively easier. Cleaning of spilled grains, routine inspection, removal of unnecessary things and general cleanliness will deny food and shelter to rodents. If there is an established rodent population inside the poultry facility it is difficult to eliminate sources of food and shelter. However sanitation measures taken outside the poultry house will prevent further infestation. These measures include cleaning of spilled grains, elimination of outside debris such as old and discarded equipment, damaged and unusable containers and other clutter that can be used as nesting place by rodents. Similarly removal of weeds around the poultry shed which provide cover, food, water and nesting material will prevent reinfestation. An uncluttered 60cm wide weed free perimeter around the poultry shed will not only deny harbourage and cover to rodents but will also enable detection of new burrows. If this perimeter is covered with 1.5cm sized gravel to a depth of 10cm it will prevent rodents from burrowing.

a. Typical layer house

b. Burrow of *B. indica* inside a layer house

PLATE 1

Inspection of infestation

The first step in rodent control in poultry is to determine the extent of infection which can be done visually. Areas of rodent activity can easily be detected by sightings, droppings, tracks, burrows, pathways, fresh gnawing and dead rodents.

Similar to other rodents, those living in poultry houses are nocturnal, active just after dusk and again shortly before dawn. If rodents are repeatedly seen during day time it indicates well established infestation. The location, distribution and severity of infestation can be determined easily by inspecting the premises before dawn or after dusk using a flashlight. The information gathered thus helps in determining control procedures, for example the ideal sites for bait or trap placement. A similar survey after the control will give a good indication of control success.

The most prominent indication of presence of rodents is their droppings with each mouse excreting 40-100 faecal pellets and a rat dropping 20-50 a day. Burrows are another sign of rodent infestation. Burrowing rodents like *B. bengalensis, B. indica, T. indica* etc burrow into dry or wet poultry manure and the ground below the slabs and along the walls outside (Plate 1). The walls, electrical insulation boxes and rafts/beams provide ideal nesting places for non-burrowing species like *R. rattus* and *M. musculus*.

Population Reduction

Rodent population inside poultry facilities are usually reduced by trapping and using rodenticides.

Trapping can be effectively used in small poultry houses to reduce rodent populations and in situations where baiting is hazardous but in a large poultry house using a large number of traps as a sole control method involving baiting and removing the captured rats becomes unmanageable. But traps are ideal to eliminate survivors of control measures and when baits are not accepted.

Baits recommended for trapping include peanut butter, nuts, dry fish, fried preparations, apples, coconut pieces etc. Prebaiting i.e. the baited but traps unset for 2-3 days enhances trap success. It is important to use enough traps to succeed. It is always better to use large number of traps for a few days than using few traps over extended periods. Placement of traps should be along or across the path of rats and mice. Mice traps should be set every ten feet and rat traps at every 25-50 feet. If the population is high traps can be set both inside and outside the poultry house.

Rats in the poultry houses are familiar with human odour hence it may not be necessary to prebait or to make traps odour free. Dead animals should be removed immediately.

Traps should be serviced regularly. By maintaining a map to mark trap placement and trap success, area of greatest rodent activity can be identified. Subsequently control measure can be concentrated at these points.

The three commonly used trap types are:

Snap traps: Simple, inexpensive and are very effective when used properly. They should be used only in layer houses where birds are caged safely.

Multiple-capture live traps are very effective as a supplemental method for controlling rodents in a poultry house. These traps can capture up to a dozen or more mice and rats every day. These are ideal in all types of poultry-layer, broiler, etc. as there is no danger to the birds. Rats and mice being colonial and social, capture of a single member of the colony facilitates trapping of other members of the social group.

Glue board is a standard sized, standard shaped cardboard or thin metal sheet on which strong glue of polybutane origin is spread. Any animal coming in contact will stick to it strongly. They are quite effective when placed along walls and are more successful in trapping mice than rats. However, they are expensive when used over a long period and larger areas. Additionally heat, cold, dampness and dust render them useless. They should not be used in places where pets, children and wildlife have chances of coming across them. If there is accidental gluing of non-target animals, the glue can be dissolved by using cooking oil.

Sherman traps are single catch traps but very efficient and safe to use in poultry. But their use is limited to small sized poultry houses as cost, time and manpower needed for large areas is prohibitive.

Rodenticide Application

Using toxicants to kill rodents in poultry facilities is the most effective, practical and popular method of pest management although it is inherent with many challenges. The infestation is heterogeneous in nature-the entire length of the poultry from the pit to the ceiling. With the constant and unlimited availability of nutritious food and water, baits offered should be undeniably attractive to lure the rodents. Then there are chances of accidental poisoning of birds and also bait contamination by dust, feathers, poultry manure and insects rendering it unacceptable to rodents.

Rodenticide are of two types-acute rodenticides and anticoagulants

Acute rodenticides comprise only three effective rodenticides namely bromethalin, cholecalciferol and zinc phosphide. Their mode of action is described while discussing rodenticides to control commensal rodents (Sridhara *et al.*, 2005). Of these only zinc phosphide is marketed. At 2% in the bait, zinc phosphide baiting for a day preceded by two-three days prebaiting using well designed, bird-proof bait stations gives good mortality of rodents.

Anticoagulant rodenticides cause death by internal bleeding and preventing blood from clotting. The older anticoagulants were multiple dose poisons and caused death after being consumed continuously for several days. These included warfarin, diphacinone and chlorophacinone. Not only they had to be fed continuously and slow in action, but also left resistant populations. But the newer anticoagulants such as broadifacoum, bromadiolone and difethiolone cause death after a single feeding. These single-dose anticoagulants have two advantages namely they are quick and ideal for animal facilities such as poultry and swine where in the presence of attractive, alternate foods, the rodents may not be attracted to baits. Rodents resistant to multiple dose anticoagulants can be easily killed by second generation, single dose anticoagulants. Of the several second generation anticoagulants evaluated only Bromadiolone is available in both wax and loose form containing 0.005% of the active ingredient.

Table 2. Rodenticides used in poultry farms

Type of rodenticides	Active ingredient	Concentration of AC	Remarks
I Anticoagulants	(a) Brodifacoum	0.005%	a) Single dose anticoagulant
	(b) Bromadiolone	"	b) Single dose anticoagulant
	(c) Diphacinone	"	c) Death after single feeding but requires 2-4 days of feeding
	(d) Difethiolone		d) Single dose
II Acute Rodenticides	(a) Bromethalin		a) Single dose
	(b) Cholecalciferol		b) 1-3 days feeding necessary
	(c) Zinc phosphide	2%	c) Single dose; but shyness when sub lethal amount is ingested, Prebaiting necessary

Bait formulations: Several forms of bait formulations are available for baiting in poultry and livestock facilities. These are pellets, wax blocks and loose baits. In poultry and livestock facilities a combination of bait formulations gives optimum reduction of pests as different areas of the facility require different type of baiting. The wax blocks are convenient to bait nests in walls, difficult to access areas like attics and damp locations. The pelletized and meal formulations are best for large areas requiring a large number of bait placements inside burrows and also when bait stations are employed. This type of baiting is more economical than using hundreds of wax blocks. However loose baits tend to get spoiled rapidly compared to wax blocks.

Bait containers can be made at home using metal or plastic PVC pipes (12-24 inches length), wooden boxes or large cans. They should be large enough to accommodate many rodents at a time and should have two rodent size openings for entry and exit at the opposite sides. If the bait boxes have to be placed amidst free roaming poultry as in broiler houses, the bait should be inaccessible to birds and the bait station should be sturdy enough to withstand the rapid, aimless movement of poultry, or blown away by the wind. If there is possibility of exposure to children, pets and other non-target animals, tamper-proof bait boxes should be secured to walls and ground. With these precautions bait boxes are quite effective in rodent control in poultry and have the additional advantage of keeping baits uncontaminated, helping in monitoring bait consumption and reducing chances of accidental poisoning.

Baits and baiting : While choosing a bait for rodent control in poultry and other animal facilities it is necessary to remember that they have access to an excellent, nutritionally balanced food within the poultry and animal enclosures. Attractive bait ensures successful rodent control. To determine an attractive bait, it may be necessary to conduct a choice test of candidate foods.

Test baiting helps in choosing an attractive bait to mix poisons. The choice should involve not more than three food items. Test baiting for mice is not necessary as they are not neophobic.

Dry baits are usually locally available, preferred cereals mixed with 5-10% vegetable oil commonly used in the area. Sometimes additives like 5% sugar enhance the attractiveness of dry baits.

Water baits especially anticoagulant water baits work best in poultry houses where birds have been sold and the house is without any feed and water flow. Chick drinkers are ideal for water baiting. After every use the containers have to be cleaned thoroughly.

Mice don't need water regularly. Addition of sugar and fruit juice to liquid poisons may lure mice to poisons. Rats on the other hand readily drink water baits as they need moderate amounts of water daily.

Pre baiting is habituating rodents to selected food with which poison will be mixed. This is carried out by exposing measured quantities of a chosen food, without mixing poison and keeping for a few days at the same site where the poison bait would be kept. This continuous exposure to unpoisoned bait helps in overcoming neophobia (avoidance of new foods) and bait shyness (avoidance of food after consumption of insufficient poisoned food) of rats. To some extent it can also serve as test baiting with minor modifications. It also helps in calculating the amount of bait to be prepared and also the sites where the bait has to be placed.

Amount of poison bait to be used: It is very essential to ensure adequate bait placement so that all the rodents infesting the poultry house have access to plenty of food. Otherwise under baiting, apart from failing to kill all the rodents renders the rats consuming sub lethal doses bait shy.

The amount of poison bait needed depends on the species, the level of infestation and the rodenticides to be used. These factors are variable from poultry house to house and hence it is best decided after pre or test baiting in each situation.

Continued re-use of the same bait and rodenticides in the same location generally decreases acceptance of bait, results in bait shyness and poor control. Therefore, baits and rodenticides should be rotated frequently.

Mixing baits and poisons: While mixing acute rodenticides like zinc phosphide it is advisable to use hand gloves. Mixing should be as per the directions printed. Small quantities can be prepared in basin type plastic containers. Baits and poisons are to be mixed using a large wooden or plastic spoon or paddle.

Poison bait should be placed in clean containers. Residual scents in the bait containers reduce bait acceptance by rats and mice, especially rats. After mixing and dispersing, hands and the utensils used for mixing should be cleaned thoroughly.

POISONING WITH RODENTICIDES

(a) In Broiler Houses

It is best to poison rodents using acute rodenticides on the first or second day after the broilers have been removed. Next spots where poison has to be kept have to be identified. The best place to poison bait will be beneath the feeder lines since rats and mice are habituated to eating near them. Prior to placing the poison bait one has to make sure that no residual chicken feed is available to rodents. Shallow cups, egg cartons, cardboard boxes can be used as bait containers.

It is quite possible that some of the rats have not consumed lethal amounts of poison bait to die. These surviving rodents become bait shy and avoid the bait. Hence it is recommended that the same poison and bait should not be used three months after the acute poisoning. All the unused, unprotected poison should be carefully removed from the poultry house before chicks are brought in and destroyed/buried.

(b) In layer houses

Placing poison baits on concrete walkways beneath feeder lines is ideal to kill rodents because they are used to eating spilled feed in these areas.

Covered bait stations: Using open bait is not advisable when there are birds in the poultry house. Using covered bait stations rodents can be poisoned any time in the poultry house. Card board or metal boxes of 1x8" or 1x6" with opening on either size hinged to a wall area or kept on the ground are very effective in baiting rodents. On the ground, bait stations have to be kept in the regularly traversed pathway of rats and mice. Baiting should also be carried out in other suspected areas of rodent inhabitation such as barns, sheds and also adjacent homes. Otherwise rats from these shelters occupy the vacant niche created by the death of poultry house rodents.

Guidelines for efficient baiting

Baiting to be successful has to be placed in such a way that rodents inside poultry should encounter it repeatedly. The following guidelines will help in successful baiting:

- Direct placement of poison baits inside the burrows both within and outside the poultry facility.
- Bait containers to be used in the pathways of rodents.
- Attics, ceilings and centre beams should be baited with anticoagulant wax blocks, using a wire or string passing through the wax block to tie wherever necessary/ possible.
- Trapping should be carried out in feed, storage areas and egg storing areas to avoid contamination with rodenticides and to prevent rats dying in inaccessible areas.
- As mice are sporadic and nervous feeders, many bait placements containing small amounts should be kept.
- Permanent bait stations must be kept outside the poultry structure, well protected to withstand climatic conditions. Baits have to be replenished regularly.
- More attention should be paid to bait stations where consumption is high. Contrarily bait stations receiving no or little attention may be moved out.
- Rodent control in turkey and broiler facilities should preferably be carried out during flock turnover.

- When birds are present, tamper resistant bait stations should be either nailed/affixed to the wall or secured to the ground using heavy heights.

Killing rodents should be followed by cleaning out the area around poultry houses immediately of all trash which provide good shelter to rats and mice. It is also advisable to remove or to keep the grass and weeds around the poultry short as rodents avoid moving in exposed areas to avoid predators. If possible it is better to carry out control programmes simultaneously in adjoining farms. Control done and sanitary measures undertaken once in three months as a community programme will keep the poultry houses free from rodents.

Maintenance baiting and trapping

Even the most successful rodent control programme leaves a few rodents alive. These in no time multiply to original levels of infestation which coupled with possible migration of rodents from outside results in fairly good number of rodents in poultry facilities soon. To prevent this, permanent bait stations outside the poultry structure with regular replenishment of bait have to be kept and trapping should be carried out within the building. These measures will keep pressure on residual population slowing down its build up. After the control, the exterior and interior of the facility should be inspected at weekly intervals for fresh rodent signs. Traps have to be checked and emptied. Bait stations have to be filled with fresh bait.

CATS

When managed properly cats do keep the rodent population under check in a poultry house mainly by preventing population build up. Cats work best when they are contained within a poultry house. It is recommended to maintain one cat per 5000 laying birds to a maximum of 5 cats per house. Numbers above this affects their hunting activity. Females less than five years make best predators. For the cats to be effective hunters, they should be fed well and kept healthy and strong by keeping a bowl of constantly replenished commercial cat feed and adequate amount of water. It is advisable to keep a wooden or cardboard box with some sort of bedding and a litter box with litter and sand, for nesting and sleeping.

PRECAUTIONS

Rodenticides being mammalian toxicants are always hazardous to humans and animals other than rodents. Anticoagulants are potentially dangerous to predators and scavengers which hunt dying and eat dead rodents. It is advisable to keep poisons out of reach of poultry, pets, livestock and other non-target animals. Carcasses should be handled with gloved hands or using tongs or newspaper. All dead animals should be collected, buried. Pets should be confined during poison operations. Remaining rodenticides should be stored safely with proper labelling. In case of accidental poisoning the victims should be rushed immediately to physician who should be told about the poisons handled. All unused baits at the end of operation should be disposed safely.

INFESTATION, DAMAGE AND MANAGEMENT IN LIVE-STOCK FACILITIES

Dairy and swine facilities are similar to poultry houses in terms of providing an environment rich in food, good shelter and predator free. Similar to caged layer birds, cattle are confined but swine are mostly free ranging as the broiler chick are. There are no published accounts of rodent damage in these animal facilities in India although the author has been regularly managing rodent problem in swine and cattle facilities on request in the University of Agriculture Sciences, Bangalore (Sridhara, Personal observation). Even outside India published information on rodent damage in swine facilities is limited to 4-5 publication two of which elaborate a computer model to assess rodent control in swine facilities (Vercauteren *et al*., 2002; Hygnstorm *et al*., 2002).

R. norvegicus and M. musculus are reported to cause an annual loss of $6.33 million due to structural damage in Nebraska pork industry (Johnson & Timm, 1987). Losses due to consumption of livestock feed were estimated to be $0.75 million annually. Rats and mice are also reported to be vectors of swine diseases like swine dysentery, encephalomyocarditis, porcine rotaviruses, trinomosis, Leptospira, salmonellas and pseudo rabies (Henzler, 1997). The problem in America is all the more severe as rodent damage has increased particularly to building insulation by house mice (Hygnstrom, 1995). In the absence of information on the overall cost-effectiveness of various methods of rodent control, Hygnstrom *et al*. (2002) developed an interactive STELLA systems model by incorporating data from production models, literature and personal experience. The model generates cost-benefit analysis as well as predicts outcome of various level of control of house mice. The authors developed the website, http://rodent.swine.unl.edu/ to promote awareness on rodent problem and integrated strategies for managing them in swine facilities.

There are reports of effectiveness of pulse baiting with brodifacoum and bromadiolone (Arends *et al.*, 1986), mouse tubes impregnated with brodifacoum (Jenkins, 1990), trapping and permanent bait stations (Wang 2000) in successfully reducing rodent populations in swine facilities.

The strategy being practiced by the author (Sridhara, personal observation) in piggery and diary is to prebait the live burrows and bait station with a rice-millet mixture for two days followed by 2% zinc phosphide baiting. The surviving populations are baited with bromadiolone wax blocks. This method, done at six monthly intervals gave us 70-80% reduction in pest populations. As it is difficult to keep sanitary conditions and also exclude rodents, intensive rodent control using toxicants twice a year seems to be the only practical and effective way to keep rodent populations down in diary and swine facilities.

CONCLUSIONS

Rodents have been identified as one of the major constraints in poultry production. The major poultry industries in India maintain poultry facilities rodent free by preventive and remedial control of rodents. But it is the individual farmer who, although well aware of rodent nuisance and damage, has not adapted routine rodent control and rodent prevention measures. Most of the time the visiting veterinarian ends up as a consultant for rodent control too. Rodent control should be a part of training of poultry managers and banks financing poultry should make rodent free sheds and management mandatory. The problem in livestock and swine facilities is equally bad. The extension worker, veterinarian and agricultural assistant of local government who visit villages regularly should be imparted training in rodent control so that the technology is easily available to farmers even in the interiors of the country.

REFERENCES

AHMED, N., G. CHOPRA & M.C. SOOD. 1984. Assessment of rodent damage in poultry farms. *Poultry guide* 21(3): 57-60

ARENDTS, J.J., C.S. APPERSON., S.H. ROBERTSON & C.S. PAYNE. 1986. The effectiveness of rodenticides used for control of Norway rat and house mouse in poultry and livestock facilities in North Carolina. *Proc. 6th Intl. Biodeterioration Symp.,* Washington, D.C. (Eds.) S. Bary & D.R. Houghton. CAB international.

BHARADWAJ, D & I. PRAKASH. 1983. Rodent pests in poultry farms at Jodhpur. *Rod. Newsl.* 7(3):11.

CHOPRA, G 1992. Poultry farms in: RODENTS IN INDIAN AGRICULTURE. Vol. I (Eds.) Ishwar Prakash & P.K. Ghosh. Scientific Publishers, Jodhpur, India. pp: 309-329.

CHRISTOPHER, M.J., M. BALASUBRAMANYAM & K.R. PURUSHOTHAM. 1984. Rodent problem in poultry farms of Tirupati, Andhra Pradesh. *Poultry Guide.* 21: 21-25.

HENZLER, D.J. 1997. Rodent control and pathogen reduction: best management practices for rodent control in swine premises. In: *Proc. 5th Swine Disease Conference for Swine practioners,* Iowa State University, Ames, USA. pp: 39-44.

HYGNSTROM, S.E. 1995. House mouse damage to insulation: II. *Intl. Biodeterior. Biodegrade.* 33: 143-150.

HYGNSTROM, S.E., K.C. VERCAUTERN., R.M. TIMM., B.M. CORRIGAN, J.G. BELLER, L.L. BITNEY, M.C. BRUMM., D. MEYER., D.R. VIRCHOW & R.W. WILLS 2002. An economic model of integrated house mouse control in swine production facilities. Ref. http://rodent.swine.unl.edu/

JENKINS, D.L. 1990. Mouse tube success in pig farms. *International pest control* 32:5, 120-121.

JOHNSON, R.J & R.M. TIMM. 1987. Wildlife damage and agriculture in Nebraska. *Proc. Eastern Wildl. Damage Control Conf.,* 3:57-65.

KRISHNAMURTHY, T.R. 1990. Population dynamics and chemical methods of control of a few rodents in some villages of Karnataka. Ph.D thesis submitted to Mysore University, Mysore, India.

LOVEN, J. 2003. Controlling rodents in commensal poultry facilities. In: ANIMAL DAMAGE MANAGEMENT. Downloaded from the Website of Purdue University, USA viz. http://www.ces.pordue.edu/estmedial/ 28p.

MALHI, C.S & V.R. PARSHAD. 1993. Ridding rodent heaven of rats: pest control in hen house. *Ceres.* 144: 10-13.

PARSHAD, V.R. 1999. *Rodent management in poultry farms.* Technical Bulletin No.8. Project coordinator's cell, AICRP on Rodent Control, Central Arid Zone Research Institute, Jodhpur, India. 11p.

SAXENA, Y. 1999. Management of rodent pests in poultry farms, *J.Ecotoxicol Envt. Monit.* 9:1, 67-69.

SIDDIQUE, S.M. 1974. Rat menace and its control. *Poultry Guide,* 11(7):27-29.

SOOD, M.L & C.S. MALHI. 1980. Biology and control of rodents in poultry farms in Ludhiana district (Punjab). *Indian Poultry Review,* 11(15): 28-33.

SPILLET, J.J. 1968. *The ecology of the lesser bandicoot rat in Calcutta.* Bombay Natural History Society and John Hopkins University Centre, Calcutta, India.

SRIDHARA, S. 1999. Quarter century of research on rodent control (1973-98) in the University of Agricultural Sciences, Bangalore-highlights and recommendations. University of Agricultural Sciences, Bangalore, India. 126 p.

SRIDHARA, S. & T.R. KRISHNAMURTHY. 1992. Population dynamics of *Rattus rattus* in poultry and implications for control. *Proc. Fifteenth Vertb. Pest. Conf.* Newport Beech, California, USA. pp: 224-228.

SRIDHARA, S & T. RAVEENDRA BABU. 2005. Vertebrate borne diseases and management of commensal rodents. In: Vertebrate Pests in Agriculture- The Indian Perspective (in press), Scientific Publishers, Jodhpur, India.

SRIDHARA, S., V. SHIVAYYA & T. RAVEENDRA BABU. 2005. Vertebrate Pests in storage. In: Vertebrate Pests in Agriculture – the Indian scenario (in press), Scientific Publishers, Jodhpur, India.

VERCAUTEREN, K., S.E. HYGNSTROM., R.E. TIMM., R.M. CORRIGAN, J.G. BELLER, L.L. BITNEY, M.C. BRUMM., D. MEYER. D.R. VIRCHOW & R.W. WILLS 2002. Development of a model to assess rodent control in swine facilities. In: Human conflicts with wildlife. Economic considerations, *Proc. 3rd NWRC Special symposium.* (Eds.) L. Clark, J. Hone, J.A. Shivik, R.A. Watkins, K.C. Vercauteren & J.K. Yoder. Colorado, USA. pp: 59-64.

WANG, P.Y. 2000. *Hog farm rodent damage and its control.* Report of the Taiwan Sugar Research Institute. No. 168. pp 1-13.

❑❑❑

8

RODENT PEST MANAGEMENT IN ARID ZONE CROPS

R.S.Tripathi and Vipin Choudhary

AINP on Rodent Control, Central Arid Zone Research Institute, Jodhpur-342 003

ABSTRACT

Rodent pest management is very important in Arid Zones of Rajasthan for two reasons-the development of and increasing irrigation and the high densities of rodent population, which not only reduce crop yields drastically but also abet desertification by burrowing in the sandy soils. Although most of the species are found in more than one habitat, the pattern of rodent infestation is generally *Rattus cutchicus* and *Mus cervicolor* in rocky habitat, *Gerbillus nanus* in sandy soils while *Mus musculus* and *Mus booduga* inhabit ruderal habitats. The remaining species viz, *Funambulus pennanti, Meriones hurrianae, Tatera indica, Millardia meltada, Golunda ellioti, Mus platythrix etc*. are abundant in sandy to ruderal habitats. An important change in rodent infestation is the presence of xeric species only in 10 year old irrigated systems, presence of commensal *M. musculus, R. rattus* and *Suncus murinus* (Shrew) in fields with 30 or more years of irrigation and the new infestation by mesic *Bandicota bengalensis*, *Nesokia indica* and *Mus booduga*. Most rodents breed from March to September with minimal populations in winter and summer, peaks occurring around/after monsoons. Damage is about 6% to cereals, 3% to pulses, 4-26% to groundnut, 4-20% to vegetables with considerable damage to grasses and tree plantations. Studies indicated 2.0 to 2.5% zinc phosphide in cereal bait while 0.005% bromadiolone either as wax blocks or loose formulation mixed with cereals as effective. Thrust areas of research are: evaluation of rodent control techniques, monitoring of pest population, establishing damage thresholds, improving bait acceptability, evaluation of biocontrol and antifertility agents. Additionally epidemiological aspects *T. indica* and *B. bengalensis* in plague transmission and educatating resource persons and farmers in community rodent control are also important.

INTRODUCTION

Rodents constitute one of the largest mammalian groups in the Rajasthan desert. They exhibit great plasticity in their choice of a wide spectrum of desert habitats. As a result of their gnawing, debarking, cutting and feeding potential these tiny vertebrates are a serious threat to crops, orchards, rangelands and other desert vegetation. Rodents aggravate the problem of desertification by depredating upon the sparse vegetation and disturbing the soil through their fossorial propensity. Their role in intensification of desertification may be understood by the fact that a single species, *Meriones hurrianae*, is able to excavate about 61,500 kg of stabilized soil per km^2 per day during summers and deposit it outside its burrow openings in loose formation. This loose soil is easily blown away by strong desert winds. Thus, rodents are a prime factor for soil erosion. Various desert development programme viz., arid land farming, horticulture, range management, afforestation, sand dune stabilization etc. are difficult to implement successfully without management of rodent pests.

ARID ZONE ECOSYSTEM

Arid zone occupies nearly 3,17,090 sq. Km. area in India, of which 2,46,790 sq Km is hot arid and rest is classified as cold arid region (Krishnan, 1968). About 62% of the hot arid area of India falls in Rajasthan state alone. Hot arid areas are also distributed in the states of Gujarat (19.6%), Haryana (4.0%), Punjab (4.6%), Andhra Pradesh (6.9%), Karnataka (2.7%) and Maharashtra (0.4%) (Rao & Singh, 1998). In Rajasthan state hot arid region mainly falls in 12 western districts. Here the ecosystem is hyperthermic and is characterized by low and erratic rainfall, poor soil health and frequent droughts. The Indian arid zone forms the eastern side of Thar Desert and is one of the most densely populated deserts of the world. Human and livestock population is consistently increasing. During the last 20 years, human population has increased by 50% and that of livestock by 25%. Such an increase in population is leading to greater exploitation of meagre natural resources. Agriculture and Animal husbandry are the two main occupations of desert people. To meet the food and fodder requirements, farmers have adopted rain fed farming, development of range/*gochar* lands, arid horticulture and tree plantations. Thus, arable farming, horticulture and management of range/*gochar* lands for livestock production, and Silviculture/forestry provide sustainability to the region and form the backbone of rural economy. With the canalization of river waters through Indira Gandhi Nahar Project (IGNP), and large-scale exploitation of ground water, the area under irrigated farming has increased from 0.79 mha in 1970 to 1.02 mha in 1992 and is likely to increase further to 2.69 mha by 2010 (Singh & Saxena, 1998).

Agricultural productivity including horticulture, fodder grasses and silviculture in the state faces a great risk from vagaries of desert climate, particularly uncertainties of monsoon rains, poor soil nutrients and high velocity winds. Besides these natural impediments, the arid crops and other productive vegetations are always under threat by native rodents, which are primarily herbivorous and constitute one of the largest mammalian groups in Rajasthan desert. Due to their sheer numbers and omnipresence in all the major production systems of arid zone, these tiny vertebrates are considered as a serious pest problem and require proper management strategies.

RODENT PESTS OF ARID REGION

Extensive surveys in different parts of Rajasthan revealed occurrence of 18 rodent species belonging to three families viz., Sciuridae, Hystricidae and Muridae. Of these, former two families are monotypic i.e., represented by one species only. Gerbils, rats, mole rats and mice constitute the family muridae (Tripathi *et al.*, 1992). The species biodiversity in the region is presented in Table 1.

Distribution of rodent fauna in various habitats of desert biome of Rajasthan has been well documented by Prakash (1975) and Tripathi *et al.* (1992). Prakash *et al.* (1971) identified four types of rodent habitats viz. sandy, rocky, gravel and ruderal in Thar Desert. Most of the rodent species are found in more than one habitat, but a few viz., *Rattus cutchicus* and *Mus cervicolor* are found in rocky habitat and *Gerbillus nanus* occurs in sandy habitat only. Similarly *M. musculus* and *M. booduga* inhabit ruderal habitats i.e. crop fields and/or residential areas. Other species viz., *Funambulus pennanti, Meriones hurrianae, Tatera indica, Millardia meltada, Golunda ellioti, Gerbillus gleadowi, Mus platythrix* etc. are abundant in sandy to ruderal habitats. Following sequence of rodent abundance in different habitats of Rajasthan has been recognized;

- **Sandy habitat:** *M. hurrianae >T.i. indica >G. gleadowi > M. meltada*

- **Gravel habitat:** *M. hurrianae* >*T.i. indica* >*M. platythrix sadhu.*
- **Rocky habitat:** *R.c. cutchicus* > *M.p. sadhu* > *M. cervicolor* >*F. pennanti.*
- **Ruderal habitat:** *T. indica* >*R. meltada* > *M. hurrianae* > *G.gleadowi* > *F. pennanti*

Table 1. Rodent species occurring in arid zone (Tripathi *et al.*, 1992)

Family /sub family	Species
(a) Family Hystricidae :	(1) *Hystrix indica* (Kerr)
(b) Family Sciuridae :	(2) *Funambulus pennanti* (Wroughton)
(c) Family Muridae	
(i) Sub-family Gerbillinae	(3) *Tatera indica indica* (Hardwicke)
	(4) *Meriones hurrianae* (Jerdon)
	(5) *Gerbillus nanus indus* (Thomas)
	(6) *Gerbillus gleadowi* Murray
(ii) Sub-family-Murinae	(7) *Vandeleuria oleracea* (Bennett)
	(8) *Rattus rattus* (Linn.)
	(9) *R. cutchicus cutchicus* (Wroughton)
	(10) R. (=*Millardia*) *meltada pallidior* (Gray)
	(11) *R. gleadowi* (Gray)
	(12) *Mus musculus* (Linn.)
	(13) *M. cervicolor* (Hodgson)
	(14) *M. platythrix sadhu* (Bennett)
	(15) *M. booduga* (Gray)
	(16) *Golunda ellioti* (Gray)
	(17)*Nesokia indica* (Gray and Hardwicke)
	(18) *Bandicota bengalensis* (Gray)

Although eighteen species of rodents are encountered in arid zone, only 7-8 species may be considered as pests and a complex of 3-4 species are reported to inflict economic damage in any particular cropping system/habitat (Table 2). Both the gerbils, viz., *M. hurrianae* and *T. indica* are regarded as most predominant pests in rain fed as well as irrigated arable cropping systems and grasslands. Besides these, *R. meltada and M. booduga* prefer to inhabit irrigated croplands. *G. gleadowi* damages rainfed crops grown on dunes/interdunal spaces (Rana & Tripathi, 1999). The squirrel, *F. pennanti* occupies trees scattered in the crop fields while *G. ellioti* prefers the thickets of the bushes commonly found on the periphery of the crop fields.

CHANGES IN RODENT COMPOSITION *Vs.* LAND USE PATTERN

Availability of irrigation water consequent to canalization of river waters has totally changed the complexion of the desert in its command areas. Canal waters were first introduced in arid tracts of Rajasthan

Table 2. Rodent pest species composition in different cropping systems in arid zone

	Rodent pest species
Rain fed crops	*M. hurrianae-T. indica-G. gleadowi*
Irrigated crops (sprinkler)	*T. indica-M. hurrianae-M. meltada*
IGNP areas	*T. indica-M. meltada-N. indica-Mus sp.*
Grasslands	*M. hurrianae-G. gleadowi-T. indica*
Horticultural crops	*F. pennanti-M. hurrianae-T. indica*
Forest/silva fields	*M. hurrianae-T. indica-M. meltada-N. indica* (sandy areas)
	M. hurrianae-R. cutchicus-H. indica (Rocky area)
Forest & fruit nurseries	*F. pennanti-T. indica-M. hurrianae*

Indira Gandhi Canal

through Ganga Canal in 1927. In latter years Bhakhra Canal (1951-52) and Indira Gandhi Canal (1956-57) were also commissioned. Before that the region was a typical desert having undulating topography with sand dunes and dry deciduous degraded land with bushes and grasses as major vegetation. Canal irrigation has transformed the erstwhile dry deciduous degraded land into food basket of the desert State. Indira Gandhi Canal is the most promising irrigation project traversing the entire Thar Desert through major canals and its distributaries in the districts of Bikaner, Jaisalmer, Jodhpur, Barmer etc.

Ecological relationship between rodent species composition and changing land use pattern in the IGNP command areas of Rajasthan desert having varying durations of intensive irrigation have been reviewed by many (Prakash, 1991; Rana *et al.*, 1994 ; Jain *et al.*, 1996). Studies on species composition and trap index revealed that the rodent population was highest in 10 year old irrigation cropping systems and minimum in the 60 year old irrigated fields. Two mesic species of rodents viz., *Millardia meltada* and *Mus musculus* were found to inhabit the areas with 60, 40 and 30 years old irrigation cropping systems, whereas, two xeric rodent species viz., *M. hurrianae* and *G. gleadowi* were not found in intensively irrigated fields but reported only in fields with 10 year old irrigation systems. This showed that the xeric rodents are being replaced by mesic species (Rana *et al.*, 1992). Changes in land use pattern due to intensive irrigation through canal systems had resulted in "atavism" also in the habitat as shown by presence of commensal mammals, *M. musculus*, *R. rattus* and *Suncus murinus* in the crop fields with 30 or more years of irrigation practice (Rana *et al.,* 1992). The two mole rat species viz., *Bandicota bengalensis* and *Nesokia indica* and the mouse, *M. booduga* considered as mesic in habit, too have been reported in IGNP command areas (Prakash, 1991; Tripathi *et al.*, 1992; Idris *et al.*, 2004).

During the last few years sprinkler irrigation has become very common in the region. Studies conducted in Churu district indicated presence of seven species of rodents in the sprinkler irrigated crop fields. *M. hurrianae* was the most predominant (63.64%) followed by *T. indica* (15.63%) and *M. meltada pallidior* (12.5%). Of the trapped small mammals, *Suncus* sp., an insectivore represented 8.13%. In the IGNP areas at Lunkaransar (Bikaner district), *M. hurrianae* was completely absent in the cotton fields with highly compact soil although predominant in sandy soils. These fields recorded presence of *N. indica* hither to absent in the region. The Field mouse, *M. booduga* was predominant followed by *R. meltada* (Idris *et al.*, 2004).

BEHAVIOUR AND BIOLOGY

Most of the field rodents are fossorial and nocturnal, which protect them from direct attack of predators and climatic vagaries. The burrows could be very simple (*T. indica, M. booduga* and *M. meltada*) or complicated (*B. bengalensis* and *M. hurrianae*). The mole rats (*B. bengalensis* and *N. indica*) keep their burrow openings tightly plugged with soil for safety against natural enemies and flooding.

FOOD AND FEEDING BEHAVIOUR of rodents indicate their preference for various food items, which helps in selecting an effective carrier for poison bait preparation. Studies on bait preferences have revealed that majority of rodent species of the region prefer pearl millet, wheat, sorghum etc with slight variation (Rana *et al.*, 1992). Most preferred baits with their caloric values are detailed in Table 3. Similarly preference for oils, sugar, salts etc. have helped in selection of attractive bait additives. Based on the preference of rodents, poison bait recommended by AICRP on Rodent Control (CAZRI, Jodhpur) for Rajasthan is: pearl millet/wheat with groundnut oil (Rana & Tripathi, 1999).The effect of bait shyness induced by sub-lethal intake of acute poisons can be mitigated to some extent by changing the base and oil component of the baits.

Table 3. Most preferred baits and calorific requirements of Indian rodents (Rana *et al*, 1992)

Rodent species	Preferred bait	Calorific requirement (KJ/100g/day)
F. pennanti	Whole wheat grain or cracked pearl milt	104.97 (67.63-146.21)
Tatera indica	Millet and sorghum grains	247.45-253.31
M. hurrianae	Wheat floor or cracked pearl millet	150.48 (117.04-183.92)
Rattus rattus	Maize flour+10% arachis oil +1% sugar	121.21 (63.32-250.38)
Rattus meltada	Pearl millet flour+10% sesame oil; sorghum, maize and wheat grains	58.52-83.6
Mus musculus	Rice flour or pearl millet +1% arachis oil	52.17 (145.08-372.18)

Biology: Desert rodents exhibit a very high reproductive potential. They are prolific breeders with a litter size of up to 12 young ones per female. Such a high fecundity of rodents is countered by several biotic and abiotic factors operating in nature. Rodents sometimes regulate their population by feeding on their young-ones (cannibalism) during stress situations depending upon the carrying capacity of their habitat. In drought years the rodent population maintains a low profile due to scarcity of natural food; however the population explodes if good rainfall succeeds the drought year. Most rodents breed from March to September, although a few breed all through the year. Minimum births occur during the extreme winter season. Summers, when the conditions are largely unfavourable in desert, are also not considered as peak breeding season but both the *Gerbillus* species have shown peak breeding in the month of June. Generally the breeding peak occurs during the monsoon season corresponding with the availability of plenty of green food in the desert. Many desert plants flower and most of them produce fresh shoots and leaves in this season which might be an important factor for the optimal breeding activity seen during monsoons. Another small peak in breeding activity is again noticed during post winter months (Feb-April). Thus, both these breeding peaks coincide with the maturity/harvesting stage of *kharif* and *rabi* crops respectively. Most of the rodent species exhibit relatively lower population levels during winters and summers. Maximum pest population is encountered in monsoon and post monsoon seasons.

LOSSES DUE TO RODENT PESTS

Rodents cause immense losses to crops, fodder grasses and forestry plantations in arid regions. The monsoon season is very conducive for breeding of desert rodents with a very high population at the ripening/ maturity stage of *kharif* crops. During *rabi* season, the area under cropping is drastically reduced due to limited availability of irrigation facilities forcing the the pest rodents to migrate to the crop fields *en mass* and cause extensive damage to *rabi* crops. The losses caused by rodents in arid production systems are given in Table 4 and detailed as under:

Table 4. Estimates of damage caused by rodents in arid ecosystems

Sl. No.	Crops	Damage/ losses
1	Pearl millet	107.7 Kg/ha
2	Sorghum	5.85 - 6.28 %
3	Moong and moth bean	2.6 -3.12 %
4	Groundnut	4.1 - 25.8 %
4	Vegetables	4.0 - 19.9%
5	Grasses	1040 Kg/ha
6	Tree plantations	4.4 – 10.0%

Field crops: Pearl millet, being the staple food of desert people, is grown extensively as rain fed crop and suffers damage to the tune of 2-

5% by field rodents. The crop experiences maximum loss when harvested crop is heaped in the fields after harvest. Gerbils follow the harvested produce by digging tunnels under the heap and feed on the cobs. Jain *et al.* (1993) reported a yield loss of 107.7 kg/ha due to rodents. The arid pulses viz., *moong* and moth bean experience 3.12 and 2.80% pod damage (Tripathi *et al.,* 2004). In chillies these pests damaged 15-20 kg fruits per ha at a population density of 142 burrows/ha. Similarly wheat crop experienced 18.7-21.3% damage due to desert rodents (Advani *et al*., 1982). Mustard crop recorded maximum rodent depredation at vegetative growth stage when about 22.9-43.5% reduction in plant stand was noticed in the peripheral areas of the crop fields (Tripathi *et al,* 1992). Groundnut cultivation has suffered a major setback due to rodent devastations (30-50%) during the last 4-5 years. Irrigated cropping system of wheat-mustard-cumin recorded 5.5-11.1% damage by rodents. Besides the regular rodent depredation, an upsurge of *G. gleadowi*, during 1970, *Tatera indica* during 1989-90 and *M. hurrianae* in 1999-2000 have been reported.

Vegetable and fruit crops: Vegetables viz., tomato, carrot and radish were reported to experience 15.8-29.9% damage by desert rodents. Cucurbits recorded 4.1% fruit damage resulting in yield loss to the tune of 6.5 q/ha. Advani & Mathur (1982) reported that regular rodent control operation in vegetable crops fetches a cost: benefit ratio of 1: 900 in desert region. Like field and vegetable crops, the fruit crops viz., *ber*, pomegranate and date palm were also extensively damaged by rodents (Jain *et al.,* 1994; Patel *et al*., 1995). In orchards *F. pennanti* was the predominant pest, whereas in nurseries, a complex of *F. pennanti, T. indica, M. hurrianae* and *R. rattus* were quite common. Squirrels were reported to cause a loss of 29% of ripe pomegranate. A three year study on susceptibility of different pomegranate cultivars against squirrel attack revealed that Bassein seedless was highly vulnerable to squirrel damage recording more than 58% fruit damage while other six varieties recorded moderate fruits damage by squirrels (12.0-19.0%) with Jalore seedless and Khog cultivars being damaged least (<10%) (Patel *et al*., 1995). Date palm at ripening stage suffered 60-80% squirrel damage.

Range grasses and forestry plantations: Both these production system of arid ecosystem are mainly grown in common propriety resources, which mostly remain unattended and serve as an ideal abode for desert rodents. The gerbils cause immense losses to these perennial-cropping systems. They dig out the seeds of *Cenchrus ciliaris, C. setigerus* and *Lasiurus sindicus* and feed on them almost to their roots. The intake of grass seeds by *M. hurrianae* is much greater than other rodents. Field rodents have been reported to devastate some 40 acres of *L. sindicus* and 27 acres of *C. ciliaris*, *C. setigerus* and *L. sindicus* in an experimental pasture at Bikaner (Prakash, 1981).

Rodent damage to tree species, particularly young saplings is of two types (i) debarking and (ii) slicing. The debarking type of damage has been observed in *Albizzia lebbek, Prosopis cineraria, Acacia tortilis* and *Parkinsonia aculeatea* (Prakash, 1976; Tripathi & Jain, 1990). The debarking activity is done mainly by *M. hurrianae, T. indica* and *M. meltada.* Observations indicated that these rodents debarked *A. tortilis* and *P. aculeata* plantation up to a height of 13.6 and 11.6 cm respectively. The slicing activity of *N. indica* while burrowing underground resulted in the death of 4.4, 10.0 and 10.0 per cent plantations of *A. tortilis, P. juliflora, A. nilotica,* respectively in forest plantations due to root damage (Jain & Tripathi, 2000).

RODENT MANAGEMENT

Several methods of rodent control are listed in literature, but no single method can be effective if applied in isolation. Rodents are highly evolved mammals and their long association with the wisest mammal i.e. man, has made them wiser. Trap avoidance, bait shyness/aversion to acute poisons, resistance/cross resistance to anticoagulants are only few examples of rodent behaviour and adaptation challenging man's endeavours to control them. Therefore, an IPM approach based on sound ecobiology and ethology of pest species *vis-à-vis* population reduction technologies having economic viability and sociological acceptance need to be evolved. A few of the common and feasible methods of rodent control are discussed below:

Trapping: This method, though very old is considered very safe and effective for controlling rodents in indoor habitats (Jain *et al.*, 1992b). Sherman traps were found to be highly effective in trapping rodents live from fields and indoor habitats. Glue boards as kill type traps has shown high promise in indoor use as well as in poultry farms, and have been found superior to snap, Sherman and wonder traps (Tripathi *et al.*, 1994). The traps have a limited utility as a management tool in fields, but can be used as a follow up measure after initial knock down of pests by poison baiting. Single catch traps, like Sherman and snap traps are excellent for identification of pest species, population structure and for further surveillance of the problem in the fields. Trapping success in fields is very high for nocturnal rodents. However, *M. hurrianae,* a diurnal rodent avoids the Sherman and other box type traps to a greater extent. For this species, the trap has to be concealed camouflaged for increased trappability (Rana, 1982).

Habitat manipulation: Food and shelter are the essential and governing factors for the survival of rodents. By manipulating these factors, the carrying capacity of the rodent's habitat can be reduced (Prakash & Mathur, 1987). The reduced food and shelter increases the competition and aggressiveness among rodents and causes decrease in population through mortality/emigration. Barnett & Prakash (1975), and Prakash (1976) suggested that trimming of bunds around the crop fields and planting of *Opuntia* on the bunds is helpful in reducing the pest population. A band of crop, which is difficult to husk by rodents, i.e. cluster bean, if sown around the main crop, can reduce the entry of rodents (Prakash & Mathur, 1987). Similarly regular weed control in and around crop fields also causes stress in native rodents resulting in their migration to other areas.

Rodenticides

(a) Acute rodenticides: A number of acute rodenticides namely zinc phosphide, Sodium fluoroacetate and RH-787 were found highly potent against desert rodents (Prakash & Mathur, 1992). Zinc phosphide is the most widely used acute rodenticide. For desert rodents LD_{50} values of this acute poison is 35.0 mg/kg (*T. indica* and *M. hurrianae*) and 250.0 mg/kg (*M. musculus*). Optimum dosage of zinc phosphide found for majority of rodent pests is 2.0-2.5 per cent in baits on w/w basis which yields 60-70% control success against mixed population of rodents in arid ecosystem. Dosages above or below this level aren't effective and make the pest population bait/poison shy. Aluminium phosphide, a burrow fumigant provided a control success of 78.3-97% in *kharif* crops (Soni *et al.*, 1985). The fumigant has also been recommended as a follow up treatment in a number of trials for controlling the residual population of rodents surviving the action of acute rodenticide (Prakash & Mathur, 1987; Jain *et al.*, 1993).

(b) Chronic rodenticides: Several anticoagulant rodenticides have been evaluated against desert rodents in laboratory and fields. These are grouped into:

(i) Multiple dose anticoagulants which are commonly referred as first generation anticoagulants. e.g. fumarin, warfarin, coumatetralyl. chlorophacinone etc.

(ii) Single dose anticoagulants are commonly referred as second generation anticoagulant rodenticides. e.g. bromadiolone, brodifacoum, flocoumafen, difethialone etc.

LD_{50} values for warfarin, fumarin and coumatetralyl has been investigated as 4x19.1, 4x15.9 and 4x0.60 mg/kg for *T. indica* and

4x15.9, 4x15.9 and 4x0.53 mg/kg for *M. hurrianae*, respectively (Mathur & Prakash, 1984). LD_{50} values for most of the desert rodents for brodifacoum are between 0.083 to 0.77 mg/kg and for bromadiolone 0.11-1.00 mg/kg (Mathur *et al.*, 1992)

Warfarin and fumarin, the two multi dose anticoagulants required 8-14 days feeding to bring 100% mortality at 0.025% in baits against most of the desert rodents. Coumatetralyl baiting at 0.0375 and 0.005% yielded 100% mortality of *R. rattus,* 70% mortality in *T. indica* and 80% mortality in *M. hurrianae* (Mathur & Prakash, 1984). As tracking powder at 0.075%. it recorded 50% mortality of *M. musculus*. Chlorophacinone (0.0075%) in baits provided 100% mortality of *F. pennanti, T. indica* and *M. hurrianae*; 91.61% mortality of *R. rattus* and 50% mortality of *R. meltada* and *M. musculus* (Mathur, 1981). Cholecalciferol, yielded 89% mortality of Indian gerbils in wax blocks formulation (Rana *et al.*, 1994) and fumarin (0.025% conc.) yielded 50 and 60% mortality of *T. indica* and *M. hurrianae*. Under field conditions, warfarin (0.025%), coumatetralyl (0.0375%) and chlorophacinone (0.0375%) in cereal baits yielded over 70% control success on multiple baiting (Mathur, 1981). These rodenticides, though quite potent in action could not get wide acceptance due to their multiple feeding requirements.

Amongst second-generation anticoagulants, bromadiolone is the only rodenticide registered for common use. It is available at 0.005% in two formulations i.e. loose bait and wax blocks. It provides 100% mortality of *T. indica* and *M. hurrianae* in 1-2 days feeding; and 88% mortality of *F. pennanti* in four day feeding. Among other second generation anticoagulant rodenticides evaluated against desert rodents brodifacoum (0.005%) caused 100% mortality in *R. meltada*, *T. indica, M. hurrianae*, *F. pennanti* and *G. gleadowi*. A comparative performance of these anticoagulants is presented in Table 5. Similarly, flocoumafen (0.005%) and difethialone (0.0025 and 0.005%) were also found highly potent against desert rodents (Jain, 1986; Jain & Tripathi, 1988, Jain *et al.,* 1992a; Chaudhary & Tripathi, 2002a).

In fields bromadiolone loose baits produced 70-80% control success in groundnut, cotton, *moong*, rice, *arhar* and sugarcane crops and brodifacoum produced 80-90% control success in similar crop lands (Rana *et al.*, 1994). In grassland inhabited by *M. hurrianae*, *T. indica, G. gleadowi* and *Rattus gleadowi* brodifacoum baiting yielded 90.5% control success (Mathur & Prakash, 1984). Wax block baits of bromadiolone provided 88.8% control success in horticultural crops (Rana & Tripathi, 1999). Pellet and wax block formulation of brodifacoum recorded 80-92.0% control success within fifteen days of burrow baiting. Difethialone (0.0025%) in loose baits yielded over 77% success in bajra-moong-moth cropping system and 85.11 -87.49% in groundnut-guar cropping system. In wheat-mustard-cumin cropping system during *rabi* season 87.73 and 89.95% rodent control success was achieved by this rodenticide (Chaudhary, 2002b). Combination trials viz., anticoagulant followed by anticoagulant or acute rodenticides followed by anticoagulants in different *kharif* and *rabi* crops yielded a control success up to 90% and the rate of population rebuild up in such trials was also very slow.

Table 5. Single dose toxicity of grain baits of second generation anticoagulant rodenticides (No choice)

Rodent species	Per cent mortality and (mean days to death) with			
	Bromadiolone (0.005%)	**Brodifacoum (0.005%)**	**Flocoumafen (0.005%)**	**Difethialone (0.0025%)**
M. hurrinae	88	83-100 (4.7-8.0)	100 (7.0)	100 (5.9)
T. indica	100 (9.5)	90 (6.9)	90 (8.2)	100 (7.9)
R. meltada	100 (6.4)	80-100 (6.6-7.1)	100 (5.7)	100
F. pennanti	-	66 (7.0)	100 (8.1)	100 (6.1)
R. rattus	100 (7.5)	83-100 (8.0-10.8)	100 (8.6))	100 (5.9)
M. musculus	91 (8.1)	60-100 (4.6-8.4)	100 (6.8)	-

LACUNAE

Despite significant research progress in the studies of eco-biology and behaviour of pest rodents of Thar desert and evaluation of rodent control techniques, more research thrusts are needed on monitoring of pest population, deriving damage threshold levels, enhancement in bait acceptability, bio control techniques, anti fertility agents etc. Intensive ecological evaluation of changing faunal diversity of rodents in arid zone is also needed in view of transformation of desert into irrigated croplands due to large-scale use of canal waters in western Rajasthan. The new entrant like *B. bengalensis*, a colonizing rodent is more destructive than the xeric rodents. Moreover this species is highly susceptible to various diseases and by virtue of it's sharing the same habitat with *T. indica*, an established reservoir of plague *bacillus*, it may pose serious threat to public health. Therefore, epidemiological considerations of such changes require attention of researchers. In addition to these there is dire need to educate the farming community (including farm women) on this very important aspect of health hazard. Since there is lack of trained manpower on rodent management, the researchers have a greater role to play in imparting education and training to the officials of various line departments e.g. agriculture, health, forests, Non Government Organizations etc. The social engineering activity in rodent control initiated by All India Co-ordinated Research Project on Rodent Control has proved highly effective because farmers' fields are the work sites of the scientists. State Governments and NGOs may also take up such programmes in a farmer participatory mode to popularize the rodent management technologies. Besides research, sincere efforts are required to effectively transfer rodent management technologies to the farmers.

RECOMMENDATIONS

Considering the havoc caused by rodents to agriculture, very limited options are available to the farmer to contain them. It is, therefore advisable to integrate the available options like cultural control methods with poison baiting and trapping in an integrated way for long lasting solutions to rodent problem. Timing of control and placement of poison baits and traps play significant role in yielding desired results. Jain *et al.* (1993) advocated large scale rodenticidal baiting during summer months when most of the rodent species maintain lower population levels and the acceptance of poison bait by pest rodents is greater. Additionally the farming community is also relatively free to take up rodent control on a cooperative basis. Similarly second operation should be undertaken before sowing of *rabi* crops. Thus two control operations integrating acute and chronic rodenticides in a sequence before sowing and harvest of the crops is recommended. Social engineering activity in rodent control initiated under the AICRP on Rodent Control has recorded more than 70% success with such a schedule (Mathur, 1992). Two methods of bait placement are recommended for fields i.e. burrow baiting and use of bait stations. The former method provides better opportunity to access fresh baits by target rodents and protects the non-targets from its exposure. Similarly, bait station made of locally available materials viz.; bamboo or PVC pipes have also proved highly effective. The bait stations should be placed on the runways, aligning vertical surfaces like bunds for better success. Needless to mention, the common propriety resources like Government lands viz., rail/road sides, *panchayat* lands, forests etc. are the real breeding shelters of rodents and act as focal centres for their migration towards crop fields. Rodent control work, especially during summer months must be taken up by farmers in such areas also.

CONCLUSIONS

Indian Arid zone with 18 species of rodents presents a rich faunal biodiversity. These species occupy almost all habitats in the desert biome. About half a dozen species inhabiting the crop fields, sandy plains, inter-dunal spaces near human settlements and hamlets etc. are the major pest problems infesting arid land cropping systems. The damage caused by these pest species to various crops of the arid region is at times quite alarming and require regular intervention by farming community to manage their menace. Though enough information is available on eco-biology and behaviour of desert rodents, their management still depends largely on use of toxic rodenticides. However, judicious use of rodenticide baits integrated with cultural and mechanical control techniques on a regular basis (as a part of package and practices) is the need of the hour in defeating the challenges posed by desert rodents for enhanced productivity of arid lands. For this, farming community needs to be made aware of rodent pest management technology. Various extension agencies of the government and non-governmental organizations have a greater role to play in popularizing these technologies at grass root level.

REFERENCES

ADVANI, R. & R.P. MATHUR. 1982. Experimental reduction of rodent damage to vegetable crops in Indian villages. *Agroecosystems* 8: 39-45.

ADVANI, R., I. PRAKASH & R.P. MATHUR. 1982. Assessment of rodent damage and yield losses in standing wheat crop in a desert village complex. *Z. Angew. Zool.* 99: 257-266.

BARNETT, S.A & I. PRAKASH. 1975. RODENTS OF ECONOMIC IMPORTANCE IN INDIA. Arnold Heinemann. New Delhi and London. 175 p.

CHAUDHARY, V. 2002. Bioefficacy of Difethialone against *Tatera indica* Hardwicke and *Rattus rattus* Linneans in Arid Ecosystem. Ph.D Thesis, submitted to Jai Narain Vyas University, Jodhpur, India. 222 p.

CHAUDHARY, V & R.S. TRIPATHI. 2002. Evaluation of toxicity and acceptability of difethialone treated baits in Indian desert gerbil, *Meriones hurrianae* Jerdon. *Curr. Agric.* 26 (1-2): 31-34.

IDRIS, M., R.S. TRIPATHI., V. CHAUDHARY & B.D. RANA. 2004. Incidence of Indian field mouse, *Mus booduga* Gray in Thar desert. In: HUMAN IMPACT ON DESERT ENVIRONMENT (Eds.): Pratap Narain, S. Kathju, Amal Kar, M.P. Singh & P. Kaur. Scientific Publishers, Jodhpur, India. pp. 343-347.

JAIN, A.P. 1986. *Evaluation of second generation anticoagulant rodenticides in India. I. Bromadiolone*. Technical Report, AICRP on Rodent Control, Central Arid Zone Research Institute, Jodhpur, India. 17 p.

JAIN, A.P & R.S. TRIPATHI. 1988. *Evaluation of second generation anticoagulant rodenticides in India.* II. Brodifacoum. Technical Report, AICRP on Rodent Control, Central Arid Zone Research Institute, Jodhpur, India. 21 p.

JAIN, A.P & R.S. TRIPATHI. 2000. An Integrated approach to rodent pest management. In: IPM SYSTEM IN AGRICULTURE. Vol. 7 (Eds.): R.K. Upadhyay, K.G. Mukherjee & O.P. Dubey. Aditya Books Pvt. Ltd., New Delhi, India. pp. 275-307.

JAIN, A.P., M. IDRIS & Y.R. MEENA. 1994. Rodent damage to ber (*Ziziphus mauritiana*) nursery. *Rod. Newsl.* 18 (3): 13-14.

JAIN, A.P., M. MATHUR & R.S. TRIPATHI. 1992a. Bio-efficacy of flocoumafen against major desert rodent pests. *Indian J. Pl. Prot*. 20: 81-85.

JAIN, A.P., M. MATHUR & R.S. TRIPATHI. 1992b. Traps and Trapping. In: RODENTS IN INDIAN AGRICULTURE (Eds.): I. Prakash & P.K. Ghosh. Scientific Publishers, Jodhpur, India. Pp. 585-894.

JAIN,A.P., R.S. TRIPATHI & B.D. RANA. 1993. R*odent management: The state of art*. Central Arid Zone Research Institute, Jodhpur, India. 138p (Mimeo)

JAIN, A.P., R.S. TRIPATHI & B.D. RANA. 1996. Impact of ecological transformation on the rodent fauna of the Thar desert. In: FAUNAL DIVERSITY IN THAR DESERT (Eds.): P.K. Ghosh, Q.H. Baqri & I. Prakash. Scientific Publishers, Jodhpur, India. pp. 389-402.

KRISHNAN, A. 1968. Distribution of arid areas in India. *Proc. XXI International Geographical Congress: Symposium on Arid Zone*. Jodhpur, India. pp. 11-19.

MATHUR, R.P. 1981. Study of comparative efficacy of anticoagulant rodenticide against two gerbils, *Tatera indica indica* Hardwicke and *Meriones hurrianae* Jerdon. Ph.D Thesis, University of Agra, India, 164 p.

MATHUR, R.P. 1992. Social Engineering Activity in Rodent Control. In: RODENTS IN INDIAN AGRICULTURE, Vol. 1 (Eds.): I. Prakash & P.K. Ghosh. Scientific Publishers, Jodhpur, India. pp. 605-650.

MATHUR, R.P & I. PRAKASH. 1984. Comparative efficacy of some anticoagulant rodentidcides against Indian rodents. *Proc. Conf. Organ. Practice Vertebrate Pest Control.* (Eds.): Dubock, A.C. El-Vetham HALL Hampshire, England ICI-PPD. pp. 381-393.

MATHUR, R.P., N. KASHYAP & M. MATHUR. 1992. Bioefficacy of anticoagulant rodenticides. In: RODENTS IN INDIAN AGRICULTURE Vol.1 (Eds.): I. Prakash & P.K. Ghosh. Scientific Publishers, Jodhpur, India. pp. 517-584.

PATEL. N., A.P. JAIN., R.S. TRIPATHI., F. PARVEEN & M. MATHUR. 1995. Losses caused to pomegranate fruits by rodents in arid horticulture. *Geobios New Reports* 14: 37-38.

PRAKASH, I. 1975. The population ecology of the rodents of the Rajasthan desert, India. In: RODENTS IN DESERT ENVIRONMENT (Eds.): I. Prakash & P.K. Ghosh. Dr. W. Junk b.v. Publishers, The Hague. pp. 75-116.

PRAKASH, I. 1976. *Rodent pest management– Principles and Practices.* CAZRI Monograph No. 4. Central Arid Zone Research Institute, Jodhpur, India. 28 p.

PRAKASH, I. 1991. Rodent communities, Epidemeological and Economic Considerations. In: PROSPECT OF INDIRA GANDHI CANAL PROJECT. (Eds.): I.P. Abrol & J. Venkateshwarlu, Indian Council of Agricultural Research, New Delhi, India. pp. 110-115.

PRAKASH, I. & R.P. MATHUR. 1987. MANAGEMENT OF RODENT PESTS. Indian Council of Agricultural Research, New Delhi, India.133 p.

PRAKASH, I & R.P.MATHUR. 1992. Acute rodenticides. In: RODENTS IN INDIAN AGRICULTURE, VOL. I (Eds): I. Prakash & P.K. Ghosh. Scientific Publishers, Jodhpur, India. pp. 497-516.

PRAKASH, I., R.K. GUPTA., A.P. JAIN., B.D. RANA & B.K. DUTTA. 1971. Ecological evaluation of rodent population in desert biome of Rajasthan. *Mammalia* 35: 384-423.

RANA, B.D. 1982. Relative efficacy of two small mammal traps. *Acta oecologia*. 32: 149-153.

RANA, B.D & R.S. TRIPATHI. 1999. *Recent trends in coordinated research on rodent control*. Central Arid Zone Research Institute, Jodhpur, India. 48 p.

RANA, B.D., A.P. JAIN & R.S. TRIPATHI. 1994. *Fifteen years of coordinated research on rodent control.* AICRP on Rodent Control, Central Arid Zone Research Institute, Jodhpur, India. 141 p.

RANA, B.D., R.S. TRIPATHI & B.K. SONI. 1992. Effect of irrigation agriculture on the desert rodents. *Mammalia* 56 (2): 231-235.

RAO, A.S & R.S. SINGH. 1998. Climatic feature and crop production. In: FIFTY YEARS OF ARID ZONE RESEARCH IN INDIA (Eds.): A.S. Faroda & M. Singh. Central Arid Zone Research Institute, Jodhpur, India. pp. 17-18.

RAO, A.M.K.M & I. PRAKASH. 1981. Rodenticidal evaluation of zinc phosphide against house mouse, *Mus musculus bactrianus. Indian J. Pl. Prot.* 13: 61-63.

SINGH, Y.V & A. SAXENA. 1998. Management of limited water resources for crop production. In: FIFTY YEARS OF ARID ZONE RESEARCH IN INDIA (Eds.): A.S. Faroda & M. Singh. Central Arid Zone Research Institute, Jodhpur, India. pp. 189-213.

SONI, B.K., B.D. RANA & A.P. JAIN. 1985. Field trials of some rodenticides (Aluminium phosphide, Brodifacoum and Bromadiolone) to control rodents in various crop fields of western Rajasthan. *Pestology* 9 (1): 25-27.

TRIPATHI, R.S & A.P. JAIN. 1990. Rodent damage to desert afforestation plantations. *Rod. Newsl.* 14: 7-8.

TRIPATHI, R.S., A.P. JAIN., N. KASHYAP., B.D. RANA & I. PRAKASH 1992. North-western Desert: Special Problem Areas. In: RODENTS IN INDIAN AGRICULTURE, Vol. I (Eds.): I. Prakash. & P.K. Ghosh. Scientific Publishers, Jodhpur, India. pp. 357-396.

TRIPATHI, R.S, M. MATHUR., A.P. JAIN & N. PATEL. 1994. Relative efficacy of glue and other traps for commensal rodent management. *Ann. Arid Zone* 33 : 143-45.

9

REPRODUCTIVE MANAGEMENT OF RODENT PESTS

R.K. Parshad

Department of Zoology and Fisheries, Punjab Agricultural University Ludhiana 141 004, INDIA

ABSTRACT

The available research findings on induction of sterility in rodent pests are discussed broadly under four approaches. The primary effort is to establish the reproductive biology and breeding patterns of target species, an area in which there is considerable output for Indian rodent pest species. A second approach involves employing pheromones in two basic ways to manage rodents-namely using urine based pheromones which are sex attractants to lure rodents to traps and baits. Additionally odours originating from urine, faeces and other body secretions of predators and their synthetic analogues to interrupt reproduction at various stages are evaluated. Their role in deterring pests form attacking crops is also critically dealt. This work is awaiting identification of chemical component responsible for this effect and its commercialization. In the third line, research on several chemicals and steroid compounds to induce sterility at different stages of reproduction is assessed. But the major problem in this approach is effective delivery systems, acceptability and palatability of chemosterilants. In the fourth approach research on immunological interference with proteins on the spermatozoa responsible for fertilization and adhesion of spermatozoa to the outer wall of zona pellucida (ZP_3), using cyclomegalovirus is briefly reviewed.

INTRODUCTION

Recent developments in agriculture and urbanization, and increasing deforestation and pollution of the environment have greatly influenced the ecology and wildlife in India. The manifestations of these ecological changes have threatened the existence of important wild mammals on one hand and on the other hand have resulted in sporadic increases in rodent pests. The main reasons for the rise in rodent populations are their higher fecundity rate, adaptability, omnivorous food habits and ability to invade all types of habitats. There are 1750 species of rodents occurring in different zoogeographic regions of the world which include beavers, chipmunks, guinea pigs, hamsters, mice, porcupines, rats, squirrels, gerbils, voles, etc. The largest family is muridae (rats and mice) which contains about 1,011 species. The rodent fauna of Indian subcontinent is represented by 46 genera and 126 species (Roonwal, 1987). Of these, 18 species are pests in agriculture, storage facilities both in rural and urban areas, in villages, towns and cities. These opportunistic species cause 2 to 15% damage to agricultural crops and occasionally 25 to 100% damage during conditions of rodent outbreaks (Parshad, 1999). To combat these losses in agriculture as well as in public health, research into different aspects of biology and management of rodents received major thrust by Indian scientists and research organizations in sixties. Rodent management strategies in the last two decades were mainly orientated towards use of physical and chemical (rodenticides) methods developed through in depth studies of rodent ecology and behaviour in the changing agro-climatic conditions of the country. Of these, rodenticides are used predominantly but a very wide range of other techniques including trapping, shooting, habitat management, use of chemical repellents and physical barriers are also suggested (Parshad, 1999).

Rodents are prolific breeders and increase their populations at a fast rate generally showing typical sigmoid curve. In such populations destroying animals in great numbers with poisons or other methods is of little value unless there is some means of preventing or slowing down recovery. Therefore, for developing effective technology for long-term effects it is essential to focus on methods which regulate fertility of rodent pests. An integrated pest management strategy when it includes fertility control becomes more effective. If reproduction can be suppressed economically, this procedure becomes far superior to attempts to regulate population densities by increasing mortality. Unfortunately, so far, not a single successful method of regulation of reproductive potential of rodents is available any where in the word. However data generated during the last few decades on the reproductive biology of rodents and evaluations of antifertility agents may be of considerable value to develop suitable technology for fertility regulation.

REPRODUCTIVE STRATEGIES OF RODENTS

The development and application of technology to regulate population of a species emanates from the systematic recording of its breeding biology, but very few Indian rodent species have received attention in this regard. The information generated so far is too fragmentary. Studies carried on the breeding biology till late eighties have been reviewed earlier (Parshad *et al.*, 1989). After this several reports on the breeding pattern of wild rodents appeared from time to time but cohered effort to generate information that can form the basis of the proposed reproductive management is still lacking.

Variations have been recorded in the breeding seasons of some of the rodent species from the different regions of the country particularly Rajasthan (Jain, 1970; Prakash *et al.*, 1971; Rana *et al.*, 1983; Rana & Prakash, 1984), Punjab (Kaur & Guraya, 1983; Parshad *et al.*, 1989; Parshad & Batth, 1999; Parshad & Kaur, 2001, 2003), Andhra Pradesh (Jain, 1979), Madhya Pradesh (Patel & Dubey, 1995), Karnataka (Chandrahas & Krishnaswami, 1974; Raj & Srihari, 1991) and North-Eastern states (Chauhan, 2003) and these appeared to be correlated with the cropping pattern of that particular area.

Most of the rodent species breed throughout the year with a peak corresponding to crop maturity or harvesting when plentiful amounts of energy-rich food become available in the fields. At the same time these changes are closely linked to the seasonal changes in light, temperature and precipitation, and thus are fairly predictable. Such of the rodent species that can take advantage of the temporary abundance of food and somehow overcome periods when conditions are not optimal establish themselves and cause extensive losses to crops throughout the year. However, in some geographic zones with adverse climatic conditions, may be at higher attitudes, breeding activity is restricted to a particular favourable period of the year. A peculiar phenomenon observed in the North-Eastern hilly area of India is the periodic mass-flowering and seeding of certain bamboo species which is accompanied by a tremendous increase in the population of *Nesokia, Mus, Rattus, Golunda* and *Rhizomys* species (Chauhan, 2003).

In the seasonally breeding species, seasonal reproductive activity is exhibited by both the sexes. During the breeding months, majority of the males of *Rattus meltada* had distinct scrotal sacs, active spermatogenesis in semi inferous tubules and epididymal smears showed the presence of active motile spermatozoa (Raj & Srihari, 1991). In *Funambulus palmarum*, the spermatogenic activity in adult did not become quiescent during any part of the year (Patil & Saidapur, 1991).

Unlike *R. meltada* (Dechamma & Dutt, 1980) and *M. platythrix* (Suresh & Dutt, 1986), the epididymides of *R.r. wroughtoni* contain active spermatozoa throughout the year (Bhatt *et al.*, 1987).

The major factor that determines the size of breeding population is the rate at which young females become sexually mature and experience the first pregnancy. The studies on folliculogenesis and follicular development involving recruitment and selection of follicles for ovulation in cyclic females are limited to *R. meltada* (Guraya & Gupta, 1979; Parshad *et al.*, 1989), *R. rattus* (Parshad & Batth, 1999; Batth & Parshad, 1996, 2000a, 2000b), *Bandicota bengalensis* (Kaur & Guraya, 1983) and *Meriones hurrianae* (Saidapur & Kamath, 1993). Rodents have short oestrous cycle length which ranges from 4 to 7.4 days in most of the species (Nigam, 1981; Parshad *et al.*, 1989; Raj & Srihari, 1991; Saidapur & Kamath, 1993). Litter size varied from 6-14 (Parshad *et al.*, 1989; Parshad & Kaur, 2001) and estimated annual production per adult female ranged from 10.6 to 53.4. The mean number of embryos implanted in right and left uterine horns of *R. meltada* is observed to be 3.4 and 4.5 respectively during September and October (Patel & Dubey, 1995). The number of corpora lutea corresponded to the visible number of embryos in the uterine horns (Raj & Srihari, 1991). Gestation period varied from 12 to 47 days (Parshad *et al.*, 1989). More information on the species-specific physiological reproductive events would help to identify the sites where the interference can be of use in regulating the pest population by causing irreversible sterility.

DISRUPTION OF REPRODUCTIVE BEHAVIOUR THROUGH PHEROMONES/ODOURS

Chemical communication through pheromones is common among rodents and may affect the development, reproduction and behaviour of the recipient individuals. Failure of implantation in newly inseminated females after exposure to strange males is well documented in laboratory mice (Bruce, 1960). The block to pregnancy is estimated to be about 40% if the second male belongs to the same strain as that of stud male and over 80% if the second male belongs to a different strain. Not much success has been achieved in field and commensal rodents using rodent pheromones to manage their numbers. However, pheromones may be used to increase the acceptance of rodent bait and to make bait stations more attractive to rodents. The operative pheromone present in the urine of males (Dominic, 1966) and accessory olfactory system (Rajendren & Dominic, 1984) is suggested to be an androgenic metabolite (Rajendren & Dominic, 1988). Little is known about the genetic control of these odours and rodent responses to them but it is well recognized that this type of behavioural genetics has great potential in rodent control.

The majority of the studies on reproductive inhibition have focused on intraspecific influence of semiochemicals and how they influence reproductive output and behaviour of females. Predator odours may also directly affect the reproductive physiology and behaviour of rodents. Oestrous cycles were extended in Norway rats when exposed to mink anal sac secretions (Voznessenskaya *et al.*, 1992). Similarly, the duration of oestrous cycles was extended in bank voles (*Clethrionomys glareolus*) exposed to weasel (*Mustela nivalis*) odours (Koskela *et al.*, 1996). Recently it has been shown that the female rats exposed to cat urine during pregnancy had reduced litter sizes at parturition (Voznessenkaya *et al.*, 2003). Reduction in litter size was correlated with resorption of embryos and decline in plasma progesterone when exposed to predator odours (Naidenko *et al.*, 2003). Efforts are being made to encapsulate active ingredient from the predator urine and to use it for reducing reproductive output. The method plans to utilize naturally derived compounds that pose no environmental hazard while achieving the goal of reducing rodent population.

CHEMOSTERILANTS AS REPRODUCTIVE INHIBITORS

Chemosterilants are agents that interfere with the physiological events related to reproduction. Interference may be at the level of gamete formation, gamete interaction, prevention of implantation and removal/ manipulation of the implanted embryos. Though well exploited from human birth control perspective, these processes are still in the experimental stage and are not available for routine rodent control. Because of the great potential of chemical inhibition of reproduction in regulating rodent population, a number of chemicals have been evaluated against some species of Indian rodents (Table 1).

An ideal chemosterilant should possess some degree of specificity to avoid secondary effects on the agro-ecological system, should be effective orally, quick in action and should affect both the sexes particularly the females as the male sterilization in polygynous species will not produce the same inhibiting effect on the number of offspring produced as females. None of the chemicals tested so far against rodents meets all these requirements and there is a need for more efforts in this direction. It is expected that one day there may be many chemosterilants to use just as we have a variety of rodenticides today.

Several steroidal compounds have been proved to be effective chemosterilants but only a few of them have been evaluated for the

Indian rodents. According to Sasidharan (1981), combinations of testosterone propionate and estradiol benzoate reduced the sperm concentration and also the weights of vas deferens and prostate gland in *Tatera indica*. Dehydroepiandrosterene administered to *R. rattus* interfered with the fertility of females by inducing the formation of ovarian cysts (Parshad & Batth, 1999) but the requirement of repeated dosing and method of application limits its use in rodent control. Synthetic oestrogen SC20775 showed promise as an effective chemosterilant for *R. norvegicus* and *R. exulans,* but palatability was low and the chemical was not cost effective (Garrison & Johns, 1975).

Gestation period being short in rodents any chemical treatment before birth can affect the growth and development of the offspring.

Table 1. Reproductive inhibitors tested against Indian rodents.

Chemical	Dose	Treatment duration	Species	Effects	Reference
Testos-terone pro-pionate (TP)	90μg/animal/ day	10 days	*T. indica*	No effect on genital organs, Reduced sperm concentr-ation in vas deferens	Sasidharan, 1981
Estradiol benzoate (OB)	0.015μg/ animal/day	10 days	*T. indica*	No effect on genital organs, Reduced sperm concentr-ation in vas deferens	Sasidharan, 1981
TP+OB	90μgTP+ 0.015μg OB/ animal/day	10 days	*T. indica*	Reduced weights of testis, epididymis, vas deferens and prostate, Reduced sperm concentration	Sasidharan, 1981
Glyzo-phrol	40 mg/kg b.w.	Two months of baiting	*R. rattus*	Damaged spermatogenic cells in seminiferous tubules, Degeneration of ovum	Arora *et al*, 1978
Alpha-chloro-hydrin	82 mg/kg	Single feeding	*B.bengal-ensis*	Dysfunction of testis, lesions in epididymis, low sperm concentration	Saini & Parshad, 1991, 1993
Ethyl methane sulphonate	500 and 625 mg/kg b.w.	Single dose	*R. rattus*	Affected differentiation, structure and function of spermatozoa	Kaur & Parshad, 1997
Sodium selenite	1.0, 1.5, 2.0 mg/g of food	From 9th to 12th day of pregnancy	*B.benga-lensis*	Terminated pregnancy and affected germ cell population if survived	Parshad & Kaur, 2003

Neonatal interference with the sexual development in the offspring may cause irreversible sterility or mortality, which would suppress populations. Recent laboratory studies on *B. bengalensis*, the most predominant pest species of India, have shown that exposure of pregnant females to food containing variable doses of sodium selenite (0.5 to 2.0mg/g of food) from day 9 to day 12 of gestation terminated pregnancy in all females except at the lowest dose where the gonads of the surviving neonates had reduced population of germ cells. Selenium appears to exert an anti fertility effect through the loss of pregnancies and by reducing the reproductive potential of the surviving offspring (Parshad & Kaur, 2003). However, more extensive studies are required about the effect of selenium on subsequent generations and its applicability under the field conditions. Colchicine (an alkaloid from a plant of Liliacae family) has been evaluated as a chemosterilant in *B. bengalensis* but because of its poor acceptance in the diet it was not explored further for rodent population management (Srivastava, 1966). Alpha-chlorohydrin, which is widely marketed and is known for its toxicity to rodents at high doses, also exerted antifertility effect at low doses (Parshad, 1999). Most of the male survivors which had ingested sub lethal doses of alpha-chlorohydrin became sterile permanently both in the laboratory and in the fields (Saini & Parshad, 1991, 1993). The chemical is considered as the most effective toxicant-sterilant. Still, it has not made a major impact on rodent control. Alkylating agents seem to offer great possibility of sterilizing male rats. Ethyl methane sulphonate affected the differentiation, structure and function of spermatozoa of *R. rattus* (Kaur & Parshad, 1997). Glyzophrol, a commercial rodenticide is speculated to contain an alkylating agent, tetramine which damaged the spermatogenic cells in the seminiferous tubules in male and resulted in degeneration of ova in the female *R. rattus* (Arora *et al.*, 1978). The potential of the alkylating agents for rodent control needs to be explored further.

There are some disadvantages linked to the use of reproductive suppressive methods. Under some situations like houses, food handling and processing units etc. it is not possible to tolerate even a single rat. In these situations, the use of chemosterilants will be nearly useless. On the other hand, the use of antifertility agents is highly desirable for rat control in sewers, agriculture and forestry.

IMMUNOLOGICAL INTERVENTION

This approach of fertility regulation is broadly based on the development of vaccine and its successful delivery. This has been successfully used through baits for the control of rabies in red fox in Europe (Artois *et al*, 1993). It has the potential to provide practical method for long-term management of wild and domestic species. An immunological vaccine aimed at preventing fertilization in a species is targeted at two sites; proteins important for fertilization on the surface of spermatozoa and those proteins involved in sperm adhesion to the outer coat of the egg, particularly zona pellucida (ZP3) (Prasad *et al.*, 1996).

In many species both sites have been studied and candidate antigens have been identified and assessed for their ability to induce contraception (Hinds *et al.*, 2003). Initial studies using ectromelia virus expressing the gene for the mouse egg coat protein ZP_3 provided laboratory evidence that a delivered immunocontraceptive vaccine for the mouse caused sustained infertility (Jackson *et al.*, 1998). Several laboratory and field studies in Australia using mouse cytomegalovirus expressing mouse ZP_3, clearly confirmed the potency of virally vectored immunocontraception (Hinds *et al.*, 2003). The zona pellucida induced infertility can be the result of antibody-mediated blocking of ZP sperm receptor site and the cytotoxic T cell mediated destruction of developing ovarian follicles. This later mode of interference is being considered as highly desirable for inducing permanent sterility in problem animal populations. Additional advantage of this approach lies in the fact that zona-based vaccines are based on selective epitopes chosen in order to manipulate selectively the immune response for particular target species. Therefore, a key challenge is to identify or engineer the reproductive antigen to be species-specific.

Application of fertility control through immunological intervention for management of animal populations is still a relatively new concept. Though basic scientific knowledge to develop immunocentraceptive technology stemmed from the studies on small laboratory rodents, their potential to regulate rodent pest populations has not yet been realized to the desired extent which requires an integrated multidisciplinary approach.

As mentioned earlier in this chapter studies are underway on many of the delivery systems of vaccine utilized recombinant organisms and to use them for pest management. Consideration on the use of such organisms needs to take into account the safety constrains prior to any environmental release.

BIOLOGICAL STERILANTS

This refers to the use of organism to induce sterility of the pest species. Hepatic nematode, *Capillaria hepatica* in house mice has been found to reduce the number of litters produced and the young weaned over three months (Singleton & Readhead, 1991). Singla *et al.,* (2003) evaluated *Trypanosoma evansi* as a biocide against *B. bengalensis* and *R. rattus.* Inoculation of *T. evansi* to pregnant bandicoot rat caused mummification of foetuses within the uterus.

CONCLUSIONS

High reproductive potential of rodent pest species is the fundamental cause of their large populations and consequent severe crop damages in India and other tropical and sub-tropical countries. Fertility control techniques have a number of advantages over lethal control for pest species of high fecundity and adaptability in diverse habitats. At the same time they can also be considered as a more humane strategy of management. A number of techniques involving the manipulation of reproductive behaviour and immunological responses of animals and use of reproductive inhibitors have been developed and tested under laboratory and field conditions but none of these has so far been used for reducing rodent populations because of the associated problems dealing with the application, acceptability and efficacy except for the limited use of alpha-chlorohydrin in few countries. Immunocontraception holds promise for field efficacy. However, this technique has not been tested against any Indian pest species of rodents. Development and application of reproductive management techniques require a thorough study of the behavioural, ecological and physiological aspects of each rodent pest species. Though some attempts have been made with regard to few Indian rodent species like *Rattus rattus, Bandicota bengalensis* and *Tatera indica*, multidisciplinary approach is required for development of effective fertility control methods for different species.

REFERENCES

ARORA, K.K., B.K.VARMA, R.B. DOHAREY. & S.S. LAL.1978. Laboratory evaluation of glyzophrol – a chemosterilant. *Rod. Newsl.* 2: 4-5.

ARTOIS, M., E. MASON., J. BARRAT & M.AUBERT. 1993. Efficacy of three oral rabies vaccine baits in the red fox: A comparison. *Vet. Microbiol.* 38: 167-172.

BATTH, B.K. & R.K. PARSHAD. 1996. Quantitative assessment of changes and maturation of mast cells in the ovary of prenatal and adult rats, *Rattus rattus. Indian J. exp. Biol.* 34: 692-694.

BATTH, B.K & R.K. PARSHAD. 2000a. Ovarian cholinergic innervation of immature, cyclic, pregnant and lactating house rat (*Rattus rattus*). *Indian J. Anim. Sci.* 70: 591-593.

BATTH, B.K & R.K. PARSHAD. 2000b. Mast cell dynamics in the house rat (*Rattus rattus*) ovary during oestrus cycle, pregnancy and lactation. *European J Morph*. 38: 17-23.

BHATT, S.K., A. SUJATHA., R. ADVANI & A.S. SUKUMARAN., 1987. Population structure and breeding season in *Rattus rattus* Wroughton Hinton. *Proc. Indian Acad. Sci.* (Anim. Sci.) 96: 657-665.

BRUCE, H.M. 1960. A block to pregnancy in mice caused by proximity of strange males. *J. Reprod. Fert.* 1: 96-103.

CHANDRAHAS, R.K. & A.K. KRISHNASWAMI. 1974. Studies on the ecology of the Indian gerbil *Tatera indica hardwickei* (Gray) in Kolar (Mysore state). *Indian J. Med. Res.* 62: 971-978.

CHAUHAN, N.P.S. 2003. Observations of bamboo flowering and associated increases in rodent population in the north-eastern region of India. In: RATS, MICE AND PEOPLE: RODENT BIOLOGY AND MANAGEMENT (Eds.) Singleton G.R., L.A. Hinds, C.J. Krebs & D.M. Sprat. Australian Centre for International Agricultural Research, Canberra, Australia. pp. 267-270.

DECHAMMA, P.A & N.H. GOPAL DUTT. 1980. The testicular and epididymal changes in the soft-furred field rat, *Millardia meltada* in relation to some environmental factors. *Proc. Indian Acad. Sci.* (Anim. Sci.) 89: 303-310.

DOMINIC, C.J 1966. Observations on the reproductive pheromones of mice; I. - Source. *J. Reprod. Fert.* 11: 407-414.

GARRISON, M.V & B.E .JOHNS. 1975. Antifertility effects of SC-20778 in Norway and Polynesian rats. *J. Wildl. Manage.* 39: 26-29.

GURAYA, S.S & S.K. GUPTA. 1979. Morphology and histochemistry of ovarian changes in the field rat, *Millardia meltada. Z. ikrosk anat Forsch. Leipzig* 93: 959-973.

HINDS, L.A., C.M. HARDY., M.A. LAWSON. & G.R. SINGLETON. 2003. Developments in fertility control for pest animal management. In RATS, MICE AND PEOPLE: RODENT BIOLOGY AND MANAGEMENT. (Eds.): Singleton G.R., L.A. Hinds., C.J. Krebs & D.M. Sprat. Australian Centre for International Agricultural Research, Canberra, Australia. pp. 31-36

JACKSON, R.J., D.L. MAGUIRE., L.A. HINDS & I.A. RAMSHAW. 1998. Infertility in mice induced by a recombinant ectromelia virus expressing mouse zona pellucida glycoprotein 3. *Biol. Reprod.* 58: 152-159.

JAIN, A.P. 1970. Body weights, sex ratio, age structure and some aspects of reproduction in the Indian gerbil, *Tatera indica indica* Hardwicke, in the Rajasthan desert, India. *Mammalia* 34: 415-432.

JAIN, A.P. 1979. A note on the breeding season and litter-size in the rodents of Andhra Pradesh. *Rod. Newsl.* 3: 14.

KAUR, P & S.S. GURAYA. 1983. Body weight, sex ratio, seasonal reproductive changes in the Indian mole rat, *Bandicota bengalensis* in the Punjab. *Aust. J. Zool.* 31: 123-130.

KAUR, R & V.R. PARSHAD. 1997. Ethyl methane sulphonate induced changes in the differentiation, structure and functions of spermatozoa in house rat, *Rattus rattus. J. Bio. Sci.* 22: 357-365.

KOSKELA, E., T. HORNE, T. MAPPES & H. YLONEN. 1996. Does risk of small mustelid predation affect the oestrous cycle in the bank vole, *Clethrionomys glareolus? Animal Behav.* 51: 1159-1163.

NAIDENKO, S.V., S.V. NAIDENKO., L. CLARKE & V.V. VOZNESSENS-KAYA. 2003. Predator presence affects the reproductive success of prey in outdoor conditions. In: RATS, MICE AND PEOPLE: RODENT BIOLOGY AND MANAGEMENT. (Eds.) Singleton G.R., L.A. Hinds., C.J. Krebs & D.M. Sprat Australian Centre for International Agricultural Research, Canberra, Australia. pp: 148-150.

NIGAM, P.M. 1981. Mating behaviour, gestation period and breeding season of the lesser bandicoot, *Bandicota bengalensis* Gray. *Rod. Newsl.* 5: 20.

PARSHAD, V.R. 1992. State-of-the-art technology for preventing major crop and grain losses due to rodents–a scenario for the 21st Century. *Rod. Newsl.* 16: 6-11.

PARSHAD, V.R. 1999. Rodent Control in India. *Integ. Pest Mgmt. Rev.* 4: 97-126.

PARSHAD, R.K & B.K. BATTH. 1999. Cystogenesis of antral follicles induced by dehydroepiandrosterone (DHEA) stimulates mast cell proliferation and maturation in the house rat (*Rattus rattus*) ovary. *Indian J. exp. Biol.* 37: 933-935.

PARSHAD, R.K & R. KAUR. 2001. Breeding performance of *Bandicota bengalensis* under laboratory conditions. *Rod. Newsl.* 25: 10-11.

PARSHAD, R.K. & N. KAUR. 2003. Antifertility effects of sodium selenite in the Indian rat, *Bandicota bengalensis*. *Indian J. Anim. Sci.* 73: 903-906.

PARSHAD, V.R., P. KAUR & S.S. GURAYA. 1989. Reproductive cycles of mammals: Rodentia. In: REPRODUCTIVE CYCLES OF INDIAN VERTEBRATES (Eds.): S.K. SAIDAPUR. Allied Publishers Limited, New Delhi. pp. 347-408.

PATEL, R.K. & O.P. DUBEY. 1995. Reproduction biology and population fluctuation of *Millardia meltada* in Madhya Pradesh. *Rod. Newsl.* 19: 7.

PATIL, S.B & S.K. SAIDAPUR. 1991. Kinetics of spermatogenesis in the wild squirrel *Funabulus palmarum* (Linnaeus) *Acta. Anat.* 141: 352-363.

PRAKASH, I., A.P. JAIN & K.G. PUROHIT. 1971. A note on the breeding and postnatal development of the Indian gerbil. *Tatera indica indica* (Hardwicke, 1807) in Rajasthan desert. *Sauget. Mitt.* 19: 375-380.

PRASAD, S.V., B. WILKINS & B.S. DUNBAR. 1996. Molecular biology approaches to evaluate species variations in immunogenicity and antigenicity of zona pellucida proteins. *J. Reprod. Fert.* (Suppl.) 50: 143-149.

RAJ, G.G & K. SRIHARI. 1991. Breeding aspects of the soft-furred field rat, *Rattus meltada* in and around Bangalore, Karnataka. *Rod. Newsl.* 15: 10-11.

RAJENDREN, G. & C.J. DOMINIC. 1984. Role of vomeronasal organ in the male-induced implantation failure (the Bruce effect) in mice. *Archs. Biol.* 95**:** 1-9.

RAJENDREN, G. & C.J. DOMINIC. 1988. Effect of cyproterone acetate on the pregnancy blocking ability of male mice and the possible chemical nature of the pheromone. *J. Reprod. Fert.* 84**:** 387-392.

RANA, B.D., R. ADVANI & B.K. SONI. 1983. Reproductive biology of *Rattus rattus rufescens* in the Indian desert. *Zeit. angew. Zool.* 70**:** 207-216.

RANA, B.D & I. PRAKASH. 1984. Reproductive biology of soft-furred field rat, *Rattus meltada pallidior* (Ryley, 1914) in the Rajasthan desert. *J. Bombay nat. Hist. Soc.* 81**:** 59-70.

ROONWAL, M.L. 1987. RECORDS OF THE ZOOLOGICAL SURVEY OF INDIA: RECENT ADVANCES IN RODENTOLOGY IN INDIA. Zoological Survey of India. Miscellaneous Publication No.105. Calcutta, India. 126p.

SAIDAPUR, S.K & S.R. KAMATH. 1993. Morphometric study of ovarian follicular growth during the oestrous cycle of the laboratory maintained Indian desert gerbil, *Meriones hurrianae* (Jordon). *Acta. Anat.* 148**:** 189-196.

SAINI, M.S & V.R. PARSHAD. 1991. Control of the Indian mole rat with alpha-chlorohydrin : laboratory studies on bait acceptance and antifertility effects. *Ann. app. Biol.* 118**:** 239-247.

SAINI, M.S & V.R. PARSHAD. 1993. Field evaluation of alpha-chlorohydrin against Indian mole rat: Studies on toxic and antifertility effects. *Ann. appl. Biol.* 122**:** 153-160.

SASIDHARAN, T.O. 1981. Effect of steroid hormones on reproductive system of male Indian gerbil, *Tatera indica indica. Rod. Newsl.* 5**:** 1.

SINGLETON, C & T. REDHEAD. 1991. Future prospects of biological control of rodents using micro- and macro-parasites. In: RODENTS AND RICE (Ed.): G.R. Quick. International Rice Research Institute, Los Banos, Philippines. pp. 75-82.

SINGLA, N., V.R. PARSHAD & L.D. SINGLA. 2003. Potential of *Trypanosoma evansi* as a biocide of rodent pests. In**:** RATS, MICE AND PEOPLE: RODENT BIOLOGY AND MANAGEMENT. (Eds.) Singleton G.R., L.A. Hinds., C.J. Krebs & D.M. Sprat. Australian Centre for International Agricultural Research, Canberra, Australia. pp. 41-46

SRIVASTAVA, A.S. 1966. Evolution, development and standardization of technique for the eradication of the field rats. *Proc. Indian Rodent Symp,* Calcutta, India. pp.283-290.

SURESH, R & N.H. GOPAL DUTT. 1986. Seasonal changes in the annual reproductive cycle of the male wild mouse. *Folia Morphol.* 34**:** 230-237.

VOZNESSENSKAYA, V.V., C.J. WYSOCKI & E.P. ZINKEVICH. 1992. Regulation of rat oestrus cycle by predator odors: role of the vomeronasal organ. In: CHEMICAL SIGNALS IN VERTEBRATES (Eds.): Doty, R.L & D. Muller-Schwartze. Plenum Press, New York. pp. 281-284.

VOZNESSENSKAYA, V.V., S.V. NAIDENKO., N.Y. FEOKTISTOVA., G.J. KRIVOMAZOV., L.A. MILLER & L. CLARK. 2003. Predator odours as reproductive inhibitors for Norway rats. In: RATS, MICE AND PEOPLE: RODENT BIOLOGY AND MANAGEMENT (Eds.): Singleton G.R., L.A. Hinds., C.J. Krebs & D.M. Sprat. Australian Centre for International Agricultural Research, Canberra, Australia. pp. 131-135.

10

RODENT MANAGEMENT IN GROUNDNUT IN GUJARAT

P.G. Butani, H.J. Vyas and M.N. Kapadia

Department of Entomology, College of Agriculture, Junagadh Agricultural University, Junagadh – 362 001

ABSTRACT

The predominant rodent species attacking groundnut is *Bandicota bengalensis*. However, *Millardia meltada meltada, Tatera indica indica, Mus booduga* are also found damaging standing crops, while *Rattus rattus* and *Mus musculus* cause losses during storage. Estimation of yield reduction in groundnut due to rodents revealed 4-95% at seedling stage and 4-10% at seedling stage and the extent of loss in isolated fields was as much as 85.42 per cent. Zinc phosphide poison bait and bromadiolone wax cake were found effective against rodents generally. Bromadiolone application first at pod formation stage and next at pod maturity stage proved effective for management of rodents in groundnut crop. Alternatively three applications of rodenticides one each at germination, pod formation and pod maturity stages using 2% zinc phosphide at any one of the stages and 0.005% bromadiolone at the other two stages effectively controlled rodents in groundnut. Use of Zinc phosphide should be restricted to one time due to possibilities of development of bait shyness in surviving populations of rodents. As rodent management in barren land had great impact on the reduction of rodent activity with consequent increased yield of groundnut in adjacent fields, the farmers are advised to control rodents both in groundnut fields and fields surrounding it during germination, flowering and pod maturity stages.

INTRODUCTION

India is one of the largest importers of edible oil because groundnut, *Arachis hypogea* the predominant oil seed is grown as a rain fed crop under various biotic and abiotic stresses resulting in low yield in comparison to the world average (Rao, 1988). Besides moisture stress at critical stages, pests (insect and non-insect) and diseases have been recognized as major limiting factors for groundnut production in India. Although a good deal of research work has been done on insect pests of groundnut, very little information is available about the species composition, population dynamics and measures of rodent control in spite of the fact that the damage to groundnut crop by field rodents has been well established (Bindra & Sagar, 1968; Hopf *et al.,* 1976; Vyas *et al.,* 1985; Parshad *et al.,* 1987; Prakash & Mathur, 1987; Sridhara, 1999).

Fig. 1. *Bandicota bengalensis* burrow in groundnut

Rodent damage in groundnut occurs from the stage of sowing to harvest both in *kharif* as well as summer seasons (Table 1). Rodents inhabit bunds and are also found in barren/waste land around groundnut cultivation and in stored fodder for cattle use during off season. Soon after sowing of groundnut, rodents migrate to the sown area, dig out the seeds and consume them. This loss is comparatively more when there is less rain and the fields are in a dry condition after the sowing operation. If the rains continue for a long time after sowing, the seed/seedling damage by rodents decreases due to water logging of the field. Later rodents damage the groundnut crop by cutting the branches and remain in the burrows on the bunds. Farmers avoid the interculturing operations after peg formation in groundnut. This undisturbed condition of groundnut fields enhances the rodent activity and at the same time crop canopy also plays an important role as it provides shelter to them. Thus, the activity of field rodents increases after pod formation in groundnut and is found at a maximum level at pod maturity stage of the crop. Rodents dig burrows profusely inside the fields apart from bunds because of the undisturbed dense crop and also due to less rain during these periods. The population of field rodents and damage due to them also increases, when the unthreshed groundnut crop is kept in the field for drying after harvest. At this time rodents feed on the groundnut pods and/or store them in the burrows.

After the harvest of groundnut crop in the *kharif* season, rodents start feeding on wheat crop and later they migrate back to the summer groundnut fields. In isolated fields of groundnut crop in summer, rodent damage is severe as very few patches of cultivated crops are available.

RODENT SPECIES COMPOSITION AND RELATIVE ABUNDANCE

Table 1. Predominant rodent pests of groundnut in different States of India

State	Predominant rodent species	Source
Gujarat	*Bandicota bengalensis*	Anon (1989)
	Millardia meltada meltada	
	Tatera indica indica	
Karnataka	*Bandicota bengalensis*	Sridhara (1999)
	Tatera indica	
	Millardia meltada	
	Mus platythrix, Mus booduga	
Madhya	*Millardia meltada*	Anon (1984)
Pradesh	*Bandicota bengalensis*	Anon (1985)
	Funambulus pennanti	Anon (1986)
Punjab	*Mus musculus*	Sood & Ubi (1975)
	Mus booduga	Chopra & Sood (1983)
	Tatera indica	Gupta *et al.* (1983)
	Rattus meltada	Parshad *et al.* (1987)
	Bandicota bengalensis	

The species composition of rodents in groundnut varies from one habitat to another. The same species is often found in two different habitats but their relative abundance may vary in the two locations. The species that attack groundnut in India are *Bandicota bengalensis, Millardia meltada meltada, Tatera indica indica, Rattus rattus, Mus booduga* and squirrels in Tamil Nadu (Hopf *et al.,* 1976). Prakash & Mathur (1987) reported that *R. meltada, T. indica, Mus musculus* and *M. booduga* damaged groundnut. About 85% of the groundnut in the country is grown in the states of Gujarat, Andhra Pradesh, Tamil Nadu,

Karnataka and Maharashtra. The rodent species composition in groundnut fields of these states is listed in Table 1.

Soil types, seasons, irrigation facilities and other ecological conditions affect the relative abundance of rodent species and rodent associations (Prakash, 1975, 1981). Of the different murid species reported to be present in groundnut crop in Punjab, *M. meltada* was predominant followed by *M. musculus, M. booduga, T. indica* and *B. bengalensis* (Sood & Ubi, 1975). Later studies reported the composition of *Mus* sp> *B. bengalensis*>*T. indica* >*R. meltada* in irrigated groundnut fields while the composition of *Mus* sp.> *T. indica*> *R. meltada* characterized non-irrigated groundnut fields (Parshad *et al.*, 1987). *B. bengalensis* was found to be the predominant species followed by either *M. meltada* or *T. indica* in groundnut growing regions of Gujarat (Anon, 1990a).

Rodent species in different soil types of Gujarat is presented in Table 2. *B. bengalensis* was predominant in almost all types of soils in groundnut growing areas of Gujarat during *kharif* and summer seasons except in coastal alluvial medium black and shallow black soil in the *kharif* season and in mixed red and shallow black soil in the summer season. *M.m. meltada* and *T.i. indica* were found to be the next abundant species in the above habitats. The species composition was *B. bengalensis* followed by *T. indica* and *M. meltada* in shallow black+ medium black + coastal alluvial and shallow black type of soils and in medium black soil during summer. On the other hand *T.i. indica* was predominant in coastal alluvial + medium black + shallow black type of habitat during *kharif* season followed by *B. bengalensis* and *M. meltada.* In mixed red and shallow black soil, *M.m. meltada* was present relatively in large numbers compared to *B. bengalensis* and *T. indica* during summer.

Irrigation type influenced the relative abundance of rodents in Gujarat. *M.m. meltada* was more than *T.i. indica* in canal irrigated areas whereas the population of *T.i. indica* was greater than that of *M. m. meltada* in well irrigated fields during summer.

In Punjab rodent composition in the order of *M. meltada* > *M. musculus* > *M. booduga* > *T. indica* > *B. bengalensis* was present at the seedling stage of the groundnut crop (Sood & Ubi, 1975). Rodent infestation was quite high in the second week after sowing. The highest infestation was seen at the pod maturity stage and the species composition was similar to that of seedling stage.

Table 2. Relative abundance of field rodents associated with groundnut in different types of soil in groundnut growing area

Soil type	Per cent occurrence of rodent species			
	Bandicota bengalensis	***Millardia meltada meltada***	***Tatera indica indica***	
***Kharif* season**				
1. Medium black	58.2	24.5	17.3	Bb>Mm>Ti
2. Medium black + shallow black	68.5	20.4	11.1	Bb>Mm>Ti
3. Mixed red & black + medium black	52.2	26.4	21.1	Bb>Mm>Ti
4. Shallow black + medium black + coastal alluvial	54.9	17.9	27.2	Bb>Ti>Mm
5. Shallow black	40.2	25.3	34.5	Bb>Ti>Mm
6. Coastal alluvial + medium black + shallow black	40.0	11.3	48.7	Ti>Bb>Mm
Summer season				
1. Medium black	61.0	17.9	21.1	Bb>Ti>Mm
2. Medium black + hilly soil	61.5	23.3	15.2	Bb>Mm>Ti
3. Mixed red & black + medium black	63.6	31.8	4.6	Bb>Mm>Ti
4. Mixed red & black + shallow black	42.2	55.0	2.8	Bb>Mm>Ti

In Madhya Pradesh, the soft furred field rat, *M. meltada* was the predominant species in groundnut. The first peak was observed during the third week of August and the second peak at the beginning of maturity stage of the crop during *kharif* (Anon, 1984). In another study, *B. bengalensis* and *M. meltada* infestation appeared in groundnut fields soon after sowing. The population of *B. bengalensis* increased continuously during August and September reaching the peak in October when the crop was at podding and/or maturity stage. The maximum population of *M. meltada* was recorded in August which declined in September but continued till harvesting (Anon, 1985).

In Karnataka infestation was higher at pod formation and harvesting stages. *B. bengalensis* was the major species followed by *M. meltada* and *T. indica* (Sridhara, 1999).

In Gujarat *B. bengalensis* and *M. m. meltada* were more common from sowing to harvest stage of groundnut crop (Mittal & Vyas, 1992).

POPULATION DYNAMICS OF RODENTS IN GROUNDNUT FIELDS OF GUJARAT DURING 1991-94 (Table 3)

The rodent population density and diversity were not significant during germination, vegetation growth and flowering stages of groundnut, as estimated by trapped rodents and live burrow count. Animal numbers captured ranged 7 to 28.0 and live burrows 4 to 17.7 during *kharif*. There were only two species viz. *B. bengalensis* and *M. meltada* at germination, vegetative growth and flowering stages (Table 3). The rodent population started increasing from pod formation stage and reached maximum at harvest stage. Live burrow count also exhibited the same trend. Species infesting rose to three i.e. *B. bengalensis*, *M.m. meltada* and *T.i. indica* at the later three stages. A similar trend of density increase along with crop growth of groundnut was seen during summer also but the density was less compared to *kharif* as estimated by trapping and counting live burrows. Species composition did not change between summer and *kharif*. Sex ratio was biased towards males at all stages of groundnut growth during both the seasons (Table 3).

Table 3. Population dynamics of field rodents in groundnut during *kharif* and summer of in Gujarat (mean values of 1991, 92, 93 & 94)

Crop stage	Density				Species composition		Sex ratio	
	Rod./100 traps/ day		LBC/ha.					
	Kharif	Summer	*Kharif*	Summer	*Kharif*	Summer	*Kharif*	Summer
Sowing germination	10.0	1.3	4.7	1.0	Bb>Mm	Bb	64:36	83:17
Vegetative growth	7.8	3.16	4.0	2.2	Bb>Mm	Bb>Mm	69:41	67:33
Flowering	10.7	6.0	6.0	4.0	Bb>Mm	Bb>Mm	62:38	60:40
Pod formation	15.2	10.2	7.3	7.3	Bb>Mm>Ti	Bb>Mm>Ti	62:38	60:40
Pod maturity	22.8	16.0	13.2	11.0	Bb>Mm>Ti	Bb>Mm>Ti	62:38	58:42
Harvest	28.0	21.7	17.7	14.67	Bb>Mm>Ti	Bb>Mm>Ti	59:41	60:40

DAMAGE ESTIMATES (Table 4)

The losses due to rodents in groundnut at various stages are presented in Table 4.

Damage to groundnut seeds by the squirrel *Funambulus pennanti* was reported to be 2 to 3% during *kharif* 1984 and 5.65 to 25.51% during *kharif* 1985 in Madhya Pradesh. The damage was from the second day to the sixth day of sowing (Anon, 1985, 1986). In Punjab rodent caused damage varied from 30 to 76kg/ha, the mean being 50kg/ha (Bindra & Sagar, 1968). In the same state, a later study showed 3.86% loss in yield worth Rs. 190.18/ha due to complete and/or partial damage due to cutting and hoarding of pods (Parshad *et al.*, 1987).

Table 4. Production losses in groundnut due to rodents

State	Stage	Damage (%)	Source
Andhra Pradesh	Seedling	6.00 to 9.00	Jotwani & Beri, 1968
	Plant	18.20 to 25.80	
	Pod	4.10 to 7.60	
	Plant	14.30	Rajasekharan & Dharmaraju, 1975
	Pod	47.22	
Gujarat	Plant	3.00 to 10.24	Anon. 1989
	Pod	4.89 to 9.97	-do-
	Seedling	4.00 to 51.86	Anon. 1990b
	Plant	2.90 to 26.78	-do-
	Pod	8.23 to 14.64	-do-
	Seedling	4.37 to 95.3	Anon. 1991
	Plant	3.25 to 7.40	-do-
	Pod	4.99 to 8.53	-do-
Karnataka	-	2.00 to 5.00	Hopf *et al.*, 1976
	Seedling	30-40	Sridhara, 1999
	Peg formation	2	-do-
	Harvest stage	4.5	-do-
	Hoarding	3	-do-
Madhya Pradesh	Pod	4.62 to 28.57	Anon 1985
	Plant	8.78 to 17.72	-do-
	Pod	12.66	-do-
Punjab	-	25.80	Hopf *et al.*, 1976.
	Pod	7.10	Anon 1984b.
	Pod	3.86 to 18.97	Parshad *et al.*, 1987
Rajasthan	-	18.00	Anon. 1984a
Tamil Nadu	-	1.00 to 1.50	Hopf *et al.*, 1976.

(-) Unspecified.

In Karnataka, 30-40% seedlings were damaged. Further 2% damage was seen at peg formation stage and 4-5% damage at harvest stage. In addition 3% produce was hoarded inside burrows (Sridhara, 1999).

Apart from consuming, rodents hoard groundnut pods inside their burrows. Deoras (1966) recovered 6.6 kg of groundnut from 30 burrows of bandicoot rats in erstwhile Bombay. *B. bengalensis* hoarded 17.1g of groundnut seeds per day in laboratory experiments (Srihari *et al.*, 1979). In Gujarat, field studies recorded 320g/burrow hoarding by bandicoot rat (Patel & Nayak, 1987).

Rodent damage to groundnut ranged from 4.1 to 25.8% at national level (Prakash & Mathur, 1987).

Outbreaks of rodent population occurred in Gujarat during 1975-76 (Shah, 1979) and again in 1989-90 (Anon, 1990b). Rodent damage in field crops increased in intensity from *kharif* 1988 to *rabi* 1989-90. In an isolated field of groundnut surrounded by barren/fallow land the extent of loss was as much as 85.42% to during summer 1989. Because of heavy damage due to rodents in *rabi* crops of 1989-90, farmers avoided sowing the summer groundnut during 1990 in most of the groundnut growing areas of Saurashtra region.

RECOMMENDATIONS FOR RODENT MANAGEMENT IN GROUNDNUT

Both the acute rodenticide, zinc phosphide (2%) and the second generation anticoagulant, bromadiolone (0.005%) effectively controlled rodents in groundnut fields. Evaluation of control at different vulnerable stages of groundnut growth using these two rodenticides in different sequences revealed the following schedule of rodent control to procure optimum yields:

(a) Control at sowing and pod formation stage by baiting with 0.005% bromadiolone.

Or

(b) Control at germination and pod formation stage. At germination either 2% zinc phosphide or 0.005% bromadiolone poisoning followed by 0.005% bromadiolone application at pod formation stage.

Or

(c) Control at flowering and pod formation stage: Either zinc phosphide or bromadiolone poisoning at flowering stage followed by bromadiolone baiting at pod formation stage.

Or

(d) For effective and economic rodent control in groundnut crop three applications of rodenticides is suggested viz., at germination, pod formation and pod maturity stages. Either bromadiolone can be used at all the three stages or bromadiolone poisoning at germination and pod maturity stage with zinc phosphide application at pod formation stage. Alternatively bromadiolone poisoning can be at the first two stages of groundnut growth i.e. germination and pod formation. The third application of rodenticide can be at pod maturity stage using zinc phosphide.

Zinc phosphide should be used only once at any of the above stages in view of development of bait shyness in surviving population after poison baiting.

RODENT MANAGEMENT IN BARREN LAND AND ITS IMPACT ON RODENT DAMAGE IN CULTIVATED AREA

In groundnut crop, when the rodenticides were applied in both cultivated field and its surrounding barren land compared to rodenticides applied only in cultivated field, plant and pod damage decreased significantly resulting in higher yield of groundnut pods. Similarly, significant decrease in crop damage in wheat crop and higher yield were recorded when the rodenticides were applied in both wheat fields and its surrounding barren land as compared to the area in which rodenticide was applied only in wheat fields. Thus rodent damage in cultivated fields was significantly less when surrounding barren land was also subjected to rodent control compared to rodent control only in crop fields.

REFERENCES

ANON, 1984. *Annual Progress Report (1983-84) of AICRP on Rodent Control*, Jawaharlal Nehru Krishi Vishwa Vidyalaya, Jabalpur, India.

ANON. 1984a. *Annual Progress Report (1983-84) of AICRP on Rodent Control*, Central Arid Zone Research Institute, Jodhpur, India.

ANON. 1984b. *Annual Progress Report (1983-84) of AICRP on Rodent Control*, Punjab Agricultural University, Ludhiana, India.

ANON, 1985. *Annual Progress Report (1984-85) of AICRP on Rodent Control*. Jawaharlal Nehru Krishi Vishwa Vidyalaya, Jabalpur, India.

ANON, 1986. *Annual Progress Report (1985-86) of AICRP on Rodent Control*. Jawaharlal Nehru Krishi Vishwa Vidyalaya, Jabalpur, India.

ANON. 1989. *Annual Progress Report (1987-88 & 1988-89) of AICRP on Rodent Control*, Gujarat Agricultural University, Junagadh, India.

ANON, 1990a. Rodent control measures in Gujarat-Report of Expert Team. Ahmedabad: Directorate of Agriculture, Gujarat State, India.

ANON, 1990b. *Annual Progress Report (1989-90) of AICRP on Rodent Control*. Gujarat Agricultural University, Junagadh, India.

ANON. 1991. Annual Progress Report (1990-91) of AICRP on Rodent Control, Gujarat Agricultural University, Junagadh, India.

BARNETT, S. A & I. PRAKASH. 1975. RODENTS OF ECONOMIC IMPORTANCE IN INDIA Arnold-Heinemann New Delhi and London.

BINDRA, O. S & P. SAGAR. 1968. Study on the losses to wheat, groundnut and sugarcane crops by the field rats in Punjab. *Proc. International. Symp. Bionomics and Control of Rodents*. Kanpur, India. pp. 28-31.

CHOPRA, G & M.L. SOOD. 1983. Comparative field efficacy of RH-787 (vacor), zinc phosphide and aluminium phosphide. *Pesticides* 17(4): 17-20.

DEORAS, P.J. 1966. Some observation on the probable damage caused by rats in Bombay. *Indian J. Ent*., 28: 543-547.

GUPTA, B.K., N. AHMED N & V.R. PARSHAD. 1983. Rodent activities in mature *kharif* crops in PAU fields, Ludhiana (Punjab); *Rod. Newl.* 7(1 & 2) 1-2.

HOPF, H.S., G.E.J. MORLEY & J.R.O. HUMPHRIES. 1976. *Rodent damage to growing crops and to farm and village storage in tropical and sub tropical regions*. Centre for Overseas Pest Research and Tropical Products Institute, London. pp. 23.

JOTWANI, M.G & Y.P. BERI. 1968. Economic losses due to rats. *Proc. Intl. Symp. Bionomics and Control of Rodents* held at Kanpur, India pp. 9-16.

MITTAL, V.P & H.J. VYAS. 1992. Groundnut In: RODENTS IN INDIAN AGRICULTURE (Eds.). Prakash, I. & P.K. Ghosh, Scientific Publishers, Jodhpur, India. pp. 249-264.

PARSHAD, V.R., C.S. MAHLI., N. AHMAD & B. GUPTA. 1987. Rodent damage and control in peanut fields in India; *Peanut Science* 14: 4-6.

PATEL, N.C & N.P. NAYAK. 1987. Survey of various stored grain and field pests for all crops: Final Report of RPF III, All India Co-ordinated Scheme for Harvest and Post Harvest Technology: Gujarat Agricultural University, Junagadh, India.

PRAKASH, I. 1975. The population ecology of the rodents of the Rajasthan desert, India. In: RODENTS IN DESERT ENVIRONMENTS. (Eds.): Prakash, I & P.K. Ghosh: Dr. W. Junk Publ. The Hague. pp. 75-116.

PRAKASH, I. 1981. *Ecology of the Indian desert gerbil, Meriones hurrianae*. Monograph No.10. Central Arid Zone Research Institute, Jodhpur, India.. 87 p.

PRAKASH, I & R.P. MATHUR. 1987. MANAGEMENT OF RODENT PESTS. Indian Council of Agricultural Research, New Delhi, India.

RAO, M.V. 1988. Short-term and long-term strategies for making India self-reliant in vegetable oils; *Proc. National Seminar on Strategies for making India self-reliant in vegetable oils,* Hyderabad, India. pp. 8-9.

RAJASEKHARAN, M.R & E. DHARMARAJU. 1975. Studies on field rodents in Andhra Pradesh; *Proc. All India Rodent Seminar*, Ahmedabad, India. pp. 54-56.

SHAH, M.P. 1979. Population explosion of rodents in groundnut fields in Gujarat, *Rod. Newsl.* (3): 19-20.

SOOD, M.L & B.S. UBI. 1975. Population fluctuations of sympatric murids occurring in the fields of PAU Ludhiana: I Effect of biotic factors. *Proc. All India Rodent Seminar*, Ahmedabad, India. pp. 76-81.

SRIDHARA, S. 1999. QUARTER CENTURY OF RESEARCH ON RODENT CONTROL (1973-98) IN THE UNIVERSITY OF AGRICULTURAL SCIENCES, BANGALORE–HIGHLIGHTS AND RECOMMENDATIONS. University of Agricultural Sciences, Bangalore, India. 126p.

SRIHARI, K., S. SRIDHARA & N. DESHPANDE. 1979. Laboratory studies on the food hoarding of Indian mole rat, *Bandicota bengalensis*; *Indian. J. Rodent*. 1(1 & 2): 46-50.

VYAS, H.J., H.N. VYAS & P.M. SINGH. 1985. Evaluation of Bromadiolone-a single dose anticoagulant-against field and domestic rats. *Indian J. Ent.* 47: 45-48.

❑❑❑

11

PEST STATUS OF THE INDIAN CRESTED PORCUPINE, *Hystrix indica*

Chakravarthy, A.K., A.C. Girish and Shakunthala Sridhara*

Department of Entomology, University of Agricultural Sciences, G.K.V.K., Bangalore – 560 065;
**AINP on Rodent Control*

ABSTRACT

The Indian crested porcupine, *Hystrix indica* because of degradation and fragmentation of its scrub, rocky, hilly and forest habitat is being frequently reported as a pest on crops. Data on its feeding behaviour and pest status in India is scarce. Comprehensive studies to evolve management strategies on its foraging activity using radio collared animals and evaluation of ecofriendly crop protection methods have been carried out only in the hilly and coastal zones of Karnataka. Based on this investigation, burrows were categorized into big and small depending on number of openings and animal/s living inside. Density was 8.8 to 12.9 animals per square km. Thirty six plant species, both cultivated and wild were consumed by porcupines. Debarking was the major damage in forest nurseries of Rajasthan and coconut palms in Karnataka. Palms in the vicinity of forests were damaged more than the distant ones. Active feeding was limited to seedlings of coconut. Encasing coconut and areca nut seedlings with porcelain pipes of suitable diameter and thatches of bamboo protected them from porcupines. Smearing seedlings and adult palms with coal tar prevented porcupine debarking. Telemetry has proved to be a good tool to understand the population dynamics, behaviour and feeding ecology of porcupines for both conservation and for developing crop protection measures.

INTRODUCTION

The name ‘Porcupine’ is derived from the Latin word “porous” meaning swine and the French ‘epine’ derived from the Latin word

"Spina" meaning thorn. Literally it means "the irritable back". *Hystrix indica* is called the Indian crested porcupine because of the presence of very long stiff bristle-like hairs along the neck and upper back region. The species is characterized by relatively short tail the tip of which bears specially modified quills which are hollow tubes. When threatened the tail is shaken making a hissing noise.

SCIENTIFIC STATUS

Class	: Mammalia
Order	: Rodentia
Sub order	: Hystricognathi
Family	: Hystriacidae
Genus	: *Hystrix*
Specific name	: *Indica*
Binomial name:	: *Hystrix indica*

Fig. *Hystrix indica*

DISTRIBUTION: The porcupine, a fossorial and nocturnal animal is distributed from sea level to 2800m msl throughout India. In India, two genera and three species of porcupines are found. The species are: The Indian crested porcupine, *Hystrix indica,* the crestless Himalayan porcupine, *Hystrix hodgsoni grey* and the Asiatic brush tailed porcupine, *Atherurus macrourus assamensis*. The latter two species are restricted in geographical distribution to Himalayas and North east India. The place of origin of the genus *Hystrix* is still unknown.

MORPHOLOGY: Porcupine is a thickset rodent with short powerful limbs. Forefeet have four digits and a vestigial thumb. There are five digits on the hind limb. The animal is plantigrade in gait and sole of the hind foot is naked. The head terminates in a broad blunt muzzle with the upper lip being divided. The area around the nose appears fleshy due to inflation of the nasal area by pneumatic cavities. The eyes and ears are comparatively short. The vibrissae are black and the longest measures 20cm in length. The lower part of the body is thinly covered with dark brown and black, short glossy brown bristle-like hairs. The face is also covered with similar but smaller hairs. There is an indistinct crescent of grey and white quills on the back. These are frequently shed during their lifetime. Tail is short, covered with short white quills mixed with hollow rattling quills.

BIOLOGY AND BEHAVIOUR: Published information on Indian crested porcupine is meagre. The animal is nocturnal. Day times are spent in deep subterranean retreats which usually have more than one exit hole. It also rests under the cover of tall grass and digs burrows close to its foraging site. Diurnal activity of the animal is rare and is restricted to the vicinity of their burrows. Burrows can be quite deep, sometimes extending up to 18.5m in rocky hillsides (Prater, 1993). Two to eight animals live in the burrow. They come out after sunset. They have a keen sense of smell and display high intelligence in evading traps. Grain, vegetables, fruit and roots are their main food. When irritated or alarmed, porcupines erect their spines, grunt, puff and rattle their quills.

Porcupines breed throughout the year with females littering from March to December (Prakash, 1971). The gestation period ranges from 109-112 days. Females excavate a separate burrow in which two or four young ones are born with their eyes open and soft quills covering the body. Both parents are reported to occupy the same burrow along with their young, usually two or three (Prater, 1993). Life span ranges from eight to twenty years (Roberts, 1977).

BURROWS: Porcupines dig extensive and deep burrows with several side entrances. Burrows are excavated in the embankment of river channels, under walls and hedges so as to gain easy entry into a garden or cultivated fields (Agarwal & Chakravarthy, 1992). In the Thar desert, porcupines dig burrows in the intertidal plains, far away from crop fields. The burrow had two side openings with the main entrance slightly arched. The burrow sloped down steeply four meters downwards expanding into a round chamber. The width of the burrow opening was 15 x 20cm^2 and 20 x 30cm^2. Being nocturnal, porcupines in the desert were seen spending the day inside the burrows, venturing out during the night to raid crops like wheat and other food grain (Bhargava *et al.*, 2004).

Burrow on plain surface

Dehusked coconut

Burrows under bamboo shrub

Debarked coconut tree

Burrows under concrete structure

Toppled coconut seedling

(a) Burrows of *H. indica*

(b) Damage to coconut

Plate - 1

Porcupines made burrows in different habitats of Western Ghats in Karnataka ranging from uplands to rocky plains (Chakravarthy *et al.,* 2003a). Burrows had one to several openings and were categorized into big and small based on size. These animals prefer patches with thick dense vegetation, rocky undulating niche, strong cover, abandoned concrete structures and base of trees for digging burrows (Plate 1a). They were also found dwelling in caves. Huge amount of soil up to 60-70kg was found thrown out of the entrance of the burrow. Burrows upon digging showed four compartments–one for storing food, a second one for rearing young and two for other activities. Two to eight animals were found sharing a burrow. They seem to be partially social but each individual occupied a separate part of the burrow complex but with interconnecting tunnels to facilitate social interactions. Each individual is known to use its own entrance and exit hole (Kingdon, 1974). Sometimes the animals were found on the surface resting under bushy cover near the burrow. Once the food source around the burrows was exhausted or when threatened by predators porcupines abandoned their burrows. There was a positive correlation between the burrow size and number of porcupines occupying it.

A lone study on population based on pellet group density and direct observations in Sariska Tiger Reserve, Rajasthan revealed differences between direct and indirect methods of density estimates. The over all density was 8.8 ± 2.4 animals per sq.km. by direct method while it was 12.9 animals per sq.km when indirect method was employed. Available literature is insufficient to state trends in porcupine population throughout the year. However, an increase in population is expected during the post breeding period i.e. January to March. The density at a given area depends on multiple factors like quality and quantity of food, presence and absence of natural enemies, availability of niches for burrowing and rainfall patterns (Sharma & Prasad, 1992).

The animals become ferocious and highly vocal when agitated/ threatened.

FEEDING BEHAVIOUR: Porcupines are herbivorous with a wide range of food which includes selected roots, bulbs, succulent tubers and bark of certain tree species. They also feed on ripe fallen fruits. Foods are chosen based on availability, nutritive value and palatability. They exhibit different feeding behaviour in cultivated and natural ecosystems. Within a restricted spatial range, their foraging areas and foraging time also get restricted because of availability of select foods.

Thirty six plant species, both cultivated and wild flora were recorded to be consumed by porcupines in Western Ghats of Karnataka (Table 1; Chakravarthy *et al.*, 2003b; Girish, 2005). The major plants which were eaten by porcupines in the hilly and coastal districts of Karnataka are: coconut, sweet potato, bamboo, tapioca, cashew nut, cocoa, cane, cactus, wild turmeric, *Colocasia, Acacia, Garcenia xanthochymus, Diascorea oppositifolia, D. bubbifolia,* banana, *Gloriosa superba* and *Caria pinnata* (Girish, 2005). Feeding preference differed from location to location depending on the availability and accessibility to the animal and distance from its burrow. This explained the significant differences in porcupine damage to coconut palm from one month to another in coastal Karnataka.

PEST STATUS

Porcupines preferentially ate bamboo, tapioca, cane, sweet potato, wild turmeric, *colocasia* and *cassia* roots. They also fed on fruits of mango, cashew and coconut fallen on the ground. The nuts of cashew and areca were eaten by cutting open the fruit with incisors.

Systematic observations on crop damage by porcupines in India are very few. Idris & Rana (2001) recorded girdling and debarking activities of *H. indica* in the forest nursery of arid region in Aravalli hills of Rajasthan. They found stems of *Euphorbia caducifolia* damaged in the nursery. The species caused 30% damage to seedlings of neem, *Azadiracta indica* and 12% to *Eucalyptus* sp. by cutting young plants two to three inches above the ground.

In coastal and hill regions of Karnataka, porcupine is considered a major pest on coconut. Damage to coconut plantations depended upon age, distance of plantation from forest, availability of alternate foods and the season. The combination of burrowing, feeding on stem tissue bark stripping and browsing on ground vegetation should be collectively considered to assess the impact of porcupine feeding in cultivated palm ecosystem. In the absence of underground roots and tubers, porcupines consumed fruits and nuts (Girish, 2005). Damage to coconut was two pronged. Seedlings and young trees were attacked differently (Plate 1b). Seedlings were cut at the base, coconut seedling was dehusked by incisors followed by breaking of nut and consumption of endocarp. Porcupines preferred young coconut seedlings aged less than two years (Fig. 1). Feeding severally and repeatedly resulted in the death of young as well as old palms. Generally the roots, bole region and trunk portion of coconut were attacked. The second type of damage was by debarking palms, preferably less than 15 years old. During this stage there is translocation of nutrients to the nuts from the roots and porcupine was likely to feed on this nutrient. However damage to the tree was insignificant because the debarked area was less than 75cm^2 per palm, which constituted a very small proportion of total stem area. Debarking was seasonal occurring during September to January when the palms were at the bearing stage (Fig. 1). Further, cultivated palms adjacent to forests were heavily damaged compared to palms situated away from the forest habitat of porcupines (Fig. 2). The debarking resulted in 10 to 46% mortality of young palms (Fig. 2; Girish, 2005). Three coconut varieties were damaged by porcupines although the difference in damage to the three varieties was non-significant (Chakravarthy & Girish, 2002).

Table 1. Food range of Porcupine in hill and coastal regions of Karnataka

Common name	Scientific name	Damage (%)
Sweet potato	*Ipomoea batatas* Lamk.	20 (10 tubers)
Bamboo	*Bambusa arundinacin* (Retz.)	12 (15 tillers)
Tapioca	*Manihot esculenta* Crantz.	50 (68 tubers)
Alocasia	*Alocasia indica* Schott.	2 (5 tubers)
Cane	*Calamus tenuis* Redt.	25 (45 tillers)
Diascorea	*Diascorea esculenta* Burm.	7 (42 tubers)
Ananus	*Ananas comosus* (L.) Merr	10 (250 plants)
Banana	*Musa* spp.	25 (1500 suckers)
Cactus	*Agave americana* L	35 (850 plants)
Wild turmeric	*Zingiber zerumbet* Rose ex.Sm	15 (250 plants)
Colocasia	*Colocasia esculenta*	10 (150 plants)
Gauri gedde	*Gloriosa superba*.L	5 (100 plants)
Byne palm	*Caryota urens* L.	20 (45 seedlings)
Acacia	*Acacia catechu* Wild	35 (45 fruits)
Ridge gourd	*Luffa acutangula* (L.) Roxb.	2 (75 plants)
Bottle gourd	*Lagenaria vulgaris*	20 (10 plants)
Muskmelon	*Cucumis melo*	25 (5 plants)
Mango	*Mangifera indica* L.	1 (108 fruits)
Groundnut	*Arachis hypogea* L	1 (one acre)

Common name	Scientific name	Damage (%)
Pumpkin	*Cucurbita moschata* L.	2 (one acre)
Beans	*Dolichos lablab* L	1 (one acre)
Rubber kai,	*Ficus elastica*	5 (10 plants)
Thare mara nuts,	*Terminalia bellerica*	2 (20 plants)
Elachi mara fruits,	*Ziziphus zizuba*	1 (5 plants)
Kakke mara roots,	*Cassia fistula*	5 (45 plants)
Kare mara roots	*Kanthium* sps	2 (4 plants)
Thavere gedde	*Nelumbo nucifera* Gaertn.	10 (45 plants)
Kadu irulli	*Crinum defixum ker*-Gawl.	5 (10 plants)
Bhela fruits	*Ferronia elephantum*	2 (3 plants)
Coconut	*Cocos nucifera* L.	15 (seedlings)
Areca nut	*Areca catechu* L	3 (seedlings)
Ragi (finger millet)	*Eluesine coracana* L	1 (one acre)
Paddy	*Oryza sativa* L	2 (one acre)
	Garcenia xanthochymus	10 (45 plants)
	Diascorea oppositifolia	5 (10 plants)
	D. bulbifolia	2 (3 plants)
Hard metallic substances		

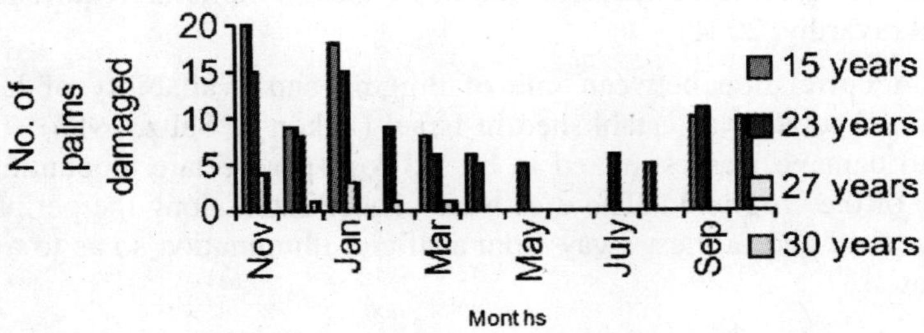

Fig. 1. No. of Coconut palms damaged in Coconut plantations of different ages at Subramanaya, 2002-04

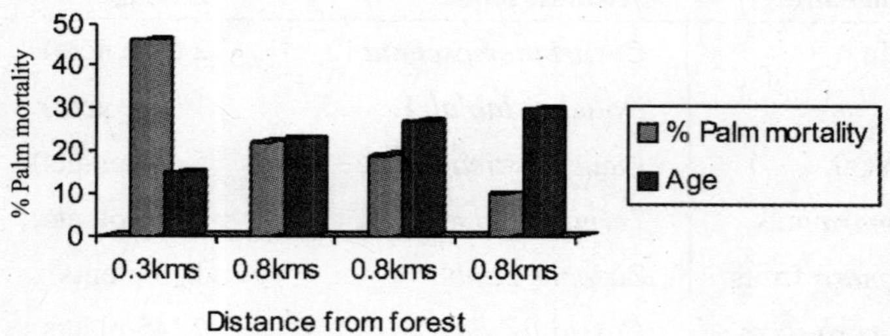

Fig. 2. Mortality of Porcupine damaged coconut palms

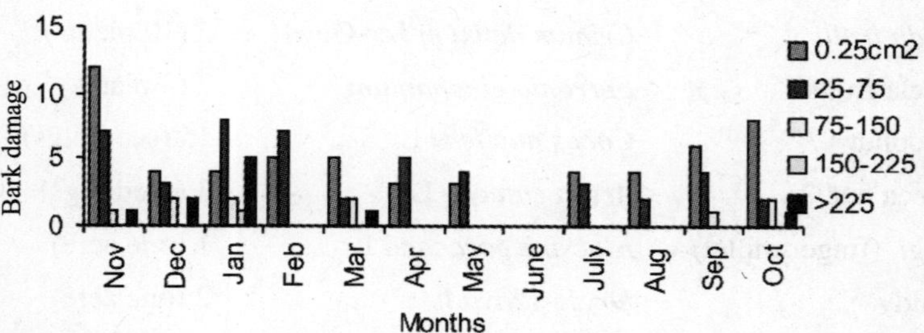

Fig. 3. Distribution of Porcupine debarking in Coconut palms at Subramanya

Porcupines also fed on fallen cashew nut kernels. Damage was maximum during April when it was at fruiting stage. Consequently there was less consumption of fallen coconut and areca nut during this period (Girish, 2005). In addition they were also seen nibbling at shed antlers of deer, metallic substances like iron ploughs, pickaxe etc possibly to check the growth of incisors and/or to obtain mineral requirements (Chakravarthy, 2004).

A correlation between rate of digging and availability of larger tubers of potato was established in Israel (Alkon & Saltz, 1985). Total potato damage was estimated to be 1-3 tons per hectare amounting to 0.6% of the crop yield. The digs were concentrated along the periphery of the field and farthest away from artificial illumination so as to avoid predators.

BEHAVIOURAL ECOLOGY

Using radio collars for the first time in India, Girish (2005) observed porcupines leaving their burrows around 6 to 7pm, spending 10 to 11 h in the night foraging, scanning for food, walking, running, etc. They were found foraging in pairs and sometimes alone but always in the vicinity of another animal. Rain deterred activities in the flatlands of Mysore zoo but not in the heavy rainfall, hilly areas of Nagarhole, Subramanya and Bhadra reserve area. Similarly moonlight affected foraging in the open area but not in the forest habitat where moon light seldom penetrates the thick canopy of trees. Porcupines were found adapted to living in a wide range of temperatures. Radio collared animals exhibited inter and intra individual variations of activity in relation to ambient temperature and precipitation. Generally activity of porcupines decreased in cooler habitats and environments (Chakravarthy, 2004).

The faecal pellet size varied from 1x0.5cm to 2.5x0.75 cm and colour from grey to dark brown. Undigested cellulose content was the major component of faecal pellet of porcupines indicating the herbivorous nature of feeding.

CROP PROTECTION

Schemnitz (1994) suggests several methods to reduce or prevent porcupine damage. Principally these include removal, repelling and excluding.

Animal removal or translocation from problem areas can stop plant damage completely. This can be accomplished by live trapping porcupines using cage trap or wooden trap with a minimum dimension of 30cm height, 25cm width and 80cm length. Preferably traps should be set at the base of a large tree or leaning against a damaged tree. Salt soaked pieces of leather or wood, salt mixed with edible oil or pieces of carrot, turnip, etc are the suggested baits (Schemnitz, 1994).

A stockade trap, which is a mini wooden fortress, usually of jungle wood is effective in capturing porcupines alive. When the animal steps on the trigger platform, the door drops down and the animal is trapped within the stockade. Coconut kernels or coconut or palm oil seedlings are attractive baits while using stockade traps (Chakravarthy, 2004).

Steel leg hold traps where legal, can be used to catch porcupines. They can be baited with salt soaked material. Care must be taken to avoid non target animals.

Snaring is more opt to capture porcupines which accidentally enter human habitat. A pole snare manoeuvred over the head of the porcupine can be easily used to lift the animal and put it into a transport box/cage.

Repellents: Thiram sprayed on vegetation was found to repel porcupines due to its bitter taste and offensive smell. Lacing the stem for about half a meter above the ground with coal tar, waste oil or paste of cement deterred porcupine attack on trees (Chakravarthy, 2004).

Exclusion: Encasing oil palm or coconut seedlings with a wire guard netting protected them from porcupine attack (Chakravarthy, 1996). Similarly small areas of high economic loss as well as nurseries can be protected with electric or non-electric fencing at 1-5 above the ground. Farmers in Western Ghats of Karnataka use a mat of bamboo thatch to protect seedlings of coconut, areca nut, cashew, coffee, etc for the first two years from porcupine damage. Metallic tree trunk guards can also be deployed.

Predators: Natural predators can be encouraged. Predators include lion, leopard, striped hyena, black bears, and great horned owls.

Ecological methods: The affinity of porcupines for crops along the margin of the fields indicates that changes in field shape that reduce edges will also reduce the area of crop field vulnerable to porcupine depredation. Single large fields would have fewer perimeters than several separate fields of the same total area. Circular shapes have proved to be optimal to deter crop raiding by porcupines. The behaviour of porcupines to avoid moonlight and distant-artificial illumination in order to be invisible to predators may be utilized to keep them away from crop fields (Alkon & Saltz, 1985).

Covering of the base of palm with bamboo thatch was found to be economical and ecofriendly in preventing porcupine damage to coconut (Chakravarthy & Girish, 2001).

Lethal Methods: Since the animal is a protected species, lethal approach to manage porcupines should be the last option and should be resorted only when damage levels are extreme and other methods of damage prevention fail.

Chemical control: A prerequisite for all methods of porcupine control is to have a map with location of porcupine dens well marked. Fumigation and baiting with rodenticides are effective in killing porcupines, the details of which are as under.

Fumigation can be by cyanogas or aluminium phosphide tablets preferably during day time. After locating porcupine dens, all openings, except one are to be closed and cyanogas pumped through the single, open tunnel and then sealed. Alternatively 10-15 tablets of aluminium phosphide can be inserted into the active den using a long-handled shovel and sealed afterwards.

A two ingredient pyrotechnic has been fabricated and used successfully in Pakistan (Khan, 1990). This is a gas cartridge containing

65% sodium nitrate and 35% ground charcoal. The ignited device is kept inside the burrow and sealed. The burning chemical releases carbon monoxide which is lethal to the porcupines.

Poison baiting: In Pakistan sodium monofuoroacetate (1080) is smeared on cut potatoes or apples and then introduced inside porcupine burrows. The method was very effective. Permanent bait stations with strychnine treated salt blocks were also successful in forest areas (Khan, 1990). Wheat flour mixed with either 0.0375% coumatetralyl or 0.005% brodifacoum, when baited for 5-7 and 3 days respectively gave considerable reduction of porcupines (Khan, 1990)

CONCLUSIONS

Studies in coastal and hill regions of Karnataka (South India), and in Rajasthan and review of literature revealed that porcupines inflict economic damage to coconut plantations in areas or patches adjacent to forest or wild vegetation tracts in the former and forest nurseries in the later. A system for monitoring porcupine in cultivated and natural ecosystems is needed as the animals appear to exhibit different patterns in feeding behaviour. Currently the animal numbers appear to be declining in certain areas. But in certain other patches, it's habitat range is expanding. Hunting for meat by tribals in spite of porcupines coming under Wildlife Protection Act (1972), degradation and fragmentation of its habitat and attempts by farming community to kill it to protect their crops are the factors contributing to declining numbers of porcupines.

The economic and ecological role of porcupine in its habitat is not yet fully established and information on the impact of its feeding in cultivated ecosystems is also not conclusive. Therefore telemetry as a tool to study bioecology, feeding impact on plants in natural and cultivated ecosystems and population ecology should be encouraged in the country. Better density estimates and population indices are also required. As scat analysis in porcupines reflect the dietary habits and gut contents, faecal pellet analysis need to be conducted through out the year to reveal the animal's dietary patterns. This knowledge coupled with measures in preserving its natural habitat will go a long way in conserving this less common rodent as well as mitigating damage to agriculture by non-lethal methods.

REFERENCES

AGARWAL.V.C & S. CHAKRAVARTHY. 1992. The Indian crested porcupine, *Hystrix indica* (Kerr). In: (Eds.): Ishwar Prakash & Ghosh,

P.K. RODENTS IN INDIAN AGRICULTURE. Scientific Publishers, Jodhpur, India. pp. 25-30.

ALKON, P.G. & D.SALTZ. 1985. Patterns of crested porcupine (*Hystrix indica*) damage to cultivated potatoes. *Agri. Ecosystem. Environ*. 14: 171-183.

BHARGAVA, R.N., L.S. RAJPUROHIT., BISSA PRASHANTH & S. MADAN. 2001. On the porcupine (*Hystrix indica*) in Western Thar Desert. *Tiger Paper* 28(4): 1-3.

CHAKRAVARTHY, A.K. 2004. Annual Report. DST Research Project on Porcupine feeding behaviour and ecology in cultivated palm ecosystems. New Delhi, India. 32p.

CHAKRAVARTHY, A.K & A.C. GIRISH. 2001. Protection of coconut (*Cocus nucifera*) palms from porcupines (*Hystrix indica Kerr*) damage. *Rat-tattle-(RISCINSA News letter)*. 2(1):2.

CHAKRAVARTHY, A.K & A.C. GIRISH. 2002. Porcupine (*Hystrix indica Kerr*) damage to three coconut varieties in coastal Karnataka. *Rod. Newsl.* 26(1-2): 2.

CHAKRAVARTHY, A.K., B.B. HOSETTI & A.C. GIRISH. 2003a. Observations on burrows of Indian crested porcupine, *Hystrix indica kerr*. *Rod. Newsl.* 27(1-2):4.

CHAKRAVARTHY, A.K., B.B. HOSETTI & A.C. GIRISH. 2003b. Porcupine (*Hystrix indica*) foraging activity in cultivated ecosystems is Western Ghats of Karnataka. *Rat-a-tattle,* (*RISCINSA Newsletter*) 3(1):18.

GIRISH, A.C. 2005. Feeding behaviour of porcupines in cultivated ecosystems of Western Ghats of Karnataka, South India. Ph.D thesis submitted to Kuvempu University, Shimoga, India.

IDRIS, M & B.D. RANA. 2001. Some observations on infestation of porcupine, *Hystrix indica* Kerr in the forest nursery. *Rod. Newsl.* 25(1-2):5

KHAN, A, A. 1990. Control of porcupine and Murree vole in crops, orchards and forests In: VERTEBRATE PEST MANAGEMENT (Eds): Brooks, J.E., E. Ahmad., I. Hussain. S. Munir & A.A. Khan. Pakistan Agricultural Research Council, Islamabad, Pakistan. xiii+206p.

KINGDON, J. 1974. EAST AFRICAN MAMMALS: AN EVOLUTIONARY ATLAS, vol. 2. Academic press, New York. 404p.

PRAKASH, I. 1971. Breeding season and litter size of the Indian desert rodents. *Z. angew. Zool.* 58(4): 441-454.

PRATER, S.H. 1993. THE BOOK OF INDIAN ANIMALS. III edn. Bombay Natural History Society, Bombay, India.

ROBERTS, T.J. 1977. THE MAMMALS OF PAKISTAN. Ernst Benn Ltd., London. 361p.

SCHEMNITZ, S.D. 1994. Porcupines: Damage prevention and control methods. In: PREVENTIONS AND CONTROL OF WILDLIFE

DAMAGE. Co-operative Extension Service, University of Nebraska, Lincoln, USA. pp: B-81 – B-92.

SHARMA, D & S.N. PRASAD. 1992. Tree barking and habitat use of porcupine in Sariska National Park in Western India. *Mammalia* 56(3): 351-361.

❑❑❑

12

PHEROMONES IN RODENT PEST MANAGEMENT

G. Archunan* and S. Achiraman

Department of Animal Science, Bharathidasan University
Tiruchirappalli-620 024 Tamil Nadu, INDIA
** e-mail: garchu56@yahoo.co.in*

ABSTRACT

Conventional methods of rodent control are mainly poison based and to some extent trapping. The major snags in effective reduction of rodent pests are development of bait shyness to acute rodenticides and resistance to anticoagulant rodenticides. The phenomenal success of pheromones in insect pest management and also the significant role of olfaction through which pheromones act have prompted evaluation and exploitation of pheromones to manage rodent pests. Some pheromones act as sex attractants, a character which has the potential to lure rodents to traps, poison baits and to suppress neophobia and bait shyness. Sex specific volatile compounds has been identified from the scent glands of *Rattus rattus* and *Bandicota bengalensis* and used with considerable success to trap field rodents. The volatile compounds of preputial/cheek glands and urine of *R. rattus* when added to poison baits are reported to increase mortality. 1-nitropentane was more effective than 1, 10 dibromodecane and 1-chlorodecane. The role of urinary proteins in carrying these pheromones and their potential role in rodent control is discussed. Proposals are made to use pheromones to increase the efficacy of traps, baits, poison baits and to disrupt mating so as to reduce population.

INTRODUCTION

Rodents have long been recognized as detrimental to human health and agriculture. They also cause losses to animal facilities, human dwellings and storage structures. They cause direct damage to various commodities by gnawing, feeding, and indirect damage by spoilage, contamination and deterioration by enhancing vulnerability to fungal and bacterial infestations during pre and post–harvest stages (Fall, 1991). Many attempts have been made to manage rodent population but rarely the effort to reduce the population levels to eco-safe level was successful. Traditional pest control methods, especially those involving lethal agents are quite unfavourable to the environment and human beings by possessing more demerits than merits. Hence, there is a need to develop modern, safe, effective technique for managing rodent population. This technique must be derived from natural system in the hope that it will be effective, pest specific and will not have any negative impact on the environment.

Rodents depend mainly on chemical signals to attract mates and for social interaction. The chemical substances involved in conveying messages between individuals are termed as semiochemicals (Regnier, 1971). The semiochemicals that are used in intra-specific communication are known as pheromones. Karlson & Butenant (1959) were the first to define pheromone as air-borne chemical substances that are secreted externally by an insect and cause a specific reaction in a receiving individual of same species; the reaction involves either the release of a specific behaviour or physiological change in the recipients. Although originally the word "pheromone" was introduced in insect communication, it has been extended to mammalian behaviours also (Doty, 1980; Dominic, 1987; Keverne, 1999; Vandenbergh, 1999; Rekwot *et al.,* 2001). Establishment of pheromonal trap in insect pest management is a landmark in pheromone technology. Presently pheromones are commercially available for mass trapping of most of the insect pests. Parellaly sufficient evidence is available to indicate the importance of rodent pheromone in behavioural interactions (Dominic, 1987; Johnston, 1990). In view of the great role played by pheromones in the life of rodents particularly in reproduction it is possible to develop pheromonal trap for managing rodents.

This chapter deals in detail about rodent pheromones and reviews its possibility for application in rodent pest management programme. Investigations on rodent communication for the past few decades have yielded results of fundamental importance which have the potential to become foundation for new chemosensory technology. Both laboratory as well as some field observations are included in this review. However, the emphasis of this chapter is on results of research at the interface between chemistry and biology of olfactory communication. The pheromonal studies in rodents may significantly contribute to investigations on other mammals for various purposes such as artificial breeding in farm/domestic animals, wild life management and control in human reproduction.

RODENT PEST MANANGEMENT

(i) Pest status

Rodentia, which make up almost 40% of mammalian species, represent the largest order of mammals, comprising some 1700 species under 389 genera and 35 families (Macdonald & Fenn, 1994). Among the 1700 rodent species, nearly 150 species have been defined as pests, though only 20 are economically important (Barnett & Prakash, 1975). However, the diversity of rodent fauna in the Indian subcontinent is remarkable; 46 genera, 135 species and about 300 subspecies have been reported (Ellerman, 1961). Among rodents, rats are the most important pests of agriculture and also affect human health. However, not all rodent species damage agricultural crops and human properties. The following rodents are considered to be of economic importance in India: House mouse (*Mus musculus*), Indian field mouse (*Mus booduga*), lesser Bandicoot rat (*Bandicoota bengalensis*), house rat (*Rattus rattus*), soft furred field rat (*R. meltada*) and Indian gerbil (*Tatera indica*) (Barnet & Prakash, 1975).

The house rat, *R. rattus* and the house mouse, *M. musculus* are the major commensal pests. *R. rattus,* also called roof rat, is one of the most abundant and widely distributed cosmopolitan rodent species in our country. The extraordinary ability of the rat to adapt itself to every environment, its great fecundity, and natural cunningness has made it one of the most successful of all animals. The classical case of the lesser Bandicoot rat, *Bandicota bengalensis*, adapting to the urban environment in Howrah and Bombay, and replacing *Rattus* sp. is well known (Seal & Banerji, 1966). The Indian gerbil is also reported to be one of the important pests in agriculture becoming a commensal in Rajasthan.

(ii) Damage caused by rodents

Losses by rodents to various crops across the country are well established by several studies (Prakash, 1988; Prakash & Mathur, 1987). The available data on the damage and economic losses caused by rodents in rice, wheat, sugarcane, maize, pearl millet, sorghum, oil seed, cotton, legume and vegetable are well documented (Parshad, 1999). Reports further show that the chronic damage ranges from 2 to 15%

throughout the country and sometimes leads to 100% loss in the field crops (Parshad, 1999). Rats damaged growing corn, melons, tender coconut, tomatoes, pumpkin and fruits. They destroyed feed, feedbags and also transmitted deadly pathogens in poultry farms. Rats are responsible for damage to buildings, houses and warehouses. They gnaw paper, books, upholstery, roof and other materials for making their nests. They also gnaw hard substances such as wooden doors, windows, electric wires, aluminium and lead, similar materials to maintain their incisors at proper size. In addition to the above direct losses, they also cause indirect losses through wastage and contamination of edible items. Tropical countries like India have been witnessing massive epidemics of rodent mediated diseases. In 1994, Surat (Gujarat) was badly hit by plague epidemic. Rats are also responsible for spreading murine typhus and rickettcial pox through their ectoparasites i.e., mites and fleas, leptospirosis through their urine, rat-bite fever through their salivary secretion and amoebic dysentery through the consumption of food contaminated with rat excreta.

(iii) Control Measures

(1) Trapping is one of the oldest methods of rat control and various types of traps are in use. The '*chitre*' or 'wonder' trap has been found to be very efficient to trap *R. rattus* alive. Glue traps (ecosafe traps) are also quite effective in controlling the population of roof rats in stores, warehouses, godowns and other sites (Urs, 1996). However, trapping is impractical if the affected areas are too large with high rat population (Fielder & Fall, 1994)

(2) Other methods: Other physical control measures like hunting, clubbing, use of flame-throw and flooding of burrows and frightening methods such as electronic devices are occasionally being employed in rodent control programmes (Fall, 1991). But such techniques are unsatisfactory for rodent control measures.

(3) Burrow dwelling rats and their young can be eliminated by fumigation using a combination of sodium nitrate and charcoal. This method is not a permanent solution because the rodents can escape by migration.

(4) In predatory control cats are occasionally used to control rat populations. Kites, owls and snakes are the natural predators of rodents in the environment (Campbell *et al.*, 1987). Pythons were used to reduce rat numbers in the Islands of Australia, granaries in Singapore and riversides in Bangkok (Whitaker & Bhaskar, 1978).

(5) Chemosterilants: Use of chemosterilants to produce sterility among rodents in order to control them has not been successful so far.

(6) Electric fencing around the crops was found to prevent the entry of rodents in the oil-palm fields in Africa (Toovey, 1953), rice fields in India (Prakash & Sarwal, 1958) and North America (Fitzwater, 1972). As it is dangerous to other animals and humans, this method is not recommended.

(7) Repellents: Repellents have been classified as visual, acoustic, chemical and mechanical. Visual group of repellents prevent animal damage through scares that disturb the animals through visual senses. A group of chemicals categorized as 'area repellents' and 'contact repellents' have come into vogue that could also be applied to rodent pest management (Fall, 1977). But effective chemicals suitable for field use are yet to be developed.

(8) Chemical means: Rodent elimination by chemical means has been a popular method throughout the world. It is usually an economical way of eliminating rats, particularly if numbers are large. The earliest form of chemical control was probably through the use of arsenic compound in rodenticide formulation. Of the many rodenticides used, zinc phosphide, arsenic trioxide and sodium mono fluroacetate were more common during the last quarter of this century. Zinc phosphide was widely used during the Second World War by Federal and State agencies through pest control operators in Europe, UK and USA.

Many of the commonly used rodenticides have become ineffective as a result of resistance acquired by rodents. Several reports on the resistance in rodents against rodenticides are available (Greaves *et al.*, 1994). Since rats have well-developed sense of smell and taste, they easily discriminate the poisoned and unpoisoned baits (Bhardwaj & Prakash, 1982). Hence, rats can easily escape from chemical control operations. Unless proper care is taken, the preparations and distribution of poison baits may be hazardous to human health and environment.

Bait Shyness: The phenomenon of bait shyness is extensively studied in a number of species. Bait shyness is reported in almost all the Indian rodents such as *Tatera indica, Meriones hurrianae, Gerbillus gleadowi, Rattus rattus, Rattus cutchicus, Rattus meltada, Mus booduga, Mus musculus, Mus platythrix, Bandicota bengalensis* and *Bandicota indica* (Prakash & Mathur, 1987). There is almost no alternate to zinc phosphide, the only effective acute rodenticide. As such it is essential to overcome bait shyness so that the poison can be used for the second time in the same operation. Information on the obliteration of bait shyness of rodenticides is scanty. An alternative technique is needed to manage the rodent population. Thus the use of pheromone seems to be a positive approach in rodent pest management in this direction.

PHEROMONAL COMMUNICATION IN RODENTS

Like most mammals rodents also mark their territories with urine or other bodily secretions. These secretions carry chemical signals between the members of the same species. Mice and rats also leave scent markings around their burrows and nest entrances to help them relocate their shelters. Pheromonal markings also assist colony members in social interactions. Scents left behind can communicate family relationships, colony status, reproductive status and aggressiveness.

(1) Sources

Urine and faeces are the two most primitive as well as common sources of mammalian pheromones (Dominic, 1976; Prakash *et al*., 1998). Apart from urine and faeces, a number of specialized cutaneous gland secretions elicit pheromonal responses in mammals including rodents. There are about 40 different specialized scent glands identified in mammals (Balakrishnan & Alexander, 1985). Based on the specific body sites, they are named as armpit, flank, tarsal, ventral, anal, caudal, digital and inter-digital glands (Mary & Balakrishnan, 1991). In rats and mice, the secretion of preputial glands is reported to act as a sex attractant (Bronson & Caroom, 1971). Vaginal secretions are also the source of pheromones responsible for sexual attraction and arousal in male hamster (Singer *et al*., 1984).

(2) Classification

Based on the type of response by the recipients pheromones are classified as two major types i.e., "primer" and "releaser". Releaser pheromones induce a rapid behavioural response in the recipients generally mediated through the central nervous system. The pheromones involved in sex attraction, evocation of aggression, recognition, alarm behaviour and mother-young interactions are the examples for releaser pheromones. Primer pheromones induce a delayed response to prolonged stimulation mediated through central nervous system and endocrine system. The following are some of the best examples for primer pheromones: i) disruption of oestrous cycle in group-housed mice ii) Male-induction of oestrus in group-housed females iii) Male induced acceleratory influence on the onset of puberty and iv) pregnancy block by odours of strange males (Vandenbergh, 1999). The rodent sex pheromones included as releaser types are being investigated in our laboratory to use in pest management programme.

(3) Perception of pheromones

In mammals, odorants are detected by the olfactory epithelium (OE) that lines the nasal cavity. Signals generated in olfactory sensory neurons in the OE in response to odorants are relayed through main olfactory bulb (MOB) to the olfactory cortex and then to other brain areas. Through these pathways odour signals ultimately reach higher cortical areas involved in conscious perception of odours as well as limbic areas, such as the amygdala and hypothalamus that are involved in emotional and motivated responses. Mammals have a secondary olfactory sense organ called the vomeronasal organ (VNO/ Jacobson organ). The VNO is a tubular structure in the nasal septum that is connected to the nasal cavity by a small duct. Removal of the VNO or severing its connection to the brain interferes with pheromone effects but not with general odour sensing. Although some pheromones are sensed in the OE (Dorries *et al.*, 1997), many reports suggested that the VNO might be specialized to detect pheromones (Halpern, 1987; Keverne, 1999). It is likely that the two systems are specialized for receiving and responding to two different kinds of molecules; the main olfactory system primarily handles relatively volatile chemical substances that are air-borne, whereas the vomeronasal organ (i.e. accessory olfactory system) primarily deals with large, relatively non-volatile molecules that are contacted by the animal.

(4) Pheromones and behaviour

Several types of behaviour have been recorded in rodents in response to pheromonal signals. Attractive, aggressive, mother-young interaction, territorial marking are some of the behaviour exhibited by the influence of pheromonal cues. Further, these signals are assessed and have been confirmed through their manifestation in behavioural changes exhibited by conspecifics. Therefore, pheromones and behaviour are inseparable. The behaviour shown by conspecifics to releaser pheromones are interesting and have been reported in detail including farm animals.

Recognition of the individual's sex may be the first step in a series of behavioural interactions. Urine of male mice contains releaser pheromones, which attract females as well as stimulate aggressive activities in males. It is well known that in many rodent species odours of oestrous females are more attractive than that of diaestrous females

(Johnston, 1983; 1990). Similarly, male rats investigated urine odours of receptive females, although sexual experience may be necessary for this preference in some cases (Dominic, 1991). One of the thoroughly studied examples of sexually significant odours in the mammalian literature is the vaginal secretion of hamster (Pertrulis *et al.*, 1999). These secretions serve to attract and reduce aggressive behaviours as well as elicit testosterone surges in males (Pfeffier & Johnston, 1994).

House mouse (*M. musculus*) living in pens or as commensals in farm buildings establish territories that are defended by the residents. Recognition of boundaries of adjacent territories is thought to be due to the presence of odours emanating from excretory products of the resident animals. Evidence suggests that urine is the main source of odours responsible for territorial marking (Hurst, 1990). The anogenital area of Wistar pups are shown to be very attractive to the dam which is due to the preputial gland secretion in pups (Brouette-Lahlou *et al.*, 1991). This attraction is an important link in the mother-young interaction and crucial to pup survival. An animal shows a specific sexual behaviour in response to some chemical cues and the same cues may be involved in aggressive behaviour. The behaviour of animals before, during and after copulation have been studied more intensively in recent years because of interest in the mode of mechanisms and consequences of sexual selection (Ferkin *et al.*, 2001). In most cases the chemical characteristics of signalling and the metabolic pathways by which they are produced are unclear.

Self-grooming is a specific type of behaviour among terrestrial vertebrates and may serve several functions (Sprujit *et al.*, 1992). Many animals including mouse, vole, rat and squirrel engage in self-grooming during and /or after copulation and such activity have also led to a variety of suggestions concerning motivational, physiological and social functions of grooming (Spruijt *et al.*, 1992; Ferkin *et al.*, 1996). The studies suggest that self-grooming may be a tactic used by animals to communicate sexual interest by broadcasting scents that are specifically important for reproduction (Ferkin *et al.*, 1996, 2001). Behaviour studies in rodents provide evidence that certain behaviour such as sexual attraction, grooming and aggression are triggered and maintained by pheromones.

CHEMISTRY OF PHEROMONES

Pheromones have strikingly different chemical structures. It is as if this type of communication is so ancestral that many different molecules have evolved or been co-opted to serve pheromonal functions.

Pheromone is a single compound. The important property of a compound used as a chemical signal is an appropriate volatility. This limits the upper molecular weights of an olfactory pheromone to about 300 and the effective lower molecular weight to around 50. Many different chemicals have been reported as pheromones that include alcohols, aldehydes, acids, saturated or unsaturated aliphatic or aromatic compounds from non-polar molecules such as alkenes to very polar compounds which may be acidic (acids or phenols) or basic (amines) (Dominic, 1991). The list of identified pheromonal compounds in various sources of rodents and their function are given in Table 1.

Table 1. Rodent pheromonal compounds

Animal	Source	Compound	Function	Reference
Hamster	Vaginal fluid	Dimethyl di sulphide	Attraction	Singer *et al.*, 1984
Mouse (♂)	Urine	a) 2-sec-butyl dihydrothiozole b) Dihydro-exo-brevicomin	Attraction, aggression and puberty acceleration	Jemiolo *et al.*, 1985
Mouse(♂)	Urine	a) Isobutylamine b) Isoamylamine	Puberty acceleration	Nishumura *et al.*, 1989
Rat (pup)	Preputial gland	Dodecylpropionate	Dam attra-ctant	Broutte-Lablow *et al.*, 1991
Albino rat (♂)	Preputial gland	2,6,10- decatrien-1-ol,3,7,11-trimethyl (Z,E)	Attraction	Kannan, *et al.*, 1998
Albino rat	Clitoral gland	a) 6,11-dihyro-dibenz-b,e- oxepin-11-one b)2,6,1dodecatrien-1-ol-3,7,11-trimethyl(z)	Attraction	Kannan & Archunan, 2001b
Albino rat (♂)	Cheek gland	1,2-benzene dicarboxylic acid (2-ethylpropyl) ester	Attraction	Kannan & Archunan, 2001a
Albino rat (♀)	Cheek gland	2,6,10-dodecatrien-1-ol,3,7,11-trimethyl (Z,E)	Attraction	Kannan & Archunan, 2001a
Lesser bandicoot rat (♂)	Cheek gland	a) 3-octen-1-ol b) Cyclopentane undecanoic acid c) 2,4,6,8-tetramethyl-1-undecane	Attraction	Kannan & Archunan, 1999

Animal	Source	Compound	Function	Reference
Lesser bandicoot rat (estrus♀)	Cheek gland	L –alanine,1,1-dimethyl ethyl ester	Attraction	Kannan & Archunan, 1999
Mouse (♀)	Urine	2,5 dimethyepyraz ine	Puberty delay	Novotny *et al.*, 2001
Mouse (♂) & (♀)	Urine	2-heptanone	Oestrus Extension	Novotny *et al.*, 2001
Mouse (♂)	Preputial gland	α- and β Farnesene	Puberty acceleration	Novotny *et al.*, 2001
Houser rat (♂)	Urine	a) Ethanol,2-(octylthio) b) 1,3,5-triazone-2,4-diamine c) 1-Chlorodecane	Attraction	Selvaraj & Archunan, 2002b
House rat (♀)	**Urine**	a) Hydroperoxide b) 1-nitropentane c)4-azidoheptane	Attraction	Selvaraj & Archunan, 2002b
Mouse (♂)	Urine	a) 3-amino-s-triazole b) 4- ethyl phenol c) 3-ethyl-2,7-dimethyl octane	Attraction	Achiraman & Archunan, 2002a

In addition to this, lot of workers believe that lipids and proteins are actively involved in olfactory communication in most of the mammalian species. Sebaceous glands of rat, rabbit and fishing bat have many kinds of fatty acids, glyceroids that appear to be involved in individual recognition and scent marking behaviours. Mice and rats excrete great amount of protein (non-volatile) in their urine but their function remained unknown for many years. The apparent waste of large quantity of protein in urine first suggested the idea that it would be involved in some physiological function; its unique presence in the males was an indication that it would be involved in pheromonal communication (Pelosi, 1994; Achiraman & Archunan, 2002b).

RODENT PEST MANAGEMENT WITH PHEROMONES

There are effective pheromones for a variety of insect pests of stored products and now they are being developed for flies, fabric pests, cockroaches and others. Eventually, there may be pheromones for numerous pests. If one can understand the behaviour that is induced by chemical stimuli, then it can be exploited for human advantage. The most important properties of pheromones are: i) requirement in minimal quantity, ii) species-specificity iii) effective even during night and iv) effective over long distance. By using pheromones we can not only monitor pests but also use them as attractants for other control measures. The available literature reveals that rodents mostly depend on chemical communication for their survival. Hence, applying the pheromonal technology in rodent pest management would definitely be a successful one.

LABORATORY FINDINGS

A. Efficacy of releaser pheromone

1. Pheromones as sex attractants

Sex pheromones can be utilized in two ways in rodent control. Synthetic analogues of large number of rodent pheromones are known at present. It has been demonstrated that the male house rats are more attracted to scent odours of oestrus female than non-oestrus female rats (Selvaraj & Archunan, 2002a). It indicates that the female oestrous urine may contain specific volatile compounds, which probably are involved in the reproductive behaviour especially in the attraction of opposite sex. In addition, the oestrus female shows significant response towards male than female urine.

The preputial glands of rat contain various volatile constituents such as dibutylpthalate, decanoic acid, 3-decyn-2-ol, 1-nitropentane and n-hexylmethylketone (Kannan *et al.*, 1998; Kannan & Archunan, 1999). Male rat urine contains 1,3,5- Triazone 2,4 diamine and 1 chlorodecane; the female urine had also three volatiles viz. (i) Hydroperoxide , (ii)1–nitropentane and (iii) 4-azidoheptane (Selvaraj & Archunan, 2002b). It is further confirmed that the synthetic compounds act as sex attractants for conspecifics. Therefore, urine and preputial glands appear to be potential pheromonal source to develop attractive traps for RPM programme.

2. Pheromones in masking bait shyness

Rodents can easily learn to discriminate harmful food and harmless food through their well-developed olfactory system (Bhardwaj & Prakash, 1982). Bait shyness behaviour in *T. indica* and *M. hurrianae* is reported to have been overcome to some extent by adding conspecific urine to the poison baits (Soni & Prakash, 1988). These reports lead to the importance of scent sources and their role as phagostimulant and also in masking poison/bait shyness. Our recent study reported that the preputial gland extract of the albino and house rats have the ability to inhibit bait shyness resulting in increased poison bait consumption and mortality rate. Rats took relatively larger amount of poison bait due to the attraction of preputial gland extract, thereby masking the poison/bait shyness. Prakash & Mathur (1987) have also pointed out that mixing freshly collected urine with poison bait significantly reduced the aversive behaviour towards rodenticides in desert gerbils. Identical report has been obtained in male house rat (Selvaraj & Archunan, 2004). These observations suggest that the conditioned aversion towards poisons by rodents can be overcome by sex pheromones obtained from conspecifics.

B. Efficacy of primer pheromone

1. Pheromones in mating disruption

Primer pheromones have not been utilized in rodent pest management programme so far. However, insect synthetic pheromones introduced into the environment were found to mask the natural pheromones disrupting the normal pheromonal communication between conspecifics. Such a disruption would cause failure of mating. The Bruce effect is a similar phenomenon whereby the exposure of a recently inseminated female to the pheromones of an unfamiliar male conspecific causes the termination of pregnancy in mice (Bruce, 1959). This effect has been most extensively studied in laboratory mice. Now it has been shown to occur in prairie deer mice (*Peromyscus maniculatus bairdii*), collared lemmings (*Dicrostonyx groenlandicus*) and several species of voles such as *Microtus ochrogaster, M. pennsylvanicus, M. agretis, M. montanus* and *M. pineetorum* (Brennan, 1999). The usage of strange male odorant may be another possible application in controlling the rodent population. Similarly, the female sex odorants can also be effective in disrupting the mating behaviours of other females. Since this technique has worked well in insect pest management, it can also be effective in RPM programme.

FIELD STUDIES

a. Mode of pheromone application in the field

Considerable research has gone into methods of application of pheromones in insects, whereas not much has been done in rodents. Kannan (1998) identified sex-specific volatile compounds from the scent glands of house rat and lesser bandicoot rats and used these compounds in trapping rats in fields. In his study about 10–20% of male and 30–40% female rats were trapped with the assistance of preputial gland extract, which is considered as partial success in the field. However, control of rodents upto 10% is considered as a major achievement (Prakash & Mathur, 1987). Thus the success in these studies supports the pheromonal role in rodent control by trapping.

Selvaraj (2002) made an attempt to find out the volatile compounds of preputial/cheek gland extract and urine of house rats in suppressing bait shyness behaviour in poultry farms and rice godowns. Among the various compounds tested 1, 10, dibromo decane, 1-nitropentane and 1-chlorodecane were found to be highly effective. The field evaluation confirmed the compound, 1-nitropentane, as more effective in increasing the mortality rate followed by 1,10 dibromodecane and 1-chlorodecane. The field evaluation further confirmed that the mixing of synthetic pheromonal compounds of glandular/urinary origin with the poison bait increased mortality in house rats (Selvaraj & Archunan, 2004). Large-scale application of these pheromonal compounds may lead to increased death rate in fields also. It is possible to use these pheromonal compounds individually in trapping rats in fields, but it is believed the success rate may not be satisfactory. The limited success of rodent control is due to the volatility of these compounds, which evaporate and diffuse quickly into the environment. The methods of pheromone release in the fields for insect control are by micro encapsulation; hollow fibre and pheromone baited traps. However, in rodents these methods may not be reliable. So, there is a need for a method to slowly release sex-specific pheromone(s) in the field.

PROTEINS AND THEIR BINDING ACTIVITY IN DEVELOPING PHEROMONAL TRAP

Both mice and rats secrete pheromones in their urine and deposit these signals as scent marks throughout their home ranges. Furthermore, these pheromones are bound with a protein molecule, a member of lipocalin family. These proteins are known as Major Urinary Protein (MUP) in mice and α-2U globulins in rats. More interestingly, the time of dissociation of pheromone protein complex is very long and urinary protein can store and slowly release the odorants into air from few hours to days. In addition, the urinary proteins are usually stable to drying and heating and are not likely to be quickly denatured when released out of the body. These proteins play a crucial role in protecting the pheromones from being rapidly lost by evaporation or degradation.

They thereby extend the 'shelf-life' of the scent marks by releasing the pheromones slowly as the protein degrades (Cavagioni *et al.*, 1999; Beynon & Hurst, 2003).

Urinary pheromones are transported or carried by a specialized vehicle called pheromone carrier proteins (Lipocalin) in most of the mammalian species including mice and rat (Flower, 1996; Hurst, 1990). The identification and isolation of this specific protein and assessing its binding activity with the pheromonal compounds will surely help in developing phermonal traps for rodent pest management.

Advantages/disadvantages of pheromones in RPM

S.No	Advantages	Disadvantages
1.	Effective in minor quantities	Lack of identification of all pheromones
2.	Non-pollutant and ecosafe	
3.	Labour saving	Control of only one sex
4.	Species-specific and safe to non-target organisms	Require more knowledge
5.	Easy handling	Delay in success

TECHNIQUES FOR PHEROMONE TRAPS

Based on the techniques adapted and success in insect pest control with pheromones, the following strategy has been employed to develop pheromonal trap for rodents:

1. Pheromonal traps without baits

In this technique, the plastic lid containing a meager quantity of pheromonal compound (s) is kept inside a rodent trap in the shape of funnel with large container. Such traps can be kept in grocery shops, poultry farms, godowns etc. The attractant will be slowly released thereby rats will be trapped in large numbers.

2. Pheromonal traps with baits

The pheromonal compound (s) can be added to baits used in the rodent traps such as tender coconuts, tomato etc. This can be used in houses and small shops to trap the small population of rats infesting. Since very little amount of pheromones is used in this technique, it seems to be eco-safe in houses.

3. Pheromones with poison baits

The pheromonal compound (s) can be mixed with the poison bait and introduced in agricultural fields as well as near the rat burrows. The rats will be attracted towards the sex pheromones which will mask the bait shyness thereby the consumption of bait is increased leading to enhanced mortality rate.

4. Pheromones to disrupt mating

In another interesting aspect, the pheromonal compound (s), which is involved in pregnancy block, estrous cycle disruption and aggression are to be identified and introduced in areas where the rodent populations are more. These pheromonal compounds disrupt the reproductive function and bring about confusion; thereby the population of rodents can be managed.

LACUNAE

The role of pheromones has been mostly confined to laboratory mice and rats. Many rodent pheromones and their bioactivity have not been studied extensively in wild species. Even the identified pheromonal compounds have not been synthesized and tested in the fields so far. Moreover, the investigations are concentrated largely on mice with very few studies on rats. Field rodents are almost totally neglected. Although the discovery of insect pheromones and their application in insect pest management is well established, attempts to manage rodents using pheromones have not successful when compared to insects. The following are the possible reasons for lack of pheremonal application in RPM programme: i) the limited understanding of rodent behaviour towards pheromones ii) all chemically identified pheromones are not suitable for pest control and iii) factors like stability, volatility and reactivity of the pheromonal compounds may affect their function.

RECOMMENDATIONS

- The isolation and identification of pheromonal compounds and their binding proteins in a number of rodent pests are required. This may be a huge task but it is possible.
- Elaborate behavioural studies in rodents need to be carried out.
- Numerous field evaluations are needed. Even though the success is very slow the results will be fruitful after a long period.

- Adequate funds should be allotted for research on rodent pheromones.
- Government and non-government organization are to be educated about the biological role of rodent pheromones.

CONCLUSIONS

The progress in understanding pheromonal communication among rodents in the past four decades is simply astounding. However, most of the knowledge about application of pheromones in rodent control is still at preliminary stage. Even though there is lack of work on rodent pheromones this review provides broad picture to consider pheromones as an important tool in rodent pest management. Our laboratory is one of the prime labs in India making efforts in identifying rodent pheromones and in developing pheromonal traps for rodents. The usage of pheromones in rodent control is eco-safe and cost effective. Thus, it is important to concentrate on identification of pheromones, bioassay and field evaluation with the objective of designing an effective RPM programme. This pheromone technology along with other control techniques will bring a significant input for RPM programme.

ACKNOWLEDGEMENTS

The pioneering work carried out by Prof. C.J. Dominic Banaras Hindu University, Varanasi and Prof. Ishwar Prakash of Central Arid Zone Research Institute, Jodhpur on rodent pheromones in India are gratefully acknowledged. The preparation of this Chapter was partially defrayed from Indian Council of Agricultural Research and University Grants Commission, New Delhi.

REFERENCES

ACHIRAMAN, S & G.ARCHUNAN. 2002a. Urinary proteins and pheromonal communication in mammals. *Indian J. exp. Biol.* 40: 1077-1078.

ACHIRAMAN, S & G.ARCHUNAN. 2002b. Characterization of urinary volatiles in Swiss male mice (*Mus musculus*): bioassay of identified compounds. *J. Bio. Sci.* 27: 679-686.

ADAMS, M.G. 1980. Odour producing organs of mammals. *Zool. Soc. London.* 45: 57-86.

BALAKRISHNAN, M & K.M. ALEXANDER. 1985. Sources of body odour and olfactory communication in some Indian mammals. *Indian Rev. Life Sci.* 5: 277-313.

BARNETT, S.A & I. PRAKASH. 1975. RODENTS OF ECONOMIC IMPORTANCE IN INDIA. Arnold-Heinemann, London. 75 p.

BEYNON, R.J & J.L. HURST. 2003. Multiple roles of major urinary proteins in the house mouse, *Mus domesticus*. *Bioche. Soc. Trans*. 31: 142-146.

BHARDWAJ, D. & I. PRAKASH. 1982. Poison aversion in *Rattus rattus* through acclimatization. *Indian J. exp. Biol.* 20: 396-398.

BRENNAN, P. 1999. Bruce effect. In: ENCYCLOPEDIA OF REPRODUCTION Vol. 1 Academic press, New York. pp 433-438.

BRONSON, F.H. & D. CAROOM. 1971. Preputial gland of the male mouse: Attractant function. *J. Reprod. Fert.* 25: 279-282.

BROUETTE-LAHOU, I., R. AMOUROUX., F. CHASTRETTE., J. COSNIER., J. STOFFELSMA & E. VERNET-MAURY. 1991. Dodecynyl propionate, Attrcatant from rat pup preputial gland: Characterization and Indentification. *J. Chem. Ecol.* 17 : 1343-1354.

BRUCE, H.M. 1959. An exteroceptive block to pregnancy in the mouse. *Nature*. 184: 105.

CAMPBELL, R.W., D.A. MANUWAL & A.S. HARESTAD. 1987. Food habits of the common barn owl in British Columbia. *Can. J. Zool.* 65: 578-586.

CAVAGGIONI, A., C. MUCIGNAT-CARETTA & R. TIRINDELLI. 1999. Pheromone signalling in the mouse: role of urinary proteins and vomeronasal organ. *Arch. Ital. Biol.* 137: 193-200.

DOMINIC, C.J. 1976. Role of pheromones in mammalian fertility. In: NEUROENDOCRINE REGULATION OF FERTILITY. (Eds.): T.C.A. Kumar. Karger, S. Basal, Switzerland. pp. 236-245.

IBID 1987. Primer pheromonal influence on reproduction in rodents. *Indian Rev. Life Sci*. 7: 21-49.

IBID 1991. Chemical communication in animals. *J. Sci. Res.* (Banaras Hindu University). 41: 157- 169.

DORRIES, K.M., E.ADKINS-REGAN & B.P. HALPERN. 1997. Sensitivity and behavioural responses to the pheromone and rosterone are not mediated by the vomeronasal organ in domestic pigs. *Brain. Behav. Evol.* 49: 55-62.

DOTY, R.L. 1980. Scent marking in mammals. In: COMPARATIVE PSYCHOLOGY: RESEARCH IN ANIMAL BEHAVIOUR. (Eds): M.R.Denny. Wiley Pub., New York. pp. 385-399.

ELLERMAN, T. 1961. THE FAUNA OF INDIA, (MAMMALIA, RODENTIA), VOL III (Pts I and II). Government of India, New Delhi. 849 p.

FALL, M.W. 1977. *Rodents in tropical rice*. Technical bulletin no. 36. Rodent Research Center, University of the Philippines at Los Banos College, Laguna, Philippines.

FALL, M.W. 1991. Controlling rice losses to rodents in rural communities. In: RODENTS AND RICE. (Eds.): G.R. Quick. IRRI, Los Banos, Philippines. Pp. 7-16.

FEILDER, L.A & M.H. FALL. 1994. Rodent control in practice in tropical field crops. In: RODENT PESTS AND THEIR CONTROL. (Eds.): A.P. Buckle & R.H. Smith. CAB International, Wallingford, U.K. pp. 313-318.

FERKIN, M.H., E.S. SOROKIN & R.E. JOHNSTON. 1996. Self grooming as a sexually dimorphic communicative behaviour in meadow voles, *Microtus pennsylvnicus*. *Animal Behaviour.* 51: 801-810.

FERKIN, M.H., S.T. LEONARD, L.A. HEATH & C.G. PAZ-Y-MINO. 2001. Self grooming as a tactic used by prairie voles, *Microtus ochrogaster*, to increase scent marking detection by conspecifics. *Ethology.* 107: 939-949.

FITZWATER, W.D. 1972. 'Barrier fencing in wildlife management'. *Proc. Fifth Vertbr. Pest Conf.*, Fresno, CA. pp 49-55.

FLOWER, R.D. 1996. The lipocalin protein family structure and function. *Biochem. J.* 318: 1-14.

GREAVES, J.H., A.P. BUCKLE & R.H. SMITH. 1994. Resistance to anticoagulant rodenticides: RODENT PESTS AND THEIR CONTROL. (Eds.) A.P. Buckle & R.H. Smith. CAB International, Wallingford, UK. Pp 197-217.

HALPERN, M. 1987. The organization and function of the vomeronasal system. *Ann. Rev. Neurosci.* 10: 352-362.

HURST, J.L. 1990. Urine marking in population of wild house mice, (*Mus domesticus*): communication between the sexes *Anim. Behav.* 40: 233-243.

JEMIOLO, B., D.J. GUBERNICK., M.C. YONDER & M. NOVOTNY. 1994. Chemical characterization of urinary volatile compounds of *Peromyscus californicus*, a monogamous biparental rodent. *J. Chem. Ecol.* 20: 2489-2499.

JOHNSTON, R.E. 1983. Chemical signals and reproductive behaviour. In : PHEROMONES AND REPRODUCTION IN MAMMALS. (Eds.): J.G. Vandenbergh. Academic Press, New York. pp. 3-37.

JOHNSTON, R.E. 1990. Chemical communication in golden hamsters: From behaviour to molecules and neural mechanisms. In: CONTEMPORARY ISSUES IN COMPARATIVE PHYSIOLOGY. (Ed.): D.A. Dewsburry. Sinauer Press, New York. pp. 381- 409.

KANNAN, S. 1998. Histomorphology, Biochemistry and Chemical identification of Rat scent glands in view of rodent pest management: A new light in IPM programme. Ph.D. Thesis. Bharathidasan University, India.

KANNAN, S & G. ARCHUNAN. 1999. Identification of volatile compounds from cheek glands of lesser bandicoot rats and assessment of behavioural response for identified compounds. *Ind. J. Exp. Biol.* 37: 798-802.

IBID. 2001a. Chemistry of clitorial gland secretions of the laboratory rat: Assessment of behavioural response to identified compounds. *J. Biosci.* 26: 247 - 252.

IBID. 2001b. Rat cheek gland compounds: Behavioural response to identified compounds; *Indian J. Exp. Biol.* **39** 887-891

KANNAN, S., K. RAMESH KUMAR & G. ARCHUNAN. 1998. Sex attractants in male preputial gland: Chemical identification and their role in reproductive behaviour of rats. *Curr. Sci.* 74: 689-691.

KARLSON, P & A. BUTENANDT. 1959. Pheromones (Ectohormones) in insects. *Ann. Rev. Entomol.* 4: 39-58.

KEVERNE, E.B. 1999. The Vomeronasal organ. *Science* 286: 716-720.

MacDONALD, D.W & M.G.P. FENN. 1994. The natural history of rodents: Pre-adaptation of pestilence. In: RODENT PESTS AND THEIR CONTROL (Eds.). A.P. Buckle & R.H. Smith. CAB International, Wallingford, UK press. Pp.1-21.

MARY, E & M. BALAKRISHNAN. 1991. Specialized cephalic skin glands of *Suncus murinus viridescens*. *Acta. Theriol.* 36: 187-192.

NISHIMURA,K., K. UTSUMI., M. YUHARA & Y. FUJTANI. 1989. Identification of puberty accelerating pheromones in male mouse urine. *J. Exp. Zool.* **251**: 300-305.

NOVOTNY, M.V., B. JEMIOLO., D. WIESLER., W. MA., S. HARVEY, F.XU., T.M. XIE & M. CARMACK. 1999. A unique urinary constituent, 6-hydroxy-6-methyl-3-heptone, is a pheromone that accelerates puberty in female mice. *Chem. Biol.* **6**: 377-383.

PARSHAD, V.R. 1999. Rodent control in India. *Integrated Pest Management Reviews* 4: 97-126.

PELOSI, P. 1994. Odorant –binding proteins. *Rev. Biochem. Mol. Biol.* 29: 199-228.

PETRULIS, A., M. PENG & R.E. JOHNSTON. 1999. Effect of vomeronasal organ removal on individual odour dicrimination, sex-odour preference and scent marking by female hamsters. *Physiol. Behav.* 66: 73-83.

PFEIFFER, C & R.F. JOHNSTON. 1994. Hormonal and behavioural responses of male hamsters to females and female odours: Role of olfaction, the vomeronasal system and sexual experience. *Physiol. Behav.* 55: 129-138.

PRAKSAH, I. 1988. Rodent problems in India. *FAO. Plant Prot. Bull.* 36: 145-148.

PRAKASH, I & R.P. MATHUR. 1987. MANAGEMENT OF RODENT PESTS. Indian Council of Agricultural Research, New Delhi, India.

PRAKASH, M & S.S. SARWAL. 1958. Electric fencing. *Indian Forester.* 84 : 70-73.

PRAKASH, I., MD. IDRIS & S.KUMARI. 1998. Scent marking behaviour of three desert rodents. *Proc. Indian Natl. Sci. Acad.* B 64: 319-334

REGNIER, F.E. 1971. Semiochemicals structure and function. *Biol. Reprod.* 4 : 309-326.

REKWOT, P. I., D. OGWU., E.O. OYEDIPE & V.O. SEKONI. 2001. The role of pheromones and biostimulation in animal reproduction. *Anim. Reprod. Sci.* 65: 157-70

SEAL, S.C & K. BANERJI. 1966. Changing pattern of rodent population in Calcutta and Howrah. In: *Proc. Indian rodent Symp.* (Ed.) K.L. Harris Calcutta, India Pp. 69-83.

SELVARAJ, R. 2002. Pheromonal identification and its involvement in the control of house rat (*Rattus norvegicus*). Ph.D. thesis, Bharathidasan university, Tiruchirapalli, India.

SELVARAJ, R & G.ARCHUNAN. 2002a. Chemical identification and bioactivity of rat (*Rattus rattus*) urinary compounds. *Zool. Stud.* 41: 127-135.

IBID. 2002b. Role of male scent glands in improving poison bait acceptance in female rats, *Rattus norvegicus. Indian J. exp. Biol.* 40: 53-57.

IBID. 2004. Efficacy of male scent glands and urine in masking poison bait shyness of female house rats, *Rattus rattus. Pest. Biochem. Physiol.* (Communicated).

SINGER, A.G., A.N. CLANCY., F. MACRIDES & W.C. AGOSTA. 1984. Chemical studies of hamster vaginal discharge: Male behavioural responses to a high molecular weight fraction require physical contact. *Physiol. Behav.* 33 : 645-651.

SONI, G. R & I. PRAKASH. 1988. Mitigation of poison shyness in desert gerbil by adding conspecific sebum of ventral scent marking gland and urine in poison baits. *Indian. J. exp. Biol.* 26: 476-478.

SPRUJIT, B.M., J.A.R.A.M. VANMLOOF & W.H. GISPEN. 1992. Ethology and neurobiology of grooming behaviour. *Physiol. Rev.* 72: 825-852.

TOOVEY, F.M. 1953. Field planting of oil palms. *J.W. Afr. Inst. Oil Palm Res.* 1: 68-75.

VANDENBERGH, J.G. 1999. Pheromones in Mammals In: ENCYCLOPEDIA OF REPRODUCTION. Academic Press. UK. pp. 764-769.

WHITAKER, R. & S. BHASKAR. 1978. Rodent control in the Lakshadweep islands. *Rod. Newsl.* 2(4): 1.

URS, Y. L. 1996. Control of *Rattus rattus* in an aquarium through eco-safe traps and ultrasonic repeller- A new approach. *Rod. Newsl.* 20: 2

❑❑❑

13

CONSERVATION STATUS OF RODENTS IN INDIA

E.A. Jayson and K.M. Jayahari

Division of Forest Ecology and Biodiversity Conservation, Kerala Forest Research Institute, Peechi, Thrissur, Kerala-680 653

ABSTRACT

India has ninety three species of rodents out of which twenty are endemic. Among the endemic species, *Ratufa indica dealbata* is extinct, eight are in vulnerable category, one is in the critically endangered list (*Biswamoyopterus biswasi*), one is endangered (*Artherurus macrourus*), one is data deficient, two are considered as data deficient and many of them are treated as pests. Totally there are seven endangered species namely *Berylmys bowersi, Cricetulus migratorius, Diomys crumpi, Marmota bobak, Mus famulus, Niviventer brahma* and *Ratufa macroura dendolena.* There are 19 vulnerable species and 22 data deficient species. In view of the large number of vulnerable, endangered, and threatened and data deficient species of rodents in India, it is strongly advocated that all large scale rodent control programmes should precede studies on species composition and population density. Other suggested measures are non-lethal approaches to rodent control such as barriers and predators. In addition wildlife census should include small mammals and steps should be initiated to detect and reduce bioaccumulation of rodenticides.

INTRODUCTION

The concept of pest control is an economic approach to a biological problem and to a conservation biologist the 'management' of rodents sounds better than their 'control'. Rodents, which are small mammals, are an integral component of many ecosystems because they act as prey, predators and insectivores (Martin & McComb, 2002). More than being a myth, serving as source of food and medicine for humans, and being pests, rodents are an essential part of the ecosystems serving as the connecting link between the biotic elements in an ecosystem. Their '*r*-reproduction strategy allows them to be an important component of the trophic level to which they belong, by transferring energy from green plants and insects to small carnivores, reptiles and birds. Different species of rodents at different ecological niches contribute to the food web according to their ecological status, an act which is crucial for the transfer of energy. Alteration of small mammal communities by pest control programs will affect the various components of a stable ecosystem and ultimately the whole system (Sieg, 1987). Additionally bioaccumulation of poisons used for rodent control is a serious problem in rodent pest management programs.

Rodents by their '*r*'-reproductive strategy and their connecting link position between the trophic levels of an ecosystem have a crucial role in the maintenance of biodiversity and also in the functioning of the ecosystems. Detailed studies on the species composition and population structure of rodents in an area can reveal the series of ecological processes that prevailed in the past. In the Indian subcontinent, where alternate wet and dry geographic zones with highly diverse environments exist and where many of the animal groups show disjunct distribution, examination of the existing species relationship will confirm the various biogeography hypotheses and also will help to understand the ecological history of earlier periods. Rodents will serve as the best field tools for such studies and if a certain rodent species becomes extinct it will show as missing link in such studies. The above facts demonstrate that the conservation of rodents is crucial in many aspects.

RODENT SPECIES OF INDIA (Fig. 1 to 3)

IUCN RED LIST CATEGORIES

According to the IUCN red list categories (2000) the taxa are preliminarily divided into two broad categories namely Evaluated and Not Evaluated. The Evaluated category is further subdivided in to Data deficient and adequate data available groups. The adequate data available groups are further categorized into Extinct, Extinct in wild, Threatened and Low risk categories. The threatened group is further classified into Critically Endangered, Endangered and Vulnerable categories. The Low risk group is further sub grouped into Conservation dependent, near threatened and least concerned categories.

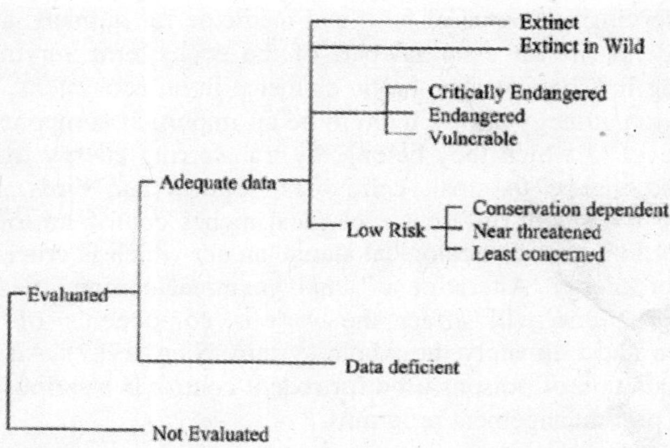

Fig. 1 IUCN Red list categories

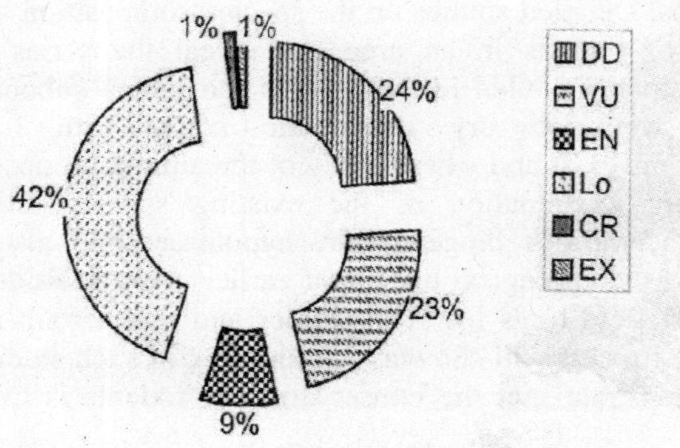

Fig. 2 Status of rodents in India in IUCN Red list categories

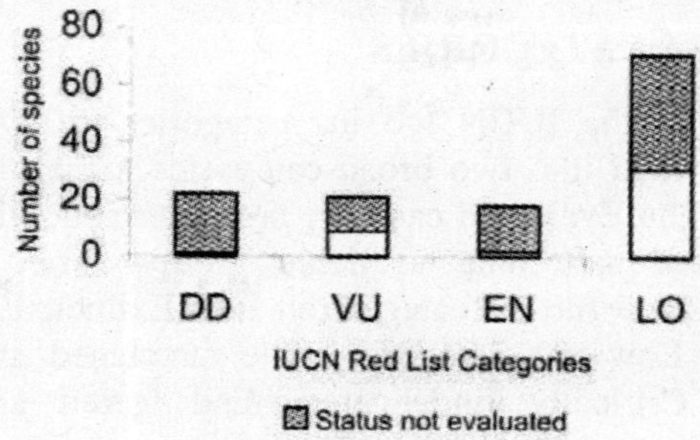

Fig. 3 Number of species in different categories whose status has not evaluated either globally or nationally

Ninety three species of rodents are present in India and out of these 20 species are endemic. Forty three per cent fall in Low risk category (IUCN 2002) *i.e.* more than fifty per cent of the species are in threatened category or status unknown (Data Deficient). A total of 72 per cent of the species lack proper IUCN status evaluation either globally or nationally due to deficiency of data, but pest control measures are applied even on such species for economic gains. Thirty-seven species are in the Low-risk category and 9 species are in the vulnerable category. Eleven species are considered as vulnerable since their status is not yet evaluated (CAMP, 2002). Among the twenty endemic species *Ratufa indica dealbata* is extinct. Eight of the endemic species are in vulnerable category, one is in the critically endangered list, one is endangered, one is data deficient and two species are considered as data deficient and many of them are treated as pests. This unfolds a scenario where, relentless control measures are implemented in the absence of information about the conservation status of many rodent species.

Conservation status of different threatened species is given below.

EXTINCT SPECIES (EX): One species

1. ***Ratufa indica dealbata*** **(Erxleben, 1777)** [EX]-**The species is endemic to India,** was distributed at a single location in Gujarat State. Arboreal and inhabiting dry deciduous forests. It is now categorized as extinct.

CRITICALLY ENDANGERED SPECIES (CR): One species

1. ***Biswamoyopterus biswasi*** **(Saha, 1981)** [CR] Namdapha Flying Squirrel. **The species, endemic to India**. Arboreal, nocturnal and distributed only in Arunachal Pradesh. The distribution is restricted to a location in Namdapha with a declining extent of occurrence. Human interference and the loss of habitat are assessed as the main threats. Recently discovered species. Home range<10km.

ENDANGERED SPECIES (EN): 1+7 Species

Atherurus macrourus **Thomas,** [EN]. Brush tailed porcupine. **The species is endemic to India** and the distribution is confined to two isolated locations in Arunachal Pradesh and Meghalaya.

1. ***Berylmys bowersi*** **(Anderson, 1879)** [EN]. Bower's Rat is both fossorial and arboreal, distributed in the rain forests of Arunachal Pradesh, Meghalaya, Mizoram, Manipur and Nagaland in North-east India.

2. *Cricetulus migratorius* (Pallas, 1773) [EN]. Little Gray Hamster. This is an Endangered species. This species is distributed in Jammu & Kashmir. Even though the species has a wide distribution, its presence is restricted to certain locations in the given region. Distribution extends from the dry cultivated valley to the arid montane steppy of its range.

3. *Diomys crumpi* (Thomas, 1917) [EN]. The distribution of Manipur Rat is restricted to a single location which is not confirmed. Habitat is forest tract in hilly region.

4. *Marmota bobak* [EN]. The Himalayan Marmot has a fragmented distribution in Sikkim. Terrestrial, fossorial and diurnal. Lives amidst dwarf rhododendron and dwarf juniper.

5. *Mus famulus* (Bonhote, 1898) [EN]. Bemjpte's mouse. **The species is endemic to India** and distributed at the high altitude sholas, grasslands and montane forest in southern India and Andaman Nicobar Islands. Recently reported from Eravikulam National Park and Nilgiri Biosphere Reserve.

6. *Niviventer brahma* (Thomas 1914) [EN]. Mishmi rat. This fossorial rodent is seen in the rain forest of Arunachal Pradesh. Restricted distribution at a single location and habitat destruction is the main threat.

7. *Ratufa macroura dendolena* (Pennant 1769) [EN]. Grizzled (Sri Lankan) Giant Squirrel. This is reported from the Chinnar wildlife sanctuary and Srivalliputhur Grizzled Giant squirrel sanctuary and Palani hills in India. Also reported from Muthodi Wildlife sanctuary in Karnataka. The species comes under IUCN Vulnerable category. The riverine forest of the region supports the population of the species in peninsular India. Habitat loss is the major threat.

VULNERABLE SPECIES (VU): 19 species

1. *Apodemus draco* (Berrett-Hamilton 1900) [VU]. Fukien wood mouse; population is restricted to a single location of less than 100 sq.km in Arunachal Pradesh. Occurs in and around human dwellings in tropical rain forests.

2. *Hystrix brachyura* Linnaeus, 1758 [VU]. Malayan Porcupine. Four highly fragmented isolated populations are seen in four specific locations at Central and Eastern Himalayas.

3. *Micromys minutus* (Pallas, 1771) [VU]. Harvest Mouse is distributed in Meghalaya and Nagaland in India. Arboreal, inhabiting primary and secondary forests.

4. *Ratufa indica centralis* (Erxlebon 1777) [VU]. Malabar Giant Squirrel, is **endemic to Indian sub continent.** Arboreal, inhabiting the moist deciduous, semi evergreen and evergreen forests of the Western Ghats south to Biligiri Hills, Karnataka. Hunting and loss of habitat due to fragmentation are the main threats (CAMP 2002).

5. *Ratufa bicocolor gigantea* (Sparrmen, 1778) [VU]. Malayan Giant Squirrel. The species is arboreal, seen in the evergreen, semi evergreen, and moist deciduous forests of northeast India including West Bengal.

6. *Hylopetes alboniger* (Hodgson 1836) [VU]. Particoloured flying squirrel is a nocturnal, arboreal rodent seen in the temperate forests of Sikkim, Mizoram, Arunachal Pradesh and Nagaland. The distribution is restricted to limited area.

7. *Marmota caudata* (Geoffroy, 1844) [VU]. Long tailed Marmot is distributed in Jammu and Kashmir. Terrestrial and fossorial. Loss of habitat due to war and being hunted as food are the major threats to the species.

8. *Petinomys fuscocapillus fuscocapillus* (Jerdon, 1847) [VU]. Small Travancore Flying Squirrel. The species is endemic to the Western Ghats, viz. Annamalai, Periyar Tiger Reserve, Peppara Wildlife sanctuary, Srivalliputhur, Peechi.

9. *Chiropodomys gliroides* (Blyth, 1856) [VU]. Pencil-tailed Tree Mouse is distributed in Cherrapunji and Khasi Hills of Meghalaya State. Arboreal, inhabiting primary and secondary forests especially wherever bamboo is present.

10. *Rattus palmarum* (Zelebor, 1869) [VU]. Nicobar Rat. **This is endemic to India**. The species is distributed in the mangrove and tropical rain forest of Nicobar Islands.

11. *Hyperacrius wynnei* (Blanford, 1881) [VU]. Muree Vole is fossorial and distributed in the pine forest and associated grasslands in Jammu and Kashmir.

12. *Rattus tiomanicus* (Miller, 1990) [VU]. Malaysian Wood Rat. Arboreal and is found in the scrub, garden, plantation and secondary forests of Andaman and Nicobar group of Islands.

13. *Rattus stoicus* (Miller 1902) [VU]. Andaman Rat. **This is endemic to India** and distributed in the Andaman Islands, Henry Lawrence islands

Fummumbutus pennanti *Ratufa indica*

Platacanthmays lasiurus

14. *Dacnomys millardi* (Thomas, 1916) [VU]. Millard's (Large toothed) Rat is distributed in the forested tracts of the hilly regions of West Bengal and Arunachal Pradesh.

15. *Niviventer eha* (Wroughton, 1916) [VU]. Smoke bellied rat is fossorial and distributed in the high altitude, wet forests of Sikkim and West Bengal. The restricted distribution is confined to locations.

16. *Cricetulus alticola* (Thomas, 1917) [VU]. Laddakh Hamster. This species is seen in the cold rocky deserts of high altitude. Human interference is the main threat to the population.

17. ***Cremnomys elvira*** **(Ellerman, 1947)** [VU]. Ellerman's Rat. Reported to occur in Tamil Nadu**. This is an endemic species** found in the Eastern Ghats of India. This rat inhabits thorny forest and semi arid forest. The population is restricted to a small region of 11 km^2 from where the type specimen was collected (CAMP 2002).

18. ***Rattus ranjiniae*** **(Agarwal & Ghosh, 1969)** [VU]. Ranjini's Rat. **Endemic to India** reported only from Thrissur, Alleppy and Thiruvananthapuram Districts of Kerala.

19. ***Millardia kondana*** **(Mishra & Dhanda, 1975)** [VU]. The type specimen was collected from Pune District of Maharashtra. Distribution of the species was recorded only from the type locality. Fossorial, inhabits scrubs and bushes, forest marshes, rocky areas and around human dwellings.

DATA DEFICIENT SPECIES (DD): 22 species

1. ***Alticola albicaula*** **(True, 1894)** [DD]. Baltistan Mountain Vole which **is endemic to India** and is found in the Jammu and Kashmir region of the Indian Sub Continent. Fossorial, inhabits steep rocky areas near cultivation.

2. ***Alticola montosa*** **(True, 1894)** [DD]. **The species is endemic to Jammu & Kashmir region** and is found in Central Jammu & Kashmir and in the Pakistan occupied Jammu & Kashmir. Fossorial occurring in rocky uplands covered with coarse grass. The political problems like war and international terrorism prevents a detailed study.

3. ***Alticola roylei*** **(Gray, 1842)** [DD]. Royle's mountain Vole. The species is found at Khumaoun in Uttaranchal. Fossorial, inhabiting uplands and rocky grounds covered with coarse grass and also near human habitation.

4. ***Alticola stoliczkanus*** **(Blanford, 1875)** [DD]. Distributed in Jammu & Kashmir and Sikkim. Fossorial occurring around boulders at extremely high altitudes, preferably close to streams.

5. ***Apodemus sylvaticus*** **(Linnaeus, 1758)** [DD]. Thomas Wood Mouse is distributed in the Himalayan regions. Fossorial. Living around human dwellings and also in jungles at high altitude.

6. ***Berylmys manipulus*** **(Thomas, 1916)** [DD]. Manipur rat is distributed in Nagaland, Manipur and Assam States in India. Fossorial. occurring in hilly region with bamboo forest.

7. ***Eothenomys melanogastor*** **(Milne-Edwards, 1871)** [DD]. Peer David's vole. Distributed in Arunachal Pradesh.

8. ***Funambulus layardi*** **(Blyth, 1849)** [DD]. Layard's Stripped Squirrel. This is a diurnal arboreal species distributed in India and Sri Lanka. Habitat is hilly forest.

9. ***Funambulus sublineatus*** **(Waterhouse, 1838)** [DD]. Dusky Stripped Squirrel. Diurnal and arboreal. This is distributed in India and Sri Lanka. In India the species is distributed in the Western Ghats, from Coorg and South wards.

10. ***Hadromys humei*** **(Thomas, 1886)** [DD]. Hume's (Manipur bush) Rat. This is distributed in the states of Manipur and Assam in India.

11. ***Hylopetes barberi*** **(Blyth, 1847)** [DD]. Kashmir Flying Squirrel is found in Kashmir.

12. ***Hyperacrius fertilis*** **(True, 1894)** [DD]. True's Vole is recorded from Jammu & Kashmir and Punjab. Colonial, fossorial. Inhabits sub alpine, scrub and meadows.

13. ***Leopoldamys edwardsi*** **(Thomas, 1882)** [DD]. Edward's Rat. The species inhabits the rain forests of hilly region of west Bengal, Nagaland, Arunachal Pradesh and Meghalaya.

14. ***Microtus leusurus*** **(Blyth, 1863)** [DD]. This species is distributed at the grass lands and high mountain steppe region of Jammu Kashmir and Himachal Pradesh.

15. ***Mus pahari*** **(Thomas, 1916)** [DD]. Gardner's shrew mouse, Sikkim Mouse. The species is terrestrial and fossorial recorded from the forested hills in North-East India.

16. ***Niviventer langbianis*** **(Robinson & Kloss, 1922)** [DD]. Langian Rat. This is distributed in the North- East India.

17. ***Niviventer niviventer*** **(Hodgson, 1836)** [DD]. White Bellied Rat is distributed in many locations starting from Himalayas to Mishmi Hills in North-East India. Fossorial and forest dwelling. Occasionally near human habitat.

18. ***Niviventer tenaster*** **(Thomas, 1916)** [DD]. Tenasserim Rat. This species is distributed in the North-Eastern India.

19. ***Rattus nitidus*** **(Hudgson, 1845)** [DD]. The Himalayan Rat is distributed in the Northern and North Eastern India. Terrestrial, fossorial, living close to villages, cultivated fields and rocky areas.

20. ***Rattus sikkimensis*** **(Hinton, 1919)** [DD]. Sikkim Rat. This species is distributed in Sikkim, Meghalaya and Arunachal Pradesh. Terrestrial, arboreal, living near cultivated fields and adjacent forests.

21. ***Platacanthomys lasiurus*** **(Blanford W.T. 1891)** Malabar spiny dormouse, distribution restricted to Coorg, Mysore districts of Karnataka and northern Kerala.

22. ***Rattus turkestanicus*** **(Satunin, 1903)** [DD]. Turkstan Rat. This species is distributed in the Indian Himalayan region. Fossorial/ terrestrial, living in and around houses, rocky river beds and fields.

23. Sicista concolor **(Buchner, 1892)** [DD]. Chinese Birch Mouse is seen in the high altitude rocky areas of Jammu and Kashmir. Fossorial, inhabiting rocky areas and high altitudes

CONCLUSIONS

Rodent control programmes without proper evaluation of the status of the pest populations and species composition may lead to severe ecological imbalance in the nature. For example *Rattus ranjiniae,* categorized as vulnerable species (IUCN category) has a very narrow, restricted distribution, limited to a single location in Thiruvananthapuram District, Kerala. No measures are being taken to protect the species from destruction when rodent pest control is implemented. There is a dire need to take up steps to protect the endemic, endangered, threatened species before initiating rodent pest eradication programme. Some criteria to be observed is given below.

1. Detailed studies should be conducted before any massive rodent pest control measures is implemented, so that the species composition and population density of each rodent species can be understood and thereby the rare species can be protected.
2. Massive killing of rodents can be avoided by erecting physical or chemical barriers, which will prevent the entry of pests into the field.
3. Small mammal surveys should be an integral part of wildlife census conducted regularly so that the rare and endangered rodent species can be located and monitored.
4. Measures should be initiated to reduce the bioaccumulation of toxins in rodent populations.
5. Rodent control can be achieved by enhancing predatory pressure by increasing the density of small carnivores, birds of prey and other natural predators.

ACKNOWLEDGEMENTS

We are grateful to Dr. J.K. Sharma, Director, Kerala Forest Research Institute, Peechi, Kerala for valuable advice and suggestions. The authors are thankful to the Department of Science & Technology, Government of India for sponsoring the studies on rodents. Dr. C. Renuka, Programme Coordinator, Division of Forest Ecology and Biodiversity Conservation is acknowledged for the encouragement.

REFERENCES

CAMP workshop report, 2002. Rodentia, Insectivora and Scandentia CAMP Handbook: Planning for the Rodentia CAMP Review-2002. Zoo Outreach Organization, Coimbatore, India. 142p

MARTIN, K.J & W.C. McCOMB. 2002. Small mammal habitat associations at Patch and Landscape scale in Oregon, *Forest Science* 48(2): 255-264.

SIEG, C.H. 1987. Small mammals: Pests or vital components of the ecosystem. *Proc. 8th Wildlife Damage Control Workshop*, 1987. Rapid city, South Dakota, USA.

❑ ❑ ❑

14

MANAGEMENT OF CROP RAIDING ELEPHANTS

Shakunthala Sridhara

AINP on Rodent Control, University of Agricultural Sciences, G.K.V.K., Bangalore – 560 065

ABSTRACT

The Asian elephant *Elephas maximus* occurs as four distinct populations in the forests of South, Central, Northwest and Northeast regions of India. Elephants forage extensively, grasses, bamboo, bark of various trees and reed forming the bulk of their food. Being both grazers and browsers, grasses of *Thelmeda* sp., *Cymbopogon* sp., seedlings, clumps, lateral shoots of bamboo, twigs and leaves of several species of trees are the preferred food in nature. Severe crop damage by raiding elephants is reported for South India, Nilgiri biosphere, around Dalma Wildlife Sanctuary in South Bihar and in south West Bengal. Damage was severer to cereals like ragi (finger millet) in South India and paddy in West Bengal and Bihar. Only 5.5 to 9.1% of the cultivated area was raided. The major factor responsible for crop raiding is shrinkage, fragmentation and degradation of their habitat. The attraction of nutritive food in crop fields also contributes. The two ultimate factors responsible for crop raiding by elephants are optimal foraging strategy and enhancement of reproductive success by feeding on nutritionally rich cultivated crops. The chief strategies of crop protection namely prevention of raiding by barriers, guarding, chasing by various methods, monetary compensation, translocation, capture etc have limited success. Ultimately elephant nuisance can be prevented effectively only by making elephant habitat self sustaining, preventing its degradation and immediately stopping all cultivation around it. Parallely conversion of forest for monoculture plantations, illegal grazing, collection of firewood, fodder & unregulated fire should be stopped as also mining, construction of new roads, railway tracks etc. To some extent tourism should also be strictly limited. Growing suitable vegetation for elephants, development of water shed projects inside forests, swamp protection, rehabilitation of hamlets inside the elephant–habitat protect crops and conserve elephants. Elephant corridors are already planned/ executed. Adapting agro forestry as a buffer zone between forests and villages as well as changing the pattern of land use by growing non-edible crops as done elsewhere in Southeast Asia to manage crop raiding elephants are worth evaluating in local problem areas. Providing alternate fuels to wood, growing fodder, encouraging non-agricultural activities, compensating for not growing edible crops etc. are the suggested actions for people affected.

INTRODUCTION

The association of man and elephant is from time immemorial. In Hindu mythology and religion elephant is an object of worship, a faithful companion, a giant with a heart of gold. Elephant is a mammoth animal. Although shy by nature it is capable of becoming wild and fatally aggressive. It's requirements for food and living space is enormous. It's once vast natural habitats are reduced to small, fragmented refuges consequent to growing human population with increasing demands for land for agriculture and industry. As a result elephants not only come more and more in contact with human beings but compete with him for resources resulting in regular conflicts. Although crop depredation and killing of humans is historically documented, such conflicts which were far and infrequent are increasing alarmingly from past few decades. Crop losses due to the Asian elephant, *Elephas maximus* are well documented for Palamau National Park (Mishra, 1971), South India (Sukumar 1990; Nair *et al.*, 1980), North-eastern states (Choudhury, 1989), Nilgiri Biosphere (Balasubramanyan *et al.*, 1995), Dalma Wildlife Sanctuary (Datye & Bhagwat, 1995) and South West Bengal (Singh *et al.* 2002). Crop raids and losses are also documented for Africa (Buechner, 1965; Allway, 1979), Srilanka (Mckay 1973), Malaysia (Olivier, 1978; Blair *et al.*, 1979), Indonesia (Nyhus *et al.,* 2000), Tanzania (Newmark *et al.*, 1994), Uganda (Naughton-Treves, 1998) and Kenya (Thoules & Sakwa, 1995).

SCIENTIFIC STATUS

Class : Mammalia

Order : Proboscidea

Sub order : Elephantoidea

Family : Elephantidae

Genus : *Elephas*

Species : *maximus*

Binomial name: *Elephas maximus*

DISTRIBUTION AND HABITAT: Earlier the range of elephant distribution extended from Tigris and Euphrates valleys of Syria and Iraq to yellow river of China and South to Sumatra. At present it is found in varying numbers in India, Nepal, Bhutan, Bangladesh, China, Mayanmar, Thailand, Kampuchea, Laos, Vietnam, Malaysia, Sri Lanka and Indonesia.

In India elephants occur as four distinct populations distributed in South, Central, Northwest and Northeast regions.

The South Indian population of elephants is distributed over the forested hills of Western Ghats and Eastern Ghats of Karnataka, Kerala, Tamil Nadu and recently Andhra Pradesh. Geographically the area falls between 8° 15'N – 15° 30'and 74° 15 E – 78° E. The annual rainfall ranges from 50-500cm. The altitude of this habitat varies from 100-2000 msl. Vegetation is diverse from dry scrub through deciduous forest to wet evergreen forest. There has been considerable degradation of forests due to mining, hydroelectric projects, plantations of teak, rubber, coffee, tea and spices, numerous patches of cultivation and human settlement.

The central Indian elephant corridor consists of dry deciduous, moist deciduous and semi evergreen forests of Eastern Ghats in Bihar and Orissa states. The habitat is deteriorating due to felling of trees, shifting cultivation, replacement of natural sal (*Shorea robusta*) with teak (*Tectona grandis*) plantations and destruction of forests.

The Northwest Indian habitat is spread over the Terai forest region of Uttar Pradesh along the foothills of Himalaya. Reservoirs, power channel and industries are deteriorating this habitat also.

The North East Indian elephant habitat is highly fragmented extending from the Himalayan foothills of Bhutan and northern West Bengal to the states of Assam, Arunachal Pradesh, Nagaland, Manipur,

Mizoram, Tripura and Meghalaya. The habitat includes moist deciduous and evergreen forests of the Himalayan slopes as well as the swampy grasslands of flood plains of Brahmaputra river.

MORPHOLOGY: An average male elephant is 2.5 to 3 meters high and weighs 5000 kg. Body colour is greyish black. An elephant's most distinguishing feature is its long flexible muscular trunk. Appearing first at the age of two, the trunk continues to grow throughout the life of the elephant and weighs 60 kg. Elephants have a pair of ivory tusks which are larger in males. The ears are large and flappy. The size of the ear reflects its habitat; the hotter the climate, larger the ears. Legs have thick bones and a broad foot, which is supported by thick fatty material.

BIOLOGY AND BEHAVIOUR: Life span of elephants is 65-70 years in both the sexes. Gestation is 21 months with 1-2 young born at a time. Babies at birth weigh around 100 kg.

Elephants are social animals, normally living in herds of 5 to 20 animals, usually under a dominant bull. They live in a close knit family group. Within the group there is a hierarchical relationship. The herd size is dependent on season and availability of resources like food and water. The social life is matriarchal. There can also be all male herds, as well as solitary elephants or bulls. Communication is by emitting rumbling sounds or trumpeting and by touching each other with trunks. Pheromones too play an important role.

Elephants are emotional animals greeting each other with friendliness, comforting the weak members of the group, pampering and teaching the young, looking after the sick and protecting the family in times of danger. They even grieve the death of a family member, covering the dead body with leaves and staying with the body for several days.

Elephants are diurnal and are active from two hours before sunrise to two hours after dusk, sleeping only 4-5 hours a day. The hottest part of the day is spent under shady trees. They travel hundreds of kilometers and are good swimmers.

Elephant home ranges extend over extensive, continuous stretch of forests seeking food, water and shelter. The size of home range varies from region to region, season and availability of resources. A minimum range of 650 km^2 is reported for elephants in Kerala.

FOOD AND FEEDING BEHAVIOUR (Table 1): Elephants are known to spend 70% of the time on feeding. They are reported to consume food between 1.5% of their body during dry season to 1.9% of body weight during wet season. In captivity the food requirement is about 125 kilogram of hay and grains per day. Elephant being a polyphagous animal feeds on a number of plant species belonging to different species. Grasses, bamboo and reed form the major bulk of elephant's food. Generally there are seasonal variations in the consumption of grass/plant species and parts of plants. They also feed on the bark of various tree species. Bark feeding seems to fulfil the nutritional requirements of essential fatty acids like linoleic acid, minerals such as manganese, iron, copper, boron, calcium and sodium.

Detailed studies have been carried out on the feeding behaviour of South Indian elephants in Eastern Ghats and South India. Sukumar (1990) reported that although 112 plants were eaten by *E. maximus* in Eastern Ghats, 85% of their food came from just 25 species belonging to the families of Leguminosae, Palmae, Cyperceae and Graminae. He considers elephants as both grazers and browsers. They grazed mostly on tall grasses of *Themeda cymbaria, T. triandra* and *Cymbopogon flexosus*. Different parts of the grass were preferentially consumed in different seasons. Seedlings, clumps and lateral shoots of bamboo species, *Bambusa arundinacea* and *Dendrocalamus strictus* were eaten with relish. Twigs and leaves of several species of trees and varieties of fruits were also eaten. Overall higher consumption of grass in all the seasons and in all the habitats was seen in Mudumalai Wildlife Sanctuary of South India except thorn forests where browse formed 98.5% of the diet (Sivaganeshan & Johnsingh, 1995). In these forests during wet season II and dry season fruits were also consumed to a greater extent because of their high protein content compared to grasses. Amongst grasses *Thelmia* species was consumed relatively more across the seasons because of its greater availability. Additionally the palatability of its young leaves in wet seasons and succulancy of basal tussock in dry season favoured their preferential consumption. Palatability, protein content, softness, digestibility, greenness, carbohydrate content of grasses influenced food selection of elephants along with their availability.

TYPES OF MAN-ANIMAL CONFLICT

Broadly the conflict between humans and elephants can be categorized into four types.

a. **Killing/injuring people** occurs due to two reasons. Primarily it is accidental. Nervousness and confusion on the part of both humans and elephants result in elephant attacks on humans and subsequent death and injury to the later. On the other hand there are confirmed "rogue" elephants that have learnt the vulnerability of humans which deliberately chase and kill people. Usually these "rouges" are solitary bulls.

b. **Crop raiding** is a major destructive activity of elephants, indulged by almost all wild elephants when they find favourable opportunities to do so. An interesting observation is the fact that the crop raiding by herds is seasonal coinciding with natural stages of crop growth as against the wrought by solitary bulls all round the year.

c. **Damage to property** is by way of breaking into houses. Elephants raid houses in search of food grains, salt, sometimes for country liquor and very rarely to retrieve the calves which have accidentally strayed into villages. A large number of housebreaking occurs routinely in the tea gardens of Assam and West Bengal. As in crop raiding, there are habitual housebreakers similar to habitual depredators of crops.

d. **Loss of livestock** is rare. Usually buffaloes are the victims.

CROP RAIDING BY ELEPHANTS (Table 1)

Of the two serious damaging activities of elephants namely human kills and crop raiding, this chapter deals mainly with crop losses inflicted by elephants in India. Although there are records in the past about crop losses due to elephants, the continuous habitat loss due to expanding agriculture, industry, mining and roads inside the forests has lead to more and more cases of elephant raids on crops throughout the country. Such losses have been quantitatively estimated in detail for Bihar (Mishra, 1971; Datye & Bhagawat 1995), South India (Sukumar, 1990; Nair *et al* 1980), Nilgiri biosphere (Balasubramanyan *et al.*, 1995), North Eastern states (Choudhury, 1989) and South West Bengal (Singh *et al.*, 2002). It was Sukumar's (1985) detailed observations on the crop raiding behaviour of elephants, quantification of losses and correlating crop raiding behaviour with elephant's life history strategy in some villages of South India during 1981-83 that pioneered serious studies of quantification of crop losses in the country. He found that finger millet was the most damaged food crop. Fields with rich growth and high density at the stage of flowering or grain filling were selectively attacked. Only terminal part of the plant with inflorescence/ grain was consumed and the basal stem along with root was discarded. When finger millet was harvested, elephants foraged on paddy, little millets, sorghum and jowar. The stalk bearing inflorescence was again selected in the last two cases. While rachis from coconut tress was pulled out and consumed, in case of banana, stems were split and the fibrous pith was eaten. Often inflorescence or fruit bunch was eaten. Sugarcane stalks were broken and eaten completely. In mango tree damage was restricted to flowering branches till fruiting; jackfruits were either plucked or shaken off the trees, smashed by trampling and consumed.

With the onset of rains and larger areas coming under cultivation, the frequency and intensity of crop raiding increased reaching a peak during October-December coinciding with crop maturity. In some villages, fields were raided every night. The recorded intake of crops was 5472.2 kg by 297 herd members on 33 nights and 4450.5 kg by 131 bulls in 77 nights. The frequency of raiding was correlated with the size of the herds and the size of the enclave. Cultivated crops formed 9.3% the annual food requirement of males and 1.7% of family herds (Sukumar, 1990). Similar results were obtained later in 26 villages of Nilgiri Biosphere, distributed in the states of Karnataka and Tamil Nadu during the 1991-93 crop seasons (Balasubramanyan *et al*, 1995). These villages were within or bordering moist deciduous, dry deciduous and thorn forests. Peak of raiding coincided with crop maturation with two raiding peaks when crops were raised in two seasons and a single peak of damage if crop was raised once a year. Of the fields visited 9.1% were raided and damage occurred in 2.1% of the raided fields. Crops were damaged by both trampling and eating. Overall 3.45% of paddy, 1.5% of ragi, and 1.4% maize were damaged. Ginger was not damaged but 1.7% of the crop was trampled. Damage to other crops namely sunflower, groundnut, horse gram and cotton was nil or negligible.

Table 1. Foraging Behaviour of Elephants

Location	Plants eaten/crop raided & damage	Reference
S.INDIA	*Themeda cyambaria* *Cymbopogon flexusus* grasses	Sukumar ' 90
NILGIRI BIOSPHERE	*Thelmia sp.* Grasses	Sivaganeshan & Johnsingh'95
S.INDIA	5472.2 kg crops in 33 nights (Herds) 4450.5 kg crops in 77 nights (bulls) Mostly Finger millet, Paddy, Sorghum, Jowar, Coconut rachis, Banana pith, Sugarcane stalks	Sukumar' 85
NILGIRI BIOSPHERE	Paddy- 3.45% Ragi-1.5% Maize 1.4%	Sivaganeshan & Johnsingh '95

	Ginger – 1.7% (trampled)	
S.BIHAR	Paddy worth Rs. 1.8 million	Datye & Bhagawat' 95
S.WEST BENGAL	5.5% cultivated area mostly paddy raided average loss 40% =Rs. 3.2 crores. Paddy>Potato>Vegetables>Wheat>Maize	Singh *et al.*, 2002
MEGHALAYA	Annual losses of Rs. 24000/farmer	Williams & Johnsingh' 96

Not all elephants raided crops although both males and females of certain clans raided. There was temporal and spatial over lapping of home ranges of raiding and non raiding elephants. No significant difference was seen between male and female groups in the frequency with which they raided crops. Males caused greater per capita damage, but females inflicted greater total damage. Although the rate of raiding between states did not differ significantly, there was variation in the extent of area damaged and crops damaged in different areas (Balasubramanyan *et al.*, 1995).

Similar temporal pattern of raiding was seen in ten villages in and around Dalma Wildlife Sanctuary in South Bihar (Datye & Bhagawat, 1995). The elephants living inside the sanctuary started expanding their range after the rains. Initially, when paddy shoots appeared in the month of July, adult bulls descended to the crop fields sporadically. The family herds came down a little later in the month of August. By September-October both adult bulls and herds indulged in maximum number of raids. Once the elephant population migrated to Midnapore of West Bengal raids decreased in the subsequent months of November-December. The herd size varied significantly in mid and final stages of paddy season. The smaller herd of five members in the early stages of crop rose to thirteen later. There was an age dependent raiding pattern in males with adult bulls ravaging the crop individually while sub adult males visiting crops as members of a herd or in the company of other sub adults. All elephants, irrespective of sex exhibited fidelity to the cropping site.

Around 150 villages were raided around the sanctuary and all along the migration route of elephants to West Bengal. However the damage decreased as the villages became more distant from the sanctuary. The average number of raids was 2.53/village for adult bulls and 2.83/village for herds. Thus the latter showed a slightly higher rate of raiding.

In the migrated districts of Purulia and Midnapore, elephants were totally dependent on paddy causing heavy damage. Herds caused more damage in both Bihar and West Bengal. Damage to paddy was both by consumption and trampling. Paddy loss for the ten villages was worth 0.18 million rupees in 1989 and 1990, and 0.16 million during 1991. Hay lost ran into thousands of rupees per village but was meagre compared to grain lost.

Singh *et al.* (2002) studied the crop raiding behaviour of elephants in West Bengal in detail. Their major findings were: Since 1980 the elephant population of South West Bengal has increased consequent to increase in forest cover. Conflict between humans and elephants began from 1987 onwards consequent to migration of elephants from the neighbouring Dalma Wildlife Sanctuary in Bihar. These elephants covered a distance of 70 km in one single night starting in the months of August and left South West Bengal after paddy harvest in the month of January–March (Singh *et al.*, 2002). The elephant population was about 65 animals inhabiting 3368 km^2 distributed in four localized sub groups. Only 5.5% of the total cultivated area in the range of elephant movement was raided. Highest crop depredation occurred when the ratio of crop land: forest was 1.2:1. Crop depredation was lower when cultivated areas were larger than forest cover and vice versa. It was concluded that larger cultivated areas are beneficial to humans and moderate to high density forest cover is strategically important not only to reduce elephant–human conflict but also to conserve them. Intensity of crop damage was maximum during maturity of crop and was greatest for crops grown during *kharif* (51.8%, September-December) followed by *rabi* (33.6%: January-April) and summer (14.5%: May to August). The average crop loss was 40% resulting in annual losses of Rs. 3.2 crores. Apart from crop raiding and human deaths, 110-133 houses per annum were damaged by resident elephants. Maximum house damage took place between May and August when crop availability was poor.

The crops damaged in the decreasing order of severity were paddy (66%) >potato (16%)> vegetables (10%) > wheat (5%) > maize (1%). Damage also occurred to a lesser extent in orchards and granaries except during the months of September-December. At provisional and national level, the degree of economic loss was not very significant but at the local level it was severe as most of the affected farmers were marginal and subsistence level cultivators.

During the study period it was found that the migratory elephants from Bihar caused more damage to crops and life in the forest division of Midnapore. The better juxtaposition of croplands and forests contributed to this activity. In contrast to this, the marginal forest habitat was adequate for males to localize and raid crops because of their smaller ranges (Choudhury *et al*., 1985).

West Garo hills in Meghalaya is a terrain of patchy forests in various stages of degradation and *jhum* fields on accessible hill slopes and valleys. Since the habitat of elephants is sub optimal, the elephants are compelled to move from one patch to another in search of food. During such searches they come across the newly cultivated *jhum* fields which offer highly palatable nutritious and easily available food. The cost of loss due to elephant raiding was Rs. 24,000 (US $ 720) compared to the annual income of a Rs. 11,000 (US $ 320) of subsistence farmer calculated on the assumption that an average of 2000 elephant raids on agricultural fields occur in a year (William & Johnsingh, 1996).

REASONS FOR CROP RAIDING

The problem of crop raiding by elephants is a complex one with several factors contributing to the phenomenon. First and foremost, crop raiding is an age old habit of elephants only its severity has become greater. Sukumar (1985) attributes both *proximate and ultimate factors* responsible for this conflict with human interest.

Proximate factors have mainly to do with habitat alteration of elephants. It could be shrinkage, fragmentation or degradation which is a result of several other factors complimenting and supplementing each other. These factors are:

(a) *Reduction in natural habitat*: With ever-increasing human population, the primary necessity of higher food production is met mostly by encroaching forest areas. Although modern agriculture with the help of hybrid seeds, fertilizers, pesticides, improved irrigation and of late using genetically modified crops has gone a long way albeit almost entirely contributed to surplus food production in the country, horizontal expansion of agriculture has played a very crucial role in the earlier stages of enhanced food production and does so even now. Throughout the country almost all forest and protected areas have been encroached legally or illegally for cultivation. In most of the country the land encroached is used to cultivate food grains while in valleys of North and North West India fruit crop and vegetable cultivation take precedence. In contrast the forests in eastern, North-Eastern India are converted into tea gardens in the north and those in Eastern and Western Ghats into coffee, tea, rubber, coconut, cocoa, cashew, areca nut and spices. With economics of cultivation and price of agriculture produce deciding the cropping pattern, advertently or inadvertently, the elephant habitat of forests and mountains are being reduced throughout the country and subsequently converted/encroached for cash and food crops.

(b) ***Degradation of habitat***: Degradation is defined as any exploitation of the habitat that reduces the elephant's food resources (Sukumar, 1990). Such degradation would definitely drive elephants to easily available greener pastures. An important reason for depletion of elephant's natural food is due to intense grazing inside elephant habitat by the enormous cattle populations living around protected areas. A second lethal impact of cattle forages into elephant habitat is transmission of several diseases, an aspect much overlooked by all concerned.

Another major reason for deterioration of forests is felling trees and implementing some practices which are detrimental to the quality of elephant habitat. These include felling of large tracts of forests and conversion of natural forests into monocultures of teak, eucalyptus and other non-fodder species.

Lastly there is what Barua (1995) calls biotic interference in the form of illegal collection of fodder and firewood by humans from the forest, construction of reservoirs, roads, mining and forest fires deliberately set by villagers.

(c) ***Fragmentation of habitat*** is best illustrated by conversion of forest interiors into tea gardens and increased human settlements by way of hamlets inside elephant habitat. This leads to "pocketing" of elephant herds as well as resource depletion and obstruction to routes of elephant migration.

(d) ***Palatability and nutritive value of food crops*****:** The very same characters for which cereals are cultivated by man namely taste, digestibility, absence of toxins and nutritive value attract elephants to food crops. The fact that they are far superior in palatability and nutrition compared to the grasses, browser parts of trees, bark and fruits normally eaten by elephants in their habitat is again an influencing factor. Protein content, levels of calcium, sodium, sugars and amino acid are in far greater amounts in ragi and paddy, the two food crops predominantly raided by elephants in India (Sukumar, 1990).

Ultimate factors that induce crop raiding are two fold-namely optimal foraging strategy and enhancement of reproductive success in males.

(a) ***Optimal foraging strategy***: Being a massive animal with fodder requirement of 250-300 kg per day, an elephant has to spend 16-20 hours daily to procure such huge quantities of food. In contrast cultivated fields offer the same amount of food in lesser time and smaller area. The energy expended in terms of travelling shorter distances and lesser time foraging on a highly nutritive food is far less compared to the demands of their natural foraging in forests. In case of herds the predatory pressure by way of human retaliation is nil. For "rogue" bulls, although they face the risk of human attacks to repel them from crop fields, elephants being highly intelligent learn very soon that men with all his worldly weapons and modern fireworks can be easily bullied into retreat and leave the crop fields to them. They risk human attempts to drive them away and return to crop fields all around the year. Thus crop raiding behaviour of Asian elephant fits nicely into the optimal foraging theory in terms of more gains nutritionally compared to energy expended and risk involved.

(b) ***Enhancement of reproductive success*** seems to explain the higher frequency with which males raid crop fields. Elephant is a polygynous animal and thus needs better nutrition than available in degraded and fragmented natural habitats to increase its reproductive success (Sukumar & Gadgil, 1988). As such there is a selection pressure operating on them to obtain nutritious food even if it is risky. This may explain the crop raiding behaviour of bulls throughout the cropping season in spite of constant human attempts to chase them back to forests.

MEASURES TO MITIGATE CROP RAIDING BY ELEPHANTS (Table 2)

Elephants are remarkably intelligent animals capable of adapting to all kinds of attempts by man to mitigate crop losses caused by them. They cannot be contained by any single method compelling the scientists/managers/farmers to device suitable plan of action keeping in mind the behaviour, ecology and habitat of the crop raiders. Basically the entire issue of managing crop raiding elephants is three phased: Firstly passive methods to prevent elephants leaving their habitat by digging trenches and erecting electric and non-electric fences. Secondly, measures to be vigilant about approaching elephant, the primary aim being to spot the encroachers. The activities include maintaining buffer zones, creating watch towers, whistles to alert other villagers, continuous fire at the edges of fields where the elephants are likely to move in and cow bells placed regularly along a string fence to alert farmers when elephants foray into crop fields. Finally, employing active methods like throwing fire and flame balls, pepper sprays, shooting to scare, shouting, beating drums and other noices.

Measures devised so far fall into two types: short term and long term. The short term measures are mostly planned and resorted to give immediate protection against elephant raiding while long term measures aim at removing or reducing factors that cause elephant raiding of crop and to create living conditions for elephants within the forest self sustaining.

SHORT TERM MEASURES

These include use of repellents, deterrents, night vigils, capture, capture and translocation, destruction of "rogue" elephants, use of trained elephants, wildlife squads and monetary compensation.

1. Repellents and Deterrents: A repellent is a substance or an action or stimuli that wards off or repulses and a deterrent hinders or discourages elephant raiding of crops.

(a) ***Repellents*** are substances that are activated by humans and should immediately turn an animal away in a close approach or attack while a deterrent should prevent undesirable behaviour by turning an animal away before a conflict occurs and they need not be monitored or activated by human (Hunt, 1983).

Table 2. Existing measures to alleviate crop losses to elephants

SHORT TERM MEASURES	LONG TERM MEASURES
1. Repellents & Deterrents	1. Habitat improvement
(a) Repellents	
(i) Chemical repellents	(a) Preventing encroachment
"HATE-C4",	(b) Preventing deforestation
Secretions from temporal gland	(c)Preventing conversion of forests into
Pepper spray (*Capsicum oleroresin*)	monocultures
(b) Deterrents	(d) Growing fodder grass inside elephant habitat
(i) Passive deterrents	
Trenches	(e) Swamp protection
Electric fencing	2. Eco-development works
Non – electric fencing	(a) Use of alternatives to firewood
(ii) Active deterrents	(b) Preventing illegal grazing
Guarding	(c) Resettlement of tribals
Driving away using sound or light	3. Elephant corridors
(iii) Integrated approach using sound & light	
2. Capture of elephants	4. Agro-forestry as a buffer between forests and villages

3. Capture & Translocation
4. Use of trained elephants (*Koonkie*)
5. Wildlife squads
6. Monetary compensation

5. Changed land- use (cultivating non-edible crops)

Chemical repellents: A successful chemical repellent was manufactured by a German firm called "HATE-C4". It had no effect on elephant crop depredation in Malawi (Bell, 1984) but in South Africa, positive results were obtained when the chemical was used (La Grange, 1989). An earlier study indicated the effectiveness of temporal gland secretion of African elephants in repelling conspecifies (Gorman, 1986). The most promising results of using pheromones to attract/repel and contain cows were obtained for the elephant by Ramussen *et al.*, (1993, 1996, 1997). The atomized secretions collected from the temporal glands of *musth* bulls elicited avoidance behaviour in female Asian elephants and had the potential of repelling non-*musth* bulls (Perrin *et al.,* 1996). Elephants from both Africa and Asia are known to secrete chemical signals at the time of stress such as child birth, sickness and disruption of social organization. If identified and synthesized, such chemicals along with the *musth* chemicals have a great promise in rendering crop fields aversive to elephants.

Pepper spray to deter elephants after further evaluation is suggested in recent times (Santiapillai & Suprahman, 1986). A latest study in Zimbabwe (Osborn, 2002) indicates that a strong spray of capsicum oleoresin momentarily froze the crop raiding elephants, followed by an audible exhalation of air, a rumble or roar followed by, disoriented retreat emitting a series of excited trumpets, rumbles and roars. Compared to the traditional methods of guarding, sling shots, drums, dogs, throwing burning sticks employed by either a single farmer or a group of farmers, capsicum oleroresin repelled elephants more quickly. The capsicum oleoresin spray at the time of crop raiding results in elephants associating the unpleasant effects of the spray with crop destruction and thus the chemical may act as a secondary repellent. Crop raiding may be a learned behaviour by young males from a limited number of older bulls (Osborn & Welford, 1997). Thus it is possible to alter the behaviour of crop raiding adult bulls by aversive conditioning using capsicum oleroresin. Once the adults learn to avoid crop fields it is likely that young ones may imitate them. But we still need further evidence for this. The high cost of resin imports is justified by arguing that saving the food crops of subsistence farmers is more important than calculating the economics of deterring elephants. But compensating farmer with food grains and firewood or fuels is far more easier, practical and economical than spending huge amounts of money and efforts in deterring a mighty animal from crop fields. The psychological stress of threatening, deterring and chasing elephants is an often overlooked factor while suggesting such face to face measures of reducing elephant damage to crops.

(b) ***Deterrents*** can be further categorized into passive and active systems. *Passive* approaches are permanent in nature and need not be carried personally. These include one time major effort like digging trenches and electric fencing. *Active* systems include chasing the elephants from crop fields using traditional methods or killing individual "problem animals".

Under active systems guarding and driving away elephants is included. Guarding cultivated fields day and night will lessen damage and increase yields. Driving or repelling the elephants from fields using search lights, crackers, noisy explosives, beating drums, and loud noises is the age old method. Either elephants soon get habituated to these non-lethal methods and return to the crop or the bulls may charge at the noise makers when enraged by these methods. Or they shift from one cultivation field into the next one. There is also the possibility of keeping elephants away from crops by exploiting the low frequency distress calls emitted by elephants while calving and play back recordings of *Massi* cattle noice. The later is based on the assumption that elephants associate the sounds of *Massi* cattle with *Massi* tribesmen who periodically hunt and injure elephants. Elephants are believed to retreat on hearing the noice (Kangwana, 1995)

Night vigils: Villagers construct a temporary platform, *machan* on a sufficiently tall tree and sit out the night, keeping vigil on the crop. Mostly they are armed with bows, arrows and spears, torches and search lights. The animal moves away when confronted thus. More often, it comes back after a temporary retreat.

The passive system of deterring elephants is mainly by digging trenches and erecting fences.

Use of barriers includes elephant-proof trenches and electric fencing.

(i) Trenches: The dimensions of an ideal elephant trench should be at least 2m deep, 2m across at the top and 1.5m across of the base to effectively prevent elephant movement (Sukumar, 1989). In some places trenches are filled with pointed sticks to further deter elephants from crossing. Digging elephant proof trenches is enormously expensive and requires constant monitoring and maintenance leading to high recurring expenses. Further in hilly terrain trenches get filled up by the litter and debris from the run off water within no time and thus become ineffective.

(ii) Electric fencing: If the area to be protected is small or plots are of cash crop and raiding is sporadic, efficiently functioning electric fences prevent crop damage by elephants. However several factors have contributed to their failure in West Bengal and Bihar. These include theft of electric wires, grazers, indifferent maintenance, and lack of monitoring and wide gaps in the fence wherever it passes over water bodies. Although a strong fence with higher current and better maintenance may work better, elephants have learnt a few ingenious tricks to render electric fencing ineffective. They have been reported to mow down fence posts, breaking fence wires using tusks, throwing tree branches over the wire and even crawling under. Some solitary elephants just don't bother about the non-lethal shock and break through the fence. The other drawbacks of electric fencing are: they can never act as a physical barrier to migrating populations of elephants for long if the food becomes scarce within their habitat; they create new areas of depredation; they reduce the net forest available thus shrinking the already reduced habitat further. Further trying to confine elephants to a degraded habitat using electric fencing is highly unethical and cruel. A combination of fence and ditch has been tried in Kenya where a ditch is dug inside the fence. But the barrier was too expensive for farmers (Osborn & Welford, 1997).

(iii) Non-electric fences have been used successfully to restrict elephant in many parts of Asia and Africa. These fences are usually built with strong wooden/steel/railroad tracks driven vertically into the ground. Heavy gauge wire or cable is strung between the poles and drawn tight. Although these fences are successful, they are expensive to install and maintain (Bell, 1984).

(c) *Integrated approach of scaring elephants*: Siege & Baldus (1998) have found that an integration of sound and light is effective in making elephants to withdraw from crop fields in Africa. The strategy involved a shooter firing a blank shotgun at close range i.e. within 20-50 m of the elephant. When the animal starts running away, a flare thrown scares the elephant further. Finally a thunder flash hurled at the retreating elephants from a distance of 20-50 meters, increases the impact. When the method was slightly modified i.e. firing of shot gun and thunder flash were used at regular intervals followed by flares, elephants ran away to about 200-300m distance from the crops. Breeding elephants were more easily scared and kept away from crop fields for a minimum of two weeks when scared with shot guns and flares. It was easier to drive away solitary bulls than breeding herds using the integrated approach. On the contrary when traditional methods of shouting, making noice were used solitary bulls were not only unaffected, but they charged at the peasants trying to scare them. A flash light directed at such bulls evoked a charge and the animals attempted to follow the position and direction of the light (Seige & Baldus, 1998).

2. Capture of elephants to be effective should be limited to crop raiders and "rogue" elephants. It is highly impossible to identify raiders from non-raiders because of the overlapping of home ranges. On the other hand random capture has little effect on crop losses due to raiding elephants.

It was suggested that by taking away most of the solitary or all-male herds, losses due to crop raiding could be reduced (Sukumar, 1990). However, comparatively higher raiding by females more than males causing greater damages is becoming more common due to habitat loss and degradation (Datye & Bhagawat, 1995). The authors opine that before removing males, it is better to evaluate the damage caused by the two sexes. Additionally, sex ratios also should be taken into account because it is not advisable to remove males if their numbers are low. In such situations, they advocate paying compensation for crop damages as it is the only workable solution to alleviate the plight of farmers.

Capture and domestication of few elephants would bring down the crop damage in the migration range of elephants. Captured elephants can be used for tourism rides or kept in the parent sanctuary.

3. Capture and translocation: Domestication may work to some extent if the elephant population is excessive in a given habitat. But the possibility of excess population is rare. Carefully capturing after identifying the "rogue" elephant followed by tranquilizing and translocating him may offer some benefit. But translocation of males and herds without taking sex ratio into consideration will adversely affect elephant populations which are highly female biased with tuskers being killed indiscriminately by poachers.

Destruction of "rogue" elephants: Habitual man killing elephants are termed "rogue" which invariably happen to be bulls. Although there is provision to eliminate elephants that kill human beings under the Wild life Act (1972), habitual crop raiders and house breakers cannot be easily killed by any means. Normally once an elephant is declared a rogue, license is given to a competent person to kill it. However it takes a lot of time to identify the rogue. Coupled with the inordinate time taken by authorities to declare an elephant as a "rogue" this situation leads to more and more loss of lives and property making people resentful of both the bureaucracy and the rogue elephant. Additionally it is also not advisable to remove solitary males as the genetic viability of the herd is maintained by them.

4. Use of trained elephants (*Koonkie*) to chase away wild elephants: Specially trained elephants (*kunkis*) and Mahouts (*phandia*) chase and drive wild elephant herds into their original habitat. These operations were initiated in early 1980 in West Bengal and is even now resorted by forest department frequently. But the operation is not successful because of several reasons such as lack of availability of sufficient number of *koonkis* and *phandia*, the operation disturbing the seasonal movement and distribution of elephant herds, not being effective against solitary bulls and finally the entire operation becomes a failure because the elephants loose fear in 'chase without capture' action.

5. Wildlife squads: Many state governments have set up both permanent and temporary wildlife squads to deal with elephant menace of crop raiding, house breaking and attacks on humans. These squads are trained and equipped with vehicles, fire arms and Radio transmitter sets. They drive away raiding elephants using spot lights, crackers and gun shots. They do provide some relief. But they are not a total solution to elephant menace because many a times they lack proper training, are ignorant of animal behaviour, lack foresight and often drive away elephants from one place to another indiscriminately (Barua, 1995). They are most often pressurized by locals to open fire or chase elephants against their better judgment. Sometimes the squads reach the affected village too late or never due to unavoidable reasons or apathy. As a result villagers become hostile and impatient.

6. Monetary compensation: Thamba (1995, 1996) defines compensation as a method of raising threshold of the affected people for animals like elephants which cause lot of economic damage. Earlier such a threshold was quite high but of late people facing habitual elephant menace have become less tolerant. The major reasons for this changed attitude are increasing and regular crop raids which devastates a large portion of poor and marginal farmer's crops, frequent, violent deaths of farmers while protecting crops and in accidental confrontations and elephant raids on village houses. As per the guidelines of the Wildlife Protection Act (1972), the victims are legally eligible for compensation. The Kerala government has enacted a legislation to pay Rs. 10,000 for loss of life and handicap, and Rs. 5000 for crop losses depending on the degree of damage. In Karnataka the government pays full compensation to crops lost and suitably for loss of life. In Meghalaya Rs. 12,130,805 (US$ 391300) was paid as compensation for elephant damage for attacks between 1985-1993. Death outside the jurisdiction of forest department entails Rs. 10,000/- compensation but no compensation is offered if death or injury occurs within forests under the control of forest departments. The other state governments have also made similar provisions but the scheme was blatantly abused in some states with an estimated 26% bogus claims (Williams & Johnsingh, 1996). The overall compensation offered and paid seems to be meagre. Not all deserving cases get justice. In many cases full compensation does not reach the affected due to several socio-political and bureaucratic snags. Although the implementation and the amount of compensation is inadequate, in the absence of alternatives, it is advisable to continue the scheme with a humanistic approach to compensate the unfortunate victims suitably and quickly, and also raising the amount in view of escalated cost of living. Mandatory provision of sufficient budget for compensating elephant damage to crops and human life contributes to the conservation of elephants in an indirect way namely mitigating the hostile attitude of farmers to elephants and instilling in them interest in conservation. However, it is essential to document economic losses to assess the extent of damage. This is possible by monitoring crops and their value every season. In most cases damage is based on information supplied by the affected people which are more often an exaggerated amount. The only way to remedy the situation is following claims and instilling a sense of accountability in claimants.

LONG TERM MEASURES

Basically the problems of crop raiding by elephants and elephant-human conflicts in India are due to non-availability of adequate resources, primarily food for elephants in their natural habitat. This in turn is brought about by two major human activities namely shrinking of habitat of elephants and its degradation. The ever-increasing need for food by growing population all around natural reservoirs has naturally lead to encroachment of forest for agriculture. The degradation of the habitat is brought about by illegal cattle grazing depriving the already scarce fodder resources, illegal collection of firewood and bamboo, advertent or inadvertent forest fires by villagers etc. Often severe droughts, sometimes continuous for two to three years make water a scarce commodity. Thus ultimately and simply stating elephant nuisance can be prevented by making the habitat of elephants self supporting and preventing its degradation. The long-term measures can be four fold namely habitat improvement, eco-development work, elephant corridors and changed land use around the elephant habitat. Most of the ***'donts'*** come under habitat improvement and changed land use while ecodevelopment works and elephant corridors have the ***'do'*** components.

(1) Habitat improvement: The first major step in reducing human-elephant conflict is to mercilessly stop encroachment of land belonging to elephant habitat be it reservoirs or sanctuaries or national parks. Cultivation in a perimeter of three kilometres from the boundary should be banned totally. Vigilance should start at the beginning of crop season so as to prevent crop rising from the stage of land preparation itself. Strict warnings should be issued to villagers encroaching forest reservoir habitat. In spite of the warning, if they continue cultivation practices by sowing, all nurseries/germinating crops should be destroyed. Once the villagers understand that the forest department means its business they will soon follow the rules.

The second step is to stop such of the activities that degrade the habitat and take up measures that improve the sustainability of the elephant habitat. The activities that should be stopped to improve the habitat include conversion of forests into monoculture plantation crops, illegal grazing, collection of firewood, fodder and unregulated fire. Also to be stopped are mining, construction of new roads, and railway tracks, and to some extent to limit tourism too. Since the aviation industry in the country in poised for a leap making air travel reachable for even middle class people, all future plans of railway lines in the forests of elephant habitat have to be stopped. Instead existing roads/tracks are to be upgraded.

On the other hand habitat can be improved by growing fodder grass of elephants and bamboos, by creating water bodies, conserving water by adapting water shed projects, enriching the forest flora by planting suitable trees wherever the vegetation has thinned. Growing vegetation suitable for elephant diet has already been taken up by several state governments.

Swamp protection is also important for elephant conservation (Sivaganeshan & Johnsingh, 1995). Fire control strategies have to be evolved to protect swamps during dry season. Preservation of critical microhabitats like *nullahs*, dry streambeds, valleys and gallery forest, are also important.

In many an elephant habitat there are hamlets in the remote interior. People living here depend on forest for their requirements and grow some edible crops. Their living inside but depending on outside world for few necessities will definitely clash with the life of elephants. Settling these people outside the sanctuary will make sanctuaries much less disturbed for the elephants.

(2) Existing eco development works are to be implemented vigorously in villages around the national parks and sanctuaries. Basically the objective of such plans is to reduce the dependency of villagers on forests for grazing, firewood and other requirements. Meeting energy and fodder requirements outside the forests goes a long way in preventing degradation of elephant habitat. As some NGOs are doing in Karnataka, villagers have to be taught to use alternate sources of energy like cooking gas, which should be highly subsidized around sanctuaries. Villagers should be encouraged to grow adequate quantities of fodder for their livestock. These measures are not easy to implement and are slow to yield fruits. But the Indian farmer has always been receptive to practical solutions and as elsewhere in the world is down to earth when it comes to his crops and livestock. If he is educated how these long term measures will prevent elephant damage with some teaching, some coaxing, some warning and some rewarding he will definitely respond appropriately but such a transformation can take place only by the coordinated efforts of forest and agriculture departments. Though it looks far fetched, compensating farmers not to cultivate crops around sanctuaries in the pattern of American government paying cash to its farmers not to grow crops during some seasons to maintain prices seems a very practical and ideal solution for not only preventing crop damage by wild animals, but more importantly in conserving threatened and endangered animals. In fact it may be more economical and practical coupled with improvement of habitat and eco development works than methods like trenches, electric fencing, compensation, squads etc, which are immensely expensive and far from yielding satisfactory results.

(3) Elephant corridors: Corridors are known to exist that link one forest inhabited by elephants with another forest. By improving these corridors and keeping them free of human interference, natural and seasonal movement of elephants can be facilitated. Settlements and agriculture around the elephant corridors have been associated with crop raiding by elephants and their attacks on humans. The seasonal movement of elephants in definite routes is an instinctive behaviour. Disturbance of its natural route and nonavailability of food on its route is due to recent settlements and agriculture. It is not the fault of elephants if they feed on available crop during their journey. Maintaining elephant corridors free of human interference is one of the most vital steps in reducing human-elephant conflict.

(4) Agro forestry: Buffer zones are frequently established around the protected areas with the twin objectives of maintaining the integrity of the fauna and flora within the core area and to reduce the impact of people on the biodiversity inside (Mackinnon *et al.,* 1986). Putting simply, it means that buffer zones extend the available habitat for plants and animals i.e. extension buffering as against socio-buffering by providing resources and services to people. Agro forestry seems to have fulfilled both these roles in Southeast Asia particularly in Indonesia. These agro forestry systems are biologically richer than crop fields and monoculture plantations and not as rich as the core areas. Nonetheless they expand the habitat available for biological diversity beyond the boundaries of primary forests. Simultaneously they provide significant economic benefit to people living around protected areas. Small holders in Sumatra, Indonesia grow strands of trees as component of their diverse farming system resulting in the country becoming the world's second largest producer of natural rubber, third largest producer of coffee and the fourth largest producer of cocoa (from Nyhus & Tilson, 2004). Maintaining buffer zones around protected areas is a government policy here. Crop raiding in Sumatra was highest where secondary growth forest bordered villages and lower where grasslands and wet lands acted as buffer between forest and cultivated fields. (Nyhus & Tilson, 2004). But many more area-specific studies are needed before large scale operations of this nature are undertaken.

(5) Changed land use: Geographical Information System (GIS) analysis of data on elephant habitat availability and utilization in south West Bengal revealed that elephants ranged intensively only 243 km^2 out of 3368 km^2 overall area of their habitat. Thus habitat utilization of crop land was significantly less in proportion to their availability. Crop depredation showed a decreasing trend in response to increase in crop field or forest cover from the 1:1 ratio. When ratio of forest and crop land against depredation /unit area were plotted (Singh *et al*., 2002) highest depredation took place when the ratio was 1:1.2. Keeping this in view it was suggested that by growing non-edible crops like lentils and jute (*Corchorus capularis*) in very interspersed areas it is possible to reduce crop depredation. In Sumatra, Indonesia villagers affected by crop raiding elephants planted less valuable crops such as cassava or grasses near the border of the sanctuary to reduce the risk of elephant damage. Other villagers changed the schedule of crop raising (Nyhus *et al*., 2000). Although not much is known about crop preferences of elephants, there is definite information on crops, at least a few of them, which are unpalatable to elephants. By growing crops such as tea or oilseed which are not consumed by elephants, the attractiveness of cultivated fields can be reduced. Santiapillai (1990) suggests that a barrier of distasteful crops of at least 1 km width be grown around food crops. Attempts were made to grow the Mauritius thorn, *Caesalpinia decapetala*, cactus and sisal as barrier crops to prevent elephants raiding crops in Africa but little systematic research has been done on their effectiveness. Additionally farmers do not seem to be enthusiastic in having such barriers fearing that it will prevent them and their livestock entering the forests (Osborn & Anstey, 2002). In Sri Lanka steps taken in land use pattern include growing homestead gardens closer to villages where the likelihood of elephant damage is less. The secondary growth of hedges of forests not only provides food to elephants but gives them good cover before invading crop fields. By removing these hedges the visibility of approaching elephants becomes more. While cultivating crops it is also useful to take into consideration the fact that elephants are most likely to raid fields next to patches of jungle and crop fields that are away from the village (Thouless, 1994).

Another important factor is that elephant density is related to landform i.e. more elephants live in 'valley' landscapes than 'plateaus' resulting in farms in valleys being raided more (Bell, 1984). Agricultural lands along streams were found more vulnerable to crop depredation by elephants (Osborn & Welford, 1997).

HUMAN ELEPHANT CONFLICT (HEC)

It is an established fact that in almost all cases of people living around elephant habitats are at a disadvantage with crops routinely destroyed, property and buildings damaged, humans injured and killed. Under such life threatening activities it is unreasonable to expect affected people to be tolerant of elephant nuisance and be positive towards elephant conservation. Actually the reverse is the case with affected villagers protesting, agitating and in extreme cases favour killing the offending elephants. People become resentful of conserving a resource not owned by them and over which they have no control in spite its destructive activities. Such resentment is detrimental to the cause of elephant conservation.

Although losing an entire season's cultivation to elephants is devastating to a farmer at individual level, and heart breaking too, the total loss when analyzed at regional and national levels is meagre compared to other pests. To worsen the matters, data on crop damage and patterns of damage is poor. Long term studies in problem areas are almost nil. As a result comparison between sites and crops becomes difficult. There is a likelihood of researchers extrapolating results from damaged pockets to entire cropped area. Rarely claims of farmers are tallied with actual figures from field observations. But facing a situation of total crop loss within a night in the absence of alternate sources of income, the plight of a subsistence farmer is pitiable. The overnight destruction leaves a great impact on his survival and as well as emotions. He hates the elephant and at the same time he is afraid of the gigantic animal and its rage. He is frustrated because he cannot do anything as he is legally prevented from taking any action. Lastly the compensation he is entitled to is not only inadequate, but comes belatedly or not at all. The response of a peasant starts from complaining to local authorities, changing the cropping pattern, putting up obstacles in elephant's path, trying to frighten and chase the animal and finally seeking monetary compensation. When everything fails he resorts to killing the offending elephant.

On the other end of the spectrum, we have a majestic animal, unparalleled in its strength, beauty and intelligence, an animal that kept to itself most of the day foraging in well organized social groups where care of the group is more important than social ranking and is never a threat to any other animal including humans unless provoked. Such an animal is deprived of its natural share of food and living space because man wants more and more land to feed his ever increasing numbers and that of his livestock. As any living creature struggling to survive it resolves the issue by raiding crops and to hell with whoever interferes in its quest for food. In fact elephant with its enormous strength can do much more than raid crops given its potent rage and inherent intelligence. Man has a lot of choices to survive even if he happens to be a subsistence farmer living in villages around the protected area but the elephants pushed into or restricted to a fragmented and degraded habitat have absolutely no choice but to feed on alternate, available food in its range. It so happens that cultivated crops become an easily accessible resource.

Hitherto it was not a problem predicting temporal patterns of human-elephant conflict based on their migration pattern and route, availability and non-availability of food and water. But spatial distribution and variation of such conflict was difficult. Sitati *et al* (2003) attributed this inability to the unpredictable foraging behaviour of male elephants as well as variations in data resolution of earlier studies. Using a grid based geographical information system (GIS) with a 25-km^2 resolution combined with simple statistical tools, they could identify spatial predictors of HEC. It was possible not only to cluster crop raiding into distinct conflict zones but also predict the occurrence and intensity based on area under cultivation and for male elephant groups. Although elephant induced death and injury were less predictable, they could be correlated with proximity to roads. They claim that the method is transferable to all sites of HEC. Using this data a vulnerability map can be drawn which will enable to develop and deploy strategies to mitigate conflicts. However, it is advisable to integrate land use and livelihood strategies with other methods.

CONCLUSIONS

With all the available methods at our disposable to deter elephants from raiding crops like the first phase of preventing the animals by trenches and power fences, raising alarms by maintaining guard posts, chasing crop raiding elephants by light, sound and blank cartridges, using chemical deterrents and finally growing inedible crops, we are still unable to make even a dent in the seasonal pattern of crop raiding by elephants. The long term projects like improving the habitat of elephants and creating/restoring and maintaining elephant corridors are time consuming though theoretically the best possible way of keeping elephants contained to their natural foraging areas. Meanwhile we have to think of the villagers around protected areas and their problems. It is advisable to 'translocate' villages inside and around sanctuaries that are without permanent infrastructure like schools, hospitals, water supply. When translocation of elephants is feasible and planned, translocating villages to better areas is not difficult. We have the history of people rehabilitated when their lands were/are being submerged while constructing dams for irrigation and power generation. In villages having basic civic facilities, the life style has to be changed. Gradually with the help of officials of forest and agriculture departments they have to be oriented to grow cash crops and raise agro forestry wherever possible. Instead of food crops, sericulture, poultry and dairy should be encouraged. Village level craftsmen should be encouraged and centres for handicrafts be set up. Most importantly if the land is unfit for profitable agriculture and agriculture based activities, small scale industries keeping in view the raw materials available around should be set up. The Indian farmer is resilient, intelligent and highly adaptable as evident by our revolutions in food industry - Green, White, Blue. What is needed is a pragmatic plan to change the life style and livelihood patterns of villagers facing elephant problem. Extension workers of agriculture department and universities have brought about revolutions in cultivation. A similar approach will definitely work in converting villages around protected areas into hubs of agro forestry, dairy, poultry, sericulture, and small scale industries. What every one ultimately wants is purchasing power. Agriculture is not the only way to achieve it.

Spending on such projects of self sustenance is far more productive and beneficial to farmers and government than spending huge amounts on trenches, power fences, forest squads, firearms, etc. Once people become economically independent of cropping, their dependence on forests wanes and elephant habitat remains productive and elephant corridors assure them of safe journeys, adequate food and other resources. Ultimately once we take care of people, people will not exploit nature and the wild fauna including elephants are conserved naturally.

REFERENCES

ALLAWAY, J.D. 1979. Elephants and their interactions with people in the Tana River region of Kenya. Ph.D. thesis, Cornell University, USA.

BALASUBRAMANYAM, M., N. BASKARAN, S. SWAMINATHAN & A.A. DESAI. 1995. Crop raiding by Asian elephant (*Elephas maximus*) in the Nilgiri Biosphere reservoir of South India. In: A WEEK WITH ELEPHANTS (Eds.): Daniel, J.C. & Hemanth Datye. Bombay Natural History Society & Oxford University Press, Bombay, India. pp. 350-367.

BARUA, PARBATI. 1995. Managing a problem population of elephants. In: A WEEK WITH ELEPHANTS (Eds.) Daniel, J.C. & Hemanth Datye. Bombay Natural History Society & Oxford University Press, Bombay, India. pp. 150-162.

BELL, R.H.V. 1984. The Man-Animal interface: an assessment of crop damage and wildlife control. In: CONSERVATION AND WILDLIFE MANAGEMENT IN AFRICA. *Proc. Workshop organized by US peace corps at Kasungu Natl. Park*, Malawi. (Eds.) Bell, R.H.V. & E.M. Caluzi.

BLAIR, J.A.S., G.G. BOON & N.M. NOOR. 1979. Conservation or cultivation: The confrontation between the Asian elephant and land development in peninsular Malaysia. *Land. Dev. Digest*. 2:27-59.

BUECHNER, H.K. 1965. *Report on the potential conflict between elephants and agriculture in the proposed Tana River irrigation scheme*. In: UNDP/FAO. 1968. Survey of Irrigation potential of the lower Tana River Basin, Kenya. Vol. 7.

CHOUDHURY, C.L. 1989. The Indian elephant in a changing world. In: CONTEMPORARY INDIAN TRADITION: VOICES ON CULTURE, NATURE, AND THE CHALLENGE OF CHANGE. (Ed.) C.M. Bordeon. Smithsonian Institute Press, Washington, D.C. pp. 301-302.

CHOWDHURY, S., S.P. SHAHI & J.C. DANIEL. 1985. *Report of the Asian Elephant Specialist Group – Central India Task Force, Bihar & Orissa*. Unpublished document presented in Asian Elephant Specialist Group Meeting, Bandipur.

DATYE, HEMANT.S & A.M. BHAGAWAT. 1995. Estimation of crop damage and the economic loss caused by elephants and its implication in management of elephants. In: A WEEK WITH ELEPHANTS (Eds.)

Daniel, J.C. & Hemanth Datye. Bombay Natural History Society & Oxford University Press, Bombay, India. pp. 375-387

GORMAN, M.L. 1986. The secretion of the temporal gland of the African elephant, *Loxodenta africana* as an elephant repellent. *J. Trop. Ecol.* 2: 187-190.

HUNT, C.L. 1983. *Deterrents, aversive conditioning and other practices*. An annotated bibliography to aid bear management. National Park Service, USA, 1-136 pp. typescript.

KANGWANA, K. 1995. Human–Elephant Conflict: the challenge ahead. *Pachyderm* 19:11-14.

LA GRANGE, M. 1989. *Problem animal control*. Game Management Africa, Harare, Zimbabwe, 81pp. Typescript.

MACKINNON, J., K. MACKINNON., G. CHILD & J. THORSELL 1986. (Eds.) MANAGING PROTECTED AREAS IN THE TROPICS. IUCN, Cambridge.

McKAY, G.M. 1973. The ecology and behaviour of the Asiatic elephant in southeastern Ceylon. *Smithsonian contributions to zoology*. 125:1-113

MISHRA, J. 1971. An assessment of annual damage to crops by elephants in Palamau district, Bihar. *J. Bombay. nat. Hist. Soc.* 68: 307-310

NAIR, P.V., R. SUKUMAR & M. GADGIL. 1980. The elephant in South India. A review In. A WEEK WITH ELEPHANTS. (Eds.) J.C. Daniel & Hemanth Datye. Bombay Natural History Society & Oxford University Press, Bombay, India. pp. 9-19.

NAUGHTON-TREVES, L. 1998. Predicting patterns of crop damage by wildlife around Kibale National Park, Uganda. *Conservation Biology*. 12: 156-168

NEWMARK, W.D., D.N. MANYANANZA, D.G.M. GAMASSA & H.I. SARIKO. 1994. The conflict between wildlife and local people living adjacent to protected areas in Tanzania: human density as a predictor. *Conservation Biology*. 8:249-255.

NYHUS, P & R. TILSON. 2004. Agroforestry, elephants and tigers: Balancing conservation theory and practice in human-dominated landscapes of South-east Asia. *Agriculture, Ecosystem and environment* (2004).

NYHUS, P.J., R. TILSON.M & R.T. SUMIANTO. 2000. Crop raiding elephants and conservation implications at Way Kambas National Park, Sumatra, Indonesia. *Oryx* 34(4): 262-274.

OLIVIER, R.D.C. 1978. *On the ecology of the Asian elephant*. Ph.D thesis. University of Cambridge, Cambridge, UK.

OSBORN, F.V & L.A.WELFORD. 1997. *Living with elephants: a manual for wildlife managers in the sadc region. uaaid/sadc tcu* publication 5. US agency for International development, Lilongwa, Malawi.

OSBORN, F.V & S. ANSTEY. 2002. Elephant/human conflict and community development around the Niassa Reserve, Mozombique. Consultancy report for WWF/SARPO, Zimbabwe.

OSBORN, F.V. 2002. Capsicum oleoresin as an elephant repellent: Field trials in the communal land of Zimbabwe. *J. Wildl. Manage.* 66(3): 674-677.

PERRIN, T.E., L.E.L. RASMUSSEN & R.GUNAWARDENA. 1996. A method for collection, long term storage and bioassay of labile volatile chemo signals *J. Chem. Ecol. 21:207-221.*

RASMUSSEN, L.E.L., D.L. HESS & A.HALL-MARTIN. 1996. Chemical profiles of African bull elephant (*Loxodanta africana*); Physiological and ecological implications. *J. Mammal*. 77: 422-439.

RASMUSSEN, L.E.L., T.D. LEE., A. ZHANG. W.L. ROELOFS & G.D. DAVES. 1997. Purification, identification, concentration and bioactivity of (2) -7-dodeoen-1-yl-acetate: Sex pheromone of the female Asian elephant (*Elephas maximus*). *Chemical senses* 22: 417-438.

RASMUSSEN, L.E.L., T.D. LEE., G.D. DAVIS & M.J. SCHMIDT. 1993. Female-to-female sex pheromones of low volatility in the Asian elephant (*Elephas maximus*). *J.Chem. Ecol.* 19: 2115-2128.

SANTIAPILLAI, C. & H.SUPRAHMAN. 1986. The ecology of the elephant (*Elephas maximus L.*). In: THE WAY KAMBAS GAME RESERVE, Sumatra. WWF IUCN Final Report, 3133. WWF IUCN, Boger, Indonesia.

SANTIAPILLAI, C. 1990. *Why do elephants raid crops in Sumatra*? The Asean workshop on wildlife research and management and the meeting on the establishment of Asian Wildlife Society. Ministry of forestry, Boger, Indonesia.

SIEGE, L & R.D. BALDUS. 1998. Assessment of crop damage and application of nonlethal deterrents for crop protection east of the Selous game reserve In: Siege, L.E & R.D. Baldus, Tanzania Wildlife Discussion Paper NR 24. (Eds.) Serengeti Wildlife Research Institute, Tanzania.

SINGH, A.K., R.R.SINGH & S. CHOUDHURY. 2002. Human Elephant Conflicts in changed landscapes of West Bengal, India. *The Indian Forester*. 128(10): 1119-1132

SITATI, N.W., M.J. WALPOLE., R.J. SMITH & N. LEADER-WILLIAMS. 2003. Predicting spatial aspects of human elephant conflict. *J. Appl. Ecol.* 40:667-677.

SIVAGANESHAN, N & A.J.T. JOHNSINGH. 1995. Food resources crucial to the wild elephants in Mudumalai wildlife sanctuary, South India. In: A WEEK WITH ELEPHANTS (Eds.) Daniel, J.C. & Hemanth Datye, Bombay Natural History Society & Oxford University Press, Bombay, India pp: 406-423.

SUKUMAR, N. 1985. Ecology of the Asian elephant (*Elephus maximus*) and its interaction with man in South India. Ph.D thesis. Indian Institute of Science, Bangalore, India.

SUKUMAR, R. & M. GADGIL. 1988. Male-female differences in foraging on crops by Asian elephants. *Anim. Behav*. 36: 1233-1235.

SUKUMAR, R. 1989. THE ASIAN ELEPHANT: ECOLOGY AND MANAGEMENT. Cambridge University Press, Cambridge, UK.

SUKUMAR, R. 1990. Ecology of the Asian elephant in Southern India. II. Feeding habits and crop raiding patterns. *J. Tropical. Ecol.* 6: 33-53.

TCHAMBA, M. 1995. The problem elephants of Kaele: A challenge for elephant conservation in Northern Cameroon. *Pachyderm* 19:26-31.

TCHAMBA, M.N. 1996. History and present status of the human-elephant conflict in the Waza-Logone region, Cameroon, West Africa. *Biol. Conser*. 75: 35-41.

THOULESS, C.R & J.S AKWA. 1995. Shocking elephants: fences and crop raiders in Laikipia district, Kenya. *Biol. Conser.* 72:99-107.

THOULESS, C.R. 1994. *Conflict between humans and elephants in Sri Lanka.* Unpublished report for the GEF. EDG, Oxford, UK. 33p.

WILLIAMS, A.C. & A.J.T. JOHNSING. 1996. Status survey of elephants (*Elephas maximus*), their habitats and an assessment of elephant human conflict in Garo hills, Meghalaya. Final Report, Wildlife Institute of India, Dehradun, India.

❑❑❑

15

WILD BOAR, *Sus scrofa*, AN UNDERESTIMATED CROP PEST OF INDIA

Shakunthala Sridhara and A.K.Chakravarthy*

*AINP on Rodent Control, * Department of Entomology*
University of Agricultural Sciences, G.K.V.K., Bangalore-560 065.

ABSTRACT

Wild boar, *Sus scrofa*, found throughout India is an unrecognized serious pest of crops wherever wilderness borders cultivation more so around national parks and hilly regions. The animal is an omnivore but its diet is mainly vegetarian with protein supplements from animal sources. Although damage to sugarcane, wheat and maize is both qualitatively and quantitatively reported for Pakistan (Brooks *et al.*, 1990), and even though there are strong indications that wild boar is a serious pest of crops in India too, studies on its behavioural ecology, life history, foraging and pest status are negligible. Of the four to five studies carried out mostly around national parks and forests, wild boar was one of the several vertebrate depredators damaging crops. Rice, lentil, wheat, kidney bean, potato, sugarcane, maize, groundnut and cardamom were depredated in different parts of the country. The reviewed methods of containing wild boar damage to agriculture are: poisoning mostly with rodenticides, repelling with insecticides like temik, trapping, hunting, shooting, fencing, netting, hounding, habitat destruction, changing crop patterns and raising less attractive cultivars. By its omnivory, intelligence, tenacity, aggressive attitude, adaptability, prolific breeding and extreme capacity to survive, wild boar has all the attributes of becoming a successful crop pest or it already is, but not studied in depth. Studies on the pestiferous activities and management techniques for wild boar management are required urgently.

INTRODUCTION

Wild boar, *Sus scrofa* is ubiquitous, living in wild, consuming ground vegetation, soil arthropods, carrion and raiding crops if their habitat is close to crop fields. In these situations its lifestyle is almost commensal. Agricultural lands in the vicinity of forests and national parks are prone to wild boar damage in India and probably throughout the world. In spite of their almost pest status, studies on crop losses due to wild boar foraging in India are very few (Schultz, 1986; Chhangani & Mohnot, 2004). In neighbouring Pakistan detailed studies have been carried out on the biology, behaviour, pest status and management of wild boar to prevent damage to cultivated crops (Brooks *et al.*, 1990). This chapter is intended to provide an overall picture of wild boar damage to crops drawing information worldwide so as to focus on the nature of damage the animal is probably and capable of causing in India unnoticed and unmeasured by scientists. The various methods of containing damage elsewhere will also be dealt with summarily.

SCIENTIFIC STATUS

Class	: Mammalia
Order	: Artiodactyla
Family	: Suidae
Genus	: *Sus*
Species	: scrofa
Binomial name	: *Sus scrofa*

There are 21 subspecies of *Sus scrofa* worldwide.

DISTRIBUTION

Family suidae of five genera and nine species is native to old world. Suids are medium sized mammals living south of 58°N latitude throughout Asia, Eurasia, Africa, Madagascar and Taiwan (Walker, 1975). *Sus scrofa* is widely distributed in Asia, being very common in India, Pakistan, Ceylon and Malaysia (Prater, 1986).

HABITAT

Wild boars live generally under dense cover in forests and woodlands or high crops/grasses. They also live in burrows, which are self excavated or are abandoned by other animals. These animals prefer wet habitat patches so that they can scoop out soil in search of food. They are good swimmers and fond of mud baths.

HABITS

Wild boars are gregarious, ranging in bands of 40 to 50 individuals. Members of this family are rapid runners. When injured, wild boars are courageous and will fight back. There are authentic reports of boars killing camels and tigers. These hoofed animals are active at night. They use their snouts to dig food and are omnivorous. They not only eat fungi, leaves, fruits, roots, bulbs, tubers, but also consume snails, earthworms, reptiles, young birds, rats and eggs of vertebrates. Feeding is in the early morning and late evening, chiefly at night. No wild animal is more destructive to crops in cultivated areas adjacent to wilderness than wild boar. It also feeds on offal or carrion. Although nocturnal, wild boar moves around in agricultural fields even during day when man is not around.

MORPHOLOGY

The wild boar is a bulky animal weighing 100 to 140 kg and measuring 100-150 cm. Body colour varies from grey to pinkish brown and is covered with coarse black and brown bristles. The bristles on the nape and shoulder are longer, forming a crest (Prater, 1986).The body is heavy, neck is thick and legs are short and slender. The head is elongated with an abruptly truncated mobile snout which terminates into a flat disc with nostrils. The large ears are pointed, set forward and covered with hair. Eyes are small, situated at the back of the head, near the base of the ear. The short tail ends in a tuft of long yellowish-brown bristles. There are bristles on the cheeks also.

BIOLOGY AND BEHAVIOUR

Resting and feeding in small groups, wild boars are social animals although occasionally some adult males forage individually. Days are spent in dense foliage near some water body as the animal has very few sweat glands and is extremely sensitive to sun. The wild boar has a strong, characteristic acrid smell which seems to play an important role in their social life.

On being disturbed, wild boar emits rapid low grunts. When angry, the adult male clutters its two tusks in an audible and threatening manner. The strong and pointed head is used in fighting, digging and bulldozing vegetation. Wild boar has the reputation of being dangerous and aggressive, the characters that make it a favourite target for hunting. Its sense of smell is acute, hearing is moderate but eye sight is poor. In spite of these sensory attributes a crop raiding wild boar would be totally oblivious to the shouts, prodding and other human activities to chase him out of crop fields. In the otherwise social groups of wild boar, there is usually a lone tusker behind the group who has more courage, intelligence and fortitude. With high temper he is capable of becoming savage and vindictive on being wounded. Even tigers are believed to keep away from such a male. If attacked by hunting dogs or a man on horse back, wild boar is reputed to fight back like a true professional.

Wild boar breed throughout the year with a peak during and after monsoon; gestation lasts 112-115days, litter size ranges 2-14 piglets (Walker, 1975). In central India, the majority of young are born at two periods, shortly before and after the rains. Using grass, crop parts and stems the female constructs a well hidden nest into which the young are delivered. After breeding, the big bears live alone or in company with one or two sows (Prater, 1986). In hill region of Karnataka, peak breeding was observed during post monsoon (September-October) with average litter size of four (Chakravarthy, Pers. observations, 1983-1990).

FEEDING BEHAVIOUR AND PEST STATUS

Wild boars are omnivorous and mostly vegetarian in food habits. Generally a wide variety of seeds, fruits, young leaves, tubers and succulent stems form the bulk of their diet which is supplemented with fungi, carrion, and eggs of ground nesting birds, reptiles and insect larvae.

Wild boar dwelling close to crop fields invariably become crop pests causing significant damage. However, quantification of damage and yield losses and its economic impact have been studied sparsely. Brooks *et al.* (1990) consider wild boar as the most important vertebrate pest in sugarcane, wheat and maize in Pakistan causing more damage than other vertebrate pests namely rats, porcupines and rose-ringed parakeet. Estimated damage was 11.3% in sugarcane, 3% in wheat, 6-7% in maize and 0.9% in groundnut resulting in a cumulative loss of 130.9 million Pakistan rupees. Damage was by consumption of grain and by trampling plants to form bedding or to gain access to the interior of field. Sugarcane rind, germinating and milky stage of wheat, maize cobs and groundnut pods were the main parts of the crop eaten by wild boar. Similar pattern of damage was reported for wheat, sorghum, maize, barley and oats, the losses amounting to millions of dollars in Australia (Choquenot *et al.,* 1996). In Italy 25-62% of total production of wheat, maize, sunflower and grapes were lost to foraging by wild pigs (*Sus scrofa*) and fallow deer (*Cervus doma*) (Genov *et al.,* 1996). In France, apart from cereals, meadows and oilseeds were damaged by wild boar (Baratte, 1998). A review of diet of wild pig in Western Europe revealed that the animal though omnivorous ate more vegetative food than animal food. Overall the vegetative food was composed of roots, green plant matter and agricultural crops with at least one energy rich food such as nuts, pine seeds, or cereals, insects, earthworms and ground nesting birds. Small mammals also comprised its diet (Schley & Ropert, 2003).

The Indian studies on crop damage by wild boar are usually limited to areas around forests, national parks and hilly regions (Jayson, 1999; Chakravarthy & Srihari 2002; Rao *et al.*, 2002; Chhangani & Mohnot, 2004). In all these studies wild boar was the most damaging in the vertebrate pest complex that foraged on cultivated crops. A three year study in the forests of Kerala revealed that a total of forty five species of crops were destroyed by wild animals. The commonly affected crops were paddy (*Oryza sativa*), coconut palm (*Cocos nucifera*), areca nut (*Areca catechu*), coffee (*Coffea arabica*), oil palm (*Elaeis guineensis*), pepper (*Piper nigram*), jack tree (*Artocrpus heterophyllus*), mulberry (*Morus alba*) and mango (*Mangifera indica*). Of the principal vertebrate pests namely elephant (*Elephas maximus*), gaur (*Bos gaurus*), sambar (*Cervus unicolor*), wild boar (*Sus scrofa*), bonnet macaque (*Macaca radiata*), common langur (*Presbytus entellus*), black napped hare (*Lepus nigricollis*) and pea fowl (*Pavo cristatus*), elephant and wild boar caused maximum damage. Damage was also reported for pineapple, sweat potato, tapioca, alocasia, beans and plantain. In the continued mountain ranges of Western ghats and four districts of hilly region in Karnataka where cardamom (*Elettaria cardamomum*) is cultivated twelve vertebrate species damaged cardamom of which the lesser bandicoot rat (*Bandicota bengalensis*), squirrel (*Funambulus palmarum*), wild boar (*Sus scrofa*) and monkey (*Macaca radiata*) were considered as the major pests (Chakravarthy & Srihari, 2002). Further studies (Chakravarthy, 1991; 1994) identified wild boar as an important pest on rice in the hilly regions of Western Ghat in Karnataka causing an average of 12% to cent per cent losses in rice fields adjacent to forest tracts. Fields with early maturing tall rice cultivars sustained heavy damage than late maturing dwarf cultivars. Earlier similar damage to tall varietal rice by *S. scrofa* in areas adjacent to bamboo forests in Madhya Pradesh was observed (Thomas & Naidu, 1995).

There are three Indian and one Nepalese study on wild life damage to crops around national parks (Studsrod & Wegge, 1995). In all sanctuaries except Kumbalgarh Wildlife Sanctuary, Rajasthan wild pig was the major depredator but there were other mammalian herbivores too. In Nepal cheetal and wild boar caused 47% loss to lentil and 24% to wheat (Studsrod & Wegge, 1995). In Sariska Nilgai and wild boar more so the later raided crops (Sekhar, 1998) and in Nanda Devi Biosphere although damage was by porcupine, monkey, musk deer and patridge, monkeys and wild boar were responsible for 50-60% of total crop loss which was Rs. 5,38,620 (US$ 15,389) during the study period (Rao *et al.*, 2002). The major crops damaged were kidney beans and potato. The wild boar in Kumbalgod Wildlife Sanctuary fed on 39 species of cultivated plants. *Zea mays* (maize), *Saccharum officinarum* (sugarcane), *Arachis hypogea* (groundnut) and some vegetable species formed the bulk of the animals food. The cumulative loss in all the farms was calculated as US$ 2500-3000 (Chhangani & Mohnot, 2004). In almost all these studies there was inverse correlation between extent of damage and distance from the reserve.

THE ECOLOGICAL ROLE OF WILD BOAR

Wild boar using their strong snout and neck muscles dig and overturn soil in search of food viz. insects, grubs, mast, root, bulbs, plant material and fungus. Such rooting has two ecological consequences. In Hawaii by digging forest ground cover of delicate and rare species of plants, wild boars disturbed the ground cover facilitating invasion of exotic weeds (Stone & Anderson, 1988). In Malaysia, soil rooting and seed predation by wild pigs reduced species richness, growth and survival of woody plants in under story vegetation in a forest (Ickes *et al.*, 2001). They also depredated voraciously on native fauna such as eggs and hatchlings of green sea turtle, giant tortoises and dark rumped petrels in Hawaii (Stone & Anderson, 1988) and terrestrial small vertebrates and eggs of ground nesting birds in California threatening their survival (Barret & Teitije, 1993).

Direct predation on small vertebrates may suppress their population which is advantageous if the animal happens to be a pest. Feral pigs themselves are predated upon by large carnivores some times beneficially. In Europe they facilitated the recovery of wolf (*Canis lepis*) which was on the verge of extinction. They are an important prey for the Amor tiger, *Panthera tigris,* for the Australina dingo (*Canis familiaris dingo*), Florida panthers (*Felis concolor corgi*) and mountain lions (*Puma concolor*) in USA (Barrett & Tieje, 1993).

Wild pigs carry a number of diseases and parasites which can be transmitted to humans and livestock such as Brucellosis, Cholera, Leptospirosis, Tuberculosis (Bovine, Avian & Swine), Q-fever, Trichonosis, Toxoplasmosis, Pseudorabies and Plague.

MANAGEMENT OF WILD BOAR

The finest consolidated approach to wild boar management is available in Brooks *et al.* (1990). A total of nine methods for wild boar control are suggested. These are poisoning, pig sticking, trapping, shooting, fencing, netting, explosive bomb, hounding and habitat destruction; all of them are fraught with practical difficulties. For an average farmer in India pig sticking is a rich man's game more of a past time of a privileged few than to manage a pest problem. Bait placement and setting of traps can be quite successful to reduce wild boar numbers. There are four types of traps to capture wild boar namely panel trap, site trap and two types of steel traps. The best bait is wheat, sorghum, rice, apples, nuts and bread.

Shooting has to be extremely intensive to be effective if the population is small and isolated.

Electric fencing has been found effective by Chakravarthy (1991) but Brooks *et al.* (1990) felt that for wild boar management it needed some modifications and evaluation.

Netting is a successful method to capture wild boar in Pakistan and is ideal to collect animals for experimental purpose and not for large scale management of the pest. Hounding by trained dogs is also successful.

Thick vegetation such as reeds and mesquite plantation type growth near swamps and marshes are the ideal places for wild boar to hide. By eliminating such growth and draining the swamps harbourage for wild boar can be reduced.

Chemical control of wild boar mostly involves rodenticides and a few insecticides. Sodium mono fluroacetate (1080) is used as a grain coating of wheat. After soaking the grain in the poison solution (25g in 5 litres of water), poison bait is exposed to wild boar at bait points or bait stations. Prebaiting is essential. Though effective, the method is hazardous to non-target wildlife. A modified method eliminating prebaiting is to fill capsules with 50-60 mg of sodium monofloroacetate along with 100mg of plasil and encase in flour dough ball, the size of a squash ball. This is dusted with full cream, dry milk powder and dried Cyprus tubers and placed on the path of wild boar.

Arsenic oxide is cheap and is required in small amounts (half tea spoon) although death is delayed. In Australia two methods of application are used. Carcass are dusted with poison and placed in the tracks of the feral pig. Alternatively half tea spoon of poison is filled into the drilled hole of an apple, the hole closed with apple pulp and placed along the wild boar's path.

Zinc phosphide baiting resembles that of compound 1080. Capsules containing 500mg of rodenticide are dipped in paraffin wax of very low melting point and then covered with flour dough or put inside fruits like guava, apple or in cobs of maize at milky stage. However, it is recommended to change the bait every time bait shyness occurred (Brook *et al.*, 1990). Strychnine in spite of high lethality (10mg/kg) was not found effective in Pakistan. Anticoagulants are used effectively in some countries but delivery systems still need standardization.

Aldicarb (Temik) is used as an exposed bait in Pakistan. About half kilogram of boiled maize is put as a heap on the tracks of wild boar and covered with 15-20g of Temik and 50g of full cream dry milk powder mixed with Cyprus tubers and brown sugar. The second method is to obtain 100 per cent aldicarb by dissolving Temik in acetone, filter and the filtrate is evaporated to get pure aldicarb which is then encapsulated as mg capsule which in turn is coated with flour dough and a layer of cream milk powder and baited.

In southern Karnataka punctured polythene bags containing thimet granules (log) mixed with equal amount of sand are hung along the boundary of rice fields at 3m intervals at a height of 0.6 to 1m. The method kept away wild boar from paddy fields for 30-50 days (Chakaravarthy & Srihari, 2002).

In Madhya Pradesh, after observing severe damage to local tall varieties of rice, farmers started growing dwarf rice with long auricles on panicles to reduce wild boar damage (Thomas & Naidu, 1995).

Feral pigs are a serious problem in Hawaii and Australia. Trapping, snaring and hunting with dogs was effective in reducing wild pig problem in Hawaii (Anderson & Stone, 1993). The New South Wales government is reported to have spent $4 million to control feral pig by shooting from helicopters, trapping and poisoning in 1984 (Choquenot *et al.*, 1996). Guarding the fields by hired personnel was considered most economical and successful to contain wild boar damage in Rajasthan (Chhangani & Mohnot, 2004). Change in cropping and crop composition by cultivating non-edible crops such as medicinal plants was also suggested (Rao *et al.,* 2002). People in Indian villages primarily raise crops for family consumption. As such raising cash crops in place of food crops is not practical and not preferred by the marginal farmer.

CONCLUSIONS

Wild boar is a crop depredator worldwide and has all the characters of a successful pest. Being a crop consumer, predator of smaller vertebrates and less of a prey, omnivorous, prolific breeder and adaptable, *Sus scrofa* has evolved into a formidable pest of crops. Since it is protected under Wildlife act (1972) of India, with a provision to kill after getting permission from the government if and when it becomes a serious nuisance to human life, health and crops, managing wild boar problem is not easy. At present the crop losses are overlooked except around national parks. Although cropping is less intense here, even moderate losses make an impact on the economy of the forest dwelling, small farmer. There are complaints from farmers about wild boar problem specially during sowing. To evaluate and implement management techniques, initially we have to quantify the damage in various agro-ecological conditions of the country and establish damage patterns, frequency and extent. The various approaches of mitigating wild boar damage are to be tested to evolve location specific and as far as possible non-lethal methods to contain wild boar damage to agriculture.

REFERENCES

ANDERSON, S. & C. STONE 1993. Snaring to control feral pigs, *Sus scrofa* in a remote Hawaiien rain forest. *Biol. Conser.* 63: 195-201.

BARATTE, Y. 1998. Agriculture, wildlife and hunting. 1998. Agriculture, faun saurage et chasse: séance specialisee du II mars 1998. comptes- Rendus-de- I' Academie-d' Agriculture-de- France. 84(2): 111-124.

BARRET, R.H. & W. TIETJE 1993. The wild pig in California Oak Woodland: Ecology and Economics. *Conference presentation summaries,* University of California, Berkley, USA.

BROOKS, J.E., E. AHMAD, I. HUSSAIN & M. H. KHAN 1990. The Agricultural importance of the wild boar (*Sus scrofa*) in Pakistan. *Tropical Pest Manage.* 35(3): 278-281.

CHAKRAVARTHY, A.K. & K. SRIHARI 2002. Vertebrate pests of cardamom (*Elucine cardamomum Maton*) in hill region of Karnataka, South India. *Pest management in Hort. ecosystem.* 6(2): 139-148.

CHAKRAVARTHY, A.K. 1991. *Vertebrate pest management in cardamom (Elettaria cardamomum Maton) and other crops in Malnad region,* ICAR Annual report, University of Agricultural Sciences, Bangalore, India. 16p.

CHHANGANI, A.K. & S.M. MOHNOT 2004. Crop raiding by wild boar (*Sus scrofa*) in and around Aravalli and its management in Rajasthan, India. *Tigerpaper* 31 (2): 1-4.

CHOQUENOT, D., J. MILLROY & T. KORN 1996. MANAGING VERTEBRATE PESTS: FERAL PIGS, Bureau of Resource Sciences, Australian government publishing services, Canberra, Australia. 163p.

GENOV, P.V., L. TONINI, G. MASSEI & N. BOTEV 1996. Crop damage by wild ungulates in a Mediterranean area. *Proc. International Union of Game Biologist's 22nd Congress.* Sofia, Bulgaria. pp: 214-215.

IBID. 1994. *Vertebrate pest Management.* ICAR Annual Report, University of Agricultural Sciences, Bangalore, India. pp.34-35.

ICKES, K., S. DEWALT & S. APPANAH 2001. Effect of native pigs (*Sus scrofa*) on woody under story vegetation in a Malaysian low land rain forest. *J. Tropical Ecol.* 17: 191-206.

JAYSON, E.A. 1999. *Studies on crop damage by wild animals in Kerala and evaluation of control methods.* Research Report No. 169, Kerala Forest Research Institute, Peechi, Kerala, India. 48pp.

PRATER, S.H. 1986. THE BOOK OF INDIAN ANIMALS. Bombay Natural History Society, Bombay, India. 326 p.

RAO, K.S., R.K. MAIKHURI, S. NAUTIYAL & K.G. SAXENA 2002. Crop damage and livestock depredation by wildlife: a case study from Nanda Devi Bio-sphere Reserve, *Indian J. Envt. Mangmt.* 66(3): 317-327.

SCHLEY, L. & T.J. ROPERT. 2003. Diet of wild boar, *Sus scrofa* in Western Europe, with particular reference to consumption of agricultural crops. *Mammal Rev.* 33(1): 43-56.

SCHULTZ, B.C. 1986. The management of crop damages by wild animals. *Indian Forester* 112 (10): 891-899.

SEKHAR, N.V. 1998. Crop and livestock depredation caused by wild animals in protected areas: the case of Sariska Tiger Reserve, Rajasthan, India. *Envt. Conser.* 25: 160-171.

STONE, C. & A. ANDERSON. 1988. Introduced animals in Hawaii's natural areas. *Vertebr. Pest Conf.* 13: 134-140.

STUDSROD, J.E. & P. WEGGE. 1995. Park-people relation ship: the case of damage caused by park animals around the Royal Bandua National park, Nepal. *Envtl. Conser*. 22(2): 133-142.

THOMAS, M. & A.K. NAIDU. 1995. Indian wild boar, *Sus scrofa* L.: a rice pest in Madhya Pradesh, India. IRRN 20(1):28.

WALKER, E.P. 1975. MAMMALS OF THE WORLD. The John Hopkins University Press. III Ed. Baltimore & London 1500 p.

16

CROP DAMAGE BY BLUE-BULL (*Boselaphus tragocamelus*) AND ITS MANAGEMENT

L.S. Rajpurohit, A.K. Chhangani and S.M. Mohnot

Department of Zoology, Faculty of Science, Jai Narain Vyas University, Jodhpur-342 001, INDIA

ABSTRACT

The indigenous ungulate (*Boselaphus tragocamelus*) (*Nilgai*, Blue bull), a normally shy and quite herbivore, of late has become a serious pest of crops in the states of Haryana, Punjab, Madhya Pradesh and Rajasthan. The limited studies on the pestiferous activities which can be 50 -100% damage, are based on surveys. Cereals and pulses grown in the affected areas were the preferred food while cotton crop was used only as cover. The major reasons for the increase in numbers of *nilgai* and resultant crop loss is attributed to lack of natural predators, deforestation to expand agriculture, overgrazing of range lands by livestock and most importantly the religious protection given to it by the *Vaishnoi* sect of Hindus. At present farmers guard the crop fields themselves or hire people to guard. Recently the governments of Punjab and Haryana have relaxed the ban on hunting to reduce the pest populations. Suggested methods for their management include protection of *nilgai* habitat, fencing of *nilgai* populations or crops, guarding, using trained dogs, sterilization, translocation, culling, power fencing and enclosures or corrals.

INTRODUCTION

The Blue bull (*Boselaphus tragocamelus)* is a large and common antelope inhabiting areas with open scrub and scarce vegetation, in and around wildlife sanctuaries but avoiding dense forests. Blue bull is a clumsy-looking ungulate with a massive body and long legs. The name "blue bull" refers to the iron-grey coat of the mature male. It is the largest antelope in India. It is a hardy animal and survives very well in arid conditions and habitats where water is scarce (Roonwal, 1987). It's ranges are fast declining because of biotic interferences, mainly the spread of agriculture and exotic weed species. But, local people quite often blame this animal for raids on agricultural fields and orchards. Villagers have also assumed that the population of Blue bull has increased manifold in the recent past. In fact the animal is becoming a serious sporadic crop pest in many parts of the country (Prakash, 1986; Rajpurohit & Mohnot, 1988, 1997; Chauhan & Sawarkar, 1989; Chauhan & Singh, 1990: Chhangani, 1994; Goyal & Rajpurohit, 2000; Chhangani *et al.*, 2002).

SCIENTIFIC STATUS

Class	: Mammalia
Order	: Artiodactyla
Family	: Bovidae
Genus	: *Boselaphus*
Species	: *Tragocamelus*
Generic name	: *Boselaphus tragocamelus*

DISTRIBUTION

Nilgai is exclusively indigenous to India and its range starts from the foot hills of Himalayas in the north to Karnataka in the south, with Haryana, Rajasthan, Gujarat, Madhya Pradesh and Uttar Pradesh supporting good populations. However, it is not found in east and North-East parts of the country, Western India and Ceylon (Prater, 1986).

MORPHOLOGY

The body is greyish with a white ring below each fettock and two white spots on each cheek. Males have a pair of horns which are small, smooth, and triangular at the base. Horns are 20-22 cm long, becoming roundish at the tip. Males measure 1.95 to 2.1 m lengthwise and stand at shoulder 1.3 to 1.55m (Ellerman & Morrison Scot, 1952).

GROUP STRUCTURE

Nilgai lives in two types of herds or as solitary bulls. A bisexual herd has 3 to 20 individuals of both sexes and of all age classes; the all-bull herds may comprise 2 to 8 adult and sub-adult males (Goyal & Rajpurohit, 2000). Generally there is a single adult bull in a bisexual herd (i.e. unimale bisexual group) who leads the group but sometimes two or more adult and sub-adult bulls are observed in a bisexual herd (i.e. multi-male bisexual group). Adult and leader bull of bisexual herd have been observed fighting with a solitary bull or all-bull herd males for territories and/or other resources such as mates. The *nilgai* have been observed shifting from one area to other, depending upon the availability of crops (Prakash, 1986; Rajpurohit, 1988).

BEHAVIOUR

Earlier studies indicated blue bulls to be very shy and keeping away from human habitation. They used to avoid mixing with livestock. But, in recent past they are regularly coming in contact with rural livestock and human beings due to increased populations of blue bull and livestock. This has rendered them bold and aggressive with little fear and flight. Besides their interaction with goat, sheep, camel, cow and buffalo, they frequently come in close contact with wildlife such as black buck, chinkara, langurs and a variety of birds including peacocks, partridge, Demoiselle crane etc. Wolves and dogs have been found chasing young blue bull. In the normal situation they do keep certain distance from human beings which range between 50 to 100 feet minimum (Chhangani, 2000).

In and around Jodhpur and Kumbhalgarh Wildlife Sanctuary (KWS), Rajasthan blue bull has a peculiar association with the livestock and langurs during feeding. Langurs usually feed on variety of flowers, fruits, seeds and leaves from trees which are generally out of reach for blue bull. Langur being a selective feeder, feeds on particular plant parts in specific seasons of the year throwing away uneaten parts. Such rejected portions are consumed by blue bulls which usually follow the langurs on ground specifically for this purpose (Chhangani, 2000).

PEST STATUS

The species is a highly adaptable herbivore and an opportunistic feeder. Of late crop depredation by *nilgai* has become a serious problem in Rajasthan, Haryana and Punjab. In spite of significant losses to farmers, the limited studies on crop raiding behaviour are based only on surveys and lack statistically designed data collection and analysis as is done with other agricultural pests. The first published report of *nilgai* depredation on crops was reported for several districts of Haryana and the area around the Great Indian Bastard (G.I.B) sanctuary at Karera in Madhya Pradesh (Chauhan & Sawarkar, 1989). The agricultural belt extending over Sirsa, Hissar, Bhiwani and Mahendragadh districts of Haryana were reported to be sustaining a 7000 strong *nilgai* population. Although capable of devouring most agricultural crops, *nilgai* exhibited a definite preference for gram, wheat seedlings and *moong*. Cotton crop served as an excellent cover to hide. Damage reached as high as 75% and was never less than 10%. The authors attributed the drastically reduced natural habitat and lack of protected areas for *nilgai* as the reasons for pestiferous activities of the ungulate in Haryana.

In the G.I.B sanctuary at Kerara, Madhya Pradesh serious damage was reported for gram, *moong*, wheat seedlings, mustard and linseed. Groundnut, *musoor*, sugarcane, soybean, *til, arhar*, jowar and vegetables were consumed to a lesser extent. The crop damage was attributed to explosion in *nilgai* population beyond the carrying capacity of its shrunken and fragmented habitat. With the ban on hunting and trapping of *nilgai* by tribals and the protection bestowed on the animal under schedule 1 of Wildlife Act, the ungulates increased in numbers rapidly. In certain areas the damage was so severe that cultivation had to be abandoned.

Similar increased population of *nilgai* was reported in crop fields around Jodhpur and Kumbhalgarh Wildlife Sanctuary in Rajasthan (Goyal & Rajpurohit, 2000). The population explosion was due to several reasons, the main ones being lack of natural predators, deforestation to expand agriculture, overgrazing of rangelands by livestock and the religious protection to the animal given by *Vaishnoi* Hindus. The shrubs and trees found in this area do not provide sufficient food for *nilgai* but offer sufficient day time cover. Faced with poor food resources, the animal learns to raid cultivated crops. It often browsed on shrubs, small trees and grazed on grasses and herbs. When these were insufficient to meet its food requirements it easily manoeuvred the 6-7 feet stone wall, barbed fencing, fences of dead or live thorny plants or any barrier out up by farmers to reach the edible crops. Crop raiding usually occurred during morning, evening and night. Although they devoured almost all the crops grown during *kharif* and *rabi* with the exception of Raira (*Brassica juncea*), they seem to prefer moth (*Pennisetum aconitifolious*), *gawar* (*Cyamopsis tetragonoloba*), bajra (*Pennisetum typhoidenum*), *moong* (*Phaseolus aureus*), *til* (*Sesamum orientale*), wheat (*Triticum vulgare*), *jeera* (*Carum nigrum*), *dhania* (*Coriandrum sativum*) and a variety of vegetables grown. Amongst vegetables *dhania* was the most preferred. Fruits devoured were grafted *ber* (*Ziziphus maritiana*), *nimbu* (*Citrus medica), papita* (*Carica papaya*), *amrud* (*Picidium guajava*) and *anar* (*Puncia granatum*). The adult animal consumed about 13-15 kg plant material a day.

Damage in jowar fields

Studies in Haryana (Chauhan & Singh, 1990) reported severe *nilgai* depredation of crops in the southern districts of Haryana namely Rohtak, Jhajjar, Hissar, Sirsa, Bhiwani and Mahendragadh. Preferentially gram, wheat seedlings and *moong* were attacked. Cotton was used only as a cover. The estimated crop loss was 50%- 58% in some areas which went up to 100% when fields were invaded by a big group of *nilgai*. They cause lot of nuisance as well as heavy crop losses by invading in groups of 1000 and more.

FACTORS CONTRIBUTIONG TO PESTIFEROUS ACTIVITIES OF NILGAI

In many parts of the country like Jodhpur blue bull have no natural predators except in some protected areas. Several national parks and wildlife sanctuaries have potential predators of *nilgai* like lion, tiger and leopard as revealed by scat analysis (Prater, 1986; Choudhary, 2000). But in Thar desert in Western Rajasthan, Punjab, Haryana, part of Gujarat and Madhya Pradesh, where only feral dogs which killed young blue bulls were the natural predators. This lack of natural predator seems to be the major factor for population increase (Chhangani *et al.*, 2002). Other mortality factors include shooting, road killing etc.

Studies conducted so far suggest that the blue bull have very wide moving ranges and their herds move from one place to another and shift their lodgings. Often such herds are mistaken for different herds. This gives the impression that the blue bull population is increasing and causing lot of damage to a variety of crops.

After the onset of monsoon the fairly abundant vegetation in the sanctuaries is adequate to meet their nutritional needs and blue bull probably do not raid crops as they can survive on natural plant material. Net land cover changes in agriculture were very fast in the last 10 years in all the habits where *nilgai* has become a crop raider. Thus it seems that the extension of agriculture in the recent past has intercepted the blue bull's traditional moving routes and foraging areas on the periphery of the sanctuary. This seems to have forced them to invade the crop fields occasionally while they are on their foraging journeys (Chhangani *et al.*, 2002). During and after monsoon their visit to the crop fields decreases.

Their population size in many parts of the country suggests an alarming situation. Their feeding adaptation to crop fields, availability of exotic weed species and low natural predation along with religious attitude of people provides them better survival. This seems to be the major factor of population increase and imbalance in the ecosystem in which blue bull is thriving. The increasing population of blue bull has increased the man-wildlife conflict many times in the recent past. Wildlife conservation and crop damage by blue bull thus becomes an obvious management issue of high priority.

CROP PROTECTION METHODS

The commonly used method for crop protection is guarding the cultivated field. For successful guarding people are required to be in the fields day and night during the seasons when the crops are most vulnerable. Obviously, this is not practical. Many a times 4-5 farmers hire a person or persons (depends on farmer's groups) to guard their crop fields, and share the cost. This practice is the most common crop protection strategy in Rajasthan (Chhangani *et al.,* 2002). Although culling is suggested, it is opposed by *Bishnoi* Hindus who live in the same habitat as *nilgai* in Punjab, Haryana and Rajasthan. In spite of the animals being protected under Wildlife Act, recently the governments of Punjab and Rajasthan have relaxed the rules of hunting in view of strong demands from farmers who had suffered devastating crop losses.

MANAGEMENT STRATEGIES AND RECOMMENDATIONS

The crop raiding activity by blue bull poses a significant threat to agriculture and horticulture and thus the livelihood of farmers is at high risk. Reduction of animosity towards wild animals and prevention of agricultural and horticultural loss by blue bull needs immediate attention. All measures of preventing crop damage should precede a survey to generate adequate information on the population, ecology and behaviour of *nilgai*, land use pattern, crops grown, their cover value and the agricultural economy of the affected areas. Following are the recommendations for preventing crop damage and management strategy for *nilgai*:-

1. Habitat protection: Government and people must respect the natural habitats while planning developmental activities like agriculture, horticulture, construction of roads and dams within or on the periphery of the forestlands or protected areas. Such activities cause the destruction of forest with serious impact on wildlife, which forces wild herbivores to feed in the cultivated fields. There is an urgent need to understand the impacts of such activities and to stop these interventions in the most critical wildlife habitats. For this a social, political and legal monitoring cell is needed and if need be, to stop such of the human activities that cause loss of biological diversity.

2. Fencing with local material like *Prosopis juliflora* branches or *Euphorbia caducifolia* are successful against blue bull and act as bio-barriers. This is the cheapest and most effective method to protect the crops and as such boundaries can be made more effective without much cost. Stone wall is more effective but not economical.

3. Guarding: Constant vigilance during the crop season employing sticks, noise, shouting etc is the cheapest way of crop protection. This is also the most successful strategy for preventing crop and horticulture damage by different vertebrates particularly blue bull.

4. Trained dog: Use of trained dogs can scare the blue bull. Barking of the trained dog definitely deters the blue bull and also warns the farmer about the crop raiding activities of blue bull.

5. Biological control: Sterilization of male blue bull is theoretically effective because socially they live in a harem system in which one adult male and several adult females live in bisexual groups. This method definitely controls the population of vertebrate pests which live in social groups. However, there are no effective chemosterilants available to affect this strategy.

6. Translocation: Translocation is considered as the best wildlife conservation and management tool. Through this the problem creating vertebrates can be trapped from cultivated area and translocated to their original habitat or they can be rehabilitated to such other habitats where their population is small.

7. Culling is hard to suggest in a country like India where many wild animal species are an integral part of religious, cultural and environmental heritage. Selective reduction of excessive wildlife pest populations is a normal, logical, long range and globally used wildlife management tool as in the case of man-eating carnivores like tiger or panther. Although hunting is prohibited and legally banned, realizing the severity of the damage to food crops and horticulture, culling of selective and identified animals with the acceptance of local community and by trained wildlife experts of forest department or professional hunters hired by forest department seems to be a workable solution. The overall objective of hunting is to keep the blue bull numbers within a reasonable size so that the crop loss can be decreased and farmers can reap their harvest undamaged. It is recommended that at least 45% of the adult males and 20% of the adult females need to be culled to maintain population at low levels. Monitoring of population is essential to arrive at effective rates of culling (Chauhan & Singh, 1990).

8. Power fencing is potentially an effective method of protecting crops from *nilgai* depredation. Such a method, although successful to keep away *nilgai* from fields of high value crops, creates the problem of 'transferring damage' to fields that are not surrounded by power fencing. Transporting the problem of crop raiding blue bulls to neighbouring farms does not prevent yield losses (Chauhan & Sawarkar, 1989). Moreover erecting an electric fence is costly for a marginal farmer.

9. Enclosures or corrals: Corrals are constructed to confine a large proportion *nilgai* and blackbuck in selected patches of forest. The population thus isolated should be subjected to fertility control to keep the numbers down. The details of erecting such corrals and maintaining them are described by Chauhan & Sawarkar (1989). But there are several problems to effect such a confinement, the major one being the availability of large tracts of open, uncultivated, non forest area with sufficient natural vegetation. Cost of erecting 3 m high fencing, post fencing care of the animals which includes not only feeding but also measures to suppress their reproduction are the other factors not so conducive for restraining animals using corrals (Chauhan & Singh, 1990). Given the socio-economic situation of our farmers, fencing of *nilgai* to protect crops is beyond their means and can be executed only by government or government agencies.

CONCLUSIONS

Nilgai, *Boselaphus tragocamelus* is a classical case of an unconventional pest. Destruction of its natural habitat and acquisition of crop raiding behaviour especially around the sanctuaries and national parks where cultivation is fragmented but rich in providing nutritious cereals and pulses is a serious problem which is compounded by the protection status it enjoys both legally and religiously. One major lacuna of the problem is lack of reliable data on crop losses and behavioural ecology of the pest. Until we have information on these aspects, damage reduction measures are limited to culling, guarding and using dogs. The other methods suggested such as containing them in enclosures, fencing crop fields, power fencing, translocation, sterilization are not practical for the farmers of a developing country like India. Translocation and sterilization is possible for government agencies like the department of forest which has the resources, expertise and manpower required.

ACKNOWLEDGEMENTS

Authors are grateful to late Prof. Ishwar Prakash, Emeritus Scientist, Zoological Survey of India, Desert Regional Station, Jodhpur for his regular encouragement. They are also thankful to Prof. S. Johnson, Former Head and Dr. G.R. Jakhar, Head, Department of Zoology, J.N.V. University, Jodhpur for providing logistic support during field-work. Thanks are also due to Mr. Bundu Khan for computation work.

REFERENCES

CHAUHAN, N.P.S & V.B. SAWARKAR 1989. Problems of over abundant populations of *Nilgai* and Black Buck in Haryana and Madhya Pradesh. *Indian Forester.* 115 (7): 488-493.

CHAUHAN, N.P.S. & R. SINGH. 1990. Crop damage by overabundant populations of *Nilgai* and black buck in Haryana (India) and its management. *Proc. 14th Vertebr. Pest Conf.* (Eds.): L.R. Davis & R.E. Marsh. University of California, Davis, USA. pp. 218-220.

CHHANGANI, A.K 1994. Crop raiding behaviour and status survey of Nilgai (*Boselaphus tragocamelus*) around Jodhpur. M.Sc. Dissertation submitted to J.N.V. University, Jodhpur, India.

CHHANGANI, A.K., L.S. RAJPUROHIT & S.M. MOHNOT 2002. Status survey and Ecology of *Nilgai* (*Boselaphus tragocamelus*) around Jodhpur, Rajasthan. *Cheetal.* 41(1 & 2): 6-14.

CHOUDHARY, S.S. 2000. RANTHAMBOR, BEYOND TIGER. Himanshu Publication, Udaipur & Delhi, India.

ELLERMAN, J.P & T.C.S MORRISON-SCOT 1952. CHECK LIST OF PALAERTIC AND INDIAN MAMMALS, 1758 to 1946. British Museum, London.

GOYAL, S.K & L.S. RAJPUROHIT. 2000. *Nilgai* (*Boselaphus tragocamelus*)- A mammalian crop pest around Jodhpur. *Uttar Pradesh J. Zool.,* 20(1):55-59.

IBID. 1997. Depredation of crops by *Nilgai* (*Boselaphus tragocamelus*) around Jodhpur, India. *Him. J. Env. Zool.* 11:129-132.

IBID. 2000. Ecobehavioural Diversity of Hanuman langur, *Presbytis entellus* living in different ecosystems. Unpublished Ph.D. Thesis submitted to J.N.V. University, Jodhpur, India.

PRAKASH, I 1986. Wildlife Resource and its management. In DESERT ENVIRONMENT: CONSERVATION AND MANAGEMENT. (Eds.): K.A.Shankarnarayan & Vinod Shanker. Central Arid Zone Research Institute, Jodhpur, India. pp. 19-22

PRATER, S.H 1986. THE BOOK OF INDIAN ANIMALS. Bombay Natural History Society, Bombay, India.

RAJPUROHIT, L.S & S.M. MOHNOT 1988. The Nilgai *Boselaphus tragocamelus* an agriculture pest around Jodhpur (Rajasthan). *Proceeding of National Seminar on Advances in Economic Zoology,* December 20-22, 1988, University of Jodhpur and Zoological Survey of India, Jodhpur, India. pp. 57.

RAJPUROHIT, L.S. 1988. *Nilgai*, *Boselaphus tragocamelus* a serious crop-pest between Jodhpur and Osian (Rajasthan). *Cheetal.* 29(2): 10-13.

ROONWAL, M.L. 1987. Fauna of the great Indian desert. IN: DESERT RESOURCES AND TECHNOLOGY. Vol. 1. (Ed.): Alam Singh. Scientific Publishers, Jodhpur, India. pp 1-86.

17

RHESUS MONKEY (*Macaca mulatta*) PROBLEMS IN INDIA AND THEIR MANAGEMENT

Ekwal Imam and Iqbal Malik*

Department of Wildlife Sciences, Aligarh Muslim University, Aligarh 202 002
**Founder Director —VATAVARAN-NGO, 540, Hawa Singh Marg, Asian games village, New Delhi- 110 049*

ABSTRACT

Amongst the three commensal monkeys in India viz., *Macaca mulatta, Macaca radiata* and *Presbytis entellus*, the Rhesus monkey, *M. mulatta* is the most aggressive. Forty eight per cent of 0.3 million Rhesus monkeys of Northern India live in human habitats. Apart from being potential transmittors of several viral, bacterial, parasitic diseases and helminth parasites, they create lot of nuisance, destroy property and vegetation as well as bite women and children. Their destructive activities and harassment have been well documented for Delhi, Jodhpur, Jaipur and temple towns of North India. Harassment includes entering houses, stealing food, clothing and other articles, uprooting vegetables and garden plants, pulling out electric wires and TV antennae, threatening, attacking and biting people. Three measures are suggested to manage monkeys namely cessation of feeding by people, translocation and fertility control. It is impossible to stop feeding monkeys because this act is a cultural and religious habit of people of India from time immemorial. The option of reproductive suppression is not possible for lack of suitable technique. Thus only translocation is viable and proved practical and effective in the studies carried out so far. The IUCN guidelines for translocating monkeys are outlined as well as a need for a monkey home for captured simians is highlighted.

INTRODUCTION

Of the sixteen species of non-human primates present in India three species are considered as commensal, most of the time living close to human beings in villages, towns, cities, temple sites, parks, gardens, orchards, etc. These are: Rhesus monkey, *Macaca mulatta*, Bonnet monkey, *Macaca radiata* and the Hanuman langur, *Presbytis entellus.* Rhesus is the most commensal amongst them, most aggressive and destructive. Out of 0.3 million monkeys in Northern India, approximately 48.5% live in human habitations (Southwick & Siddiqi, 1994). It is also the most studied species.

Originally forest living, the Rhesus monkey has successfully invaded human habitat. The following are the major reasons for their pestiferous activities:

1. Habitat related factors are mainly two namely an overall decrease in forest cover, which in turn decreases the number of plant species that provide food to non human primates and unplanned replacement of tree species. The later results in change in canopy size unsuitable for Rhesus monkeys.

2. An increase in human activity: Developmental projects, hunting, trapping and illegal trade for exports, bio-medical research, meat and pet trade are the negative human activities that are adversely affecting these animals.

3. Chaoting fissioning of groups: To meet the demand of pharmaceutical companies, monkeys of specific age and sex are trapped, which affects the age and sex ratio of monkey social groups leading to chaotic fissioning of groups. More groups mean increased home range, which invariably includes human habitat. This in turn leads to enhanced competition between humans and monkeys for resources.

4. Depleted forest cover: Forests being their natural habitat, monkeys are basically shy of human beings. Unfortunately depleted forest cover has adversely affected the stability of monkey populations. As a consequence they encroach upon human habitat.

The problem of commensal monkeys is strongly linked to Hindu culture of revering monkeys as God's incarnation. This is also seen in countries where Hindu mythology has strong influence on socio-cultural life of people viz. Nepal, Thailand, parts of Indonesia etc. However this reverential attitude towards monkeys sparks off a chain reaction. On their arrival into human habitation, monkeys are initially fed by people. When they keep returning, they are shunned away. When they are shunned, they look for unguarded food inside and outside homes leading to man-monkey conflict. The Gods become pests!

SCIENTIFIC STATUS: There are three species of commensal monkeys in India. Their scientific status is as follows:

1. The Rhesus monkey, *Macaca mulatta* (Zimmerman, 1780)

Class	: Mammalia
Order	: Primates
Sub order	: Haplorhiri
Super Family	: Catarrhiri
Family	: Circopithecidae
Genus	: *Macaca*
Species	: *Mulatta*
Binomial name	: *Macaca mulatta*
Synonymous	:*Cercopithecus mulatta, Simia rhesus*

Conservation status: WPA: Scheduled II; **CITES**: Appendix II

2. The Bonnet macaque, *Macaca radiata*

Status	: Lower risk
Class	: Mammalia
Order	: Primates
Sub-order	: Haplorhiri

Superfamily	: Catarrhiri
Family	: Cercopithecidae
Genus	: *Macaca*
Species	: *Radiata*
Binomial name:	: *Macaca radiata* (Dufreshe, 1979)

3. The Hanuman Langur, *Semnopithecus entellus*

Status	: Lower risk
Class	: Mammalia
Order	: Primates
Sub-order	: Haplorhiri
Superfamily	: Catarrhiri

Family	: Cercopithecidae
Genus	: *Semnopithecus*
Species	: *Entellus*
Binomial name	: *Semnopithecus entellus* (Dufreshe, 1979)

Similar species

Macaca assamensis: Characterised by hind quarters of body being less brightly coloured than the forequarters, tail being half to more than half of head and body, and the skin around callosities is hairy.

Macaca radiata: Crown hairs forming whorl or bonnet on head.

Intraspecific variation

M. mulatta mulatta: Characters as mentioned above; Nepal, North India, coast of Vietnam and island of Hainan.

M. mulatta mcmahoni: Largest of all, longer fur and dull colour; NE Afghanistan, Pakistan.

M. mulatta villosa: larger or heavier than *mulatta*; brow ridges pronounced; Kashmir, Himachal Pradesh and North Uttar Pradesh.

M. mulatta sancti johannis: Hong Kong

M. mulatta lasiotus: S. China

M. mulatta tcheliensis: N. China

The last three sub species differ mainly in the length of tail and body colour. The present chapter deals with the Rhesus monkey *Macaca mulatta.*

DISTRIBUTION

India: West, North, North East and Central India. In the south beyond Godavari to the Krishna delta (Fig. 1).

Extralimital: North eastern Afghanistan, Pakistan, China, Bhutan, Nepal, Bangladesh, Myanmar, Hong Kong, Kampuchea, Laos, Thailand and Vietnam.

Estimated population: Wild: approximately 300,000 to 350,000 in India. *Captive*: common in Zoos and in some research institutes.

Trade: Trained and used for roadside shows by street performers. Rhesus is a very common pet. It is extensively used in medical research, was exported in large numbers from India till the export ban in 1978. Still illegally traded.

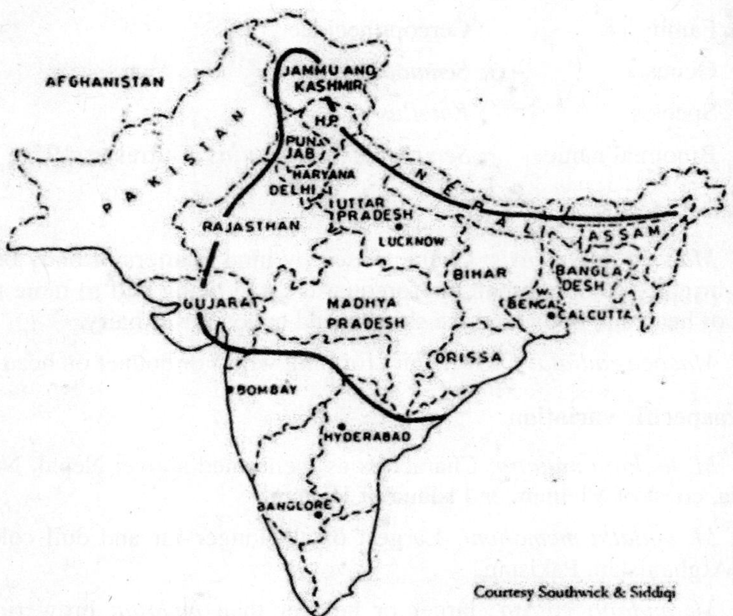

Fig. 1. Approximate distribution of rhesus monkey, *Macaca mulatta* in India

MORPHOLOGY

General appearance: A medium sized macaque, with hindquarters of the body brighter (orange red hue) than forequarters and a short tail.

Head body length: Male: 455-635 mm, Female: 370-580 mm, Tail: Male: 200-315 mm, Female: 160-285 mm.

Body weight: Male: 5.6 – 14.5kg, female: 3.0 – 10.7kg.

Head: crown hairs directed backwards from the brows and without a median parting, whorl or cap.

Face: Bare; light pink to reddish; cheek hairs short, forming a whorl located inferiorly.

Coat: Upper back olive to greyish brown; loins, rump and base of tail orange red, more brightly coloured than the forequarters; skin around ischial callosities (rump) naked; undersurface sparsely haired and paler than above.

Sexual dimorphism: Females are smaller and lighter in weight than males.

Infant: The new-born is dark brown, almost black; crown hair with center parting.

PEST STATUS: The problem of commensal monkeys is three fold:

1. Possible transmission of fatal diseases
2. Nuisance to daily life of people
3. Pestiferous activities

DISEASES TRANSMITTED BY MONKEYS

Monkeys are susceptible to several fatal infections that are also dangerous to human beings but not to lower animals. Although this susceptibility has enormous practical value in biomedical research, the very same attribute is a potential hazard to human beings who come in contact with infected monkeys. The diseases are categorized into viral, bacterial and parasitic.

(a) Viral diseases are the most dangerous as it is very difficult to diagnose and treat diseases of viral origin. The following are the major viral diseases infecting monkeys:

Herpes virus: Herpes simiae (herpes B) produces a mild disease in some species of monkeys that is analogue to cold sores caused by *Herpes hominis* in human beings. But in human beings it is fatal causing acute ascending myelitis. It is found only in macaques under natural conditions. Transmission is primarily by monkey bites and aerosolization.

Pox viruses cause four diseases in non-human primates all of which are infectious to man. *Monkey pox* which is serologically close to smallpox, is the most common infection.

Yaba virus is a rare viral infection seen in monkeys and other non-human primates.

Rabies: Monkeys are as susceptible to rabies as human beings. Since monkeys from forests are exposed to wildlife, chances of rabies in monkeys are high.

(b) Bacterial diseases: Similar to human beings, monkeys are susceptible to a wide variety of bacterial pathogens. In general macaques are more susceptible to tuberculosis and enteric bacteria. Important bacteria infecting monkeys and humans are:

Mycobacteria (responsible for tuberculosis). Amongst the non-human primates, macaques and apes are most susceptible to tubercu-losis;

Shigellosis and *Salmonellosis* are caused by two separate organisms but the symptoms, signs and treatment are similar. Reports of transmission to humans are rare.

Camphylo bacteria cause severe entero colitis in a variety of mammals and birds including monkeys and humans.

(c) Parasitic diseases

Protozoa giardia and *Entamoeba histolytica,* causing diarrhoea and dysentery, respectively are the two important protozoan parasites of primates including monkeys.

(d) Worms

Nematodes: The principal infective nematode from primate to man belongs to *Strongyloides* sp.

Cestodes: The only zoonotic cestode from non-human primates to humans is *Hymenolepsisnana.*

Ringworm: *Trichaphyton mentagrophytes* is the common ringworm affecting all primates.

(e) Arthropods

A few species of lice, mites and fleas infect monkeys causing skin diseases characterized by pruritus and scaling. However, arthropods are highly dangerous to man in their role as intermediate hosts for parasitic diseases and mechanical vectors for several infectious organisms.

Infected monkeys and latent carrier monkeys are potentially hazardous to human health when they come in close contact. This threat makes it imperative that public is not only educated by medical and veterinary professionals about health risks from commensal monkeys but also about the need to manage them scientifically.

NUISANCE TO PEOPLE

(a) Rhesus macaque, ***Macaca mulatta*****:** Basically the problem of monkey nuisance lies in their attempts to procure food and space in human habitats. This in turn is a fall out of destruction of forests, their natural habitat. With shrinking forests, changed microhabitat, decreased availability of food and water, and decreased human tolerance to increasing number of monkeys, not only there is conflict between humans and monkeys, but also mutual hostility. Attracted by food, water and cover, monkey troops invade croplands, settlements and often destroy property, gardens, household furnishing and parked cars. They are over-abundant in temples where devotees feed them out of religious sentiments, in hospital premises and schools. Angered by their destructive activities, irate humans resort to hitting monkeys with stones and sometimes even shooting them. In response monkeys become over-aggressive. They threaten people with snarls, snatch food boxes, spectacles, and handbags and very frequently bite human beings.

Although there are no countrywide surveys on people bitten by monkeys and harassed, the instances of monkey bites are increasing in recent times. Roughly around 100 people are injured by monkey bites every day in New Delhi alone. The magnitude of harassment can be imagined if one realizes that about 50 people were bitten per day fifty years ago (Malik, 2001).

Table 1. People harassed/ bitten by monkey

Category	% Harassed	% Bitten
Resident (Delhi & Vrindavan)	95.5	19
Pilgrims (Vrindavan)	30.5	3
Visitors (Tughlaqabad fort)	5	2
Office goers (North block, Delhi)	88	2
All India Institute of Medical Sciences		
Staff	80	2
Relatives of patients	20	2
School Students (Delhi)	97	2

A survey of monkey menace in four monkey infested areas of New Delhi during 1996 revealed that although monkeys harassed residents, office-goers, students and visitors, the degree of harassment varied between these groups (Table 1) (Malik, 2001). Residents were the worst affected as monkeys raided their homes and gardens. This lead to a vicious circle of aggression between residents and monkeys resulting in maximum number of bites. Office-goers were equally harassed, but were less likely to be bitten as most commuters were adult males. It appears woman and children were harassed and bitten more as monkeys seem to perceive these two categories to be weaker, least aggressive and incapable of retaliation. A similar behaviour was observed in crop raiding Hanuman langurs, *Semnopithecus entellus* in Rajasthan (Chhangani & Mohnot, 2004). This view of weaker individuals being targeted by monkeys is further supported by the fact that in schools infested by monkeys, children moving in groups accompanied by *lathi* (long bamboo stick) wielding guards are not attacked by monkeys.

Table 2. Attitude of people towards Rhesus monkey in Vrindavan

Type of harassment	Response (%)		
	Positive	Negative	Low or No response
Felt harassed	95.5	4.5	
Degree of harassment	74.5	20	5%
Monkey bites	19		
Spectacle snatching	17		
Stealing house hold items	28		
Miscellaneous	36		
Response to shifting	69	26.5	4.5

(Based on 270 respondents)

The nuisance and pestiferous activities of an over-abundant population of Rhesus monkeys was studied elaborately in Vrindavan (Malik & Johnson, 1994). Ninety five per cent of local people were harassed by monkeys by way of their entry into houses, stealing food, clothing and other articles, uprooting vegetables and garden plants, pulling out electric wires and TV antennae, threatening and attacking people and often biting severely.

Of the economic impact of diseases being transmitted to humans from monkeys, taking the average bites of 100 per day, each patient requires 10 injections of anti rabies vaccine, which costs Rs. 250/shot. This would amount to Rs. 2.5 lakh/day. If the other monkey transmitted/ borne diseases are also taken into account, the increased number of commensal monkeys not only threatens the welfare of monkeys but also poses a greater problem to public health.

Commensalism of Rhesus monkeys is systematically traced by Southwick & Siddiqi (2001) based on the long term studies carried out in Aligarh (Southwick, 1989), Tughlakabad (Malik, 1989) and Jaipur (Mathur & Lobo, 1988; Mathur & Manohar, 1990). The population of Rhesus monkeys in Aligarh showed different trends. The initial population of 337 monkeys in 17 groups in 1959 rose to 403 numbers in 22 groups by 1962. Over the next few years the population declined to 163 monkeys in 10 groups (Southwick, 1989). The decline was generally attributed to export of large number of juvenile Rhesus monkeys for biomedical research. With reduced exports, monkey population stabilized in 1970 and started showing slight increases. With the total ban on monkey export in 1978, the Rhesus population in

Aligarh doubled (Southwick & Siddiqi, 2001). Similar increases of monkey population by several hundred percent was observed in other parts of Uttar Pradesh and Tughlaqabad at the southern edge of New Delhi (Malik, 1989). This spurt in population was attributed to high reproductive and low mortality rates.

Great numbers of Rhesus monkeys also lived in Jaipur, more than in Vrindavan which later showed a natural decline (Mathur & Manohar, 1990).

(b) Hanuman langurs, *Semnopithecus entellus* are also commensal, but do not reach pest status in spite of their wide geographic distribution. Their habitat ranges from the montane forests of Himalayas to the plains of Gangetic basin, deserts of Rajasthan and the tropical forests of southern India. Conflicts with humans are rare because of their preference for natural vegetation. Most appropriately they are the true sacred monkeys of Hindus, revered and respected. Being less aggressive, not depending much on human resources for food, with low fertility and high mortality rates, Hanuman langurs rarely reach pest status. Tourist locations, temples, parks, sanctuaries etc are their preferred habitat. Lately their population is exhibiting signs of increase.

(c) Bonnet macaque, *Macaca radiata*: Although studies on Bonnet macaques are scarce compared to Rhesus monkeys and Hanuman langurs, they are as troublesome as Rhesus monkeys. For instance Bonnet macaque living in Elephanta caves near Mumbai, are reported to harass tourists and shop keepers similar to the pattern Rhesus monkeys do elsewhere. Occasional crop raiding by them is reported for south India (Southwick & Siddiqi, 2001).

PROBLEMS OF RHESUS MONKEY

We have studied troublesome Rhesus populations at Vrindavan, Gurgaon and Aligarh, the summary of which is as follows:

(a) Vrindavan (1995): This temple town of Hindus measuring 4.4km^2 is located 160km from Delhi. The town has about 1000 temples, *ghats* (bathing places along the bank of river Yamuna), some tree grooves and tanks. Although the area had thick lush green forest sixty years ago, now the only green cover available is in the temple orchards.

In the absence of any previous records on the Rhesus monkey population of Vrindavan, it was almost impossible to estimate the percentage increase in their total population during the study period. The presence of 1,338 monkeys in 4.4 km^2 with a density of 304 monkeys / km^2 was quite alarming. The observed density was much higher than the population of Aligarh and Tughlaqabad Fort of Delhi. Such high population is considered overabundant and beyond the carrying capacity of the habitat, and has no scope for any expansion. Consequently prolonged aggression was observed, mostly at feeding sites. In such situations sick monkeys get deprived of food. In addition, their frequent contact with humans may enhance the chances of transmission of potential diseases from monkey to human and vice versa (Shah & Southwick, 1965; Tiwari & Shukla, 1984), although no disease outbreak within any human community in India has ever been unequivocally linked to monkey-human contact. Nevertheless, the potential for such mishaps is undeniable.

The monkeys of Vrindavan were creating problems such as stealing clothes, food items, uprooting vegetables and garden plants, damaging television antennae and electric wires etc. The monkeys at the site of temples were expert tricksters and within no time, they would snatch spectacles. These monkeys seemed to have developed this behaviour just to obtain more food from the pilgrims because whenever food was offered they would drop the spectacles. In the absence of large trees, monkeys were using television antennae, electric wires and electric poles for playing. Whenever monkeys vigorously shook the antennae and electric poles, TV reception and electric supply were disrupted. Monkey bites, which may cause rabies, were also very common in Vrindavan. At a local hospital rabies vaccination for monkey biting, especially for children, was reported almost daily. The harassed people strongly supported translocation almost *en mass*.

With the increase in activities of monkeys, the traditional bond between man-monkey was eroding rapidly. Their activities were becoming intolerable and they were viewed more as vermin than as wildlife. This was not only true for Vrindavan, but for all monkey-infested areas of the country. It is quite possible that public dislike of Rhesus could extend to other primate pest species in India. In this worst possible scenario, public support for all conservation projects, involving monkeys might suffer. As problems created by monkeys of Vrindavan had already attained serious proportions, some initiative was required to reduce the Rhesus population by scientific management. The translocation appeared as the only solution and it was possible, due to the initiative of the local people and *Swamis* (Caretaker of Ashrams).

(b) Gurgaon (1998): The Gurgaon air force station (GAFS) is situated 28km south of New Delhi. It had a population of 27 nuisance monkeys.

(c) Aligarh Muslim University Campus, Aligarh (2000): The Central University is a residential one, situated at Aligarh in Uttar Pradesh, India. The campus of Aligarh Muslim University has 15 residential hostels and approximately 2000 quarters, accommodating more than 17,000 people. The campus has well maintained greenery. Food wastes dumped at the backyard of houses and hostels attract the monkey groups, while planted trees along the roads and other parts of the campus provide shelter to them. In addition students and residents also fed the rhesus monkey groups.

MANAGEMENT

Southwick & Siddiqi (2001) suggest three measures to contain commensalism and accompanying nuisance to public. These are, reducing supplemental feeding, translocation from trouble areas and fertility control if and when practical methods can be developed. However, the first and the third are not feasible in India. Religious sentiments do not allow people to stop feeding commensal monkeys. Fertility control is theoretically a long term, permanent solution but right now we do not have tested and proven methods of sterilizing monkeys effectively, economically and successfully. That leaves translocating as the only feasible and practical method to manage the problem of monkeys.

Translocation can be defined as a wildlife management tool, whereby wild animals are shifted from the problem area to a new place where they have adequate natural resources to settle themselves for the purpose of conservation or management. Parallel rescue/welfare reintroduction involve movement of wild primates from one area to another to resolve human-primate conflicts or the release of captive primates, such as orphaned or surplus animals in order to improve their welfare (Baker, 2002). This approach was taken by Malik & Johnson (1991, 1994) to mitigate the problem of nuisance monkeys in Vrindavan, Delhi and Tughalaqabad.

The following are the precautions drafted by IUCN to be adhered to while translocating monkeys from nuisance areas to a new, natural habitat and were followed in our translocation programmes:

1. A multidisciplinary group consisting of primatologists, representatives from natural resources agencies of government, local communities, funding agencies, veterinarians and Zoo personnel should be involved in translocation.
2. It is essential to determine the suitability of the habitat in which monkeys are going to be translocated. The new habitat

- Should satisfy the requirements of the introduced species in a sustainable way.
- Should keep in view the possibility of a major population explosion of translocated primates.
- Has to ensure the presence of corridors that connect the release site with other areas.
- Most importantly the release site should be within the taxon's historic documented range. For long term population management in the released site, a population and habitat viability analysis (PHVA) may be helpful.

3. The existing wildlife in the area of translocation should not be disturbed drastically by introducing non-residents. It is important to gauge the impact of translocated monkeys on resident population.
4. Because translocation presents a risk of disease transmission, social disruption and introduction of alien genes to wild populations, it is advisable to select sites with no resident or minimum resident population.
5. It is essential to confirm availability of sufficient food and their seasonal variability of the new location. The monkeys to be transported have to be provided with foods similar to those they will encounter in the release site and also during their confinement in enclosures transporting them. It is also important not to give foods that are grown adjacent to the release area.
6. The predator risk in the translocated area should be taken into account while introducing nuisance monkeys.
7. The survival prospects of the released animals should be ensured.
8. Translocations are usually long-term operations and require committed and continuous public, political and financial support. Socioeconomic surveys should be carried out to assess the impact, costs and benefits of reintroduction.
9. The translocation programme should be understood, accepted and supported by local people prior to initiation. If there is a risk of human-wildlife interaction at a later date, an action plan to resolve such situations should be drawn up in consultation with concerned authorities.

10. While transporting and translocating commensal monkeys to wild, the possibility of transmission of anthropozoonotic and zoonotic diseases is very high. Hence, a strict quarantine and veterinary programme must be planned and executed with the help of qualified and experienced veterinarians.

TRANSLOCATION OF RHESUS MONKEY IN INDIA

The utility of translocation in the conservation of Rhesus monkeys was first demonstrated in 1984 (Southwick *et al.,* 1984) when 60-70 monkeys from a population of 140 at Chatatri (Aligarh) were translocated. In 1989, a Rhesus group of 21 was translocated from Tughlaqabad Air force station, New Delhi to a rural area of Meethapur 10 km away (Malik & Johnson, 1991, 1994). In 1988 a subgroup of 20 Rhesus monkeys was trapped from a mango grove and translocated to a canal-bank forest patch (Siddiqi & Southwick, 1993). In February 1995, five groups of Rhesus monkey comprising of 40 individuals were successfully translocated from Central Zoological Park, New Delhi and rehabilitated in Tughlaqabad fort, New Delhi (Imam & Malik, 1997) and in the same year 7 individuals trapped from Friendicos, New Delhi were rehabilitated in Tughlaqabad fort, New Delhi.

During translocation, the monkeys were baited and trapped with the help of a portable iron cage measuring 2m x 2m x 2m with a sliding door. The trapped monkeys were transferred to a holding cage of 0.5m X 0.5m X 0.5m, then finally released into a releasing cage of 5m X 2.5m X 2.5m size. These were loaded on to a mini truck and transported to pre-selected suitable sites and released in the vicinity of trees during day time.

The post-translocation study in Vrindavan revealed that the monkeys were comfortable at the new sites. Time spent in different activities by the translocated monkey group was found to be similar to the groups of northern India (Seth & Seth, 1986), Tughlaqabad fort (Malik & Southwick, 1987) and Nepal (Southwick *et al*., 1982). No stress was observed. The local people too accepted the new guests. Inhabitants, passer-by and passing trunk drivers gave the monkeys food and eatables.

After translocating 45% of the nuisance population of monkeys from Vrindavan, the operation was suspended as alternate, suitable forest patches were not available. However, the earnest request of people of Vrindavan to continue translocation of monkeys is a sure pointer to people's co-operation for the measure undertaken which is significant in view of the town being a Hindu pilgrim centre from time immemorial.

In Gurgaon twenty two of the monkeys were trapped and translocated to Firozpur-Thirka, which had a pond and perennial *nullah* with a fairly good forest patch. This translocation was a small attempt to reduce man-monkey conflict and rehabilitate the urban monkeys in their natural habitat. The post translocation visit to GAFS and Firozpur-Jhirka revealed that translocation of rhesus macaques was a successful attempt in this case also. They were doing well even 20 months after translocation.

In Aligarh trapping was carried out from January 2000 to March 2003. Out of seventy two monkeys fifty-one were trapped from four places located in the University campus and consisted nineteen adult males, twenty adult females, eight juveniles and four infants. They were trapped over two years from different localities of the University campus. The trapped monkeys were shifted to forest patches situated 30 km away from Aligarh.

A post-translocation visit revealed that most of the monkeys were settled in the semi-natural forest of canal area and no crop raiding and monkey bites were reported by nearby villagers.

THE DILEMMA OF MANAGEMENT OF PESTIFEROUS MONKEYS

In dealing with the rich and diverse primate population of India two types of problems arise. First and foremost are the conservation priorities of rare and endangered primates which form more than fifty percent of total primate species. The three major contributors to this alarming state are loss of habitat, degradation of habitat and hunting /capturing for trade. Loss as well as degradation of habitat is due to clear cutting of forests for agriculture, felling forest trees for timber and firewood and raising monoculture plantations of teak, eucalyptus, oil palm and cloves in forest cleared areas. Grazing pressure from livestock also contributes to degradation. Unlike Africa and South-east Asia, primates of India enjoy religious status and are not subject to hunting for pleasure although some tribals may hunt for food. In India the residual population of endangered primates are limited to North-east and South. These include Rhesus, Assamese, Pig-tailed and stump tailed macaques, capped, golden and phayre's langurs, slow loris and Hoolock gibbon in north–east of India and lion tailed macaque, Nilgiri langur and slender loris in South India.

The second issue of concern is the pestiferous activities of Rhesus and Bonnet macaques and Hanuman langurs. Living mostly in forest edges they depend to a considerable degree on human environment for food, becoming pests in the process. Thus we find them at the opposite end of the distribution span necessitating management of nuisance populations where the emphasis is on driving them back to their habitat. In extreme cases of overabundant numbers reduction of numbers becomes a necessity. Long term studies on Rhesus macaques in Aligarh (Southwick, 1989; Southwick & Siddiqi, 2001) and in Tughalaqabad in Delhi (Malik, 1989) have shown rapid increase in numbers due to high reproductive rates and low mortality of infants. Thus ecologically Rhesus monkeys qualify for "*r*" selected category exhibiting vigorous population growth and efficient utilisation of commensal habitats. Richard *et al.,* (1989) have termed them as "weed macaques" in view of their aggressive commensalism. Majority of them obtain their food from human habitat by way of handouts from people, thieving from roadside shops and houses, snatching food from people as well as raiding orchards. Although 50% of the monkey population of Vrindavan was successfully translocated to forest patches, it is not considered a permanent solution (Southwick & Siddiqi, 2001). There is a strong possibility of them turning into pests in the new environment. The only deviation to growing numbers of Rhesus macaques is seen in Jaipur where they have actually shown a declining trend in population (Mathur & Manohar, 1999).

Hanuman langurs, though commensal rarely reach the pest status of rhesus macaques. They are less aggressive and not considered as pests by many. Birth rates are lower and infant mortality is higher compared to Rhesus macaques. They tend to depend more on natural vegetation and less on crop raiding and thus are a lesser pest than Rhesus monkeys. Although their populations fluctuated, they were more stable than Rhesus monkeys.

Bonnet macaques though less studied than Rhesus and langurs are inherently capable of aggressive commensalism, similar to Rhesus monkeys, both in terms of harassing people and raiding crops.

The management approaches suggested for Rhesus are: reduction in supplement feeding, translocation from troubled areas and fertility control provided successful methodologies are arrived at (Southwick & Siddique, 1994). Of these only translocation has been successfully carried out so far as explained earlier in this chapter. Fertility control is still a theoretical option.

NEED TO SET-UP MONKEY TRANSIT HOME AT DELHI

There are around 5000 rhesus monkeys in Delhi. Around 95 percent of these were living in and around human habitations during 1980's when, approximately 30% people were being harassed and 30% were bitten. By 2003 these figures increased to 60% and 16% respectively (Malik Pers. Obser.).

Every second day we receive innumerable calls from the residents of Delhi facing monkey problems in their colonies. We also receive complaints from hospitals, schools, offices, institutes and residential areas about the havoc monkeys are creating. This increased man-monkey conflict is detrimental for both monkeys and humans. It is time Delhi sets in motion measures to mitigate monkey problem. The first step in this long term solution is to set up a transit home for the problematic rhesus monkeys of Delhi. We would like to call the transit house as Monkey Home where the captured problem monkeys are transported, treated, quarantined and finally handed over to research labs or rehabilitated. This endeavor has to be a joint effort by the government, non-government organizations and experts. We suggest establishing a Transit home as a first step towards management of nuisance monkeys of urban area.

Site for setting up of Transit Home should fulfill the following conditions

i. The site of the Transit Monkey Home for Delhi monkeys should be in or around Delhi so that the monkeys would not have to travel far off which can lead to stress and fatalities.

ii. The area should be at least 15 acres so that around 2000 monkeys can be housed there at any given time.

iii. The site should be green, rough and non-manicured with a potential to be developed into a near natural habitat for monkeys.

Facilities to be provided in Monkey Home:

Monkey Home would have three major components namely Holding Facilities, Hospital and Quarantine Center and two minor components – Disposal and Miscellaneous Area (Fig. 2).

1. Holding Facilities

Three fourth of the area should be open green area and one fourth should be built up area. It should have 5 parts: (i) large, green and open area where monkeys would be housed as soon as they come. The area should be able to hold 50 to 100 monkeys at any one time. From here monkeys should go for their health check-up to (ii) X-ray room and (iii) Lab facilities. After checkups monkeys would go to either house for Monkeys suffering from non-contagious disease (iv) or to house for Monkeys with contagious diseases (v).

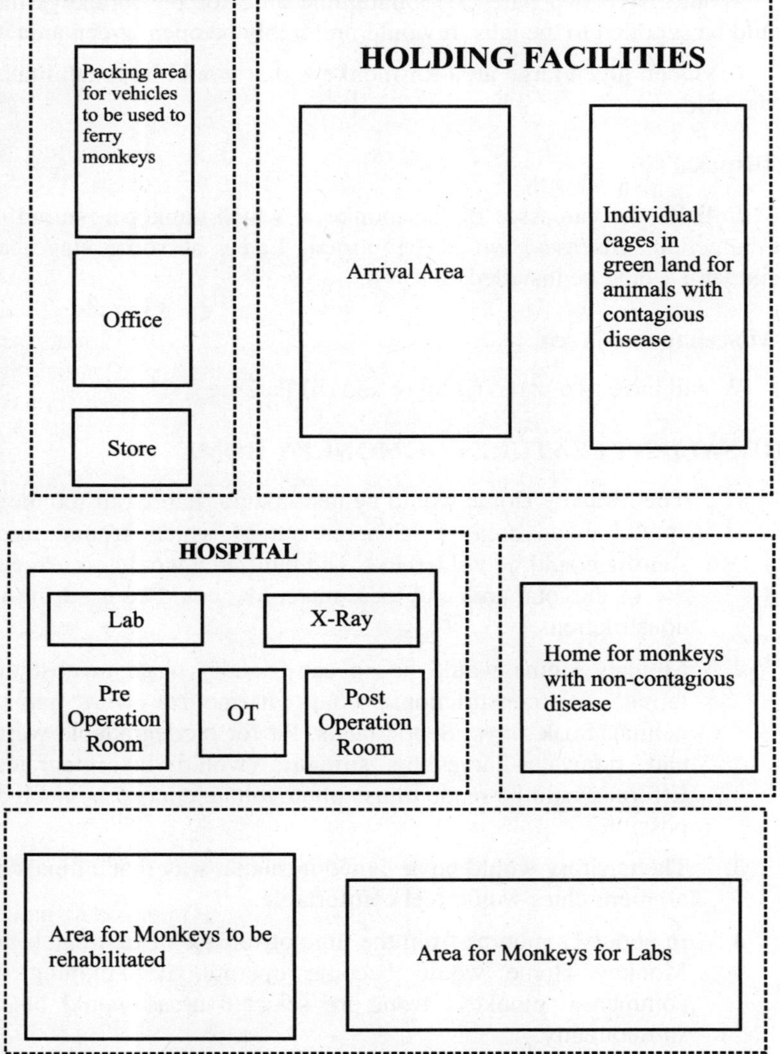

Fig. 2. Flow chart of the facilities to be provided in the transit home for monkeys

2. Hospital

It would have 3 parts: (i) Pre-operative room, (ii) Operation theatre (iii) Post-operative room.

3. Quarantine Center

Would have two parts: (i) Quarantine area for the monkeys that would be supplied to the labs. It would preferably be open, green area.

(ii) Open green large area for monkeys that would be rehabilitated in the wild.

4. Disposal site

Initially the carcasses of the monkeys, which cannot be cured of the infectious diseases, would be buried. Later, at some stage, an incinerator would be installed.

5. Miscellaneous Area

Would have two parts: (i) Store and (ii) Parking

THE SALIENT FEATURES OF MONKEY HOME

1. The Monkey Home would be near natural habitat of monkeys spread in an area of 15 acres out of which around three fourths would be wilderness. The built area would not exceed 3% of the total area and local materials would be used for all constructions.
2. Monkey Home would be an eco-friendly place in absolute terms. All constructions would harmonize with natural habitat. Soak-ways, Septic tanks, Pit for biodegradable waste and rainwater-harvesting structure would be integrated. Efforts would be made to use photovoltaic energy as much as possible.
3. The territory would be designed in such a way that animals of all hierarchies would feel comfortable.
4. In about 12 months from the time of release of first batch the Monkey Home would become operational. Trapping of commensal monkeys from pre selected areas would begin subsequently.
5. High standard of personal hygiene of all primate keepers would be maintained to avoid any zoonotic infections.

6. The Monkeys would be provided with interesting, stimulating, fresh and tasty diet.
7. All major facilities would have double doors for people to come in and out of them. Facilities would be interconnected by overhead runs and pathways. There would be crush cages at strategic points.

CONCLUSIONS

With the increase in population of human and commensal monkeys, the existing age-old harmony between them is depleting day by day. Since rhesus monkey is one of the important primate species due to its religious, traditional, ethical and biodiversity value, it has become necessary to conserve them properly and protect them from being declared as pest/vermin species. For this we have to develop a long-term national Management Plan for our country. Following are some recommendation in this direction:

1. Identification of commensal monkey groups distributed in the country.
2. Study of various problems created by rhesus monkey.
3. Quantitative analysis of damages caused by rhesus monkey.
4. Identification of forest patches which can accommodate some monkeys.
5. Rehabilitation of commensal/refugee monkeys in their natural habitat.
6. Implementation of Management Plan as discussed above.

REFERENCES

BAKER, L.R. 2002. Guidelines for Non-human primate re-introductions. IUCN/SSC Reintroduction Specialist Group: Page 1-57.

CHHANGANI, A.K & S.M. MOHNOT. 2004. Crop raid by Hanuman Langur, *Semnopithecus entellus* in and around Aravalli Hills (India) and its management. *Primate Report* 69: 1-18.

IBID. 1994. Commensal Rhesus in India: the need and cost of translocation. *Reue' E' Cologie* (TERRE VIE): ROME, 49(3): 233-243.

IBID. 2001 Monkey menace – who is responsible In: NON HUMAN PRIMATES OF INDIA. *Envis. Bull.* (Wildlife and protected areas), (Ed.) A.K.Gupta, Wildlife Institute of India, Dehradun, India.

IMAM, E. & H.S.A. YAHYA. 1995. Population dynamics and distribution of Rhesus monkey in Aligarh district. *J. Ecobiology* 8(3): 1-9.

IMAM, E. & I. MALIK. 1997. *Translocation of monkeys from National Zoological Park, New Delhi to Tughlaqabad Fort, South Delhi*. Report submitted by Vatavaran to National Zoological Park, New Delhi, India.

MALIK, I. & C.H. SOUTHWICK. 1987. Feeding behaviour of free ranging Rhesus of Tughlaqabad. *J. Bombay nat. Hist. Soc.* 84(2): 336-349.

MALIK, I. & R.L. JOHNSON. 1991. Trapping and conservation: development of translocation in India. In: PRIMATOLOGY TODAY (Eds.) A . Ehara, T. Kimura and M. Iwamoto, Elsevier, Amsterdam. pp. 63-64.

MALIK, I. 1989. Population growth and stabilizing age structure of the Tughlaqabad Rhesus. *Primates* 30: 117-120.

MATHUR, R & A. LOBO. 1988. Density estimate of monkeys of Jaipur, India. *Primate Report*. 19: 35-42.

MATHUR, R. & B. MANOHAR, 1990. Density of *Macaca mulatta* and *Presbytis entellus* in the old city of Jaipur: A three year survey. *Appl. Anim. Behav. Sci.* 27:351-361.

RICHARD, A.F., S.J. GOLDSTEIN & R.E. DEWAR. 1989. Weed Macaques: The evaluation implications of macaque feeding ecology. *International J. of Primatol.* 10: 569-594.

SETH, P.K. & S. SETH. 1986. Ecology and behaviour of Rhesus monkey in India. In: PRIMATE ECOLOGY AND CONSERVATION. (Eds.) J.G. Else and P.C. Lee Cambridge University Press, Cambridge, UK.

SHAH, K.V. & C.H. SOUTHWICK. 1965. Prevalence of antibodies to certain viruses in sera of free-living Rhesus and captive monkey. *Indian J. Med. Res.* 53:488-500.

SIDDIQI, M.F. & C.H. SOUTHWICK. 1993. Translocation of Rhesus macaques in India. In CONSERVATION AND MANAGEMENT OF PRIMATES (Eds.). R. Mast & A. H. Mitchell. Sunny press, Binghamton, USA.

SOUTHWICK, C.H., M.F. SIDDIQI. & R. JOHNSON. 1984. Subgroup relocation of Rhesus monkeys in India as conservation measure. *American J. of Primatology* 6: 428.

SOUTHWICK, C.H., T. RICHIE., J. TEAS & M.F. SIDDIQI. 1982. Rhesus monkey population in India and Nepal: Pattern of growth, decline and natural regulation. In: ADVANCED VIEWS IN PRIMATE BIOLOGY (Eds.). A. B. Chiarelli & R.S. Corruccini. Springer Verlag, Berlin and Heidelberg, Germany. pp. 128-137.

SOUTHWICK, C.H. & M.F. SIDDIQI. 2001. Status, conservation and management of primates in India. In: NON HUMAN PRIMATES OF INDIA. (Eds.) A.K. Gupta. *Envis Bull.* (Wildlife and protected areas). Wildlife Institute of India, Dehradun, India.

SOUTHWICK, C.H. 1989. Rhesus monkey populations in India and Nepal: patterns of growth, decline and natural regulation. In: BIOSOCIAL MECHANISMS OF POPULATION REGULATION (Eds.). M.N. Cohen,

R.S. Malpass & H.G. Klein. Yale University, New Haven, Connecticut, USA. pp. 151-170.

SOUTHWICK, C.H., & M.F. SIDDIQI. 1994. Primate commensalism: the Rhesus monkey in India. *Rev. of Ecol.* (Terre vie) 49: 223-231.

TIWARI, J.P. & L.K. SHUKLA. 1984. Microbiological investigation of rhesus macaque, *Macaca mulatta*, in an urban setting in Uttar Pradesh, India. In: CURRENT PRIMATE RESEARCH, (Eds.) M.L.Roonwall, S.M.Mohnot and N.S. Rathore. University of Jodhpur, Jodhpur, India. pp. 497-499.

❑ ❑ ❑

18

CROP DAMAGE BY HANUMAN LANGUR, (*Semnopithecus entellus*) AND ITS MANAGEMENT

A.K. Chhangani, L.S. Rajpurohit and S.M. Mohnot

Indo-US Primate Project, Department of Zoology,
Jai Narain Vyas University, Jodhpur – 342 003 (India)

ABSTRACT

Many non-human primate species are reported as potential pests of crops in different parts of the country. They live in wide range of habitats, from Himalayas to peninsular forests reaching the arid zones of Thar Desert, from woodland forests to rain forests of northeast, spilling over to cultivated land, villages and towns. They have evolved into highly successful commensals and pests. particularly Hanuman langur, *Semno-pithecus entellus*, Bonnet monkey, *Macaca radiata* and Rhesus monkey, *Macaca mulatta.* Studies on the economics of crop raiding monkeys in India revealed losses due to crop raiding and crop protection measures against monkey menace to range between Rs. 5000 to Rs. 12000 per year per field. This amounted to Rs. 5,00,000 annually. The common and successful plant protection strategies employed by farmers included fencing, using dogs, throwing stones, vigilance and guarding the fields.

INTRODUCTION

Although agriculture is the backbone of Indian economy, majority of farmers thrive on subsistence farming. Cereals, pulses, oil seeds are grown comparatively more than cash crops like coffee, tea, rubber, jute and spices. Cultivated food crops, in most cases are meant for family consumption. Average incomes are quite low. Any loss of crop due to pests and diseases has a serious repercussion on the livelihood of a farmer. Losses due to insects and pathogens deprive the farmer 30-40%

of his yield. Mammalian herbivores are no less harming. Small mammals such as rodents take away anywhere from 5 to 20% of the crop yield starting from sowing to post harvest storage. The losses are slow but consistent. In comparison large herbivores like elephants or troops of monkeys or herds of ungulates may destroy large chunks of crops in a single raid. Except rodents, prevention of crop losses from mammalian depredators is nearly impossible with issues of conservation, animal welfare, religion and culture getting mixed up. Compensation from government is rarely available and involves long drawn bureaucratic procedures. In areas fringing national parks, sanctuaries which harbour a rich wildlife, crop depredation by mammals including monkeys results in significant loss of agricultural yields deteriorating the life of subsistence farmer a little more.

Crop raiding by herbivorous mammals is an age-old problem resolved by farmers by indigenous methods until now. It is only now that the problem has assumed serious dimensions. This situation is especially true around sanctuaries and fragmented natural habitats. Major reasons for the escalation of the problem are:

a. Intense agricultural practices and large scale monoculture: Not only monocrops are highly attractive to wild herbivores, they are also unsuitable to harbour predators of crop raiding mammals.

b. The expanding human population needs more and more food leading to expansion of land under agriculture which brings people and wildlife into not only greater contact but also results in competition for resources.

c. The demand for more agriculture leads to cultivation of marginal land expanding upto the borders of, sometimes inside wilderness and sanctuaries. Pest species normally inhabit the edges of their natural habitat eating not only naturally available food but learn to raid crops that grow adjacent to their habitat.

d. As more and more uncultivated land is brought under agriculture, the wild habitats of many herbivores become "islands" surrounded by crop fields, cutting off wild populations from similar populations living in other patches. Such fragmented habitats are, to some degree advantageous to the affected species. The boundaries increase and centre of the wilderness become closer to the periphery with the result these animals travel less and encounter larger areas of food availability.

e. Urbanization and city life attracts migration of youth from villages. This coupled with more and more children of farmers attending school depletes manpower to guard crops against raiding herbivores.

Crop damage by mammals such as elephant (*Elephas maximus*), gaur (*Bos gaurus*), blue bull (*Boselaphus tragocamelus*), sambar (*Cervus unicolor*), barking deer (*Muntiacua muntak*), black buck (*Antilope cervicapra*), chinkara (*Gazella gazella bennetti*), wild boar (*Sus scrofa*) and porcupine (*Hystrix indica*) have been widely reported from all over the country (Prater, 1980; Schultz, 1986; Chauhan & Sawarkar, 1989; Sukumar, 1990; Bohra *et al*, 1992; Balasubramanian *et al*, 1993; Chhangani, 1994; Chhangani & Mohnot, 1997, 2004; Rajpurohit & Mohnot, 1997, Goyal, 1999).

Worldwide non-human primate species that have been reported to cause crop losses include Hanuman langur, (*Semnopithecus entellus*), black and white colobus (*Colobus guereza*), chimpanzees (*Pan troglodytes*), olive baboons (*Papio hamadryas ahubis*), yellow baboon (*Papio cynocephalus*), rhesus macaque (*Macaca mulatta*), Assamese macaque (*Macaca assamensis*), button macaque (*Macaca brunnescens*), booted Sulawesi macaque (*Macaca orchreata brunnescens*), vervet monkey (*Cercopithecus aethiops*), red tailed quenons (*Cercopithecus ascarius*), blue monkey (*Cercopithecus mitis*) and patas monkey (*Cercopithecus patas*) (Roonwal & Mohnot, 1977; Forthman–quick, 1986; Boulton *et al.*, 1996; Pirta *et al.*, 1977; Hill, 1997, 2000; Else, 1991; Kilner, 2000; Priston, 2004; Chhangani & Mohnot, 2004).

Non human primates in India have not been thoroughly studied as "pests". Different primate species of India represent widely different conservation and management problems. Many species such as the golden langur (*Trachypithecus geei*), Phayre's langur (*Trachypithecus phyrei*) and lion tailed macaque (*Macaca silenus*) have small populations, limited distribution and face serious habitat loss (Southwick & Siddiqui, 2001). In contrast, other species especially those that are commensal and can capitalize on human habitats such as Rhesus monkey (*Macaca mulatta*), bonnet monkey (*Macaca radiata*) and Hanuman langur (*Semnopithecus entellus*) are abundant, widespread in distribution and are pests/potential pests in villages, towns, cities apart from horticultural and agricultural fields. These non human primates inflict significant damage to crops along the immediate periphery of wildlife sanctuaries and national parks.

Of the three species of pestiferous monkeys conflicting with humans in India considerable amount of information is available only for two species. The problem of Rhesus monkey, *Macaca mulatta* is covered elsewhere in this book (Imam & Malik, 2005). This chapter gives an overview of the Hanuman langur, *Semnopithecus entellus*, which is a crop raider in the true sense. Although bonnet macaques are reported to inflict similar damage to cultivated crops in South India, albeit to a smaller degree, there are no published reports of any detailed study.

STATUS AND SCIENTIFIC CLASSIFICATION

Status	: Lower risk
Kingdom	: Animalia
Phylum	: Chordata
Class	: Mammalia
Order	: Primates
Sub-order	: Haplorhini
Superfamily	: Catarrhini
Family	: Cercopithecidae
Genus	: *Semnopithecus*
Species	: *entellus*

Binomial name: *Semnopithecus entellus* (Dufreshe, 1979)

A total of seventeen species of langurs or leaf monkeys are reported to occur worldwide. Hanuman langur is one of the four species that occur in India. Depending on body mass, size and colour there may be 7-15 sub species of this monkey in the country.

The Hanuman langur (*Semnopithecus entellus),* also known as the common langur, grey langur and entellus langur, is an old world monkey and is the only member of the genus semnopithecus. It is a large and fairly terrestrial species, preferring open wooded habitats. It descends down only when it feels absolutely secure.

The species is largely grey with a black face resembling the burnt face of monkey God, Hanuman. According to Hindu mythology he goes as an emissary of Lord Rama to his wife Seetha, who was abducted and kept prisoner by the demon king Ravana of Lanka, the modern Sri Lanka. When his mission of getting back Seetha fails he burns the city with his oil soaked tail and his face gets burnt in the process. Most pious Hindus believe that the black skin of Hanuman langur, *Semnopithecus entellus* is a living proof that they have descended from the monkey God of Hindu mythology, Lord Hanuman. This belief as discussed later has a large bearing on the commensalism and management of their activities when numbers reach pest status.

DISTRIBUTION

The Hanuman langur, *Semnopithecus entellus* is the most widespread and adaptable colobine in the Indian subcontinent. These langurs live in a wide range of habitats from Himalayas to peninsular forest, in semi arid woodlands, villages, towns, and cultivated land. Amongst primates Hanuman langur is next to humans in inhabiting a wide variety of landscapes. It is distributed in India, Pakistan, Sri Lanka, Bangladesh and possibly as far as Bhutan and China.

MORPHOLOGY AND BIOLOGY

Head and body length: Female: 40-58cm, Male: 51-78cm. tail length: 69-101cm. weight: female: 11.2kgs, male: 18.3kg. The colour of the fur ranges from grey, dark brown to golden with varying amounts of black, depending on the sub species. Size also varies considerably-sub species from the southern part of their range are generally smaller compared to northern species. The average life span of Hanuman langur is twenty years. After a gestation period of 168-200 days, a single infant is born which is weaned after 20 months.

BEHAVIOURAL ECOLOGY

McCam (1928, 1933) was the first to record scientific observations on Hanuman langur to be followed by Jay (1963) who worked on their social behaviour. Later the population and behavioural ecology of the species was studied by Sugiyama (1964). Starting from 1967 Mohnot and his team have been studying the various aspects of ecology, biology, behaviour and population dynamics of *S. entellus* for the past 37 years, making it the longest ever study on a non-human primate. The species is found in various habitats as diverse as semi desert, open parks, moist and dry deciduous forests and mountains upto 4000 meters above sea level. They more often extend their home ranges to the proximity of humans near agricultural fields, roads and temples, indicating high degree of ecological adaptability. Their social adaptability renders them ideal for socioecological investigations.

Hanuman langurs live in groups with numbers ranging from eleven to sixty four with adult male residents varying from one to fourteen (Newton, 1988).

Extensive studies in and around Jodhpur point out that only resident males breed. However, approximately every 27 months a new male will oust the resident male along with most of the juvenile males. The new male usually kills all the infants (Sommer & Rajpurohit, 1989; Rajpurohit & Sommer, 1993).

FEEDING ECOLOGY

In their natural habitat, Hanuman langurs spend most of their time foraging. They feed on seeds, nuts, fruits, grasses, leaves, roots and occasionally insects. With shrinking habitat their foraging and feeding habits have also changed. In cities like Jaipur and Jodhpur they have become commensal. But factually the divinity associated with them ensures steady supply of food reverentially and affectionately by devout Hindus. It is at the fringe of forests, in the vicinity of sanctuaries and protected areas that they become agricultural pests digging up seeds and young plants, raiding cereal and tuberous crops and fruit trees.

The main reason for their distribution in such varied habitats is attributed to their omnivory and digestibility. Hrdy (1977) states that they can live on anything including strychnine. Like all colobines, Hanuman langurs have special "saccualted" stomachs which are highly efficient in breaking down all kinds of plant materials viz. leaves, unripe fruits, seeds, even bark. Even cellulose is digested with the help of anaerobic bacteria present in the stomach. The langurs feed on about 240 natural and cultivated plant species. Religious people provision most of the groups with wheat/millet, *chapattis*, vegetables, fruits and nuts. Around Jodhpur provisioning accounts for one third of the total feeding time with considerable variation among groups (Winkler, 1988).

Colobines, the group to which langurs belong have an interesting feeding behaviour. Their feeding ecology has been interpreted variously over the past few decades. Based on their multichambered stomach colobines were initially categorized as folivores. With more and more studies they are at present considered as seed-eaters. They become leaf eaters when seeds are not available. Still their diet varies for the same species from location to location, from season to season and over the years. Hanuman langurs like all colobines strongly prefer seeds, young leaves, flowers and insects to mature leaves except in winter. During this season they have no alternative but to feed on mature leaves. This winter dependency on mature leaves is seen in all langurs of the Indian sub continent including capped langurs, Nilgiri langurs and purple faced langurs. They also consume huge quantities of insects during premonsoon and monsoon in Jodhpur, Kanha and Ramnagar (Nepal) when there is an outbreak of insect populations (Srivastava, 1991). In addition fruits are consumed based on the content of complex sugars.

The seasonal variation in food availability appears to influence reproduction. The langurs exhibit a lot of variation in reproduction. Unprovisioned population in Himalayas and reservoirs are seasonal breeders. On the other hand provisioned and/or crop raiding populations are known to breed more or less round the year (Srivastava & Danbar, 1966; Harley, 1985).

SOCIAL BEHAVIOUR

Hanuman langurs live in social groups. Basically troops are of two kinds-all male bachelor groups of around 60-70 animals and larger groups, sometimes as many as hundred monkeys of females and their offspring led by one or two dominant male/s. The females are closely knit and usually adhere to the same territory throughout life. In contrast, the males are rarely in the same group beyond two years. Each group occupies its own home range of about 0.5-1.5 km^2. Home ranges of all male bands can be as large as 20 km^2. The mean troop size in Jodhpur was 38.5 members and the number of all male bands was about 13 (Rajpurohit, 1987; Rajpurohit & Sommer, 1993).

The all male bands roam constantly, looking around to take over of a harem. The strongest of them challenges the alpha male of the harem. When the alpha male responds to the challenge, a fierce battle of strength ensues with females joining the alpha male to drive away the challenging male. If the challenging bachelor is weak, he would retreat. A stronger one will drive away the original troop leader and would take over the harem. He then drives away the other bachelor males. Next he systematically kills all the infants sired by his predecessor. The biological explanation for this infanticide is that the successful and dominant male is interested in propagating only his genes. This can happen only after females come to oestrous which is possible after they stop nursing. This is achieved by eliminating infants.

CROP RAIDING BY HANUMAN LANGURS

Langurs cause severe damage to crops in many parts of the country. In Sri Lanka they eat young leaves, shoots, buds of mangoes and other fruits. Some times they are killed as crop raiders (Muckenhirn, 1972). But in India people do not kill langurs because of the prevalent religious sentiments and the Wildlife Protection Act, 1972. It is regarded not only as sacred but divine by Hindus. It is a cultural practice throughout the country to feed them especially near temples. But this reverence does not stop the farmers from threatening them, throwing stones, whistle and shout at them, use firecrackers to scare them away from crop fields when the simians indulge in destroying crops. But langurs are very intelligent and well adapted to raid crops. When they attack a crop field, all the group members do not enter the field at the same time. Two to six members of the troop enter while the others remain at the edge of the field or on trees watching vigilantly. These "watchers" give alarm calls when people or dogs approach (Chhangani & Mohnot, 2004). Crop raiding by male-bands is more frequent compared to bisexual troops. This may be due to the fact that males are physically strong and better skilled in running, threatening and fighting. Females apart from being less strong are burdened with infant care. It has been observed from bisexual troops that females with infants are the last to enter the crop fields after all adults, sub-adults, juvenile males and single females have moved in. Male bands have bigger home ranges than bisexual troops and hence they invade more number of agricultural and horticultural fields. Langurs were more fearful of men than woman and children (Chhangani & Mohnot, 2004).

In India two exhaustive studies were carried out on the crop raiding behaviour of Hanuman langur populations inhabiting Kumbalgarh Wildlife Sanctuary (KWS) and Jodhpur City (Chhangani & Monhot, 2004)

(1) AT KUMBALGARH WILDLIFE SANCTUARY (KWS)

The Kumbalgarh Wildlife Sanctuary (KWS) lies between 25°40'N and 73°21' and 30ºE in West Aravalli hills of Rajasthan, India spreading across 585 sq.km. The altitude ranges from 274 to 1155 meters above sea level. A variety of agricultural and horticultural crops are grown around the area.

Population: A total of 16 groups of Hanuman langurs including 11 unimale, bisexual troops and 5 all male bands with a population of 540 animals were observed in and around KWS. The mean troop size was 41.7 and the mean band size 15.6. The male : female ratio was 1:46. The population was monitored from 1995 to 2002 for a long-term eco-behavioural study.

Feeding behaviour: Hanuman langurs were highly adaptable consuming a variety of foods that included natural, cultivated and artificial food. The activity period of langurs was 14 hours during summer and around 11 hours in winter. The time spent on feeding activity ranged from 29.5% to 39.3% depending on the ecological and topographical features of the microhabitat as well as day length. Langurs ate about 184 types of food items, which was inclusive of natural and cultivated plant parts as well as cooked food provided by people. Natural food included leaves, flowers, fruits, seeds, grains, vegetables, flowers and parts of garden bushes. The foods offered by people were raw, cooked, fried, and roasted food grains. This apart, langurs were seen eating sand, chewing bee wax and licking plant latex and rocks. Maximum natural foods (80.75%) were consumed during the month of June and minimum (59.2%) in the month of September. The consumption of cultivated plants revealed a reverse trend with least consumption of 19.25% in June and maximum intake of 40.8% during September. The overall annual consumption of natural food was 67.62% and that of cultivated food was 32.98%.

Economic loss: Extent of crop damage depended on the number of troop members and the crop protection strategies adapted by the farmers. All the farms and orchards falling within the home range of studied populations were raided and damage inflicted. Generally the extent of spoilage was more than consumption, particularly the playful activities of juveniles and infants resulted in much spoilage of ground produce as well as on tree tops. Damage was found upto 27% of total yield representing an annual loss of about US$ 900 from a total of 102 animals living in the periphery of crop fields and orchards. In addition the cost of protection ranged between US$ 150-200 per farm totalling US$ 1800-2400 for 12 farms. Apart from this direct loss, there was indirect loss due to consumption of flowers and fruits as well as damage to branches bearing flowers and fruits by juveniles and infants.

Crop protection strategies: The most commonly used protection strategy was guarding the crop during season which was adopted by 60% of the farmers. Twenty percent used "*gophan*", a device of throwing stones at invading langurs. About 15% of farmers used dogs to chase the invading monkeys. Many a times dogs killed juveniles and infants. The remaining 5% of farmers used lethal methods like shotgun, potash bomb and high voltage electric current.

(2) JODHPUR CITY

Population: The study was carried out on 35 bisexual and 15 all male bands from 1995-2000. The total population of bisexual troops fluctuated from 1400 to 1700 while that of all male bands increased from 112 to 240 over the six-year study period. The population density varied between 10 and 13 animals per km^2 on a plateau of about 150 sq. Km. The overall population showed a gradual increase from 1995 to 2000.

Crop raiding: Out of 52 groups only 20 groups (15 bisexual and 5 all male bands with a total of 1311 langurs) regularly damaged the cultivated fields in and around Jodhpur. At many places Hanuman langurs were reported to obtain as much as 90% food from cultivated plants (Yoshiba, 1968). At Mt. Abu they were dependent on crops for much of their dietary requirements (Hrdy, 1977). Damage also occurred by way of breaking of thin branches bearing flowers and unripe fruits.

Crop damage was dependent on the success rate of langurs in invading fields. Generally, the farms located adjacent to the home range of langur groups and farms without crop protection strategies were at highest risk and suffered more losses. In Himachal Pradesh damage by Hanuman langur varied from district to district and from habitat to habitat depending on the distance of farmland from the fringe of the nearest forest where langur troops lived (Sahoo, 2004; personal communication).

Table 1. Cultivated plant parts consumed by langur troop in and around Jodhpur and seasonal intensity (1995-2000)

Local name	Botanical name	Summer	Monsoon	Winter
Crop				
Ganwar	*Cyamopsis tetragonaloba*		**** YL, ML, SE	
Genhu	*Triticum eastivum*			**** YL, ML, FL, SE
Chana	*Cicer arietinum*			* YL, ML, FL, SE
Moong	*Phaselus radiatus*		*** YL, ML, FL, UF, SE	
Rizka	*Medicago sativa*		** YL, ML	** YL, ML
Moth	*Vigna aconitifolia*		*** YL, ML, FL, UF, RF, SE	
Sarson	*Brassica campestries*			** YL, ML, FL
Jawar	*Sorghum bicolor*		** YL, ML, FL, SE	
Til	*Brassica juncea*		* YL, FL	
Vegetables				
1. Gajar	*Daucas carota*			** YL,ML,FL,RF
2. Dhania	*Coriandrum sativum*		* YL, ML	** ML, FL
3. Chandalia	*Amaranthus hybridus*		* YL, ML	
4. Kakri	*Cucumis sativys*	* YL, FL	* YL, ML	

Local name	Botanical name	Summer	Monsoon	Winter
5. Muli	*Raphanus sativus*	* YL, ML	* FL, YL, ML	* YL, ML
6. Tamatar	*Lycopersicon lycopersicum*		* YL, UF	** RF
7. Band gobi	*Brassica oleracea* L. var. *capitatal*		* YL	*** YL, ML
8. Bhindi	*Abelmoschus esculentus*	* FL, YL	* UF	
9. Baingan	*Solanum melongena*		* UF, FL	** UF, RF
10. Kachar	*Cucumis melo* var. *culta*		* FL, UF	
11. Palak	*Rumex vesicarium*		* YL, ML	* YL, ML
12. Pudina	*Mentha spicata*		* YL, ML	
13. Ful gobi	*Brassica oleracea* L. var. *botrytis*		* YL	*** FB, FL
13. Mirchi	*Capasicum annuum*		* YL, FL, DF	
14. Methi	*Trigonella foenum-graecum*		* YL, ML	
15. Tori	*Luffa cylindrica*		* FL, YL	
16. Karala	*Momordica charantia*		* FL, YL	
Cultivated flowers &fruits				
1. Amrood	*Picidium guajava*			*** UF, RF
2. Anar	*Punica GRANATUM*		* FL	* RF
3. Nimbu	*Citrus medica*	* FL, YL	* FL	
4. Papita	*Carica papaya*		* FL, FR	
5. Hajara	*Tagetes erecta*			** FB, FL

Local name	Botanical name	Summer	Monsoon	Winter
6. Hajara	*Tagetes putula*			** FB, FL
7. Tulsi	*Ocimum canum*	* YL	* YL	* YL

YL = Young leaves; ML = Mature leaves; FB = Flower buds; FL = Mature flower; UF = Unripe fruits; RF = Ripe fruits; SE = Seeds
Damage intensity* = low; ** = medium; *** = high; **** = very high

Amongst food crops ganwar (*Cyamopsis tetragonaboba*) was damaged heavily during monsoon and wheat (*Triticum aestivum*) during winter. Considerable damage was seen in *moong* (*Pheseolus radiatus*) and moth (*Vigna aconitifolia*) during monsoon while medium level damage was seen in Rizka (*Medicago sativa*) and jowar (*Sorghum bicolor*) during the same season. During winter Rizka (*Medicago sativa)* and sarson (*Brassica campestris*) were damaged to some extent (Table 1).

Cabbage (*Brassica oleracea* L. var. *capitatal*) and cauliflower (*Brassica oleracea* L. var. *botrytis*) were damaged significantly during winter. Radish (*Daucas carota*), coriander (*Coriandrum sativam*), tomato (*Lycoperisicon lycopersicum*), eggplant (*Solanum melongena)* were damaged to some extent during the same season. Damage of fruits was seen again only during winter. High degrees of damage were seen in pomegranate (*Picidium guajava*) and *hajara* (*Tagetes erecta*) (Table 1).

Economic loss: Damage was maximum at 32% and minimum at 4%. The calculated crop damage from bisexual troops and all male bands came to about US$ 1000 per anum from 1311 animals living in the periphery of crop fields and orchards. The cost of crop protection amounted to US$ 1200 per year. Indirect losses by way of damage to flowering and fruiting trees was not included in this calculation.

Crop protection strategies: The different methods used and their efficacy in determining the monkey depredation was same as that in areas around Kumbhalgarh Wildlife Sanctuary namely guarding, throwing stones, using dogs and lethal methods employing gun, firecrackers and high voltage electric current. Fencing using thorny twigs and branches as well as naturally growing *Euphorbia caducifolia* were also practiced.

MANAGEMENT OF CROP DEPREDATION – PRESENT AND FUTURE

The problem can be discussed in three phases namely extent of damage, reason for crop raiding and management strategies.

(a) Extent of damage caused by non-human primates: Several species of non-human primates raid crops. But in general terrestrial species and non folivores are more involved in crop raids than arboreal and folivores. All crop-raiding genera have some common traits, which greatly contribute to their success in depredating cultivated crops. These are: their complex social organizations, omnivory and retention of arboreal habits in spite of being primarily territorial. The most common agricultural pests belong to three genera namely *Macaca, Papio* and *Cercopithecus*. In much of Africa and Arabia baboon (*Papio* sp.) is the most pestiferous while vervets are pests in sub Saharan areas. Apart from crop raiding they also intrude into tourist resorts and houses to steal food. Macaques decrease agricultural yields considerably in India, Indonesia and Japan. In Tanzania primates specially olive baboon (*Papio anubis*), yellow baboon (*Papio cynocephalus*), vervet monkey (*Ceropithecus aethiops*), black and white colobus (*Colobus guerza),* patas monkey (*Ceropithecus patas*) and chimpanzee (*Pan troglodytes*) caused 52% damage (Zubiri & Switzer, 2001). In Uganda crop losses due to baboons averaged 25% for cassava, and 19% for maize. Sweet potato was also damaged to a great extent (Hill, 2000). The other two important conclusions of the Ugandan study were: (1) Proximity of the farm to the forest edge and presence or absence of neighbouring farms influences crop-raiding behaviour of baboons. (2) The indirect losses due to crop raiding area increased demand for labour to guard the crops and sometimes costs of replanting when stands are badly damaged also added to farmers woes. In Indonesia the Button macaque, *Macaca brunnascens* heavily predated on sweet potatoes and corn (Kilner, 2000). Later Priston (2004), working on the same species in Sumatra reported greater damage in farms closer to the forest edge and within the farms damage was higher at the margins. Most damage occurred to sweet potato, banana and corn. In Japan there was no crop-raiding by Japanese macaques until 1950s. Gradually the macaques descended from the mountains and started expanding along the river system with levels of damage increasing as years passed (Bergman, 1996). In India systematic studies on crop raiding is limited to Hanuman langur, *Semnopithecus entellus* (Chhangani & Mohnot, 2004). The authors extrapolated the losses including the amount spent on crop protection measures to an annual loss of Rs. 5,00,000/- per annum in and around Jodhpur, Rajasthan.

(b) Reasons for crop raiding: Several reasons contribute to increased crop raiding by non-human primates (Zubiri & Switzer, 2001). These include changes in agricultural methods and techniques, increase in wildlife populations, increase in human population, increase in agriculture and encroachment of forest land, increased fragmentation of wildlife habitat leading to more and more wildlife-people confrontations, urbanization and improved education. That apart primates are successful crop raiders which are attributed to their intelligence, adaptability, complex social organization, aggressive behaviour and protection by wildlife laws.

(c) Crop protection strategies: Farmers throughout the world adopt various strategies to protect their crops from raiding primates. Most of them are traditional approaches as there are almost no or very few studies on crop raiding by mammals. The methods employed by a farmer are greatly limited by the resources available to him. Farmers in developing countries have access to more sophisticated and costly equipment to repel/frequent crop depredation by mammalian herbivores. In contrast farmers in developing countries have smaller incomes and financial constraints. They have evolved over the period a range of simple, manpower based techniques to prevent crop raiding. The commonly used methods are guarding using dogs to frighten the riders, using scarecrows, plastic flags, fire works, burning materials which emit unpleasant smells, fencing using local materials, electric fencing, hunting, trapping, shooting to scare and chasing by running at them, yelling, banging drums, throwing stones etc. There is no quantitative data on the efficacy of these traditional methods.

Recent research: Compared to studies on and suggestions for the problem of crop raiding by elephants, little research has gone to mitigate crop depredation by primates. Only few noteworthy attempts are available. Taste aversion learning behaviour has been tried to prevent crop raiding by monkeys in Japan and Kenya (Zubri & Switzer, 2001). The technique is based on the phenomenon that animals become aversive to the food which has been consumed with a non-lethal aversive tasting/smelling agent. This is achieved by baiting the target food mixed with an emetic like Lithium chloride. The results are not conclusive. In Japan experiments have been carried out with electric shock treatment to prevent crop raiding by macaques. But in view of high expenditure and cruelty to animal involved, this is not a practical solution.

CONCLUSIONS

Techniques such as aversion learning and electric shocks are expensive, technologically sophisticated and as such their applicability and affordability is limited to developed countries. Fact is distribution of crop raiding primates and their habitats is mostly in developing and poor countries of Asia and Africa where farming is marginal and the farmer is a subsistence farmer. He has neither the time nor money to use expensive and sophisticated methods to keep away non-human primates from eating his crops. Evaluation and improvement of traditional methods taking into consideration the habitat of crop raiders, their behaviour, land use pattern of the given area and crops cultivated seems to be a viable proposition to manage monkey menace to crops. What is needed is an integrated crop protection strategy, which in turn is possible by combined efforts of farmers, primatologists, conservationists, personnel of agriculture and forest departments. This approach needs urgent attention as the attitude of affected farmers towards conservation has changed over the past several years. The demand for guns to shoot at crop raiding monkeys has increased manifold in the affected areas (Chhangani & Mohnot, 2004). This is one of the serious man-animal conflicts that needs immediate action from the concerned.

REFERENCES

BALASUBRAMANIAN, M., N. BHASKARAN, S. SWAMINATHAN & A.A. DESAI. 1993. Crop raiding by Elephants in the Nilgiri Biosphere Reserve, India. In: A WEEK WITH ELEPHANTS (Eds.): Daniel, J.C. & Hemanth Datye. Bombay Natural History Society & Oxford University Press, Bombay, India. pp. 350-367.

BERGMAN, C. 1996. Lovable pests. *International wildlife* (March-April) (http:// www.nwf.org/internationalwildlife/ snowmonk html)

BOHRA, H.C., S.P.GOYAL, P.K. GHOSH, & I. PRAKASH. 1992. Studies on ethology and eco-physiology of the antelopes of the Indian desert. *Annuals of Arid Zone*. 31(2): 83-96.

BOULTON, A.M., J.A. HORROCKS, & J. BAULU, 1996. The Barbados vervet monkey (*Cercopithecus acthiops sabacus*): Changes in population size and crop damage 1980-1994. *Int. J. Primatol*. 17: 831-844.

CHAUHAN, N.P.S. & V.B. SAWARKAR. 1989. Problems of over abundant populations of Nilgai and Black Buck in Haryana and Madhya Pradesh. *Indian Forester*. 115(7): 488-493.

CHHANGANI, A.K. 1994. *Crop raiding behaviour and status survey of Nilgai (Boselaphus tragocamelus) around Jodhpur*, unpublished M.Sc. Thesis, Jai Narain Vyas University, Jodhpur, India.

CHHANGANI, A.K. & S.M. MOHNOT 2004. Crop raid by Hanuman langur, *Semnopithecus entellus* in and around Aravalli hills (India) and its management. *Primate Report* 69: 1-18.

ELSE, J.G. 1991. Nonhuman Primates as Pests. In: PRIMATE RESPONSES TO ENVIRONMENTAL CHANGE. (Ed.) Box, H.O. Chapman & Hall, London, UK. pp. 115-165

FORTHMAN-QUICK, D.L. 1986. Activity Budgets and the Consumption of human foods in two troops of Baboons (*Papio anubis*) at Gilgil, Kenya. In: PRIMATE ECOLOGY AND CONSERVATION. (Eds.) Else, J.G., and P.C. Lee. Cambridge University Press, New York. pp. 221-228

GOYAL, S.K. 1999. Study on population Dynamics and Range Use in Nilgai – Bluebull (*Boselaphus tragocamelus* Pallas) in Western Rajasthan. Ph.D. thesis, Jai Narain Vyas University, Jodhpur, India.

HARLEY, D. 1985. Birth spacing in langur monkeys (*Presbytis entellus*). *Int. J. Primatol.* 6: 227-242.

HILL, C.M. 1997 Crop raiding by wild animals: The farmer's perspective in an agricultural community in Western Uganda. *Int. J. Pest Manag.* 43: 77-84.

IBID. 2000. Conflict of interest between people and baboons: crop raiding in Uganda. *International J. Primatol.* 21: 299-315

HRDY, S.B. 1977. THE LANGURS OF ABU: FEMALE AND MALE STRATEGIES OF REPRODUCTION. Harvard University Press, Cambridge, Massachusetts, USA.

IMAM, E & IQBAL MALIK 2005. Rhesus monkey (*Macaca mulatta*) problems in India and their possible management. In: VERTEBRATE PESTS IN AGRICULTURE-THE INDIAN SCENARIO (Ed.) Shakunthala Sridhara. Scientific Publishers, Jodhpur, India. (in press).

JAY, P.1963. *The social behaviour of the langur monkey*. Ph.D. thesis. University of Chicago, Chicago, USA.

KILNER,T. 2000. *The behaviour and ecology of the Buton macaque Macaca brunnescens.* M.Sc. dissertation, Manchester Metropolitan University, Manchester, UK.

McCAM, C. 1928. Notes on the common Indian langur (*Semnopithecus entellus*). *J. Bombay nat. Hist. Soc.* 33: 192-194.

McCAM, C. 1933. Observations on some of the Indian langurs. *J. Bombay nat. Hist. Soc.* 36: 616-628.

MUCKENHIRN, N.A. 1972. Leaf-eaters and their predators in Ceylon: Ecological Roles of Gray Langurs, *Presbytis entellus* and Leopards. Ph.D. Thesis, University of Maryland, Bethesda, USA.

NEWTON, P.N.1988. The variable social organization of Hanuman langurs (*Presbytis entellus*): infanticide and monopolization of females. *Int. J. Primatol.* 9: 59-77.

PIRTA, R.S., M. GADGIL & A.V. KHARSHIKAR. 1977. Management of the rhesus monkey, *Macaca mulatta* and Hanuman langur *Presbytis entellus* in Himachal Pradesh. India. *Biol. Conserv.* 70: 97-106.

PRATER, S.H. 1980. THE BOOK OF INDIAN ANIMALS. Bombay Natural History, Society. Bombay, India.

PRISTON, N. 2004. An assessment of crop raiding by *Macaca oehreata* brunnesuns in South east sulawesi: Farmer's perceptions, reality, and the impact of primate behaviour and conservation. Ph.D. Thesis. University of Cambridge, UK.

RAJPUROHIT, L.S. 1987. *Male Social Organization in Hanuman langur* (*Presbytis entellus*). Ph.D. thesis, University of Jodhpur, India.

RAJPUROHIT, L.S. & S.M. MOHNOT. 1997. Depredation in crops by Nilgai (*Boselaphus tragocamelus*) around Jodhpur. *Him. J. Env. Zool.* 11: 129-132.

RAJPUROHIT, L.S. & V. SOMMER. 1993. Juvenile male emigration from one-male natal troops in Hanuman langurs. In: JUVENILE PRIMATES: LIFE HISTORY, DEVELOPMENT AND BEHAVIOUR (Eds.). Michael E. Pereira & Lynn A. Fairbanks. Oxford University Press, Oxford, UK. pp. 86-103.

ROONWAL, M.L. & S.M. MOHNOT. 1977. PRIMATES OF SOUTH ASIA: ECOLOGY, SOCIOBIOLOGY AND BEHAVIOUR. xvii + 421 pp. Harvard University Press, Massachusetts, Cambridge, USA.

SCHULTZ, B.O. 1986. The management of crop damage by wild animals. *Indian Forester* 112(10): 891-899.

SOMMER, V. & L.S. RAJPUROHIT 1989. Male reproductive success in harem troops of Hanuman langurs (*Presbytis entellus*). *Int. J. Primatol.* 10: 293-317.

SOUTHWICK, C.H. & M.F. SIDDIQI. 2001. Status, conservation and management of primates in India. In: NON HUMAN PRIMATES OF INDIA (Ed.) A.K. Gupta. Wildlife Institute of India, Dehradun, India. pp. 81-91.

SRIVATSAVA, A. 1991. Insectivory and its significance to langur diets. *Primates* 32: 237-241.

SRIVASTAVA, A. & R. DANBAR. 1966. The mating systems of Hanuman langurs: a problem in optimal foraging. *Behav. Ecol. Sociobiol.* 39: 219-226.

SUGIYAMA, Y. 1964. Group composition, population density and some sociological observation of Hanuman langurs (*Presbytis entellus*). *Primates* 5: 7-37.

SUKUMAR, R. 1990. Ecology of the Asian Elephant in Southern India: II Feeding habits and crop raiding patterns. *J. Trophical Ecol.* 6: 35-53.

WINKLER, P. 1988. Feeding behaviour of a food–enhanced troop of Hanuman langur (*Presbytis entellus)* in Jodhpur, India. In: THE ECOLOGY AND BEHAVIOUR OF FOOD-ENHANCED PRIMATE GROUPS (Eds.) J.E. Fa & Southwick, C.H. Alan R. Liss, New York, USA. pp.3-24.

YOSHIBA, K. 1968. Local and inter-troop variability in ecology and social behaviour of common Indian langurs. In: PRIMATES (Ed.) Phyllis Jay. Holt, Rinehart & Winston, New York. pp. 217-242

ZUBIRI, C.S & D. SWITZER. 2001. Crop raiding primates: Searching for alternative humane ways to resolve conflict with farmers in Africa. People and wildlife Initiative. Wildlife conservation unit, Oxford University, Oxford, Great Britain.

❑❑❑

19

VERTEBRATE PEST MANAGEMENT IN STORAGE

Shakunthala Sridhara, V. Shivayya and T. Raveendra Babu

AINP on Rodent Control, University of Agricultural Sciences, G.K.V.K., Bangalore – 560 065

ABSTRACT

The major rodent pests of storage in India are *Rattus rattus, Mus musculus* and *Bandicota bengalensis* and to a lesser extent *Bandicota indica. Rattus norvegicus* is reported from major harbour cities while *Tatera indica* seems to infest and damage stored food grains in North-West India. Available information puts rodent density in stored conditions at 0.77 rats/sq. feet of ground area. The estimated loss varies from 3.5 to 25%. Construction of rodent proof godowns, fortifying existing ones against fresh infestations after eliminating residential rodent population and their maintenance is the well established approach to prevent rodent damage in storage. The godowns are fumigated before storage to eliminate all pests including rodents. Later infestation if any is managed by rodenticide treatment. Zinc phosphide, bromadiolone, coumatetralyl, tracking powder and liquid anticoagulants are the rodenticides recommended for godowns while only anticoagulant baiting is suggested for household situations to avoid accidental poisoning. The major thrust in villages should be use of rodent proof storage bins and rodent control on a community level. It is not practical to include public hygiene in villages due to lack of sufficient infrastructure and civic facilities. However, individual house owners can be motivated to keep inside and outside of houses clean and deny access to rodents by taking up structural repairs and maintenance. Bird damage to godowns is almost nil but for the possibility of disease transmission through droppings while feeding on spilled grains inside and outside the godowns. Bird proofing storage structures will take care of this problem.

INTRODUCTION

Agriculture produce is vulnerable to damage by pests and diseases at two stages -in the fields during growth and during post harvest storage. Food protection is a three pronged strategy with protection of field crops during their growth and maturation forming the first step. Secondly prevention of qualitative damage due to moisture, insects and pathogens during post harvest storage and thirdly, protecting food stored from quantitative damage in homes and godowns from insects and rodents. Although rodent damage to crops in the field is very severe and disturbingly visual, such damage occurs for a very short duration of one or two weeks at the vulnerable stages of sowing/sprouting, and grain maturation just before harvest. Comparatively losses at home and godown storage are less spectacular but the prolonged periods over which rodents and other vertebrates consume stored food grains results in severe losses if computed at national level. In India greatest loss of food grains for an individual farmer occurs during the 7-9 month home storage because more than 70 per cent of produce is retained at home for day to day use of the farmer and also for trade.

LOSSES CAUSED

(a) Food losses: Rodents cause losses to stored food grains in three ways-actual consumption of food, destruction, damage of storage structures and facilities, and contamination of stored food through their hair, urine and excreta. In spite of considerable losses in godowns, information on quantity of food lost is poorly documented not just in India but throughout the world.

To get a realistic estimate of vertebrate damage in storage and to plan the methods to mitigate the problem, facts about the biology, ecology, breeding, feeding, behaviour, population dynamics and movement of storage pests are necessary. Unfortunately the Indian scenario in this direction is poor. The earlier studies on rodents solely focussed on the epidemiology of plague. From the past thirty years the emphasis has shifted to manage rodent pests of agriculture in fields. With the exception of detailed studies on godown rodent populations in Calcutta (Spillet, 1966; Frantz, 1973), simulated damage in laboratory (Sridhara & Krishnamoorthy, 1979) and the exceptionally good work done by Indian Grain Storage Institute to design rodent proof storage bins and educating farmers through Save Grain Campaign of Union ministry of Food, serious work on losses in storage, management of vertebrate pests and prevention of losses in storage facilities is a neglected area of research in India. There does not seem to be any significant studies on the species infesting, their behavioural ecology and life history to design location specific control strategies. Such studies are in great need because India with her diverse ecology and crops supports three to four each of rodent and bird species as pests of stored food. In such a scenario there definitely would be problems of food losses which need regional and localized strategy to reduce the impact of pests.

Economic loss caused would be in direct proportion with levels of rodent population which in turn is dependent on food and shelter available, presence and absence of mortality factors such as diseases, predators, competition for food etc. However population estimates in storage facilities is wrought with inaccuracies. Many a times populations measured at their peak during optimal environmental conditions are taken into account. But such estimates made under favourable conditions do not reflect actual densities. Further, assessments are made in a small, localized area and then extrapolated to arrive at national figures. The National Rodent Committee rejected the available figures of rodent population of the country after assessing the various estimates. It considered most of them as extrapolated figures and guesswork without substantiation from data collected under varying environmental conditions (Pingale, 1975). Jackson (as quoted by Pingale, 1975) opined that most of the figures on damage and density of rodents is not based on scientific work but politically influenced. Losses are escalated to justify implementation of a programme. Realistic estimates involve taking into consideration many factors which is quite difficult. Although various estimates of rodent population are given for India ranging from 800,000,000 to 4800,000,000 none are based on long term observations nor the data is collected from various locations. The estimates were based on such unscientific bases as a single observation on a 45 acre plot, assumption of two to three rats per acre or 300-700 rats per village depending on the size of the village. The other projected density was 0.98 to 1.29 times the human population in villages around Hapur, the ratio was 0.34 in good housing conditions of city and 0.66 in congested areas. Rodent density was calculated to be 7.6 to 10.7 in warehouses of 100-300 tonnes capacity (Pingale, 1975). In a landmark study on rodent infestation and damage to food grains, Spillet (1966) trapped an unbelievable number of 1371 male and 809 female *Bandicota bengalensis* over a 11-month period from a single godown in Kolkata. He also caught 29 Norway rats, a single male *Rattus rattus* and a single male *Bandicota indica* and six *Mus musculus*. The mean population was 188 rats. With each rat consuming 11g rice per day, total consumption per godown came to 2068g/day. Assuming equal amount being lost to contamination and wastage, the total loss of food grains in a typical Kolkata godown was calculated to be approximately 1400 kg/year. So far the only authentic density estimation is 0.77 rat/sq. meter of floor space in the godowns of Calcutta (Spillet, 1966).

Table 1. Rodent damage in godowns of South Asia (Hopf *et al.*, 1976)

Country	Commodities stored	Type of storage	% loss
India	Wheat	Village stores	5
	Rice, Wheat, Millet, Sorghum	Villages	3.5 to 5
	Rice, Wheat, Pulses, Sorghum	Mud & Bamboo bins, Bags	2 to 5
	Rice	Improved traditional storage	3 to 5.5
	Wheat	On farm storage	8 to 25
	Millet	Farm storage	5
	Sorghum	Farm storage	5
Nepal	Maize	Sacks	....
Pakistan	Wheat	On farm storage	5.70
	Sorghum		5
Bangladesh	Rice	Bamboo bins	7

The estimated losses in godowns available for South Asian countries are depicted in Table 1. The damage is 3.5 to 25% for India, 3-5% in Nepal, 2-10% in Pakistan and 7% in Bangladesh. The food grain stored was largely wheat, rice, millet, sorghum and maize mostly in traditional mud and bamboo bins, jute bags and sacks and sometimes improved traditional storage.

According to Pingale (1975), the daily food consumption of the black rat, *R. rattus* in India is reported to be 20 per cent of body weight during growth period which decreases to 10 per cent at maturity. In case of lesser bandicoot rat, *B. bengalensis*, the rate is five per cent of body weight. Considering the various age groups of different species in a godown population of rodents, a realistic calculation puts the consumption rates at 10g/rat/day. Based on this requirement and assuming that this entire food requirement is met from grains stored in villages, the total loss is calculated to be 2-4% of the quantity stored, the average being 1.69% for godowns in a town (Krishnamurthy *et al.*, 1967). It is far less in retail food grain stores, a mere 0.23% (Pingale, 1975).

In another projection daily consumptions were extrapolated to annual losses per rat for the Norway rat, *R. norvegicus, R. rattus, B. bengalensis* and the house mouse, *M. musculus* (Table 2). Accordingly

R. norvegicus consumes 15-25g/day and 6-9 kg/year, *R. rattus* eats 8-12g/day and 3-4kg/year, the intake by lesser bandicoot is 15-25g daily and 6-9 kg per annum in contrast to daily intake of 2-3g/day and 0.7 to 1 kg/year by the house mouse.

Table 2. Daily and yearly consumption of food grains by godown rodents (Hopf *et al.*, 1976)

Rodent species	Amount cons./day (g)	Amount cons./annum (kg)
R. norvegicus	15-25	6-9
R. rattus	8-12	3-4
B. bengalensis	15-25	6-9
M. musculus	2-3	0.7-1

In neighbouring Pakistan 29% of the godowns were infested by one or more rodent species namely *R. rattus, M. musculus* and 78% were subject to damage by several species of birds. The overall losses in wheat stored in godowns were calculated to be 0.2 to 0.5% of the total grain stored. This translated into annual losses of 1.25 million tonnes. In addition farms and villages also store individually harvested wheat where a further 1 to 2 per cent grain was calculated to be lost to vertebrate pests (Brooks & Lovoie, 1990). The loss in Bangladesh was 47 kg per family per rat, in this case the species was the lesser bandicoot rat, *Bandicota bengalensis* (Brooks & Lavoie, 1990). Most losses were attributed to poor sanitation and structural quality and maintenance. Spilled grain in and at the entrance of godowns attracted rodents and birds with tall weeds and grasses offering shelter to rodents living outside. Open and defective doors, windows and ventilators provided easy entry (Ahmad *et al.*, 1994).

(b) Disease transmission: The rodent pests of godowns live in close proximity of man in many tropical and sub tropical countries. Being reservoirs and vectors of several dreadful diseases they are a high health risk to human beings apart from causing huge losses of food grain. Of the forty and odd diseases transmitted/spread the most virulent are: plague, murine typhus, leptospirosis, several types of haemorrhagic fevers, Hantavirus and salmonella. Roof rats, Norway rats and bandicoots spread plague and murine typhus through fleas. The urine of field rodents and Norway rats contain the bacteria, leptospira which causes leptospirosis, while salmonella organisms are present in the droppings of all species of rodents.

By preventing access to rodents, reducing the carrying capacity of godown environment, reducing their numbers significantly and if possible eliminating them completely from godowns and storage areas not only huge food losses are prevented but a number of fatal diseases too.

INDICES OF INFESTATION

Since rodents inhabit the interiors of godowns and storage areas, there is always indication of their presence while birds being visitors and diurnal, sightings are adequate to identify their species and approximate numbers. Although nocturnal, rodents move a lot from nest to the food source several times in the night and leave tell tale traces of their movement and activities. These signs in addition to indicating presence of their occurrence also reveal the degree of infestation. In a silent night they make enough noise by way of squeaks, growling, rustlings and fightings for the inmates of a house or watchman of a godown to be aware of their nocturnal activities. The main signs of rodent infestation are droppings, runways, imprints, burrows, gnawing signs, nests, damage, sounds and sightings.

Faecal droppings are found all along the runways, along the walls, near feeding and around their shelter. The size of the pellets indicates different age groups of a single species or different species altogether. Pellets of rat are larger, around 10-15 mm in length compared to 3-6mm long mice pellets. Not only species, even relative abundance can be gauged by collecting fresh droppings for two/three days after cleaning the previous droppings much the same way as live burrows in the field are considered as an index of population level. These droppings carry germs and bacteria and contaminate food grains along with urine and body hair of rodents.

Runways are the regular paths taken by rodents from their nests to food source. The regular movement of rats and mice leaves a greasy mark on frequently travelled path, usually on beams, pipes, loft and around the holes they make in the walls, doors, windows and on the floor.

Imprints of tail, four-toed forelimbs and five-toed hind limbs are easily visible in dust or mud.

Burrows are excavated by terrestrial and fossorial species, such as bandicoots. Norway rat and house mouse usually live outside the storage structures. Soil mounds at the burrow entrance, burrow openings and runways connecting burrows are pointers to burrows. Species identification is possible by the presence or absence of soil mounds, diameter and depth of the burrows.

Gnawing: To keep their constantly growing incisors in check all species of rodents gnaw rough surfaces. Holes to gain entry are also made by gnawing through doors and walls. At the point of gnawing tooth marks and fine particles of gnawed material are visible.

Nests of indoor species such as house rat and domestic mouse are made in the attics, loft, between walls and under the floor also by mice. Outdoor burrows have nesting chambers. The oval or round shaped nests are made of soft grass, leaves, stems, shreds of paper, cloth and jute bags.

Sightings: The nocturnal habit of rodents makes them visible only during nights. If the infestation is very high, they can be seen during day time also. Trapping confirms their presence, identification and density.

Sounds: The screechy vocalization of rodents during intra and interspecific aggressive encounters, their audible scrambling and gnawing and the ultrasonic sounds emitted while caring for the young in the silence of night unmistakably denote rodent infestations.

MAJOR VERTEBRATE PESTS

Rodent pests: Although there is some regional variation in the species inhabiting godowns and storage structures at sub species level, and sometimes at species level too in the different agro-climatic zones of India, basically three species cause damage. The two ubiquitous commensal rodents are the common house rat, *Rattus rattus* and the house mouse, *Mus musculus* throughout the country. *Bandicota bengalensis* occurs in exclusion of almost all other species in the godowns of Kolkata and to some extent in Mumbai warehouses. The Norway rat, *Rattus norvegicus*, primarily a temperate zone pest is seen in the godowns of port cities of Kolkata and Mumbai, probably coming through shipments. To a small extent, the larger bandicoot, *Bandicota indica* depends on godowns for its food requirements although it always lives outside the structures, in the open.

MORPHOLOGY AND LIFE HISTORY OF MAJOR RODENT PESTS

(a) *Mus musculus* the house mouse

Morphology: The body is slender and small weighing only 15-20g. The head and body measure 75-100 m, tail is 65-90 mm. The body is dark brownish-grey on the dorsal side and light grey to creamy white on the ventral side. The tail is dark with sparse covering of hair.

Distribution and Habitat: House mouse is universal in distribution and as such is found throughout India infesting houses, storage facilities, factories and other human premises in both urban and rural habitats. *M. musculus* are also found in rice, sugarcane and other grain fields, and garbage dumps.

Natural History: House mouse is terrestrial or arboreal and nocturnal. Inside the houses and storage facilities they usually nest in attics and roofs. In fields they dig shallow burrows and use runways. The home range is very small, around 1-3m^2. Daily food consumption is 2-3g. Apart from eating, they waste a lot as the grain is discarded after eating the kernel. They breed throughout the year. Gestation lasts 19-20 days and litter size averages 5-8 young which are weaned at 21 days. Sexual maturity is reached in about 42 days.

(b) *Rattus rattus*, the house rat

Morphology: The body is slender with a pointed snout and weighs 120-260g. Head and body measure 140-220mm. Tail is characteristically longer than body and measures 160-250mm. It is colourless. Body is brownish-grey dorsally; the colour of ventral side varies from light grey to creamy white to lemon yellow.

Distribution and Habitat: Roof rat occurs universally. The habitat is highly varied, ranging from buildings to houses, flats, shops, large food stores, barns, markets, restaurants, hotels, godowns, warehouses and all such places where food grains are stored and eatables are prepared and served. Amongst agricultural fields they occur in sugarcane, orchards, coconut and cocoa plantations. Animal facilities infested by house rat include poultry, diary and swine.

Natural history: Roof rats are arboreal, colonial and nocturnal always preferring attics, lofts and beams. Sometimes they also nest outdoors, on tree tops and tall shrubs. They eat 8-12g of cereals a day. Their social life is well organised with a single, dominant male, usually two or more female subordinates to him which in turn dominate all other members of the colony. The species is territorial and defends its nest against strangers. Breeding occurs throughout the year. Gestation ranges 20-22 days. Average litter size is 6.2 young which are weaned at 21 days. Sexual maturity is reached in about 20 days.

(c) *Rattus norvegicus*, the Norway rat

Morphology: *R. norvegicus* is a heavy rat weighing 250-600g and measures 180-260mm. Tail extends upto 150-210mm. It is bare, lighter dorsally. The ears are set closer to the body. Dorsally brownish-grey and venter is dark light to dark grey.

Distribution and Habitat: A predominant commensal throughout temperate zones, the Norway rat in India is limited to godowns of port cities of Kolkata and Mumbai, that too in small numbers. They occupy the space between walls and floor, beneath garbage and rubbish.

(d) *Bandicota bengalensis*, the lesser bandicoot rat

Morphology: Lesser bandicoots are Asian equivalents of Norway rats, being both agricultural pests and commensal. Body weight is 250-600g with body and head measuring 160-270mm and tail is 130-220mm. The dark coloured body is covered with coarse hair. Long, prominent guard hairs are present dorsally. Brownish grey to almost blackish grey on the dorsal side and ventrum is grey.

Distribution: Being primarily field rodent and pest of cultivated crops, *B. bengalensis* occurs throughout India including the arid Rajasthan. For the past 50 years the species has evolved into major urban commensal particularly in Kolkata, Mumbai, and other major cities. They infest almost all irrigated and marshy crops. Being burrowers par excellence, they dig out mounds of earth while digging elaborate underground tunnels which are interconnected. Few of them expand into nests, broods and storage chambers. They can swim efficiently but are poor climbers.

Natural History: Lesser bandicoot rats breed throughout the year as in the godowns of Kolkata or they are seasonal breeders in agricultural fields. Gestation lasts 21-23 days. Average litter size is 7-8 young which are weaned at around 28 days after birth. The young become sexually mature in around 75 days after birth. Daily consumptions average 25g/rat but they are notorious hoarders, the catch weighing several kilograms. Range is smaller in urban habitat but in crop fields home ranges are extensive. They live both as solitary and colonial and are extremely aggressive.

(e) *Bandicota indica*, the larger bandicoot rat

Morphology: *B. indica* is a very large rat. Body size ranges 200mm-366mm and weighs 500-2000g or more. Tail is shorter than head and body and is covered with rough fur which is quite long dorsally. Dark or blackish brown on the dorsal side and ventral side is grey, drabber dark.

Distribution and Habitat: It is distributed from the south of Rajasthan to the tip of South India and towards the borders of North-east India. They live close to human habitat but never inside the house or godowns. They prefer to live in proximity with human dwellings such as compounds, gardens, stables, poultry and houses.

Natural history: The species is nocturnal and fossorial. It digs burrows with 1-3 openings and multiple tunnels which may or may not be connected. Breeding is both seasonal and throughout the year. Not much is known about its breeding and biology.

(f) *Tatera indica*, the Indian gerbil

Morphology: The body weight ranges between 100-150g. It is covered with soft, silky fur, which is brownish dorsally and the belly is white. Tail terminates in a tuft of hair. Hind feet are longer than forefeet. Eyes are large and ears round. Feet are white in colour.

Distribution and habitat: The species is ubiquitous inhabiting open plains, sandy soils of desert and at the edges of cultivated fields. Burrows are simple with multiple openings.

Life History: Gerbils are nocturnal and fossorial. They breed throughout the year or seasonally where climate and vegetation are distinctly variable. Although they are basically pests of agriculture, they are attracted to the grain bags stacked on plinths in outdoor conditions.

BIRD PESTS

Bird losses to stored grain are almost nil. Few granivorous are attracted to storage godowns and warehouses to feed at the easily available spilled grain while loading and unloading sacks. Scattered grain is common on the loading docks, roadways and inside the godowns. Commensal birds like the house sparrow, *Passar domesticus*, blue rock pigeon, *Columbia livia,* collared dove, *Streptopelia decaoto,* the common myna, *Acridotheres tristis* and the common crow, *Corvus splendens* are attracted to the spilled food grains. Of these sparrows being commensal may nest inside and outside godowns. Being quite bold and small, they fly in and out through open doors and unscreened windows. They also cause spillage by pecking at the jute bags containing grains as well as contaminate the spilled grain by their droppings (Garg *et al.*, 1966). The actual damage by birds to stored grain is by this contamination, also caused by other commensal birds while feeding on spilled grains because the contaminated spilled grains are rebagged. Several zoonotic diseases are transmitted through these droppings posing a potential health hazard to godown workers. It is for this reason that granivorous, commensal birds have to be discouraged from feeding inside and outside godowns (Ahmad *et al.*, 1990).

MANAGEMENT OF VERTEBRATE PESTS OF GRAIN STORAGE

Rodent pest management: Rodents are by far the most destructive of vertebrate pests of stored grains and as such much of the discussion in this chapter is concentrated on this group of pests. The two basic necessities of rodents as for any other animal are food and shelter. By denying either or both of them considerable reduction in pest population levels can be brought about. Given the poor infrastructure and sanitary conditions in Indian villages, rodent management strategies work differently in urban and rural set ups. A systematic approach to manage vertebrate pests in grain storage should be based on a preliminary inspection and assessment of damage. This consists of looking for common entry points such as cracks or holes in the doors, walls and floor, presence of live or dead animals, spilled or burst grain bags, burrows around the godown, vegetation that could harbour rodents and offer roosting to birds etc. The second step would be sanitation and house keeping of godowns

Rodent Proofing: Of all the methods, rodent proofing the structures and anticoagulant baiting have been considered as the most successful measure to bring down godown losses due to rodents. The details of design to construct rodent proof godowns are in existence since the colonial rule of British.. Before the grain is stored, godowns should be repaired to close all entry points and possible shelter points and the structure made rodent proof. Usually the proofing consists of rodent guards, undamaged, closely shutting doors and windows, crack free walls and floor, making points of entry of pipes and electrical conduits rodent proof, drains covered with grills, installing rat guards around drain pipes, and an inner door or griller mesh to ventilate but prevent rodents. Even the surroundings should be free of rodent burrows, tall vegetation, garbage and water accumulations. Once grain is stacked, the interiors should be kept clean, with no damaged grain bags, and spilled grains (Ahmad *et al.*, 1990).

Rodent proofing and rodent control should be carried out when the infestation of godowns is low mostly when they are empty or during summer. The reduction of carrying capacity of the godown environment by keeping the godown rodent-proof structurally is safer, easier and economical compared to the tedious, systematic campaign of repeated rodenticide treatment. As a result the pests living adjacent to godowns but using the food resources of the godown face deficit food and increased competition for available resources. Ultimately population level becomes low enough to tolerate. Thus the improvement of godown facilities were expected to do more to reduce losses to rodents than any other method of rodent control (Frantz, 1975).

Continued vigilance is needed to keep these structures rodent free by good sanitation and structural maintenance. Once the godown is made rodent proof, the entire structure is to be fumigated to eliminate all insects and rodents. Rodents can be further controlled by poisoning or trapping.

Rodent control is by trapping and lethal methods. Traps are either live or snap traps and are set after prebaiting for a few days. Chemical control is by 2% zinc phosphide baiting, after 2-3 days of prebaiting or by strategically placed bromadiolone wax blocks at 4-5 meters intervals. Baits and poisons have to be kept inside bait boxes. Contact rodenticides inside bait boxes and liquid anticoagulants in poultry water dispensers are also effective in godowns and warehouses.

Management in Rural houses: As 65-70% of the country is rural with less than ideal situations of sanitation, house keeping, garbage disposal and rodent proof houses, the strategy to reduce/prevent rodent caused losses to grains stored in villages is slightly different although the basic principles are same. Because of poor civic amenities and poverty, modifying environment to deny shelter is almost impossible in villages. However, food can be denied, by making existing grain storage structures rodent proof. This step has to be accompanied by systematic rodent control measures to realize optimum benefits. A large scale village level rodent control operation in Gujarat standardized this technique in early 70s. An elaborate rodent control programme in 65 contiguous villages and the town of Sidhpur was undertaken. Initially chemical control using anticoagulants was carried out. Later the component of rodent proof, metallic bins for storing grains was added to achieve long term success to prevent rodent damage to stored food grains. At the beginning, specially designed metal bins were provided at a subsidized cost. But farmers already had traditional in house storage structures of either mud bins or brick with cemented bins and were reluctant to invest in new, metal bins. The problem was overcome by altering the already existing mud-cement bins by making them rodent proof by reinforcing the cover of the bin and by providing inlet and outlet. This integrated technology of chemical control of rodents along with usage of rodent proof storage bins was a tremendous success. Over the three year campaign period the programme became so popular that adjoining villages volunteered to adapt the technology (Madsen, 1975). This approach was successfully popularized by the Save Grain Campaign of Ministry of Food, Government of India at village level across the country since then.

Elimination of residential rodent population has two prerequisites. Because of the low level of literacy in villages, anticoagulant use and trapping should be advocated for rodent control. Secondly to have sustained levels of minimum rodent population, control methods are to be on a community level and should be carried out at least once every year. This needs educating and motivating villagers and involving the entire village in rodent control. Such a campaign is possible only with initiative from government. Annual rodent eradication on community level in villages should become a policy of the government with local administration taking up the responsibility with the help of departments of agriculture, health, public works, revenue and education.

Bird pest management: There is only one method to prevent bird depredation inside godowns i.e. bird proofing the godown. It is always best to take this fact into account while building but is rarely done. The practical way of bird proofing is by tightly fitting, undamaged doors, by providing an inner door made of grilled mesh or expanded metal or by installing a vinyl plastic strip door. The later allows people, machinery and bags to pass through but will disallow birds. Windows should be screened with nylon or metal mesh. By installing "porcupine wires" on window sills, ledges, roof peaks or any other likely roosting site, birds can be prevented from landing, resting, roosting and nesting inside or outside the building. Fixing a metal sheet at a 45 degree angle across the bottom of the ledge is equally effective in preventing pigeon roosting. It is very important to maintain all bird proof measures in good condition with periodic inspections. Sparrows can be discouraged from feeding and spoiling grains by systematic searching and destruction of nests and eggs which are fairly easy to locate. But sentiments come in the way of nest destruction (Ahmad *et al.*, 1990).

GOVERNMENT EFFORTS: A consolidated estimate of food grain losses during post harvest handling and storage due to various factors like insects, rodents, birds, moisture, mould etc is about ten per cent of which 2.5 per cent is attributed to rodents. The government of India has implemented the existing recommendations to prevent storage losses to rodents, the most important being construction of rodent proof godowns and warehouses of both central and state governments. As government is the biggest procurer and storer of food grains in the country this preventive step of constructing rodent proof godowns and warehouses and fumigating them before storing grains has prevented huge losses of grain. The food corporation of India, central and state warehousing corporations also have private godowns for storage some of which are not rodent proof. In such situations zinc phosphide and anticoagulant poisoning is carried out to eliminate the infesting rodents. In case of burrowing rodents, aluminium phosphide fumigation is done. Periodic and frequent fumigation of godowns with Ethylene di bromide and Aluminium phosphide tablets eliminated both rodents and insects in the godowns.

In addition, as already mentioned, the Ministry of Food, Government of India has launched the "Save Grain Campaign" to educate, motivate and persuade farmers and traders to use modern storage practices along with rodent control measures. This is achieved by organized demonstrations and publicity programmes in villages all over the country.

Legal steps to enforce rodent control are possible only with traders and not with the public. At the level of villages, towns and cities rodent control is a social problem. In fact it is a problem of people and therefore they have to be educated about the necessity of controlling rodents for better health and to prevent grain losses. This becomes difficult in certain situations where people are aversive to killing due to ethical, moral and religious reasons. Under such conditions emphasis should be on educating them about health hazards of rodents and measures to prevent entry and access to households. Once these partial measures become successful they will slowly accept rodent eradication by live trapping first and later by chemicals.

The government has enacted some legal measures to enforce wholesale traders, floor mill owners, and rice mill owners to undertake rodent control measures in their premises under essential commodities Act of 1955 and the Rice milling industry (Regulation) Act of 1958. The prevention of Food adulteration Act/rules prescribes limits for contamination of rodent hair and excretes in the food material. Accordingly rodent hair or excreta should not exceed 5 pieces per kg of the sample. The government has constituted a Central Rodent Advisory Board to plan and organize country wide rodent control programmes in a scientific way (Krishnamurthy & Agarwal, 1975).

CONCLUSIONS

In spite of obvious, serious damage by rodents spread over most of the year in storage conditions, rodent control has not been seriously studied in rural households although there is some information on godowns and warehouses. Since most procured grain is stored in central and state warehouses which are under the control of government, there are adequate preventive and remedial measures to prevent rodent losses. However it is in the vast majority of rural houses that there is steady and considerable loss of food grains due to poor storage, poor construction and inadequate hygiene in and around villages, and lack of regular, sustained rodent control measures. The problem of proper storage can be remedied with adaption of better storing methods including improved bins. However, population reduction of rodents in villages has been neglected awfully in spite of their role in disease transmission and substantial losses of food grains. Unless government makes it a policy to execute an annual, community level rodent control in the entire village involving local administration, the chronic loss of stored food to rodents will continue unabated.

REFERENCES

AHMAD, E., J.E. BROOKS., S. MUNIR & I. HUSSAIN. 1990. VERTEBRATE PEST MANAGEMENT IN GRAIN STORAGE. Vertebrate Pest control project, National Agricultural Research Centre, Islamabad, Pakistan. 38+8 p.

AHMAD, S., R.K. PANDIT & J.E. BROOKS. 1994. *Post harvest grain losses in farm houses in Bangladesh: Rodent population estimates and potential stored paddy losses*. DWRC Res. Report No. 11-5-03. Denver Wildlife Research Centre, Denver, USA. 13p.

BROOKS, J.E & G.K. LAVOIE. 1990. Rodent control will reduce post-harvest food losses. *Agribusiness Worldwide*/November-December. 13-17.

FRANTZ, S.C. 1973. *Behavioural ecology of the lesser bandicoot rat, Bandicota bengalensis* (Gray) *in Calcutta*. Ph.D. Dissertation, Johns Hopkins University, School of Hygiene and Public Health, Baltimore, USA. 222 p.

FRANTZ, S.C. 1975. The behavioural/ecological milieu of godown bandicoot rats: Implications for environmental manipulation In: *Proc. All. India Rodent Seminar,* Ahmedabad, India. (Eds): Krishnamurthy, K., G.C. Chaturvedi & I. Prakash. pp. 95-101.

GARG, S.S.L., J. SINGH & V. PRAKASH. 1966. Losses of wheat in threshing yards due to birds and rodents. *Bull. Grain. Tech*. 4(2): 94-96.

HOPE, H.S., G.E. MORIEY & J.E. HAMPHIRES. 1976. Rodent damage to growing crops and to farm and village storage in tropical and sub tropical regions. *Cent. Overseas Pest Res*. London, UK. pp. 115.

KRISHNAMURTHY, K. & N.S. AGARWAL. 1975. Rodent control to save food grains: Efforts made by the government In: *Proc. All. India Rodent*

Seminar, Ahmedabad, India. (Eds): Krishnamurthy, K., G.C. Chaturvedi & I. Prakash. pp. 330-333.

KRISHNAMURTHY, K., V. UNIYAL. J. SINGH & S.V. PINGALE. 1967. Studies on rodents and their control. Pt. 1. Studies on rat population and losses of food grains. *Bull. Grain Technol.* 5(3): 147-150.

MADSEN, C.R. 1975. Sidhpur rodent control and grain storage project-History, working schedules and reasons for success In: *Proc. of All India Rodent Seminar,* Ahmedabad, India. (Eds.) K. Krishnamurthy, G.C. Chaturvedi & I. Prakash. pp. 16-23.

PINGALE, S.V. 1975. Rat as a competitor of man for food. In: *Proc. of All India Rodent Seminar,* Ahmedabad, India. (Eds.) K. Krishnamurthy, G.C. Chaturvedi & I. Prakash. pp. 32-35.

SPILLET, J.J. 1966. Population studies of the lesser bandicoot rat in Kolkata. In: *Proc. Indian Rodent Symposium*, Calcutta, India. (Ed.) K.L. Harris. pp. 84-92.

SRIDHARA, S & R.V. KRISHNAMOORTHY. 1979. Grain losses by spoilage by wild rodents under laboratory conditions. *Prot. Ecol.* 1: 103-108.

❑❑❑

20

BEHAVIOURAL ASPECTS IN THE MANAGEMENT OF VERTEBRATE PESTS

Shakunthala Sridhara

AINP on Rodent Control, University of Agricultural Sciences G.K.V.K., Bangalore-560 065

ABSTRACT

There are several attributes of avian and mammalian behaviour which can be exploited to prevent or reduce damage to crops and in other situations of nuisance. The basic principles underlying prevention of herbivore depredation on cultivated crops is scaring them from entering crop fields which is achieved by using frightening devices which have proved to be one of the most important non-lethal approaches of preventing vertebrate damage to crops considerably. These devices are mostly acoustic and visual scares and chemical repellents. In addition other methods like guard animals, bird resistant varieties, power fencing are also employed for pest specific problems. Auditory scares such as explosives, pyro-techniques, recorded bird alarm cells, distress calls, and ultrasonics act by inducing pain, fear, jamming communication, general distress and confusion. Lasers, reflective ribbons, raptor models, scare crows are the common visual devices that are being used with varying degree of success. Acoustic and visual scares have been found to repel birds only temporarily with habituation occurring in no time. However, an appropriate combination of the two, with a specific way of presenting them would yield better repellency. Chemical repellents are either primary repellents like cinnamamide effecting immediate repulsion or secondary repellents which work over a period of time such as predatory odours. The primary repellents are more effective to deter birds while secondary repellents mainly predatory odours work successfully in preventing mammalian herbivores from feeding on plants and crops. The amount of data on scares and repellents is voluminous for both birds and mammals in USA, Australia and UK, probably in response to the vociferous public outcry for non-lethal methods to manage vertebrate pests. Indian work is limited to evaluation of reflector ribbons, distress calls and a sole study on anthranilate in birds and almost none in mammalian depredators, except one or two preliminary studies on repellency of thiram.

Neophobia, bait shyness towards poison baits after surviving from sub-lethal doses of acute rodenticides and highly developed local food preferences are the major constraints in rodent control. These resistances are overcome by prebaiting, using anticoagulants/fumigants and/or by trapping after single day poisoning with zinc phosphide. Formulating attractive baits to mix poisons renders rodent control more effective. Pheromones can also be exploited to enhance food and poison bait intake. In addition semiochemicals particularly predator odours are potential chemosterilants capable of disrupting reproduction at various stages.

INTRODUCTION

Being competitors for the same resources in the given ecosystem, man and herbivore vertebrates are bound to clash. When man succeeded in growing adequate food grains by encroaching into and converting a portion of habitat of wild herbivores, loss of their traditional food sources led these losers to search and adapt to available foods. Some of them became commensal like rodents, crows, vultures, stray dogs and cats which exploited and shared man's bounty of food. Others like granivorous and frugivorous birds, bats, crop raiding mammals like elephants, bears, deer such as sambhar, cheetal and nilgai, wild boar, monkeys etc learnt very soon that cultivated crops can be exploited cleverly and successfully. Rodents went a step further and learnt to live off the post harvest stored food in addition to consuming crop produce in agricultural lands. The battle of wits began with man planning, and conceiving measures to prevent crop losses caused by animals and to keep away the pests. If he failed in this endeavour or did not completely succeed, he devised ways and means of killing them paving way to pesticides, traps, hunting, etc. In response animals started evolving behavioural adaptations to escape from these measures.

Basically the measures to prevent herbivore from attacking crop fields are based on two responses-scaring and repelling. When scaring and frightening fails or is inadequate, killing becomes necessary. In case of oral poisons attractants become essential to lure the pests to poisons given the easy availability of alternate food source in their natural habitat. Over the years rodents as a result of being targeted for elimination for the past several centuries have evolved two important behavioural adaptations namely neophobia and food aversion learning, which in turn has led to investigations as to how to mitigate these processes. In this chapter an overview of animal defences against man's attempts to prevent them from depredating on his crops, stored food and the counter measures being researched and practiced by the biggest depredator of natural resources on planet earth, i.e. MAN, in this catch 22 situation is reviewed.

FRIGHTENING DEVICES (Table 1)

Frightening devices are one of the most important non-lethal components of an integrated approach to reduce wildlife damage. The ultimate goal of frightening devices is to scare the depredators away from a food source or comfort zone. Being non-lethal, eco-friendly, these methods are accepted readily by public. Vertebrate pests are deterred from their pestiferous activities using frightening devices which use a single stimulus or a combination of stimuli. The senses of the pest animals namely sight, smell, taste, touch and hearing influence the receptivity of the stimuli and its effect. Generally frightening devices are based on sound and vision and they work better for birds than mammals.

BIRDS

a. Acoustic frightening devices have high potential for discouraging birds from consuming grains and fruits as birds are endowed with acute sense of hearing. The auditory scares act by inducing pain, fear, jamming communication which results in disorientation, audiogenic seizures, internal thermal effects, general distress and confusion. For best effects in bird management sound is used to generate fear not pain as the later is against the sentiment with which birds are perceived. Sounds can be generated by mechanical, electrical and explosive devices. The frightening devices commonly used are propane exploders, bird bombs, whistles, pyrotechniques etc (Anguilera *et al.*, 1991; Cummings *et al.*, 1986). Although the repellency achieved through these devices is short lived, best results can be obtained by presenting sound at random intervals, using a wide variety of noises, moving the source of sound frequently and reinforcing sound distress calls with visual scares. The disadvantages of these devices are the necessity of an operator to fix them, cost involved, disturbance to public and most importantly the possibility of birds getting habituated to them. In India, with small holdings, birds are repelled by women labourers shouting and pelting stones as soon as they sight the birds in the field during the vulnerable stages of crop damage. The practice is effective, economical and non-lethal (Sridhara per. observation).

Table 1. Frightening agents in vertebrate pest management

Device/s	Birds	Mammals
ACOUSTIC DEVICES		
(a) Exploders & pyro-techniques	Effective (Anguilera *et al.*, 91; Cummings *et al.*, 86)	Effective (Belant *et al.*, 98)
(b) Ultrasonic	Not effective[3] (Wright '82)	Not effective (Curtis *et al.*,' 77) Partially effective (Shumake' 95)
(c) Bioacoustics		
(i) Alarm calls	Effective (Frings *et al.*,' 55)	---------
(ii) Distress calls	Effective (Conover & Perito,' 81)	Effective for rats (Sprock *et al.*,' 67; Wade, '83)
VISUAL DEVICES		
(a) Lasers	Effective but costly (Glisdorf *et al.*, '02)	Ineffective (Gilsdorf *et al.*, '02)
(b) Conventional visual scares.		
(i) Reflector ribbons	Effective (Bruggers *et al.*, '86)	Effective (Mason, 98)
(ii) Strobe lights	------	Effective (Linhart, '84)
(iii) White shining plastic	Effective (Mason *et al.*,' 93)	-------
(c) Bio visual scares		
<u>Interspecific</u>		
(i) Scare crows	Partially effective (Curio, '78)	Partially effective (Mason, '98)
(ii)Raptor models	Partially effective (Conover, 84)	—
<u>Intraspecific</u>	Effective (Murton' 70)	None
CHEMICAL REPELLENTS		
(a) Primary repellents	Cinnamamide Coniferyl benzoate Acetophenols Methylanthranilate	Cinnamamide-Effective (Gill *et al.*, '95)
(b) Secondary repellents	Effective	
	(a) Methiocarb (Curtis *et al'* 94)	

Device/s	Birds	Mammals
	(b) Antraquinone (Avery *et al.*,'95; Poche' 98)	Repelled wild pigs (Chakravarthy, '93)
	(c) Thiram (Avery & Decker '91; Schafer *et al.*, '83; Dhindsa *et al.*,'91)	Predatory odours inhibiting plant damage and interrupting reproduction (Nolte *et al.*, '74); Voznessenskaya *et al.*,'92; Feoktistova *et al.*,' 03)
	(d) Caffeine (Avery & Cummings,'02)	

Ultrasonics: In spite of claims by manufacturers about the success of ultrasonic devices to repel birds, and that some vertebrate pests are capable of detecting ultrasound and infrasound (Curtis *et al.*, 1977), there is no evidence that birds are sensitive and repulsed by ultrasound (Wright. 1982).

Bioacoustics has great potential as frightening agents. These are based on auditory communication signals of birds, often in the form of alarm or distress calls. They are primarily *Feeding calls* to attract conspecifics to food source and *Warning calls*. *Alarm calls* are emitted in response to presence of predator or an unknown sudden stimulus or when a bird is excited and angry. *Distress calls* are emitted when a bird is physically traumatised, captured or restrained or faces unfavourable conditions like cold, isolation or hunger.

The distress call may be a cry for help when it becomes an attractant or a warning to other birds when it becomes a repellent. It can also startle the predator to loosen its grip. Conversely it can bring a second predator to the site. Response to distress calls is influenced by previous experience, age, sex and the resident status of the bird. These calls have been found to be effective in diverse habitats such as airports, fish farms, grassland, crop fields and dumps. In some cases a visual scare supplements the effect of a distress call. Distress call can be recorded by birds caught in mist nets and by using electric shocks. To be effective, habituation to a repellent call and its specificity if any are important.

Alarm calls are less evaluated probably because it is difficult to record them, although they are more appropriate for repelling birds.

Habituation is the gradual learning of a bird not to respond to an unimportant repeated stimulus. A successful dispersal call should have a definite biological meaning capable of disallowing any habituation. The biological meaning can be innate or acquired. However, habituation does occur but at a slower rate compared to other sounds. Habituation can be delayed by changing the location of speaker, alternating different calls (Johnson *et al.*, 1985), by the quality of recording and broadcasting devices, reinforcing calls with visual scares (Inglis, 1980), by keeping the source in easily detectable areas and creating a super stimulus (Bremond, 1980). The super stimulus is designed by enhancing the signal value of a deterrent call. The effectiveness of the call can also be enhanced by increasing the frequency, duration, intensity, harshness and by alternating alarm and distress calls (Bremond, 1980), by 20-80% reduction in resolution, integrating alarm calls with hunting calls of predators and recording with spaces in between (Schmidt & Johnson, 1984), using computers to create super normal versions of alarm and distress calls. This is achieved by quick analysis of calls, rapid and accurate editing, creating variations of the call and exposing birds continuously to ever different calls to retard habituation (Nicholls, 1990).

Call specificity: Evidence for call specificity is contradictory with claims that birds respond only to calls of conspecifics and some times only to those from the same region and that similar alarm calls exist in a variety of species (Marler, 1956). In spite of variations in bird vocalizations, transpacific effect of alarm calls in evidently present (Frings *et al.*, 1955).

Bioacoustics is definitely superior to other types of loud acoustics which are disturbing to human and domesticated animals. In contrast alarm and distress calls are effective at low intensities (Frings, 1964) and less expensive to produce.

(b) Visual devices: Lasers, reflective ribbons to some extent and strobe to a large extent are employed either singly or in conjunction with acoustic repellents to keep birds away from depredating on crops.

Lasers: Response to laser is species specific. Starlings, cow birds and doves were not affected but Canada geese and mallards exhibited avoidance behaviour to lasers. Mallards, however habituated to the device within 20 minutes (Blackwell *et al.*, 2002). Lasers have the potential of a quite, species specific and a non-lethal tool to dispel birds from urban and rural situation. The only disadvantage is their exorbitant cost ($1000 to $7500) (Glisdorf *et al.*, 2002).

Conventional visual scares used to repel birds are basically of two types-one is based on response to novelty and the second type is based on alarm response to predators (both avian and human).

Novelty based visual scares induce withdrawal but with continuous exposure, and increased encounters birds learn that the scare is harmless and get habituated to it. The speed with which repellency declines depends on the type of device and prior exposure. Thus a scare to be effective should be an uncommon one and the rate of information acquisition about it should be slow. The novelty of the scare can be maintained by exposing it for short durations, shifting its position randomly within the field and by changing its time of exposure (Inglis, 1980).

Reflective ribbons and other devices of reflection: Shiny devices reflecting light such as reflector ribbons are employed to keep away birds from crop fields. Mylar ribbons are narrow strips of reflective tape, silver colour on one side and red on the other. When stung tightly twisted between posts, the ribbons reflect sunlight and make a humming sound when wind blows. Such ribbons when placed in parallel rows, 3-10 mts apart reduced bird damage to corn, millet, sorghum and sunflower in India, Bangladesh and United States of America but were not successful to repel gold finches and morning doves ((Bruggers *et al.*, 1986). They are cost-effective only for high value crops when grown at low level and bird damage is very heavy (Conover & Dolbeer, 1989).

White, shiny plastics repelled snow geese from rye and wheat fields (Mason *et al.*, 1993). All over India farmers use white cloth and plastic sheets hung on a stick amidst the field to keep birds away from crops (Sridhara Pers. Observation).

Bio visual scares are based on pest bird's innate fear of predators. Habituation to such scares is slow because of the genetically built in tendency to respond to cues associated with predator. Bio-visual scars are of two types: <u>inter-specific</u> and <u>intraspecific</u>. Inter-specific models derive stimuli from a natural predator whereas intraspecific devices mimic warning signals of pest species displayed on sighting a predator.

<u>Interspecific devices</u> include raptor models incorporating cues involved in predator recognition such as body outline and mode of flight while hunting (Conover, 1979).

Retarding habituation to an effective visual scare is as important as designing one. This can be achieved by eliciting mobbing behaviour which also enhanced fear of the scare (Curio, 1978). Mobbing can be elicited by designing predator models holding a prey model or a lathered prey model accompanied by periodic play back of mobbing calls. A scare becomes more effective if it is moveable using wind, mechanic or electrical means (Conover, 1979)

Generally human effigies as scare crows are not very effective. A three dimensional human effigy walking with periodically moving outstretched arms (Inglis, 1980), an inflated human effigy placed on a three-wheel cart and drawn in fields and orchards (Achiron, 1968), a pop-up inflatable human effigy (Stickly & King, 1995) have been found successful in keeping birds away from crops. Women labourers walking, shouting and pelting stones at birds during vulnerable stages of bird's depredation driving away pest birds successfully has already been mentioned (Sridhara Pers. Observation). In general, stimuli emanating from a human model working in a field are the best deterrent to repel birds.

Intraspecies devices either mimic corpse or unnatural body postures and sometimes normal signals of conspecifics. Pigeon models with open wings or even a pair of pigeon wings deterred bird attack on crops (Murton, 1970). Certain alarmed, pre-flight postures also were effective in preventing Brent geese landing in cereal fields (Inglis & Issacson, 1978).

To sum up visual scares are effective when incorporated with an appropriate auditory stimulus based on the predator avoidance and intraspecies communication about alarm situations, and food availability.

(c) Guard animals: Dogs can be used to keep away nuisance wildlife including birds. Border collies successfully chased Canada geese from a property (Castelli & Sleggs, 2000)

(d) Bird resistant varieties: Research in this line is limited to sunflower. Not strictly coming under the frightening device, varieties with high anthocyanin content (Fox, 1984) and varieties with horizontally–oriented concave heads and long head-stem distance were predicted to be effective against blackbird depredation (Lilleboe, 1992). Bird resistant varieties are not an answer but can be an important component of integrated pest management where bird damage is very severe.

(e) Chemical repellents basically act by inducing an aversion in birds towards a place or a food which is otherwise attractive and avoid it. Repellents are primary and secondary based on the speed with which they affect the birds. Primary chemical repellents are avoided because they produce unpleasant olfactory sensation on first exposure or highly distasteful or they cause pain and irritation. Secondary repellents on the other hand are slow and involve learning to associate postingestional sickness with consumption of the repellent food (Rogers, 1980).

Primary chemical repellents act nociaceptively and cause pain and irritation consequent to tissue damage (Clark, 1998). Tissue damage releases neurotransmitters which are detected by the chemoreceptive fibres of the trigeminal nerve. The animal perceives unpleasant sensations like pain, irritation etc. Subsequently the bird avoids the sources of painful stimuli, in this case the foraging area. An analogue of the neurotransmitter can also be used as a repellent. Although there is physiological adaptation to continuous exposure to irritants, habituation does not occur to primary chemical repellents. E.g. Methyl anthranilate, cinnamamide, coniferyl benzoate and acetophenols.

Most primary repellents are plant derived because most plants in response to herbivory have evolved the defence mechanism of producing toxicants or irritants which upon ingestion or sampling produce chemosensory irritation. Methyl anthranitrate, the most widely used bird repellent is of plant origin. Cinnamamide, a synthetic compound derived from cinnamic acid is again of plant origin. It deters feeding by pest birds such as wood pigeon, rock dove and chestnut-capped. Although the chemical acts both as a primary and secondary repellent in birds and mammals, in birds it is a primary repellent (Gill *et al.*, 1995).

Secondary chemical repellents actually cause illness or an unpleasant experience. Birds exhibit learned avoidance of sensory cues associated with the unpleasant stimulus. The persistence of learned avoidance is related to the strength of the unpleasant stimulus and the associated cue. Methiocarb inhibits acetycholinesterase at synapses but the action is reversible. Affected birds exhibit distress responses like retching, vomiting and temporary paralysis and learn to associate the food ingested with distress experienced and avoid that food item. Methiocarb was extensively used to protect seeded, sprouted crops, ripening grain crops and fruit crops (Curtis *et al.*, 1994).

The behaviour is also exploited to minimize predation of eggs of the endangered least term, *Sterna antillarum browni* by ravens. Quail eggs injected with methiocarb upon ingestion by ravens induced avoidance behaviour resulting in reduced predation (Avery *et al.*, 1995).

Another secondary repellent, anthraquinone, has been registered as Flight control® to repel birds from turf and grass (Blackwell *et al.*, 1999). Upon ingestion it induces vomiting. The chemical is not a taste repellent or contact irritant. It successfully deterred black bird damage to sprouting and ripening rice (Avery & Cummings, 2002) and prevented damage by American robins, cow birds and red-winged blackbirds (Poche, 1998).

Thiram (tetra methyl thiuram disulfide), a fungicide emits a sulphurous odour and induced conditioned food aversion. It is also registered as a bird repellent. Thiram treated grains were avoided by blackbirds (Avery & Decker, 1991), passerine birds (Schafer *et al.*, 1983) but there was no effect in common crow (Dhindsa *et al.*, 1991).

The two types of repellents differ in their efficacy. The primary repellents by virtue of producing an immediate response are much more effective if there is a flock of transient birds. The secondary repellents require a period of learning and are better suited to repel resident pest birds.

Future repellents: Since the research and registration process to develop and market a new bird repellent is lengthy, expensive and of uncertain results, evaluation of chemicals which are already approved for human consumption or for other uses are economical and less time consuming to evaluate(Avery & Cummings, 2002). Caffeine, a naturally occurring stimulant of nervous system, is one such potential compound. Recently it was permitted by the environmental protection Agency of USA to be used as a toxicant to manage populations of non-native tree frogs in Hawaii. There are possibilities of controlling other pests too using caffeine (Hollingswerth *et al.*, 2002). Initial laboratory trials established repellency towards rice treated with caffeine (Avery & Cummings, 2002). The response was similar to anthraquinone.

Limitations: Repellents are not the sole panacea for preventing bird damage to crops. Although effective when used singly or in conjunctions with other means of protecting crops from bird menace, repellents have several limitations in respect of timing, efficacy and variation (Rogers, 1980). Repellents should be applied before the birds develop feeding pattern on crops because once the birds establish a behavioural pattern it is much more difficult to discourage them from visiting crops. Secondly crops like rice and sunflower mature over 3-4 weeks in a staggered manner necessitating repeated applications of the repellents.

Efficacy of repellents varies with crops and bird species. For instance a repellent that is effective in repelling a comparatively smaller bird may not deter a much larger flock of birds. Similarly repellency in fields where birds roost and forage such as baya weaver bird roosting in marshy vegetation but feeding on rice may be different from sunflower fields which is used only as feeding sites by parakeets.

To sum up, chemical repellents should be employed as an important component of integrated damage management strategy.

Reduction of roosting sites is another important factor in such a programme.

MAMMALS

(a) Acoustic devices work comparatively less effectively for mammals. Gas exploders and pyrotechnics are used to disperse racoons, foxes, rabbits, beaver and some rodents. Gas exploders using infra-red sensors to detect animal movement and explode have been found to reduce foraging by these animals from a few days to few weeks (Dolbeer, 1999).

(b) Bioacoustics, Sonic and Ultrasonic devices: There are very few studies on bioacoustics for mammalian pest management. Distress call of a rat killed by a skunk showed more promise than other sounds (Sprock *et al.*, 1967). However such a sound would be repulsive to humans. The only other study revealed that coyote distress calls have limited and short term effect (Wade, 1983).

The mechanism of ultrasonic repellency seems to be by way of pain induction, interference with communication, disorientation and fear. Use of ultrasonics to deter mammals is controversial. The commercially available ultrasonic devices to repel deer from cultivated crops were either ineffective or was of limited use for less than a week (Curtis *et al.*, 1977).

Almost all ultrasonic devices recommended for rodent control are for indoor situations. They are less intense when away from the source. Their energy level is absorbed by soft textured materials like cloth, insulation and also by obstacles in its path like paper, cardboard, corners, angles etc. Thus their repellency is marginal (Shumake, 1995). At the most ultrasonics can be used in combination with other repellents like pheromones, predator odours etc.

(c) Guard animals: Dogs have been used for centuries to guard livestock from predators. Of late they are used to protect agriculture also. Guard dogs effectively deterred white-tailed deer damage to white pine plantation (Greens, 1989a) and were employed to deter monkey depredation on crops in Rajasthan (Chhangani & Mohnot, 2004). Dogs should be neutered and used in pairs to allow for social interaction. There are reports of donkeys being successfully used to protect livestock from predators. They bray, chase, try to kick and bite intruders (Greens, 1989b).

(d) Visual repellents work more effectively for birds than for mammals, because of the capacity of the former to detect colours and ultraviolet light. In contrast mammals are mostly colour blind or at the most sensitive only to blue and green. Visual repellents such as scare crows and Mylar ribbon have limited effect on mammals that too short lived (Mason, 1998).

(e) Electric shock: Electric fences can deter deer (Caslick & Decker, 1979) but is of limited use against elephants (Sukumar, 1990).

(f) Lasers: Desman laser® and Dissuader® laser failed to repel white-tailed deer from crop fields (in Gilsdorf *et al.*, 2002).

(g) Predatory odours based on faeces, urine or synthetic compounds of naturally occurring secretions of predators have been shown to be repellent to herbivores (Nolte *et al.*, 1994). The avoidance of urine of predator is due to development of conditioned avoidance response of herbivores to specific predators (Swihart *et al.,* 1991), or to innate generalized responses to common sulphurous by products of meat eating (Nolte *et al.,* 1994) or perhaps from a combination of both. Repellents derived from predators are thought to provide more effective and long lasting avoidance of food crops by herbivores (Mason *et al.*, 1994). Urine of stoat, red fox, weasel and coyote were effective in preventing small mammal mostly rodent and deer damage to crops and forestry in America and Australia (Lindgren *et al.*, 1995; Morgan & Woolhouse, 1995).

CHEMICAL REPELLENTS

The chemical senses of olfaction, ingestion and chemical irritation or pain in animals have twin basic functions: to attract, and to repel. Attractant compounds are associated with food or sex while repelling compounds signal danger. Olfaction plays an important role in social behaviour and food recognition. To a great extent the attraction and repellency of an odour is dependent on learning. Identification and avoidance of food is mainly by taste. Learning is involved in taste acquisition but to a small degree. In contrast chemical irritants indicate danger and painful stimuli. By virtue of this character chemical irritants have the greatest potential of repelling animals but there is such a great variation in their effect amongst different species it is not practical to employ chemical irritants in all situations.

(a) Olfaction based repellents: In most animal species olfaction has three distinct roles namely social, environmental and/or informative, and nutritive. The social role is in sexual, maternal and intraspecies aggression. The stimuli that have application value are the ones that provoke fear such as predatory odours and the ones that help in food selection and poison avoidance.

Predator urine reduced reproductive output in rats, mice and voles (Feoktistova *et al.*, 2001). Laboratory experiments revealed extension of oestrous cycle in Norway rats by exposure to anal sac secretions of mink (Voznessenskaya *et al.*, 1992), urine of feral cats (Feoktistova *et al.*, 2001), odours of weasel in bank voles (Kostela *et al.*, 1996), urine of feral cats and overcrowded mouse odours in *Mus musculus musculus* (Feoktistova *et al.*, 2003). A commercial repellent, deer away Big Game Repellent® (BGR) which released sulphur odours and volatile fatty acids prevented deer and rabbit browsing (Mason *et al.*, 1999). BGR contains putrescent-white egg solids as the active ingredient.

A repellent sprayed on plants to deter herbivore feeding on them may in course of time cease to repel the animals due to their adaptation. This is likely to hold good even for predatory odours. In such a situation predatory odours have to be applied taking other factors of the natural environment in which such odours are encountered by the prey species for long term effects. These include rarity and location specific action of the repellents.

Whether the response of animals to odours is inherent or learnt is not clear. Although there is no evidence to suggest innate repellency or attractiveness of an odour there could be some cases where it is so. Many studies indicate that response to odour is learnt during pre and post natal development. However, in case of repellent odours, both individual experience and genetic selection are involved in preventing long term avoidances. Learning has been shown to alter and reduce innately repellent odours. Not only that, selection also prefers animals that learn to ignore harmless, repellent odours. This phenomenon occurs quickly in rapidly breeding species (Beauchamp, 1995).

Frightening devices, auditory or visual, can work only as components of an integrated pest management (IPM) strategy for bird and mammalian pests. However, they do represent a vital non-lethal approach for wildlife problems and hence are favoured by public. Unfortunately studies are few. There is a need to test their comparative efficacy individually and as a component of IPM (Gilsdorf *et al.,* 2002).

(b) Gustatory based repellents: Based on human experience, the primary tastes are categorized as sweet, sour, salty, bitter and a few others. They are also likely to be present in most of the mammals. Taste seems to play the sole role of regulating nutrient intake to ensure consumption of adequate quantities of nutrients and avoidance of poisons (Beauchamp & Mason, 1991). With taste playing such a role in survival and success of animals it is natural to assume that taste preferences and avoidances are innate with experience playing a minor role. But enormous amount of research has established that even positive or neutral tastes become aversive and are avoided if they are associated with unpleasant postingestional consequences (Garcia *et al.*, 1974). The reverse i.e. selection of beneficial foods is also true (Provenza, 1995).

The logical compounds for repelling are bitter ones. Their avoidance is believed to have evolved to protect animals from ingesting poisonous plants which in turn was an adaptation of plants to escape from depredation. Sensitivity of the bitter system seems to depend on the feeding ecology of the species. Too much sensitivity will prevent the animals from finding acceptable and adequate amount of foods. If the animals decrease their sensitivity so as to have enough food, then avoiding lethal quantities may become ineffective. The problem becomes complicated as there is always a correlation between toxicity and bitterness. The dilemma is solved by development of aversion to the flavour components associated with the toxic plant or animals. Thus if a plant has bitter alkaloids, the animal learns to avoid or limit consumption of such foods after sampling. Two classical examples for such learning are the selective feeding on nightshade by wild guinea pigs and the successful feeding on monarch butterflies by mice in Mexico. In the former case, wild guinea pigs initially avoid unfamiliar plants followed by gradual acceptance of a few and rejection of others. But neophobia towards nightshade waned after few days but feeding was limited to stems. Later studies revealed stems to contain lower levels of alkaloids compared to other parts (Jacobs & Labows, 1979). The second case pertains to Mexican mice selectively feeding on Monarch butterflies. Although monarch butterflies are very rich in protein levels, because of the presence of a bitter tasting toxic, cardiac glycoside they are avoided by three of the four local mice in the wintering grounds of butterflies in Mexico. This fourth species of mice has adapted to eating the glycoside rich proteins by three adaptations namely decreasing its sensitivity to the taste of glycoside, increasing its detoxifying ability and most probably by not developing conditioned aversion (Glendinning *et al.*, 1990).

Generally bitter compounds are relatively poor repellents because of herbivore's innate capacity to ignore plant bitteroides, an evolutionary co-adaptation i.e. as plants evolved to produce bitteroids, herbivores co-evolved by becoming tolerant of bitteroids.

(c) Chemical irritants as repellents: The fifth trigeminal nerve transmits sensations of tingling, itching, burning, pain, cooling etc., all unpleasant experiences in response to irritant chemicals which are consequently avoided. Basically irritants can make excellent repellents. Not only they are innately unpleasant but repeated exposure does not extinguish the unpleasantness. On the contrary, oral irritants sensitize with repeated exposure. The pain increases with repeated stimulation with very few exceptions. However volatile irritants seem to decrease in their effect with repeated exposure affecting their repellency. Such volatile irritants are also capable of functioning from a distance, without physical contact with the animal. But their action across classes of animals is a big disadvantage.

Sensory irritants are generally more effective than semio chemicals in deterring mammalian depredation on crops. Their advantages include immediate avoidance, independent of learning to sustain the aversion and minimal adaptation (Mason, 1998). Such immediate and unfavourable response is due to the similarity of these chemicals to endogenous substances released during injury (Clark, 1998). The successful examples include capsaicin (the hot principle in capsicum), alkyl isothiocyanate (the active principle of mustard), the principle component of tear gas, ammonia, carbon di oxide, formaldehyde, Cinnamamide and Thiram (Gill *et al.*, 1995; Ngowa *et al.*, 2003). In the lone Indian study, thiram and copper oxy chloride repelled *Rattus rattus* and *Bandicota bengalensis* from approaching cartons sprayed with these chemicals (Parshad *et al.*, 1993).

In conclusion, it is best to synthesize a repellent that attacks multiple senses similar to the skunk odour. This unpleasant smell is irritating to the eye, smells awful and possibly equally disgusting to taste.

FEEDING AND REPRODUCTIVE BEHAVIOUR (Table 2)

The capacity of rodents to survive in spite of human efforts over the centuries to eradicate them is attributed to several behavioural adaptations mostly in feeding and reproduction. The adaptations in feeding behaviour include neophobia, acquisition of food preferences, social transmission of food preferences and bait aversion learning. The phenomenal role of pheromones in the life of rodents can be exploited to interrupt reproduction and also mitigate neophobia and bait shyness. Following is the account of such adaptations.

(a) Feeding behaviour assumes importance because as on to day the most economical and successful method of rodent control is by oral poisoning using either acute or chronic rodenticides. Hence aspects of feeding behaviour particularly those affecting food intake-preferential as well as aversive, feeding patterns, acquisition of food preferences, transmission of acquired feeding behaviour, bait aversion learning etc. become highly relevant in preparing attractive poison baits and the actual process of poisoning.

Table 2. Behavioural adaptations of rodents against rodenticides and their mitigation

Behavioural adaptations	Impediment in rodent control	Mitigation practiced/suggested
Neophobia	Initial avoidance of poison baits and traps.	(a) Prebaiting with non-poisonous bait for 2-3 days. (b) Traps baited but unset for 3-4 days.
Acquisition of food preferences	Rodents of a given habitat prefer local food grains strongly and initially avoid other baits.	(a) To use locally grown cereals and locally used vegetable oil to prepare baits. or (b) A preliminary multiple choice test of food grains conducted to select preferred grains.
Bait shyness	Rejection of poison bait by members surviving acute rodenticide baiting.	(a) Changing the poison and bait. (b) Fumigation if fossorial. (c) Poisoning with anticoagulants. (d) Addition of 0.2% conspecific urine to baits.

(i) Neophobia: Rats move in their home range even in the absence of stimuli, exploring and investigating, the behaviour being termed *stimulus hunger* or *curiosity*. Even hunger cannot extinguish the exploratory behaviour (Barnett, 1975). However, a rat's response to novelty in its familiar environment is one of caution. If a wild rat's response is always exploring the unknown, it would soon consume poisons or get trapped to death. To avoid this wild rats have evolved an entirely different behaviour to novel stimulus While poisoning rodents with acute rodenticides during Second World War in England it was observed that when an unfamiliar object was placed in their familiar environment, rats avoided both the familiar food and also the path. This avoidance persisted even if the strange object happened to be a heap of nutritious food. The avoidance of novel things in a familiar environment is also displayed towards unfamiliar noise and odour, to traps and is called Neophobia (Fig. 1) (Shorten, 1954). Neophobia is displayed by *Rattus norvegicus* (Barnett, 1975), *Bandicota indica, B. bengalensis, Tatera indica, Mus platythrix, M. booduga, Meriones hurrianae, Gerbillus gleadowii* and *Rattus rufescens* but was not seen in *R. rattus, R. cutchicus* and *G. ellioti* (Idris & Prakash, 1992).

Response to unfamiliar objects is reflective of a balance between approach and withdrawal. Eventually the novel objects are explored including food even in the presence of familiar food suggesting that the tendency to approach remains at a steady level but the tendency to avoid/withdraw gradually declines. Approach in fact is attributed to fear.

Fig 1: Neophobia in a rat

Mild fearfulness involves approach and exploration while alarm results in flight or avoidance. However in disturbed and rapidly changing environments such as garbage and godowns, prolonged avoidance of new objects is almost absent. Rats get trapped within two hours after setting the traps. Generally neophobia is a temporary phenomenon lasting few hours or few days. Ultimately habituation takes place if the situation remains unchanged. This behaviour affects the success of trapping and poisoning. The initial avoidance of traps and poisons is overcome by exposing rodents to unset traps and by prebaiting with non-poisonous foods for two to three days followed by setting the traps and/or poisoning on the third or fourth day. Not much evidence exists for the exact role of sense in neophobia although Cowan (1976) demonstrated that odours play a role, but not a significant one. Since objects are avoided without contact, visual and tactile senses are assumed to be involved. Experimental evidence indicates that neophobia can be attenuated by chlorpromazine (Mitchell *et al.*, 1977), pentobarbital (Mc Gee *et al.*, 1981) and conspecific urine (Prakash & Idris, 1992).

(ii) Bait shyness (Fig. 2): In wild rats, the initial total avoidance of new food is followed by a tendency to sample it after 2-3 days. This stage is followed by a period of few hours when the food is not eaten presumably due to a temporary decline in curiosity or increase in fear (Rzoska, 1953). If the ingested food contains poison, it begins to act causing physiological changes and further feeding stops. On recovery from sublethal doses of poison, rodents correlate disturbances experienced with foods ingested and learn to avoid the poison bait subsequently. This behaviour is variously called bait shyness, bait aversion, flavour aversion learning and conditioned food aversion. The behaviour seems to extend to all components of poison bait. However if the rodents are exposed to same poison but mixed in a different bait and still continue to reject the poison bait, then the behaviour is termed poison shyness or poison aversion. Bait and poison shyness is exhibited toward baits containing acute rodenticides, such as zinc phosphide, sodium fluroacetate and barium carbonate. Most Indian rodents have been shown to be bait shy and the behaviour persists for varying lengths of time in different species (Fig. 2). Taste plays a prominent role and also olfaction in aversion learning (Bharadwaj & Khan, 1979).

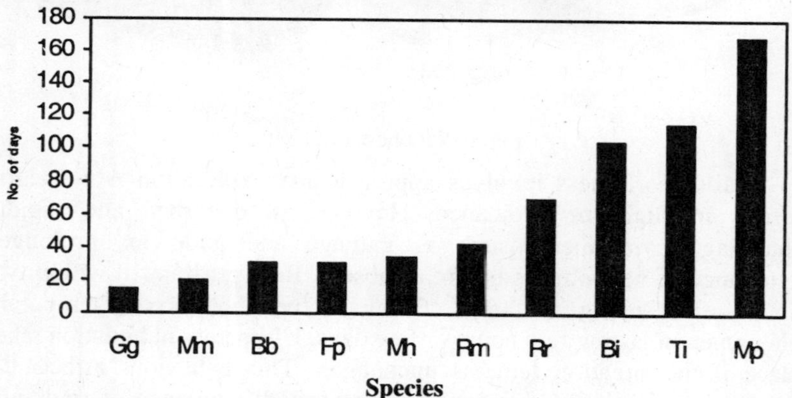

Gg-Gerbillus gleadowii; Mm-Mus musculus; Bb-Bandicota bengalensis; Fp-Funambulus pennanti; Mh-Meriones hurrianae; Rm-Rattus meltada; Rr-Rattus rattus; Bi-Bandicota indica; Ti-Tatera indica; Mp-Mus platythrix

Fig 2. Persistence of bait shyness in Indian rodents (Prakash & Mathur '87; Sridhara '99)

Suggestions to overcome bait shyness include employing a new poison in new bait after prebaiting the rodents with the new bait, by offering a choice of earlier experienced poison bait and poison in a new bait, to fumigate surviving rodents (Sridhara, 1983a, 1983b), by adding 0.4% conspecific urine (Prakash & Idris, 1992), by mixing a toxin flavour mimic in the prebait (Robbins, 1980) and by changing the texture of the bait (Bharadwaj & Khan, 1979).

(iii) Food preferences and bait formulations: Being omnivorous rodents feed on a variety of plant and animal foods including garbage. In spite of the tendency to sample all foods encountered/provided, rodents exhibit marked preference for some baits and slight preference for others. They even prefer more favourable of the two or most preferable among several food (Barnett, 1975). Thus for rodent control to become successful it becomes essential that poison be mixed with the most preferred food so that it is consumed in sufficient quantities to cause death. As wild rodents are extremely sensitive and can discriminate odours and tastes at minute levels, attractive baits can be formulated only after studies on food preferences of the target species. The most acceptable and commonly used baits are cereals which get readily mixed with poison and do not perish quickly. Even texture of the grain, i.e. whole, broken or powder form influences bait intake. Additives like 3-10% vegetable oil, molasses, sugar, salt and corn syrup increase bait acceptance (Sridhara, 1974-78). Generally rodents prefer cereals like rice, wheat and millet. Vegetable oils not only enhance bait intake but are effective in binding poison dust to grains. In view of multitude of species infesting crop fields and godowns in India, it is advisable to use a mixture of predominant cereals and vegetable oil of the area being treated to prepare baits for rodent control.

(iv) Acquisition and transmission of food preferences: In rodents many behavioural processes evolve from third week after birth onwards including neophobia. Food preference acquisition is one of the widely studied behaviour of rodents. Evidence for food imprinting is inconclusive with earlier worker failing to detect effect of infant ingested experienced foods on adult diet selection. Odour and flavours of foods experienced during preweaning or immediately after weaning were found to influence food selection in adult rats. The mechanism of such acquisition was through three ways namely (a) Gustatory cues present in mother's milk influenced the pups to identify and prefer such foods when they became adults. (b) Residual cues left at the feeding site by the mother aids in food selection and (c) Pups learn to feed on foods eaten by mothers by simple imitation. The maternal pheromone, caecotrope excreted following parturition is qualitatively influenced by mother's milk and facilitates young pup's identification and preferential consumption of foods containing cues resembling mother's milk (Leon *et al*., 1977). When faced with a choice among several novel foods, rats choose novel foods eaten by conspecifics with whom they had interacted previously (Galef & Wigmore, 1983). These effects were mediated through volatile cues present in the breath of interacted rats which in turn was a combination of the smell of ingested foods and an indigenous odour, carbon di sulphide. Carbon di sulphide when added to food facilitated its intake and also increased visits to bait stations containing such foods. The chemical also increased trap attraction and interfered with acquisition of bait shyness (Galef *et al.*, 1988). Thus carbon di sulphide seems to be a potential enhancer of rodent bait intake and effective mitigant of neophobia and bait shyness.

(b) Reproductive Behaviour

Although sexes in small mammals are morphologically identical except for the external genitals and mammary glands, they are dimorphic through odour. The extent of glandular development, the repertoire and intensity of odours produced and released, complexity and marking frequency are more in males compared to females. The pheromones associated with reproduction and sexual behaviour has relevance to rodent pest management. Releaser pheromones acting either directly on the central nervous system or through neurochemicals play a significant role in attracting opposite sexes, marking territories, evolving sexual and aggressive behaviour. Pheromones also facilitate male recognition of receptive and non-receptive females, in female recognition of normal and castrated males and female's greater behavioural responses to urine of intact males compared to that of castrated males (Carr *et al.,* 1966). Primer pheromones have slower but long lasting effects on reproduction by three major actions namely suppression, and induction of oestrous, and pregnancy block by strange males. In addition they also influence interspecific competitions, sexual selection, social status, neophobia, food intake and bait shyness (Lee & Bor, 1956; Champlin, 1971).

(i) Oestrous suppression: Housing small groups of unisexual females resulted in mutual suppression of oestrous. This change could be decreased by removing olfactory bulbs suggesting the involvement of an olfactory pheromone (Lee & Boor, 1956). Housing of a group of females in cages recently soiled by an earlier group of females also resulted in anoestrous, decreased ovarian and uterine weights suggesting pheromonal involvement (Champlin, 1971).

(ii) Induction of oestrous by exposure to males: When females are housed in a cage soiled by males or when males are present in the same room as the grouped females, there is acceleration and synchronization of oestrous, a phenomenon called Whitten effect. The pheromone responsible has been shown to occur in male urine and is released as a vapour. Several species of mice viz. *Mus musculus, M. booduga, Peromyscus maniculatus* and *Microgaster orchagaster* exhibit this effect which is diet dependent in rat (Whitten, 1966; Dominic, 1976).

(iii) Pregnancy block by strange males: When a mated, pregnant female is exposed to a strange male, her pregnancy gets terminated and she returns to oestrous within a few days. This male induced pregnancy block is called Bruce effect and it has a direct relation with the degree of unfamiliarity of the male and strain. The action is due to a pheromone emanating from male urine (Bruce, 1970).

(iv) Pheromones in intraspecies competition, sexual selection, social hierarchy, neophobia and bait shyness: There was a pheromone based sexual attractant in the urine of *M. hurrianae, T. indica* and *R. meltada.* In *M. hurrianae* sent marking played a big role in maintaining social hierarchies. Conspecific urine and sebum enhanced food intake and mitigated bait shyness in *M. hurrianae* and *T. indica* (Prakash & Idris, 1992). Similarly the extract of preputial gland of albino and house rats increased food intake and suppressed bait shyness (Selvaraj & Archunan, 2004). Sex specific volatile compounds from the scent glands of house rat and lesser bandicoot rat had positive effect on trapping. The volatile compounds of preputial/cheek gland extracts and urine of house mouse which suppressed bait shyness and increased uptake of poison bait have been identified as 2,6,10- decatrien-1-ol,3,7, 11-trimethyl(Z,E), 6,11-dihyro-dibenz-b,e- oxepin-11-one, 2,6,1dodecatrien-1-ol-3,7,11-trimethyl(z), 1,2-benzene dicarboxylic acid (2-ethylpropyl) ester, 2,6,10-dodecatrien-1-ol,3,7,11-trimethyl(Z,E), 3-octen-1-ol, Cyclopentane undecanoic acid, 2,4,6,8-tetramethyl-1-undecane, L –alanine,1,1-dimethyl ethyl ester, Ethanol,2-(octylthio), 1,3,5-triazone-2,4-diamine, 1-Chlorodecane, Hydroperoxide, 1-nitropentane, 4-azidoheptane, 3-amino-s-triazole, 4- ethyl phenol and 3-ethyl-2,7-dimethyl octane (Kannan, 1998; Selvaraj, 2002).

Rodent pheromones have thus the potential of disrupting reproduction, enhancing intake of poison and mitigating neophobia and thus capable of becoming efficient tools in rodent pest management.

CONCLUSIONS

Vertebrate pests-both birds and mammals with the exception of rodents-always create a sense of awe towards the strength and beauty of these animals in spite of the damages they cause. This sentiment makes any management with toxicants unacceptable not only to the public but also the victim of these pests, the farmer. The quest for non-lethal means of managing vertebrate pests is time immemorial but earlier they were mostly based on wisdom of the farmer and his observation on the behaviour of pest animals. It is only from the past 5-6 decades that the measures based on auditory, visual and chemical reception are being researched and evaluated to keep away vertebrate pests from crop fields. Most of the chemicals act as repellents, not as lethal pesticides, again the exception being rodents. The Indian effort in this direction is pathetically poor although our mammalian pests on agriculture are varied causing substantial losses. There is need at least to initiate work on predatory odours which does not involve much investment. On the other hand, the output on rodent behaviour and research to overcome bait shyness, neophobia, bait formulations, and pheromones especially on the desert rodents is substantial and has resulted in quite a few applications.

REFERENCES

ACHIRON, M. 1968. Building a better scare crow. *Natl. Wildl.* 26: 18.

ANGUILERA, E., R.L. KNIGHT & J.L. CUMMINGS. 1991. An evaluation of two hazing methods for urban Canada geese. *Wildl. Soc. Bull.* 19: 32-35.

AVERY, M.L. & J.L. CUMMINGS. 2002. Chemical repellents for reducing crop damage by blackbirds In: MANAGEMENT OF NORTH AMERICAN BLACKBIRDS, *Proceedings of a special symposium of the wildlife society 9th Annual Conference..* (Ed.) G.M. Linz. Bismark, North Dakota, USA. pp. 41-48

AVERY, M.L. & D.G. DECKER. 1991. Repellency of fungicidal rice seed treatments to red winged black birds. *J. Wildl. Manage.* 55: 327-334.

AVERY, M.A, M.A. PALELKA., D.L. BERGMAN., D.G. DECKER., E.C. KNITTLE & G.M. LINZ. 1995. Aversive conditioning to reduce raven predation on California Least tern Eggs. *Colonial Waterbirds.* 18: 131-138.

BARNETT, S.A. 1975. THE RAT: A STUDY IN BEHAVIOUR. Chicago University Press, Chicago, USA.

BEAUCHAMP, G.K. 1995. Chemical signals and repellency. In: REPELLENTS IN WILDLIFE MANAGEMENT (Ed.) J.R. Mason. Colorado State University, Fort Collins, Colorado, USA. pp. 1-10.

BEAUCHAMP, G.K & R. MASON. 1991. Comparative hedonics of taste. In: THE HEDONICS OF TASTE. (Ed.) R.C. Bolles. Lawrence Erlbaum Associates, Hillside, NJ, USA. pp. 159-186.

BHARADWAJ, D & J.A .KHAN. 1979. Effect of texture of food on bait shy behaviour in wild rats (*Rattus rattus*). *Appl. Anim. Ethol.* 5: 361-367.

BLACKWELL, B.F., T.W. SEAMANS & R.A. DOLBEER. 1999. Plant growth regulator (Stronghold™) enhances repellency of anthraquinone formulation (Flight Control™) to Canada geese. *J. Wildl. Manage.* 63: 1336-1343.

BLACKWELL, B.F., G.E. BERNHARDT & R.A. DOLBEER. 2002. Lasers as non-lethal avian repellents. *J. Wildl. Manage.* 66: 250-258.

BREMOND, J.C. 1980. Prospects of making acoustic super stimuli. In: BIRD PROBLEMS IN AGRICULTURE. (Eds.) Wright, E.N., I.R. Inglis & C.J. Feare. BCPC Publications, London. pp. 115-120.

BRUCE, H.M. 1970. Pheromones. *Brit. Med. Bull.* 26: 10-13.

BRUGGERS, B.L., J.E. BROOKS., R.A. DOLBEER., P.P. WORONECKI., R.K. PANDIT., T. TARIMO & M. HOQUE. 1986. Responses of pest birds to reflecting tape in agriculture. *Wildl. Soc. Bull.* 14: 167-170.

CARR, W.J., L.S. LOEB & N.R. WYLIE. 1966. Response to feminine odours in normal and castrated male rats. *J. Comp. Physiol. Psychol.* 62: 336-338.

CASLICK, J.W & D.J. DECKER. 1979. Economic feasibility of a deer-proof fence for apple orchards. *Wildl. Soc. Bull.* 7: 173-175.

CASTELLI, P.M & S.E. SLEGGS. 2000. Efficacy of border collies to control nuisance Canada geese. *Wildl. Soc. Bull.* 28: 385-92.

CHAKRAVARTHY, A.K. 1993. *Vertebrate Pest Management.* Final report of ICAR ad-hoc project, University of Agricultural Sciences, GKVK, Bangalore, India.

CHAMPLIN, A.K. 1971. Suppression of oestrus in grouped mice: the effects of various densities and the possible nature of the stimulus. *J. Reprod. Fert.* 27: 233-341.

CHHANGANI, A.K & S.M. MOHNOT. 2004. Crop raid by Hanuman langur, *Semnopithecus entellus*, in and around Aravali Hills (India) and its management. *Primate Report.* 69: 1-18.

CLARK, L. 1998. Physiological, ecological and evolutionary bases for the avoidance of chemical irritants by birds In: CURRENT ORNITHOLOGY. (Eds.): Nolan, V. & E. Ketterson. Vol. 14. Plenum Press, New York. pp. 1-37.

CONOVER, M.R. 1979. Responses of birds to raptor models. *Bowling Green Bird Control Seminar.* 8: 16-24, Bowling Green, Ohio, USA.

CONOVER, M.R. 1984. Comparative effectiveness of avitrol, exploders, and hawk-kites in reducing blackbird damage to corn. *J. Wildl. Manage.* 48: 109-16.

CONOVER, M.R & R.A. DOLBEER. 1989. Reflecting tapes fail to reduce blackbird damage to ripening cornfields. *Will. Soc Bull.* 17: 441-443.

CONOVER, M.R & J.J. PERITO. 1981. Response of starling to distress calls and predator models holding conspecific prey. *Z. Tierpsychol.* 57:163-172.

COWAN, P.E. 1976. The new object reactions of *Rattus rattus* L: The relative importance of various cues. *Behav. Biol.* 16: 31-34.

CUMMINGS, J.L., C.E. KNITTLE & J.L. GUARINO. 1986. Evaluating a pop-up scare crow coupled with a propane exploder for reducing black bird damage to sunflower. *Vertebr. Pest Conf.* 12: 286-291.

CURIO, E. 1978. The adaptive significance of avian mobbing: I-Teleonimic hypothensis and predictions. *Z. Tierpsychol.* 48: 175-193.

CURTIS, P.D., C. FITZERALD & M.E. RICHMOND. 1977. Evaluation of the yard guard: ultrasonic yard guard protection for repelling white tailed deer. *Eastern Wildl. Damage Manage. Conf.* 7: 172-176.

CURTIS, P.D., I.A. MERWIN., M.P. PRITTS & P.V. PETERSON. 1994. Chemical repellents and plastic netting for reducing bird damage to sweet cherries, blue berries and grapes. *Hort. Science.* 29: 1151-1155.

DHINDSA, M.S., P.S. SANDHU., H.K. SAINI & H.S. TOOR. 1991. House crow damage to sprouting sunflower. *Tropical pest Manage.* 37: 199-181.

DOLBEER, R.A. 1999. Overview and management of vertebrate pests. In: HANDBOOK OF PEST MANAGEMENT. (Ed.) J.R. Roberson, New York, Marcel Dekker, Inc. pp. 663-91.

DOMINIC, C.J. 1976. Role of pheromones in mammalian fertility, *Proc. Int. Symp. Neuroendocrine regulation of fertility*. Shimla, India. pp. 236-245.

FEOKTISTOVA, N.Yu. S.V. NAIDENKO., E.A. KARTASHOVA & V.V. VOZNESSENSKAYA. 2001. The role of predator odours in regulation of oestrous cycle in small rodents. SUPPLEMENTS TO ETHOLOGY (Ed.) Shakunthala Sridhara. Contributions to the XXVII International Ethological Conference, Blackwell Science, Berlin. 152p.

FEOKTISTOVA, N.Yu., S.V. NAIDENTO., A.E. VOZNESSENSKAYA., G.J. KRIVOMAZON., L. CLARK & V.V. VOZNESSENSKAYA. 2003. The influence of predator odours and overcrowded mouse odours on regulation of oestrous cycles in house mice (*Mus musculus*) In: RATS, MICE AND PEOPLE: RODENT BIOLOGY AND MANAGEMENT (Eds.): Singleton G.R., L.A. Hinds., C.J. Krebs & D.M. Sprat. Australian Centre for International Agricultural Research, Canberra, Australia. pp. 173-175.

FOX, G. 1984. Bird resistance update. *The sunflower* 10(4): 29

FRINGS, H. 1964. Sound in Vertebrate pest control. *Vertebr. pest conf.* 2: 50-56.

FRINGS, H., M. FRINGS. B. COX & L. PEISSNER. 1955. Recorded calls of herring gulls (*Larus argentatus*) as repellents and attractants. *Science.* 121: 340-341.

GALEF, B.G. Jr. & S.W. WIGMORE. 1983. Transfer of information concerning distant foods: A laboratory investigation of information centre. *Anim. Behav.* 31: 748-758.

GALEF, B.G. Jr., R.J .MASON., J.R. PRETI & N.J. BEAN. 1988. Carbon di sulphide: a semiochemical mediating socially induced diet choice in rats. *Physiol. Behav.*, 42: 119-124.

GARCIA, J., W.G. HANKINS & K.W. RUSINIAK. 1974. Behavioural regulation of the milieu interne in man and rat. *Science.* 185: 824-831.

GILL, L.G., R.W. WATKINS., J.E. GURNEY., J.D. BISHOP., C.J. FEARE., C.B. SCANLON & D.P. COWAN. 1995. Cinnamamide: A non-lethal chemical repellent for birds and mammals. In: REPEELANTS IN WILD-LIFE MANAGEMENT (Ed.) J.R. Mason. Colorado State University, Fort Collins, Colorado, USA. pp. 43-51.

GILSDORF, J.M., S.E.HYGNSTROM & K.C.VERCAUTEREN. 2002. Use of frightening devices in wildlife damage management. *Integ. Pest Manage Rev.* 7: 29-45.

GLENDINNING, J.I., L.P. BROWER & C.A. MONTGOMERY. 1990. Responses of three mouse species to deterrent chemicals in the monarch butterfly. 1. Taste and toxicity tests using artificial diets laced with digitoxin or monocrotaline. *Chemoecology.* 1: 114-123.

GREENS, J.S. 1989a. Animal and plant health inspection service animal damage control livestock guarding dog program. *Proc. Great Plains Wildl. Damage Control Workshop.* 9: 50-3.

GREENS, J.S. 1989b. Donkeys for predation control. *Eastern Wildl. Damage Control Conf.* 4: 83-6.

HOLLINGSWORTH, R.G., J.W. ARMSTRONG & E. CAMPBELL. 2002. Caffeine as a repellent for slugs and snails, *Nature* 417: 915-916

IDRIS, M & I.PRAKASH. 1992. Shyness behaviour In: RODENTS IN INDIAN AGRICULTURE (Eds.) Prakash, I & P.K. Ghosh, Scientific Publishers, Jodhpur, India. Pp. 433-444.

INGLIS, I.R. 1980. Visual bird scares, an ethological approach. In: BIRD PROBLEMS IN AGRICULTURE. (Eds.) Wright, E.N., I.R. Inglis & C.J. Feare. BCPC publications, London. pp. 121-143.

INGLIS, I.R & A.J. ISSACSON. 1978. The response of dark bellied Brent geese to models of various postures. *Anim. Behav.* 26: 953-958.

JACOBS, W.W. & J.N. LABOWS. 1979. Conditioned aversion, bitter taste and the avoidance of natural toxicants in wild guinea pigs. *Physiol. Behav.* 22: 173-178.

JOHNSON, R.J., P.H. COLE & W.W. STAUP. 1985. Starling response to three auditory stimuli. *J. Wildl. Manage.* 49(3): 620-625.

KANNAN, S. 1998. Histomorphology, Biochemistry and Chemical identification of Rat scent glands in view of rodent pest management: A new light in IPM programme. Ph.D. Thesis. Bharathidasan University, India.

KOSTELA, E., T. HORNE. T. MAPPES & H. YCONEH. 1996. Does risk of small mustelid predation affect the oestrous cycle in the bank vole, *Clethrionomys glareolus*? *Anim. Behav.* 51: 1159-1163.

LEE, V.D.S & L.M. BOOR. 1956. Spontaneous pseudopregnancy in mice II *Acta Physiol. Pharmacol.,* 5: 213-214.

LE MAGNEN, J. 1952. Les pheromones olfactosexuels chez lerat blanc, Arches. *Sci. Physiol.* 6: 295-332.

LEON, M., B.G.Jr. GALEF & H. BEHSE. 1977. Establishment of pheromonal bonds and diet choice in young rats. *Physiol. Behav.*, 18: 387-391.

LILLEBOE, D. 1992. NDSU develops bird resistant lines. *The sunflower.* 18(4): 16-17.

LINDGREN, P.M.F, T.P. SULLIVAN & D.R. CRUMP. 1995. Review of synthetic odour semiochemicals as repellents for wildlife management in the pacific North West. In REPELLENTS IN WILDLIFE MANAGEMENT (Ed.) J.R. Mason. Colorado State University, Fort Collins, Colorado, USA. pp. 217-230.

LINHART, S.B. 1984. Strobe light and siren devices for protecting fenced pastures and range sheep from coyote predation. *Proc. Vertebr. pest Conf.* 11: 154-156.

MARLER, P.P. 1956. The voice of the chaffinch and its function as a language. *Ibis.* 98: 231-261.

MASON, J.R. 1998. Mammal repellents: options and considerations for development. *Proc. 18th Vertbr. Conf.* (Eds.): R.O. Baker & A.C. Crabs. University of California, Davis, USA. pp. 325-329.

MASON, J.R., L. CLARK & N.J. BEAN. 1993. White plastic bags repel snow geese (*Chem caerulescens*). *Crop Prot.* 12: 497-500.

MASON, R.J., G. EPPLE. D.L. NOLTE. 1994. Semiochemicals and improvements in rodent control. In: BEHAVIOURAL ASPECTS OF FEEDING. (Eds.) B.G. Galef Jr., M. Mainardi & P. Valsecchi. Harwood Academic Publication, Switzerland.

MASON, J.R., J. HOLLICK. B.A. KIMBALL & J.J. JOHNSON. 1999. Repellency of Deer Away Big Game Repellent to Eastern Cotton tail rabbits. *J. Wildl. Damage* 63(1): 309-314.

McGEE, R., P. GOLUS & C. RUSSEL-JONES. 1981. Long term effect of pentobarbital anaesthesia on neophobia in the rat. *Behav. Neural Biol.* 32: 376-379.

MITCHELL, D., M. FAIRBANKS & J.D. LAYCOCK. 1977. Suppression of neophobia by chloropromazine in wild rats. *Behav. Biol.* 19: 309-323.

MORGAN, D.R & A.D. WOOLHOUSE. 1995. Predator odours as repellents to Bush tail possums and rabbits. In: REPELLENTS IN WILDLIFE MANAGEMENT (Ed.) J.R. Mason, Colorado State University, Fort Collins, Colorado, USA. pp. 241-252.

MURTON, R.K. 1970. Sophisticated "scare crow" could be the answer, *Arable former* Sept. 10-11.

NGOWO, V., J. LODAL. L.S. MULUNGU., R.H. MAKUNDI., A.W. MASSAWE & E. LEIRS. 2003. Evaluation of thiram and cinnamamide as potential repellents against maize seed depredation by the multimammate rat, *Mastomys natalensis,* in Tanzania. In: RATS, MICE AND PEOPLE:

RODENT BIOLOGY AND MANAGEMENT (Eds.): Singleton G.R., L.A. Hinds., C.J. Krebs & D.M. Sprat. Australian Centre for International Agricultural Research, Canberra, Australia. pp. 260-261.

NICHOLLS, D. 1990. Computer sound tools to aid development of acoustic signals for bird control. In: NATURAL BIRD PEST WORKSHOP PROCEEDINGS. Armidale, NSW, Australia. pp. 111-118.

NOLTE, D.L, J.R. MASON., G. EPPLE., E. ARONOV & D.L. CAMPBELL. 1994. Why are predator odours aversive to prey? *J. Chem. Ecol.* 20: 1505-1516.

PERRONE, M. 1980. Factors affecting the incidence of distress call in passerines. *Wilson. Bull.* 92: 404-408.

POCHE, R.M. 1998. Development of a new bird repellent, Flight control. *Proc. 18th Vertebr. pest Conf.* (Eds.) R.O. Baker & A.C. Crabs. University of California, Davis, USA. pp. 338-344.

PRAKASH, I & M. IDRIS. 1992. Scent making behaviour. In: RODENTS IN INDIAN AGRICULTURE (Eds.) I. Prakash & P.K. Ghosh. Scientific publishers, Jodhpur, India. pp. 445-460.

PRAKASH, I & R.P. MATHUR. 1987. MANAGEMENT OF RODENT PESTS. Indian Council of Agricultural Research, New Delhi, India.

PARSHAD, V.R., M.S. SAINI & S. JINDAL. 1993. Repellent action of two fungicides against the house rat, *Rattus rattus* and the Indian mole rat *Bandicota bengalensis*. *Int. Bodet. Biodgr.* 31: 77-82.

PROVENZA, F.D. 1995. Postingestive feedback as an elemental determinant of food preference and intake in ruminants. *J. Range. Manage.* 48: 2-17

ROBBINS, R.J. 1980. Taste aversion learning and its implication for rodent control. *Proc. Ninth Vert. pest Conf.,* (Ed.) J.P. Clark, University of California, Davis, 114.

ROGERS, J.G. 1980. Conditioned taster aversion: its role in bird damage control. In: BIRD PROBLEMS IN AGRICULTURE. (Eds.) Wright, E.N., I.R. Inglis & C.J. Feare. 1980. British crop protection council, London.

RZOSKA, J. 1953. Bait shyness, a study in rat behaviour. 1953. *British J. Anim. Behav.* 1: 128-135.

SCHAFER, E.W., W.A. BOWLES & J. HURLBUT. 1983. The oral toxicity, repellency and hazard potential of 988 chemicals to one or more species of domesticated birds. *Arch. Environm. Contam. Toxicol.* 12: 355-382.

SCHMIDT, R.H & R.J. JOHNSON. 1984. Bird dispersal recordings; an overview. Special Technical Bulletin 817, American Society for testing and Materials, Philadelphia, USA.

SELVARAJ, R. 2002. Pheromonal identification and its involvement in the control of house rat (*Rattus norvegicus*). Ph.D. Thesis, Bharathidasan University, India.

SELVARAJ, R & G. ARCHUNAN. 2004. Efficacy of male scent glands and urine in masking poison bait shyness of female house rats, *Rattus rattus.Pest Biochem. Physiol.* (Communicated).

SHORTEN, M. 1954. THE CONTROL OF RATS AND MICE. Clarendon Press, Oxford, UK.

SHUMAKE. A.S. 1995. Electronic rodent repellent devices: a review of efficacy test protocols and regulatory actions. In: REPELLENTS IN WILDLIFE MANAGEMENT. (Ed.) J.R. Mason. Colorado state University, Fort Collins, Colorado, USA. pp. 253-270.

SPROCK, C.M., W.E. HOWARD & F.C. JACOB. 1967. Sound as a deterrent to rats and mice. *J. Wildl. Manage.* 31: 729-41.

SRIDHARA, S. 1977. New food reaction in four Indian wild rodents. *Z. angew. Zool.* 64: 449-452.

IBID. 1974-78. Food preference studies in rodents and their importance in rodent control. *Plant Prot. Bull.,* XXV-XXX, pp. 39-43.

IBID. 1983a. Rodenticide induced bait shyness and neophobia in *Tatera indica cuvieri. Z. angew. Zool.* 70: 429-440.

IBID. 1983b. Bait shyness in the bandicoot, *Bandicota indica. Indian J. exp. Biol.,* 21: 2560-2563.

IBID. 1999. QUARTER CENTURY OF RODENT RESEARCH IN THE UNIVERSITY OF AGRICULTURAL SCIENCES, BANGALORE-RESEARCH HIGHLIGHTS AND RECOMMENDATIONS, University of Agricultural Sciences, Bangalore, India. 136p

STICKLY, A.R. Jr & J.D. KING. 1995. Long term trial of an inflatable effigy scare device for repelling cormorants from catfish ponds. *Eastern Wildl. Damage Control Conf.* 6: 89-92.

SWIHART, R.K., J.J. PIGNATELLO & J.I. MATTINA. 1991. Aversive responses of white tailed deer, *Odocoileus vinginianus*, to predator urines. *J. Chem. Ecol.* 17: 767-777.

SUKUMAR, R. 1990. Ecology of the Asian Elephant in Southern India. II Feeding habits and crop raiding patterns. *J. Tropical Ecol.* 6:35-53.

VOZNESSENSKAYA, V.V., C.J. WYSOCKI & E.P. ZINKEVICH. 1992. Regulation of the rat oestrous cycle by predator odours: role of the vomeronasal organ. In: CHEMICAL SIGNALS IN VERTEBRATES. (Eds.) R.L. Doty & Muller-Schwartze. Plenum Press, New York. pp. 281-284.

WADE, D.A. 1983. Coyotes. In: PREVENTION AND CONTROL OF WILDLIFE DAMAGE, (Ed.) R.M. Timm. Lincoln NE, USA: Great Plains Agricultural Council, Wildlife Resources Committee and University of Nebraska Cooperative Extension Service, USA. pp: C31-41.

WHITTEN, W.K. 1966. Pheromones and mammalian reproduction. In: ADVANCES IN REPRODUCTIVE PHYSIOLOGY I (Ed.) A. McAllen. Academic Press, New York. pp. 155-177.

WRIGHT, E.N. 1982. BIRD PROBLEMS AND THEIR SOLUTION IN BRITAIN. *Vert. pest Conf.* 10: 186-189.

❑❑❑

21

UNCONVENTIONAL AND SPORADIC VERTEBRATE CROP DEPREDATIONS

A.K. Chakravarthy, N.E. Thyagaraj and Shakunthala Sridhara*

*Department of Entomology, University of Agricultural Sciences, G.K.V.K., Bangalore-560 065; *AINP on Rodent Control*

ABSTRACT

Selected instances of unusual crop depredations by vertebrates in India are dealt with in this chapter. Wild boar, bison, frugivorous bats, monkeys in north-east India, hares, jackals and sloth bears are the unconventional pests of India about which some information is available. Wild pig is increasingly causing crop damages throughout Asia mainly because it's habitat is shrinking and in wild areas its natural enemies are decreasing. Feeding on coconut fronds of seedlings by bison is mainly incidental. Because of lack of sufficient food in the wild, langurs in the North-East are raiding cultivated fruits. In order to supplement their food with herbage or plant material, jackals incorporate contents of cocoa pods and thus become a concern for the grower. Hares are a problem at germination stage of rice, pulses and vegetable crops as they feed on germinating seedlings. Sloth bears are serious crop pests when their natural habitat and populations get entrapped amongst villages and cultivated fields. Non-chemical tools for management of such vertebrate pests is emphasized to mitigate crop depredations.

INTRODUCTION

Increase in human population has escalated the pressure on the forests and other wild areas and their resources. Due to the shrinkage of the natural areas and wildlife habitats, animals are forced to gradually change their habits of searching and usage of food which in turn has affected their feeding and living habits. Herbivores now commonly leave the natural habitat in forest areas to raid crops in adjacent patches/

villages. The wild animals are becoming dependent on cultivated crops especially in the fringe areas of cultivated ecosystems. Elephants, wild pigs, bisons, monkeys langurs, bears, peacocks, porcupines, bats were not habitual crop depredators in the past in the Indian sub-continent. But currently these animals are threatening agriculture. This activity of wild animals adversely affects the interests of the cultivators, who claim compensation for their damaged crops.

In addition, the social and economic pressures for development especially in the tropics are on the rise. The development schemes are often incompatible with the environmental and conservation concerns. With substantial habitat loss and fragmentation, the populations of wild animals are coming in direct conflict with the interests of human populations. Many species of animals are becoming commensals in human environs and pests in crop fields. It is in these situations that crop protection from the unusual animal depredators assumes urgency. These unusual wild animals cannot be branded as pests although their feeding on crops may result in economic losses. To mitigate problems of crop depredations by these unconventional species, a holistic approach is necessary. This chapter provides instances of feeding on crop in certain situations/agro ecological zone by vertebrates, which are not regular crop feeders. The review is aimed at understanding the problem of unconventional pests and in developing sound management practices. The wild animal depredators on crops discussed here are:

1. The wild bison, Bos or Bibos gaurus

Class :	Mammalia
Order :	Artiodactyla
Family:	Bovidae
Genus :	*Bos*
Species :	*gaurus*
Binomial name:	*Bos gaurus*

Coconut cultivated in coastal Karnataka is depredated by several vertebrates. Bison, while feeding on grass and other wild vegetation browse on the newly emerging fronds of the coconut palms. Thus coconut plantations at the vicinity of forest or wild tracts are highly vulnerable to accidental bison damage. Observations recorded on bison damage in one such situation are reported here. Bison feeding behaviour (Prater, 1986) and damage to palms (Fitzwater & Prakash, 1989) has been reported earlier.

The affected plot was regularly foraged and the animals were primarily attracted by the grasses and the ground vegetation. Bison damage commenced in the plot from July, after monsoon rains, when ground vegetation was dense reaching a peak during August-November and started decreasing from December to April. The damage to coconut seedlings was related to the ground biomass production. The animals while feeding on ground vegetation incidentally fed on young coconut fronds also. The number of bisons in the 300 acre farm increased at the rate of one per 18 months. On an average, there were eleven animals in the farm.

As the age of seedlings increased, the feeding pressure of bison decreased. As a result, the number of severely damaged seedlings decreased from 49 in 1990 to 3.0 in 1999 (Table 1). This may be due to regular patrolling by night staff in the experimental plot and with increase in age, the coconut leaves became tough with lignin becoming unpalatable. On an average, 53% palm seedlings showed damage during 1996 and about 5% during 1998. Severely damaged seedlings had only two healthy leaves which drastically impaired their growth and in some cases led to mortality of seedlings. Damaged seedlings did not recover and fresh fronds did not emerge. The trend in the proportion of healthy to damage fronds/seedlings was similar to the seasonal damage pattern. During 1996-97 the proportion of healthy to damage fronds was in the ratio of almost 1:2. During 1997-98, this ratio was reduced but during 1998-99, number of healthy fronds prepondered over damage fronds, i.e., with increasing age of coconut seedlings, the bison damage decreased.

The bison feeding on young (2 to 3 years old) coconut palms could not be prevented by IBEX fencing and by fencing with wires and stone pillars. Gaurding and scaring the animals during night afforded some protection. There is a need to find protection methods against bison feeding on coconut seedlings in coastal and hill regions of Karnataka. Proper regulation of ground vegetation in the young plantations may render the area less attractive to the animals and lessen the damage.

Table 1. Mean number of coconut palms in four different categories of bison damage (1996 to 1999)

Plant height (m)	Mean number of palms damaged (n=211)		
	1996-97	1997-1998	1998-1999
4	20.45	18.00	112.26
3	58.54	62.27	76.00
2	83.63	85.36	19.83
1	48.50	45.37	3.00
CD @5%	21.78	23.21	27.35

2. The Indian peacock, *Pavo cristatus*

Class : Aves
Order : Galliformes
Family : Phasianidae
Genus : *Pavo*
Species : *cristatus*
Binomial name : *Pavo cristatus*

The Indian peafowl, *Pavo cristatus* Linnaeus is a resident bird, distributed throughout the Indian sub-continent. Very common and abundant in Gujarat, Rajasthan and in places where it is protected on religious or sentimental grounds. The bird keeps in small flocks or droves consisting usually of a cock and 3 to 5 hens. In the wild, they cautiously emerge into forest clearings, ploughed fields or just sown crop fields. The birds forage in ploughed fields during early morning or late noon hours by scratching the ground for food. Peafowls roost at night in tall trees and descend to feed in early dawn. They are omnivorous consuming seeds, grains, lentils, tender shoots of crops, flower-buds, berries, insects, worms, grubs and several species of reptiles mainly snakes in and around villages (Ali & Ripley, 1987).

In coastal southern Karnataka paddy (*Oryza sativa*) was the dominant crop preferred by peacocks. Birds singly or in flocks of 5 to 6 would land in the periphery of paddy fields on open patches or on bunds.

Table 2. Peacock damage to paddy in Dakshina Kannada (Karnataka)

Village	Plot area (acres)	% damage	Yield loss (kg)	Remarks
Koyoor	06	08	80	Damage occurred along the periphery only. No. of birds ranged 2 to 20
Umedabaile	04	10	60	Damage only at the borders, grain picking, number of birds = 2 to 3
Kaliyady	10	05	80	Damage only at the borders, grain picking, number of birds = 2 to 3
Sirlal	02	18	40	Damage in patches, panicle swallowing, number of birds = 2 to 3
Durga	02	15	40	Damage in patches at all places, panicle swallowing, number of birds = 4 to 5
Mundli jarkala	01	20	120	Damage in patches, group feeding, trampling, lodging and panicle cutting: number of birds = 2 to 3
Hankibetta	0.5	6 to 8	40	Damage in patches, group feeding, trampling, lodging and panicle cutting: number of birds = 4 to 6
Kervase	02	12	100	Damage in patches, group feeding, trampling, lodging and panicle cutting: number of birds = 5 to 6

They generally preferred to forage in rice fields during morning or evening hours throughout the year. Feeding on paddy grains was by trampling or lodging the tillers. Birds were found picking grains, gulping panicles and cutting the entire panicles. After feeding on paddy grains birds were observed roosting on near by trees (Table 2). Maximum feeding by peacocks on paddy grains occurred during August and September. The density of birds, on an average per acre rice fields varied from three to eleven (n=16 fields). Besides paddy, the birds fed on cucurbit vegetables (buds/flowers/tender fruits), Brinjal, cowpeas, okra and other pulses (pods/seeds).

The damage to rice crop in eight villages varied from 5 to 12%, the yield loss varied from 40 to 100kg paddy/acre. Due to religious sentiments and as birds are protected by Wildlife Protection Act, farmers are not keen to take up any crop protection measures against peacock damage.

3. Langurs of the *presbytis* genus

Class	: Mammalia
Order	: Primates
Sub-order	: Haplorhiri
Superfamily	: Catarrhiri
Family	: Cercopithecidae
Genus	: *Presbytis*

Golden langur, *Presbytis geei*

Capped langur, *P. pileatus*

In Northeast India, three species of langurs occur, viz. phayre's leaf monkey (*Presbytis pharyrei*), capped langur (*Presbytis pileatus*) and Golden langur (*Presbytis geei*). The first two are indigenous and the last one is an exotic species. Phayre's leaf monkey was first reported in India in 1974 from Tripura (Bhattacharjee & Chakravarthy, 1992). Capped langur occurs in Assam, Tripura, Arunachal Pradesh, Nagaland, Manipur and Chittagong Hill tracts. Phayre's leaf monkey is an arboreal animal living in troops of 6 to 13 individuals. The animals live in dense forests and only under special circumstances take to ground.

Earlier phayre's monkey and capped langur were not observed in fruit orchards in villages or those adjacent to forests. Now these langurs are reported to raid crops. These two species of langurs were observed feeding on ripe fruits, green twigs, green leaves, flowers, pods, seeds, fleshy roots of mango, pigeon-pea, *Zizyphus*, *Brassica* spp. (Cabbage), citrus, guava, banana, jackfruit and *Embelica* (Gooseberry). Such feeding resulted in economic losses to some growers locally. Langur feeding was observed throughout the year. But maximum feeding occurred in April-June when alternate foods became scarce in the wild. Natural foods of monkeys included *Careya arborea, Delemix regia, Hebia brazilianensis, Bombax ceiba, Bursera serrata,* etc.

4. The hare, *Lepus nigricollis* (F. Cuvier)

Class : Mammalia

Order : Lagomorpha

Family : Leporidae

Binomial name: *Lepus nigricollis*

Rabbits and Hare: Rabbits, *Oryctolagus cuniculus* elsewhere and their counterpart in India, the Indian hare, *Lepus nigricollis* are capable of causing damage to crops and forests. Hares are nocturnal and herbivorous.

There are no published accounts of hare damage to Indian agriculture although agricultural field workers and farmers are aware of hare damage to sprouting cereals and pulses in fields adjacent to forests. One such observation was by Chakravarthy (1994) in Western Ghats where he found hares damaging rice and cowpea seedlings. They browsed over sprouts, cotyledons or fresh, tender twigs or apical portion of seedlings. Damage was concentrated at the edges and ranged from 2 to 17%. Seedlings of bamboo were damaged upto 50% necessitating replanting with bamboo varieties which were resistant to hare damage. Saplings which were damaged less than 50% regenerated and were covered with thick mat of thorny sticks to deter hares. Sown/

germinating seeds and seedlings were observed to be completely consumed by hares in fields adjacent to woodlands near Bangalore particularly during summer (Sridhara, Pers. Obs.). Ten percent thimet mixed with dry sand and suspended in perforated polythene bags around the fields repelled hares from attacking bamboo seedlings (Chakravarthy, 1994).

In the absence of sufficient published information on hare damage to agriculture and its management in India methods in vogue elsewhere to manage hare or rabbit problem may be helpful to manage hare problem in India. The traditional methods of managing these herbivores are shooting, poisoning with sodium monofloroacetate, warren destruction and exclusion fencing.

The pestiferous activities of rabbit are well documented for UK, Australia and some European countries. Rabbit damage to agriculture is by way of competing with livestock for pasture, especially during drought, damaging crops, forestry and tree plantations, soil erosion, land degradation and affecting water quality. In Scotland considerable damage was seen for crops grown in winter specially wheat, rapeseed and grass amounting to 11 million pounds annually. A survey revealed shooting as the most effective method. Other methods such as snaring, box trapping and ferreting were considered costly and time consuming (Fuchs & Neil, 1998). Fumigation by hydrogen cyanide and phosphine was found to be cost effective (Ross, 1986). The other chemicals used for rabbit management are Thiram®, a potent emetic that can be easily absorbed through skin (Stecher *et al.*, 1968), irritants like capsaicin and bitterants like denatonium benzoate (Andelt *et al.*, 1994) and a commercial repellent which releases sulphur odours and volatile fatty acids (Mason *et al.*, 1999).

The pestiferous activities of rabbit are well documented for United Kingdom (Rees *et al.*, 1985) and Australia (Williams *et al.,* 1995). By introducing the myxomatosis virus, both the countries succeeded in drastically reducing rabbit populations. But within years both the virus and host co-evolved with the virus evolving into virulent forms of different intensities and the rabbit being selected for resistance (Fenner & Ratcliffe, 1965).

An accidental disaster of mass destruction of rabbit farms in 1984 in China (Liu *et al.*, 1984) was detected to be due to the viral haemorrhagic disease (VHD), caused by a member of Calcivirus family, which later came to be known as rabbit Calcivirus. The virus was capable of inducing 90% mortality of rabbits in the laboratory. It has been successful in containing rabbit pest populations since 1995 in Australia and since 1997 in New Zealand (O'Brien & Thomas, 1998).

5. The golden Jackal, *Canis aureus*

Class	: Mammlia
Order	: Artiodactyla
Family	: Canidae
Genus	: *Canis*
Species	: *aureus*

Binomial name: *Canis aureus*

Inhabiting close to towns, villages and cultivated areas at the fringe of forests, jackals are potential depredators of crops. Widely spread throughout India, jackal is an opportunistic feeder. Although they feed on crops such as corn, sugarcane, melons and vegetables, their role in scavenging is far greater and as pests is negligible (Prater, 1986). They have also been observed feeding on cocoa pods and pineapple in Western Ghats and Coastal Karnataka, respectively (Chakravarthy, 1994). Cocoa pods nearest to the ground were attacked more. Partially eaten pods hanging from the cocoa plants is an indication of jackal damage. The average damage was around 4.57%. Chewing and sucking juice was the observed damage to pineapple. But it was negligible. A survey in Bangladesh indicated that they are a nuisance primarily on sugarcane and maize and frequently preyed on poultry (Poche *et al.*, 1987). The study revealed damage to 7 out of 8 varieties of sugarcane. Jackals bit into stalk and chewed one or several joints. There was no selection in attack, neither sugar content nor rind hardness affected the choice. The authors feel that in spite of localized heavy damage, the overall impact of jackal on sugarcane production in Bangladesh is minimal. But to a subsistence farmer, such damage can have severe economic implications. Reports of jackal depredation on crops are also available for Israel (Mendelssohn, 1972).

Although some rodenticides like sodium monofloroacetate and strychnine are used to kill unwanted and nuisance wild candies including jackals, it is not advisable to kill jackals as they play a very important role in the energy flow of the ecosystem by virtue of their scavenging activities. Removal of jackal is likely to increase the population of hare, *Lepus nigricollis.* Such an increase in hare numbers consequent to jackal killing resulted in increased damage to agriculture by hares compared to jackals (Mendelssohn, 1972). The best way to deal with them is to trap them as and when they become nuisance and remove individual animals using steel traps (Poche *et al.,* 1987).

6. The Sloth bear, *Melurus ursinus* and other bears

Class : Mammalia
Order : Artiodactyla
Family : Meluridae
Genus : *Melurus*
Species : *Ursinus*
Binomial name: *Melurus ursinus*

Sloth bear (*Melurus* sp.) is an indigenous bear species of Indian subcontinent with its distribution limited to moist and dry forests of India, Nepal, Bangladesh and Sri Lanka. There are only about 10,000 sloth bears in the world today thus making it one of the prime targets for conservation. It is included in IUCN Red Data Book, 1988 and listed in the CITES Appendix III. However, information on its behavioural ecology in the wild is scanty. Two subspecies occur in the Indian subcontinent i.e. *Melurus ursinus* in India, Nepal, and Bangladesh; *M. ursinus inoratus* in Sri Lanka (Prater, 1986).

Sloth bears live where there is sufficient forest to provide food and favourable places where out cropping of rock and tumbled boulders offer them shelter during the hot weather and rains. They come out shortly before sunset, hunt for food all night and retire in the morning. In cloudy and cool weather, their food mainly consists of fruit and insects. Fruits of banyan, wild figs, mangoes, *jamoon, ber*, etc are preferred. Bears are notorious honey stealers. They raid maize, sugarcane, date palms and toddy crops.

Though fruit is an important food item of sloth bears from April to June, termites are the staple food. Fruits, flowers, grass fibres, honey and insects were reported to be the major food items of sloth bears in Nepal (Laurie & Seidensticker, 1977). Ishwariah (1984) estimated about 18 bears which was equivalent to a crude density of one bear\8.33 km in a 150 km broken belt of rocky hills covered with scrub jungle at Ramanagaram, near Bangalore. The area was originally tropical dry deciduous forest. Reckless felling and changes in forest usage, plantation and scrub jungle replacing original forest trees, encroachment of agriculture upto the foothills not only reduced food availability and range to the bears but also fragmented their habitat due to gradual human settlements. Under these conditions bears in this area have learnt to raid crops grown particularly sugarcane, groundnut and seasonal fruits such as jackfruit and mango so much so that it is projected that 50% of their nutritional requirement is presumed to be derived from agricultural crops. Sloth bears have become major crop pests and a menace as bear-man conflicts are increasing day by day with several deaths of humans occurring due to mauling. On the other hand illegal hunting and shooting, lethal power fencing, poisoning, using explosives in favoured baits by affected farmers is affecting the bear population tremendously.

Black Bear: Although there are no published reports of Black bear damage to agriculture or forestry in India, the species has been studied fairly well because of the economic impact it has on forestry in USA and Canada. *Ursus americanus* commonly forages on Douglass-fir trees during spring. The tree's vascular tissue is bear's staple food. They usually attack trees of 15-30 years of age by removing the bark with their claws and then scrapping the vascular tissue from the heartwood with their incisors. However, only aged trees are vulnerable to bear damage. Such damage is detrimental to the health and growth of timber stand. Complete girdling is lethal while partial girdling reduces growth rates and makes the tree vulnerable to insect and disease infestation. The timber loss is substantial because bears select the most vigorous trees within the most productive stands and often damage is inflicted after implementation of stand improvements like thinning and fertilizer application (Mason & Adams, 1989). The increased attack after fertilization is attributed to dietary selection of vigorous trees by bears (Nelson, 1989). Damage is related to concentration of sugars and terpenes in the vascular tissue of Douglass-fir, the former for energy and the later induces avoidance. Trees with highest sugar : terpene ratio are selected. Apart from Douglas-fir redwoods various species of fir, spruce, cedar and pines, maple, alder, larch, aspen, cottonwood, bitter cherry, willow etc are also damaged.

Management practices include selective hunting, pruning trees which are known to contain decreased vascular tissue and sugar concentrations and are avoided by bears, selection of varieties with less sugar in vascular tissue without sacrificing growth potential and quality of wood, and by providing alternate food in feed containers placed strategically during times of shortage of natural food. These measures are known to reduce bear damage to forest trees in USA (Nolte *et al.,* 2003). Electric fencing in conjunction with other practices like growing unattractive crops adjacent to forests, aversive conditioning, lethal control of nuisance animals, converting monoculture plantations to broad leaf forests are advocated to contain the depredation of Asiatic Black bear, *Ursus thibetanus* in Japan (Mauseth,1999).

Bears were the major vertebrate pest to cause severe damage to bee keeping causing an annual loss of $623,000 in USA. The principal method of damage prevention was by electric fencing (O'Brien & Marsh, 1990).

CONCLUSIONS

Of the six species of unconventional and unusual vertebrate depredators discussed in this chapter, only two are true herbivores namely hare and bison, the rest are omnivores. All resort to crop consumption when there is cultivation adjacent to their natural habitat. The over riding factor that has made them crop consumers is inadequate food resources in their natural habitat consequent to habitat loss and/or degradation. Their foraging on agricultural crops and orchard fruits is sporadic and mostly in seasons of food and water scarcity such as severe summers and periods of drought. Crop depredation, however small, is a serious problem to the subsistence farmer cultivating food for livelihood. The problem can be tackled at two levels-making the habitat of these animals adequate in food sources and secondly repelling and deterring the crop depredators. The former is the responsibility of the forest department. Suggestions include improving forest vegetation, planting suitable fruit bearing trees, growing suitable grasses, taking care to preserve the natural vegetation and water bodies inside the forest. In worst and unavoidable situations when resources are absolutely nil or inadequate provisioning food and water at strategic points within the forest is recommended. Auditory repellents like beating drums, pyrotechnics, and explosives generating noise, fireworks, and visual deterrents like fire, scare crows and olfactory repellents such as thimet are some of the non lethal approaches that can discourage these sporadic crop raiders.

REFERENCES

ALI, S. & S.D. RIPLEY. 1987. COMPACT HANDBOOKP OF THE BIRDS OF INDIA AND PAKISTAN. Oxford University Press, New Delhi, India.

ANDELT, W.F., K.P. BURNHAM & D.L. BAKER. 1994. Effectiveness of capsaicin and bitter repellents for deterring browsing by captive male deer. *J. Wildl. Manage.* 58: 330-334.

BHATTACHARJEE, T.P. & D. CHAKRABORTHY. 1992. Group structure and home range of phayre's langur in Sepahijala Wildlife Sanctuary, Tripura. *Tigerpaper* XIX (3):31-33.

BROOKES, J.E., E. AHMAD & M.H. KHAN. 1989. The agricultural importance of the wild pig, *Sus scrofa* L. in Pakistan, *Tropical Pest Manage.* 35(3): 278-281.

CHAKRAVARTHY, A.K. 1991. Vertebrate Pest Management in cardamom (*Elettaria cardamomum* Maton) and the other crops in Malnad region (ICAR), Annual Report, University of Agricultural Sciences, Bangalore, India. 34 p.

CHAKRAVARTHY, A.K. 1994. *Vertebrate Pest Management (ICAR).* Annual Report, University of Agricultural Sciences, Bangalore, India. 54 p.

FENNER, F & F.N. RATCLIFFE. 1965. MYXOMATOSIS. Cambridge University Press, Cambridge, UK.

FITZWATER, W.D & ISHWAR PRAKASH. 1989. HAND BOOK OF VERTEBRATE PEST CONTROL. Indian Council of Agricultural Research, New Delhi, India.

FUCHS, R.M.E & G.J. NEILL. 1998. A survey of rabbit damage and control measures used in the east and north-east of Scotland. *Proc. 18th Vertebr. Pest Conf.* (Eds.): Baker, R.O & A.C. Crab. University of California, Davis, USA. pp.181-183.

ISHWARAIAH, V. 1984. STATUS SURVEY REPORT AND RECOMMENDATIONS CONSERVATION OF THE SLOTH BEAR IN RAMANAGARAM TALUK, KARNATAKA. WWI, India. 25p.

LAGRANCE, M. 1986. The mechanical control of bush pig (*Potamochoerus porcus*), in Zimbabwe. *Proc. Twelth Vertebr. Pest Conf.* (Ed.) T.P. Salmon. University of California, Davis, California, USA. pp. 215-225.

LAURIE, A & J. SEIDENSTICKER. 1977. Behavioural ecology of sloth bear (*Melursus ursinus*) in Royal Chitawan National Park, Nepal. *J. Zool.*, London. 182: 187-204.

LIU, S.J., H.P. XUE, P.Q. PU & N.H. QUIAN 1984. A new viral disease in rabbits. *Animal Husbandary Veterinary Medicine* 16: 253-255.

MASON, A.C & D.L. ADAMS. 1989. Black bear damage to thinned timber stands in Northwest Montana. *West J. Appl. For.* 4: 10-13.

MASON, J.R., J. HOLLICK., B.A. KIMBALL & J.J. JHONSON. 1999. Repellency of Deer Away Big Game Repellent to Eastern cotton tail rabbits. *J. Wildl. Manag.* 63(1): 309-314.

MAUSETH, J.D. 1999. Anatomical adaptations to xeric conditions in Maihuenia (Cactaceae), a relictual, leaf bearing cactus. *J. Plant. Res.* 112: 1107, 307-315.

MENDELSSOHN, H. 1972. Ecological effects of chemical control of rodents and jackals in Israel. In: THE CARELESS TECHNOLOGY, ECOLOGY AND INTERNATIONAL DEVELOPMENT. pp: 527-548.

NELSON, E.E. 1989. Black bears prefer urea fertilized trees. *West J. Appl. For.* 4: 13-15

NOLTE, D.L., K.K. WAGNER & A. TRENT. 2003. *Timber damage by Black Bear: Approaches to control the problem.* Document 3E32E84-Animal Damage Management, USDA, Forest service, Missoula, MT, USA. 10p.

O'BREIN, J.M & R.E. MARSH. 1990. Vertebrate pest of Beekeeping. *Proc. 1990 Fall Annual conf., Apiary Inspectors of America.* Tuscon, Arizona, USA. pp: 25-29

O'BREIN, P & S. THOMAS. 1998. Rabbit Calcivirus. Updates on a new biological control for pest rabbits in Australia. In: *Proc. 18th Vertebr. pest conf.* (Eds.): Baker, R.O & A.C. Crab. University of California, Davis, USA. pp: 397-405.

POCHE, R.M., S.J. EVANS, P. SULTANA, M.E. HAQUE, R. STERNER & M.A. SIDDIQUE. 1987. Notes on the golden jackal (*Canis aureus*) in Bangladesh. *Mammalia* 51(2): 259-270.

REES, W.A, J. ROSS, D.P. COWAN, A.M. TITTENSOR & R.C. TROUT. 1985. Humane control of rabbits. In: HUMANE CONTROL OF LAND MAMMALS AND BIRDS. Universities Federation for Animal Welfare, Potters, Ber. UK. pp: 96-102

ROSS, J. 1986. Comparison of fumigant gases used for rabbit control in Great Britain. *Proc. 12th vertebra. pest conf.* pp: 153-157.

STECHER, P.G., M. WINDHULZ, D.S. LEAHY, D.M. BOLTON & L.L. EATON. 1968. The Merack index. Merack & Company, Rahway, New Jersey. USA.

WILLIAMS, C.K., I. PARKER, B.J. COMAN, J. BORLEY & M.C. BRAYSHER. 1995. MANAGING VERTEBRATE PESTS: RABBITS. Bureau of Resource Sciences and CSIRO Division of Wildlife and Ecology, Australia Government Publishing Service, Canberra, Australia.

❑❑❑

22

VERTEBRATE PREDATORS OF CROP PESTS

A.K. Chakravarthy, N.E. Thyagaraj and Shakunthala Sridhara*

Department of Entomology, College of Agriculture, University of Agricultural Sciences, G.K.V.K., Bangalore – 560 065
** AINP on Rodent Control*

ABSTRACT

Vertebrates, although promising in certain situations, are seldom used in the biological control of pests, be it insects or rodents. Importance of vertebrates as regulating agents of natural pest population is disputed as most of them shift to alternate hosts or become pests themselves after certain period. Secondly, all vertebrate predators do not always consume all the insects they encounter. Unlike the arial birds and bats, mobility of small mammals is another limiting factor. Despite these limitations, vertebrates definitely contribute to pest suppression in forests, gardens, orchards, cultivated fields and even in urban situations. Their presence is often encouraged by the provision of suitable nesting sites, cover and by the supply of food during inclement weather or during periods of food scarcity. A critical appraisal of pisces, amphibia, reptilia, aves, chiroptera, rodents and small mammals in pest suppression in cultivated ecosystem revealed that predatory activities of many vertebrates in agroecosystems are compatible to some extent with other non-toxic, non-chemical methods of pest suppression. But much research is needed before the full potential and complete range of vertebrate depredations are utilized for crop protection.

INTRODUCTION

Though potential, vertebrates seem to have been used little in the biological control of animal pests. Vertebrates that depend on other animals for food are predators. Compared to invertebrate predators their behaviour is more complex and adaptable. However they readily turn to other sources of prey when one kind becomes scarce (Woods, 1974).

They are in fact poor bioagents when the pest population is low. Nevertheless, birds and mammals are warm blooded animals with high metabolic rates and consequently with high energy demands. This is especially true in small insectivores such as shrews which have high ratios of body surface area to volume and thus efficient predators.

In general, an insight into the history of pest management particularly in the third world reveals that before 1950, the indiscriminate pesticide use was not common and many pests were under natural control. Since 1950 the use of synthetic pesticides came into prominence and the age old environment-friendly practices were relegated to the background. Decades later frequent failures of pest control led to resurrection of biological control.

The importance of vertebrate predators in pest control is best illustrated by the case of the Indian mynah bird, *Acridotheres tristis* Linn. which has the distinction of the first internationally imported beneficial predator. The bird was introduced from India to Mauritius in 1762 for suppression of the sugarcane red locust, *Nomadacris septemfasciata* Serville. The bird effectively checked locust populations for eight years (Coppel & Mertins, 1977). Introduction of Indian Mynah bird to Hawaii in 1865 was also successful (Rao *et al*., 1971). Later it became troublesome by spreading the seeds of the noxious shrub, *Lantana camara*.

Role of vertebrate predators particularly birds in pest suppression has been reviewed by several authors (Sweetman, 1958; Buckner, 1967; Woods, 1974; Coppel & Mertins, 1977; Singh *et al.,* 1990; Zhang Zhi-Qiang, 1992; Verghese & Sriharan, 1993; Parasharya *et al.,* 1994; Rao *et al.,* 1998). The major vertebrate predator belong to the vertebrates classes of Pisces (fish), Amphibia (Salamanders, frogs, toads), Reptilia (Snakes, lizards), Aves (birds) and Mammalia (moles, shrews, bats, mice, skunks, armadillos, ant eaters, cats, mongoose and others). Much observational information has accumulated on predatory behaviour in the field as well as from stomach content examinations. However, in several situations, quantitative data on the effectiveness of the vertebrates used in insect/rodent pest suppression is lacking.

VERTEBRATE PREDATORS OF AGRICULTURAL AND OTHER PESTS

Vertebrate predators of insect and rodent pests are reviewed under the following Vertebrate classes: Fishes, Amphibians and Reptiles, Birds and Mammals (Table 1).

Gambusia affinis Common myna, *Acridotheres tristis*

Toad, *Bufo marinus* Common shrew, *Suncus murinus*

Indian mongoose, *Herpestes edwardsi* Tree shrew, *Tupais glis*

Table 1. Vertebrate predators of pests

Sl. No	VERTEBRATE CLASS	PREDA-TORS	PREY	REMARKS	REFERENCES
1.	FISHES	*Gombusia affinis*		Can predate only on aquatic, semiaquatic prey.	Woods, 1974
		Lebistes reticulatus			Chakravarthy (1983-90, Pers. Obs.)
		(Poecilla reticulatus)	Mosquito Larvae	*Gombusia* likely to prey on invertebrate predators & native fish.	
		Tilapia sp.			
		Cyprinus carpio	Larvae, insects		
		Fish	algae and weeds of paddy	Rice yield increased by 4-10%	Woods, 1974
2.	AMPHIBIA	*Bufo marinus*	*Phyllophage* sp.		Woods, 1974
		Bufo bufo	Rhinoceros beetle		Leonard, 1933
					Rao *et al.,* 1971
			Caterpillars		Woods, 1974
			Cutworms		
			June beetle		
			Mole cricket		
			Cockroach		
			Garden insects		
3.	BIRDS	*Hylochichla ustulata*	Budworms		Buckner, 1967
			Webworm	98% mortality	Coppel & Mertins,1977
		Passer domesticus	Locusts	in summer	
		Sturnus roseus	Nocturnal insects		Ali, 1979
		Nightjaws	Agricultural pests	Introduced to several countries	
		Sparrows, Mynas			Daniels, 1998
		Bulbuls, *Ibis,*	Larvae of		Rao *et al.*, 1971
		Dicrurus adsimilis	spodoptera & Helicoverpa		
		Acridotheres tristis	Orthoptera	In castor, pigeon pea, chick pea, groundnut, lab lab and cow pea	Rao *et al.,* 1998
			Lepidoptera		Chakravarthy, 1988
			Coleoptera		
			Hymenoptera	In guava & cotton	Chouhan *et al.*, 1978
			Aphids	In ragi	
		Primia socialis	Insect pests	In mixed stands of field bean and cabbage. Also diamond moth	Chakravarthy, 1988
		Primia subflara	Borer		
		Crows	Aphids		Verghese & Sriharan, 1993
		Motacella sp.			
		Tyto alba	20 species of small mammals.		Chakravarthy & Lingappa, 1978;
			House rat, lesser bandicoot	71-86% rodents, rest grey musk shrew also fish, parakeet, barbet & shrew	Jayarathnam, 1978.
		Scops owl	gerbils, mole rat,		Kanakasabai & Santhanakrishnan, 1991
		Bubo bubo	shrews, house		
		Athena brama	rat, *Mus* sp.		

				DharmaKumarsinhji, 1939
				Fulk & Khokhar, 1976
				Jain *et al.*, 1982
4a. MAMMALS	Moles, shrews & Mice *Peromyscus* sp.	Insects Cocoons of Sawfly		Coppel & Martins, 1977
		Large sawfly		Warren, 1971
	Sorex cinereus	Musk rats		Woods, 1974
	Mustela vision	Pocket gophers		
	Coyotes	*R.rattus*	In Hawaiian sugarcane	Baldwin *et al.*, 1952
	Herpestes auropunctatus	Weeds		
	Manatee, *Trichechus sp.*	Water vole, rabbits		Woods, 1974
	Mustela erminea	Rabbits		Smith *et al.*, 1995
	Mustela p. putorius	Moths, beetles, borers, bugs, mosquitoes		Walker & Molur, 2003
	Insectivorous bats	Mosquitoes		
	Myotis sp.	Rats and mice		Walker & Molur, 2003
	Megaderma lyra			
4b. PREDATORY ODOURS OF MAMMALS	Natural and synthetic odours of stout, red fox, , bobcat, weasel Mink anal sac secretions. Weasel odours	Prevented damage by snow hare, voles, pocket gophers, deer, bush tail possum, rabbit, hare, woodchuck, meadow voles. Reduced extension of oestrus cycle in rats -do-		Lindgren *et al.*, 1995 Sullivan *et al.*, 1985, 88 Belant *et al.*, 1998 Morgan & Woolhouse, 1995 Swihart, 1991, 95 Voznessenkaya *et al.*, 1942 Koskela *et al.*, 1996

FISHES

Fishes can be employed against only such pests which spend part or all of their lives in water. Their main use has been against larval mosquitoes and chironomids. Fishes can also be employed against simuliids, adults of which are important vectors of disease. Predatory fish can be used against pests of public health importance and against pest insects of paddy crop. The most important order of fishes employed for mosquito control is the cyprinodontiformes with two, useful families, Cyprinodontidae and Poeciliidae exemplified by two species,

Gambusia affinis and *Lebistes reticulatus* (=*Poecilla reticulata*). Larger fishes such as *Tilapia* species and carp, *Cyprinus carpio* have been used as predators for mosquitoes and weeds. There have been a number of successes with mosquito-eating fish in more than one continent. *P. reticulata* in particular was successful against anopheles mosquitoes in Sri Lanka and was also successful in suppressing populations of *Culex pipens fatigans* in fresh and brackish water of South India (Rao *et al.*, 1971).

Objections to the indiscriminate artificial spread of *Gambusia* have been raised by environmental groups who claim that its presence eliminates or threatens the existence of native fish species or other aquatic fauna such as predatory invertebrates (Legner *et al.*, 1974). Introduction of invasive exotic fish species in riverine tracts of hill and coastal regions of Karnataka led to declines in populations of native fish species. This is in fact the situation in Thungabhadra, Cauvery, Hemavathy and Arkavathy river systems in south Karnataka, south India (Pers. Observations, Chakravarthy-1983 to 1990).

Rapid breeding, small adult size, top-feeding habits, preference for mosquito larvae, capacity to penetrate vegetation to reach prey, hardiness and pollution tolerance, are the most desirable qualities of fish for mosquito suppression. Additionally fish in rice fields can increase the yield by 4 to 10% and keep down algal growth, restrain weeds and destroy insect larvae which could be injurious to the crop (Woods, 1974).

AMPHIBIANS AND REPTILES

Although reptiles are carnivorous and predatory, they seem to be of little use for biocontrol. The large monitor lizard, *Varanus indicus* has been introduced into parts of Asia to control rats, but with disappointing results. Several studies have shown that many lizards destroy a large number of insects including pest species. On the other hand amphibia have proved to be more useful practically.

Much of the food of both immature and mature toads consists of insects. In forest areas near water sources, tree frogs (*Hyla* species) feed on adults of pine saw fly (Coppel & Benjamin, 1965). Toads are the only group of amphibians utilized for biological insect pest suppression. The toad family Bufomidae and the genus *Bufo* contain several species of economic importance. The giant American toad, *Bufo marinus* has a successful record of insect control in different regions of the world. It has been particularly useful in sugarcane fields where it attacks *Phyllophaga* species and the rhinoceros beetle. Gardeners often introduce the much smaller toad, *Bufo bufo* into their gardens, where they effectively suppress insect pests (Woods, 1974). Earlier it helped in the reduction of swarming caterpillars and cutworms in Fiji (Rao *et al.,* 1971). Still earlier it was introduced in 1920 in Puerto Rico for suppression of the June beetles, mole cricket and cockroach (Leonard, 1933). Stomach analysis of toads collected in sugarcane growing areas showed that the diet consisted of 51% of insect pests of agriculture. However, changes in agricultural practices, drainage, insecticides, etc are making survival of populations of *B. marinus* difficult. Generally frog and toad populations are declining as their habitats are deteriorating. Road traffic is becoming a major mortality factor of amphibians particularly in India and other Asian countries (Pers. Observations, Chakravarthy, 1980-1990)

BIRDS

Habitats of birds and insects are closely interrelated and the advantages of flight place birds parallel to insects. Many birds have a high rate of insect intake with entomophagy as an adaptive strategy in birds with respect to foraging time and place (Morse, 1971).

In India as elsewhere, the problem of birds as crop pests has often been amplified since the damage done easily translates into monetary terms. However, there has not been much success in quantitatively assessing the contribution of birds in terms of pest control and pollination in agriculture. For instance, the Rosy pastor *Sturnus roseus* which is a major pest of sorghum in winter is also a destroyer of locusts in its summer breeding grounds (Ali, 1979). Unfortunately we are not in a position to compare these opposing aspects in a quantitative sense. Studies are currently underway in many parts of the country to assess the beneficial role of birds in agriculture, especially under the All India Coordinated Research Project on Agricultural Ornithology funded by the Indian Council of Agricultural Research, New Delhi.

A number of birds have been reported as insect and rodent predators in agricultural fields. Owls have played a significant role in controlling rodents at night. Nightjars are useful nocturnal insect eaters. Many species of smaller birds including bee-eaters, rollers, kingfishers, swifts, swallows, drongos, shrikes, babblers, warblers and even crows and starlings destroy insects of the agricultural fields considerably. In general, in India with the exception of pigeons, parakeets and barbets, all birds consume one form or the other of insect food during their life time. Most birds are voracious consumers of termites. The usefulness of sparrows and mynas in destroying garden and field insects led to their introduction into many other parts of the world (Daniels, 1998).

Among insectivorous birds, cattle egret (*Bubulcus ibis*), black drongo (*Dicrurus adsimilis*) and common myna (*Acridotheres tristis*) are important natural enemies of *Spodoptera* and *Helicoverpa* larvae. *Spodoptera* larvae which hide in the soil and leaf litter under plants during day time get exposed to birds during irrigation. Cattle egrets get their nutritional requirements completely from crop fields where insecticide use is nil or minimal.

The well known examples of bird predation on insect pests of agriculture include consumption of aphids, borers and other injurious insects by several bird species. Aphids are effectively predated on by Ashy-wren-warbler, *Prime socialism* in guava and cotton and by wagtail, *Motacilla* species in field bean (Chakravarthy, 1988) (Chakravarthy & Lingappa, 1978). Almost all species of pod borers were fed upon by birds. Drongo fed on the exposed larvae of pod borer, *Helicoverpa armigera* while crows were the most important predators of the borer, *Adisura atkinsoni* in mixed stands of field bean (Chakravarthy, 1988). Other reported cases of bird predation include consumption of injurious insects of ragi (*Eleusine coracana*) by the fledglings of Indian Wren-Warbler, *Primia subflara* (Verghese & Sriharan, 1993), predation on *Brevicoryme brassicae* in cabbage and on diamond moth, *Plutella maculipennis* by wagtails of *Motacilla* species (Jayarathnam, 1978).

Indian myna, *Acridotheres tristis* is the most common and familiar bird throughout India. Its importance as a successful vertebrate predator of insect pests has been well documented (Toor & Ramzan, 1974; Sengupta, 1976; Narang & Lamba, 1984; Chauhan *et al.,* 1998). Nestlings of myna were predominantly carnivorous with Orthoptera, Lepidoptera, Coleoptera and Hymenoptera forming 52 to 63 % of the diet (Chauhan *et al.*, 1998). Myna's preference for insect matter has been well exploited in several parts of India. Farmers attract mynas by perches, yellow or 'coloured rice balls' in the fields of castor, pigeon pea, chickpea, groundnut, lab-lab and cowpea. They also encourage bird predation on insects by not applying insecticides in crop fields (Chakravarthy, 1988).

Barn owl, *Tyto alba* feeds on about 20 small mammals throughout the world. The species is widely distributed and locally abundant in rural areas in India. Of the sixty species of owls occurring in India 13 feed on rodents in South India alone (Ali & Ripley, 1983). These are the barn owl (*Tyto alba*), grass owl (*Tyto capensis longimembris*), peninsular scops owl (*Otus scops rufipennisi*), Indian great horned or eagle owl (*Bubo bubo bengalensis*), forest eagle owl (*Bubo nipalensis nipalenis)*, dusky horned owl (*Bubo coromandes coromandes*), barred jungle owlet (*Glaucidium radiatum radiatum*), Malabar jungle owlet (*Glaucidium radiatum malabaricum*), Indian brown hawk owl (*Ninux scutulata lugabris*), South Indian brown hawk owl (*Ninux scutulata hirsuta*), Southern spotted owlet (*Athena brama brama*), Southern mottled wood owl (*Strix ocellata ocellata*) and the brown wood owl (*Strix leptogrammica indranee*).

Based on castings, the diet of *Bubo bubo bengalensis* was found to be composed of gerbils, painted patridge, common patridge and egrets (Dharamkumarsinhji, 1956). The scops owl (*Otus sp*) was found to feed on fish, parakeet, barbet, shrew, house rat and lesser bandicoot rat in captivity (Kajuria, 1970). Pellet analysis of *Bubo bubo* in Pakistan revealed *Tatera indica* (49.1%) as the major prey followed by *Nesokia indica* (31.6%) *Suncus sp., Gerbillus nanus* and *Rattus rattus* (Fulk & Khokhar, 1976). *Athena brama* was found to selectively feed on *Mus sp.* in Rajasthan although they were least abundant prey. This selective feeding was attributed to the smaller size of the prey and avoidance of the predator by larger prey. The owl shifted its diet from insects to rodents in winter as a preparation for breeding (Jain *et al.,* 1982).

The diet of the barn owl, *Tyto alba* in Tamil Nadu consisted of 71-86% rats and mice, the remaining diet being composed of grey musk shrew (Neelanarayanan *et al*., 1998). A single pair of barn owls and 6 chicks during the 3 month nesting period is known to consume 1000 rats. Later pellet analysis of barn owl showed *Bandicota bengalensis* forming 49.1% of faeces followed by *Rattus meltada* (23.1%), *R. rattus* (17.8%), the shrew, *Suncus murinus* (7.6%) and *Mus booduga* (2.1%). Introduction of owls from Tamil Nadu to Andaman and Nicobar islands was halted due to the apprehension that the introduced species may replace the native owl species.

MAMMALS

Predatory mammals require considerable energy, both to maintain body temperature and activities (Buckner, 1967). This energy is reflected in high potential of prey attack. Compared to birds the predation of pests by small mammals is less spectacular due to several factors. Small mammals do not always consume the insect prey they encounter. Some store for later consumption. Comparatively small mammals are not very mobile. Thus immediate numerical responses are unlikely and in fact, rarely observed. They also do not migrate with the seasons, as do many birds nor move rapidly from the area when a particular prey becomes scarce. Functional responses are difficult to assess because vertebrates take a wider range of foods than most invertebrate predators.

Small mammals were one of the important factors regulating the European spruce sawfly, *Diprion hercyniae* (Neilson & Morris, 1964). Moles and shrews (Insectivora) and mice (Rodentia) are known to utilize a large numbers of insects in their diets. Deer mice of *Peromyscus* species were the most important cocoon predators of Pine sawfly (Coppel & Mertins, 1977). As mice are capable of climbing trees, they feed on larvae in cocoons spun on needles and branches as well as those on the ground. Moles and shrews do not have arboreal habits and thus are restricted to terrestrial insects. Shrews depend completely on animal diet and their importance as natural bio control agents is increasingly realized. The masked shrew, *Sorex cinereus* is an effective predator of the larch sawfly, *Piconema erichsonii*, an important defoliator of *Larix* species. Shrews became an important predator on other forest insect pests too (Warren 1971). However, insectivorous mammals, other than bats are rarely observed in cultivated ecosystems. Consequently their value in the regulation of pests is underestimated.

Larger vertebrates are potential control agents for vertebrate pests. However, many consider that far from reducing their prey populations, the predators maintain them at higher densities and remove only the surplus populations (Howard, 1917). Mink (*Mustela vison*) preyed mainly on those muskrats (*Ondatra zibethica*) which had not been able to establish home ranges (Woods, 1974). Similar observations on other vertebrate predator-prey systems have been documented. Pocket gopher numbers bear no relationship to the density of their chief predator, the coyote.

Mongooses were released in a number of islands during the nineteenth century to control rats in agricultural areas. In Hawaii, the small Indian mongoose, *Herpestes auropunctatus* was introduced in 1880 to reduce the rat, *Rattus rattus,* a serious pest on Hawaiian sugarcane. Although they have not eliminated rats, they are still a major predator on rats in these islands (Baldwin *et al.*, 1952). Similarly in Puerto Rico, the mongoose *Mungos birmanticus* reduced the numbers of a subterranean lizard, *Ameira exsul,* but the later's prey, *Phyllophaga* larvae, became a pest. This situation was balanced later by the introduction of the giant toad, *Bufo marinus* (Leonard, 1933). The World Health Organisation (WHO) is investigating the possibility of using Japanese weasel, *Mustela sibricica itatsi* to control rats, *R. rattus* and *R. norvegicus* in the pacific islands. As weasels alone may not have sufficient impact on the rat populations, their use is being integrated with the application of rodenticide. The use of the large aquatic mammal, the manatee (*Trichechus*) for weed control has also been attempted (Woods, 1974).

Bats (Chiroptera) are considered important beneficial predators of crepuscular and nocturnal flying insects. Insectivorous bats consume many types of insects including crop pests, such as moths, beetles, borers, bugs and even mosquitoes. Research on insect consumption by bats has shown that *Tadarida brasiliensis,* the Mexican free-tailed bat can consume more than half its weight of insects every night. Colonies of insectivorous bats, often numbering in millions, are estimated to consume ten tonnes per million bats on a night basis. This amounts to 6-13 thousand tonnes of insects every summer. Rates of insect consumption are also known from studies on *Myotis lucifugus,* the little brown bat, which can eat insects equal to its own body weight per night. With a fast digestive rate and numerous foraging bouts per night, these bats consume considerable number of insects. Faeces of most *Myotis* bats contained remnants of mosquitoes (Walker & Molur, 2003).

Sinha (1986) found *Megaderma lyra* consuming rats and mice from godowns, crops fields and houses. The size of *Megaderma* colony varies from 25 to 340 individuals. Farmers in Bihar call the bat "Goddess Laxmi," the goddess of wealth and protect colonies of these bats. He observed these bats selecting amphibians, lizards, small mammals and birds and large insects like locusts, grasshoppers, beetles etc as food. The insectivorous bats have been exploited for control of insect pests in Europe by providing concrete shelters and bait boxes.

In the Netherlands, the water vole, *Arvicola terreseris terrestris* became nuisance in a large number of plantations of a temperate fruit on the island of Terchelling. A few stoats, *Mustela erminea* and weasels, *Mustela nivealis* were introduced. The weasels vanished but the stoats multiplied, reduced water vole and rabbit populations and then became pests in their own right! Similar unforeseen non target predation leading to ecological imbalance was observed elsewhere. Mangoose in Hawaii in course of time started predating on local endangered bird and small mammal fauna. The fox, *Vulpes vulpes* was introduced as a species for pleasure hunting in Australia and to control rabbits but became an important nuisance itself. Yet another case of an introduced mammalian predator to control a native pest becoming a nuisance is the introduction of European polecat, *Mustela putorius putorius* and stoat, *Mustela erminae* in New Zealand. Brought in 1880s to control rabbits in the pastures of New Zealand, they spread to forests and reached the status of pest by 1900 (Lavers & Clapperton, 1990). Although they preyed largely on lagomorphs, analysis of diet revealed them to be "opportunistic general predators" (Smith *et al.*, 1995), containing eighteen ground nesting birds and ducklings in addition to rabbits. Along with feral cats and rats they were responsible for the decline of threatened and endangered black stilts, *Himantopus noraezelandiae* (Murray, 1992) and yellow eyed penguins, *Megadyptes antipodes* (Darby & Seddon, 1990). Recent studies indicate that when 99% rabbits are eliminated, ferrets enlarged their home range and switched their diet to lizards, invertebrates and birds (Norbury & Murphy, 1996; Norbury *et al.*, 1998). Similarly *Mustela putorius furo* introduced as a pet in USA devastated populations of endangered bird species such as California clapper rail, California Least turn and snowy plover (Theylander, 1994). The threat to wild life is so severe that attempts to legalise the ferret are met with strong opposition (Moore & Whissen, 1998). In fact the Australian government has developed a risk assessment procedure to evaluate the risks and benefits of introducing exotic predators (Bomford, 1991). For this reason and because of negative results of past introductions, there have been few recent cases of the use of larger predators as bio control agents of small mammal pests.

Predator odours are also effective in managing pest populations but without killing. In the first line of research both synthetic and natural predatory odours are shown to prevent wildlife damage to forestry and agricultural crops. The urine, faeces, anal and scent glands of several predator species excrete pheromones of functional significance known as "keiromones" (Inter-specific chemical signals). Prey species perceive the odour as presence of predators, somewhere around. This perception is thought to elicit escape or avoidance response due to innate fear of predators. The cue for predator identification is the presence of sulphur compounds, the by product of meat digestion in the urine (Nolte *et al.*, 1994). Areas sprayed with predator odours repel herbivores/crop pests forcing them to alternate, safe habitats. Successful examples of predator odour deterring herbivory include urine of stoat and red fox to prevent snow hare damage to forest seedlings and inducing avoidance response in voles and northern pocket gophers in field conditions (Lindgren *et al.*, 1995), coyote urine deterring deer feeding on woody plants (Sullivan *et al.*, 1985) and food sources (Belant *et al.*, 1998), a predator urine based formulation protecting seedlings of *Pinus radianta* from browsing by Australian bush tail possum, European rabbit and the brown hare (Morgan & Woolhouse, 1995), apple trees being protected from woodchuck gnawing using urine of bobcat, (Swihart 1991), compounds in the urine of short tailed weasel and red fox protecting apple trees from voles (Sullivan *et al.*, 1988) and reduction in gnawing of apple trees by woodchucks and meadow voles by using urine of red fox and coyote (Swihart *et al.*, 1995).

Another indirect way of predation is inhibition of various stages of reproduction in rats by exposing them to predator urines. There is laboratory evidence for extension of oestrous cycle by mink anal sac secretions (Voznessenkaya *et al.*, 1992) in Norway rats and similar effect by weasel (*Mustela nivalis*) odours in bank voles, *Cleithrionomys glareolus* (Koskela *et al.*, 1996).

MANAGEMENT OF VERTEBRATE PREDATORS

Management here refers to utilisation of beneficial activities of vertebrate predators in agroecosystems, their conservation and sustenance over a long period. While published information is mainly available on birds, the information on other predators is meagre. In general, need based insecticidal applications, maintaining habitat-heterogeneity, buffer zone, multiple cropping systems, attracting beneficial vertebrates by planting/supplementing their natural foods in the cultivated patches, providing/creating seasonal or year-round water source and certain organic farming practices can attract and hold vertebrate predators for longer period.

In case of birds and bats, provision of nesting sites, resting/perching sites, landing stakes in fields and planting of broad-leaved bushes like *Acalypha* attract and facilitate predation by insectivorous birds. Many insectivorous birds like mynas, egrets, Indian roller, etc. follow the plough to pick-up the insect larvae or grubs thus exposed. Bird perches @ 8 to 10 per acre in pigeon pea, chick pea, cowpea, tomato and other crops increase bird predation on injurious insects (Chakravarthy, personal observations, 1980-90). In India there is an imperative need to shift the outlook of biological control from the existing narrow spectrum of insect parasitoids and predators to a broader spectrum embracing other invertebrates and vertebrate predators in agro ecosystems.

INTEGRATED MANAGEMENT OF VERTEBRATE PREDATORS

There are 46 beneficial species of birds and 28 species damaging crops as well as feeding on insects. For orchard crops, use of vertebrate predators can be combined with bioacoustics and botanical repellents. On millets like sorghum and pearl millet, use of neem products can be combined with predation by vertebrates. In and around plantations and agricultural fields, planting of manila tamarind (*Pithecalobium dulce*), Flame of the forest (*Butea monosperma*), Toothbrush (*Salvadora persica*), Mulberry (*Morus alba*), *Ficus religiosa* etc. would attract insectivorous birds and bats which in turn can predate on insect pests.

Vertebrate predation can be combined with other conventional and traditional methods of crop protection. Carnivores can control mammalian pests for long periods, but only after pest numbers have been reduced by other means. It can be prolonged drought, disease or human intervention. When predators become nuisance and need to be controlled, integrated pest management is suggested to avoid resurgence of prey pests. Cultivation practices also go well with vertebrate predation. Agricultural systems like sustainable agriculture and organic farming encourage useful activities of vertebrate predators.

CONCLUSIONS

Amongst vertebrates, birds, bats and shrews have been documented as successful bio agents of pests in cultivated ecosystems. Birds and bats consume large numbers of pests and can have considerable impact on localized outbreaks. Fishes, frogs, and toads have proved useful both in aquatic and terrestrial ecosystems. Beneficial activities of vertebrates can be augmented in sustainable, organic farming systems which restrict use of chemicals. As such predatory activities of vertebrates can be well combined with non-chemical and traditional methods of pest suppression. Integrated management practices for conservation of vertebrate predators or their useful activities will go a long way in reducing pesticides for insect and small mammal pest management. However careful and long term experiments are required before a vertebrate species is introduced for biological control in a new area.

REFERENCES

ALI, S. 1979. THE BOOK OF INDIAN BIRDS. Bombay Natural History Society, Bombay, India.

ALI, S. & S.D.RIPLEY. 1983. HANDBOOK OF THE BIRDS OF INDIA AND PAKISTAN (compact edition). Oxford University Press, Bombay, India. 763 p.

BALDWIN, P.H., C.W. SCHARTZ & E.R. SCHWARTZ. 1952. Life history and economic status of the mongoose in Hawaii. *J. Mammal*. 33:355-356.

BELANT, J.L., T.W. SEASMANS & L.A. TYSON. 1998. Predator urines as chemical barriers to white tailed deer. *Proc. 18th. Vertebr. Pest Conf.*

(Eds.) R.O. Baker & A.C. Crab. University of California, Davis, USA. pp.359-362.

BOMFORD, M. 1991. *Importing and keeping exotic vertebrates in Australia: criteria for the assessment of risk.* Bulletin No.12. Department of Primary Industries. Australian Government, Canberra, Bureau of Rural resources, Australia. 92 p.

BUCKNER, C.H 1967. Avian and mammalian predators of forest insects. *Entomophaga* 12:491-501.

CHAKRAVARTHY, A.K & S. LINGAPPA. 1978. Wagtails as predators of field bean aphid. *J. Bombay nat. Hist. Soc.* 54: 457-459.

CHAKRAVARTHY, A.K 1988. Bird predators of pod borers of field bean (*Lablab niger*) *Trop. Pest Manage.* 34(4): 395-398.

CHAUHAN, R.B., B.M. PARASHARYA & D.N. YADAV. 1998. The food of nestlings of Indian Mynah 138-148. In: BIRDS IN AGRICULTURAL ECOSYSTEM (Eds.): Dhindsa, M.S., P.S. Rao & B.M. Parasharya. Society for Applied Ornithology and Indian Council of Agricultural Research, New Delhi, India.

COPPEL, H.C & D.M. BENJAMIN. 1965. Bionomics of the Nearctic pine-feeding diprionids. *Ann. Rev. Ent.* 10: 69-96.

COPPEL, H.C & J.W. MERTINS. 1977. BIOLOGICAL INSECT PEST SUPPRESSION. Springer-Verlag. Berlin, Germany. 314 p.

DANIELS, R.J.R. 1998. Conservation of Birds in the Indian Agricultural context. In: BIRDS IN AGRICULTURAL ECOSYSTEM (Eds.): Dhindsa, M.S., P.S. Rao & B.M. Parasharya. Society for Applied Ornithology and Indian Council of Agricultural Research, New Delhi, India. pp. 166-171.

DARBY, J.T. & P.J. SEDDON. 1990. Breeding biology of the yellow-eyed penguins (*Megadyptes antipodes*). In: PENGUIN BIOLOGY (Eds.) L.S. Davis & J.T. Derby. Academic Press, San Diego, USA. pp.42-45.

DHARAMKUMARSINHJI 1956. BIRDS OF SAURASHTRA (India), Dilbahar, Saurashtra, India.

FULK, G.W. & A.R. KHOKHAR. 1976. Short-tailed mole rat (*Nesokia indica*), prey of the great-horned owl (*Bubo bubo*). *Pakist. J. Zool.* 8: 97.

HOWARD, L.O. 1917. The practical use of the insect enemies of injurious insects. Year book, U.S. Dept. Agri, pp: 273-288.

JAYARATHNAM, M. 1978. *Population dynamics of diamond back moth, Plutella xylossella.* Ph.D. thesis, University of Agricultural Sciences, Bangalore, India.

JAIN, A.P., R. ADVANI & I. PRAKASH. 1982. Winter food of spotted owlet, *Athena brama indica. J. Bombay nat. Hist. Soc.* 80:415-16.

KHAJURIA, H. 1970. A new leaf-nosed bat from Central India. *Mammalia* 34: 622-627

KOSKELA, E., T. HORNE., T. MAPPES & H. YLONEN. 1996. Does risk of small mustelid predation affect the oestrous cycle in the bank vole, *Clethrionomys glareolus? Anim. Behav.* 51: 1159-1163.

LAVERS, R.B. & B.K. CLAPPERTON. 1990. Ferret. In: THE HANDBOOK OF NEW ZEALAND ANIMALS. Oxford University Press, Auckland, New Zealand. Pp.320-330.

LEGNER, E.F., R.D. SJOGREN & J.M. HALL. 1974. The Biological control of medically important arthropods. *CRC Critical Rev. Environ. Contr.*, 4:85-113.

LEONARD, M.D. 1933. Notes on the giant toad, *Bufo marinus* (L) in Puerto Rico. *J. Econ. Ent.* 26:67-72.

LINDGREN, P.M.F., T.P. SULLIVAN & D.R. CRUMP. 1995. Review of synthetic odour repellents for wildlife management in the Pacific Northwest. In: REPELLENTS IN WILDLIFE MANAGEMENT. (Ed.) J.R. Mason. National Wildlife research Centre, Fort Collins, Colorado, USA. pp.217-230.

MORSE, D.H. 1971. The insectivorous bird as an adaptive strategy. *Ann. Rev. Ecol. System.* 2:177-200.

MOORE, T.G. & D.A. WHISSON. 1998. Potential risk associated with the legalisation of exotic predators such as the ferret (*Mustela putorious furo*) in California. *Proc. 18th Vertebr. Pest Conf.* (Eds.) R.O. Baker & A.C. Crab. University of California, Davis, USA. pp. 96-99.

MURRAY, D. 1992. Mackenzie Basin black stilt predator control. In: *Proc. Natl. Predator Manage. Workshop.* (Eds.) D. Vitch, M. Fitzerald, J. Innes & E. Murphy. Threatened species. Occasional Publication No.3. Craigieburn, Canterbury. pp. 52-53.

NARANG, M.L. & B.S. LAMBA. 1984. A contribution to the food habits of some Indian Mynas (Aves). *Rec. Zool. Surv. India.* Occasional paper No.44:1-76.

NEELANARAYANAN, P., R. NAGARAJAN & R. KANAKASABAI. 1998. Studying diet of Barn owl, *Tyto alba* by pellet analysis In: BIRDS IN AGRICULTURAL ECOSYSTEM. (Eds.) Dhindsa, M.S., P.S. Rao & Parasharya, B.M. Society for applied ornithology & Indian Council of Agricultural Research, New Delhi, India.

NEILSON, M.M & R.F. MORRIS. 1964. The regulation of European spruce sawfly numbers in the Maritime provinces of Canada. *Can. Entomologist* 96: 773-784.

NOLTE, D.L., J.R. MASON, G. EPPLE., E. ARONOV & D.L. CAMPBELL. 1994. Why are predator urines aversive to prey? *J. Chem. Ecol.* 20:1505-1516.

NORBURY, G. & E. MURPHY. 1996. Understanding the implications of Rabbit Calcivirus disease for the predaror/prey interactions in New Zealand. A review. Landcare Research Contract Report lC9596/61. 28 p, Auckland, New Zealand

NORBURY, G., D.C.NORBURY & HEYWARD. 1998. Behavioural responses of two predator species to sudden declines in primary prey. *J. Wildl. Manage*. 62(1) : 45-58.

PARASHARYA, B.M., J.F. DODIA, K.L. MATHEW & D.N. YADAV 1994. Natural regulation of white grub by birds in agro ecosystem. *J. Bio Sci.* 19: 381-389.

RAO,V.P., M.A. GHANI., T. SANKARAN & K.C. MATHUR 1971. *A review of the biological control of insects and other pests in South-East Asia and the Pacific Region*. Tech. Bull: 6, Common Wealth, Inst. Biol. Cont. Bangalore, India.

RAO, RANGA., J.A. WIGHTMAN & V. RAMESHWAR RAO 1998. Birds as insect control agents in the groundnut crop. In: BIRDS IN AGRICULTURAL ECOSYSTEM (Eds.): Dhindsa, M.S., P.S. Rao & B.M. Parasharya. Society for applied ornithology and Indian Council of Agricultural Research, New Delhi, India. pp.196.

SENGUPTA, S. 1976. Food and feeding ecology of the common Myna, *Acridotheres tristis* (L.) *Proc. Indian. Nat. Sci. Acad.* 42: 338-345.

SINGH, J., A.S. SIDHU., M.S. DHINDSA & H.K. SAINI 1990. The effect of fodder cuttings on larvae population, diseases incidence and bird predation of *Helicoverpa armigera* in berseem. *Ann. Biol.* 6: 153-159.

SINHA, Y.P. 1986. Bats of Bihar: Taxonomy and field ecology. *Rec. Zoological Survey of India*. Misc. Publ.77: 1-60.

SMITH, G.P., J.R. RAGG., H. MOLLER & K.A. WALDRUP. 1995. Diet of feral ferrets (*Mustelo furo*) from pastoral habitats in Otago and Southland, New Zealand. *New Zealand J.Zool.* 22:363-369.

SULLIVAN, T.P., L.O. NORDSTROM & D.S. SULLIVAN. 1985. The use of predator odours as repellents to reduce feeding damage by herbivores. II Black tailed deer (*Odocoileus hemonius columbianus*). *J. Chem. Ecol.* 11:921-935.

SULLIVAN, T.P., D.S. SULLIVAN, D.R. CRUMP, H. WEISER & E.A. DIXON. 1988. Predator odours and their potential role in managing pest rodents and rabbits. *Proc. Vertebr. Pest Conf.* 13:145-150.

SWIHART, R.K., J.J.PIGNATELLO & M.J.I. MATTINA. 1991. Aversive responses of white tailed deer, *Odocoileus virginianus* to predator urines. *J. Chem. Ecol.* 17: 767-777.

SWIHART, R.K., M.J.I. MATTINA & J.J. PIGNATELLO. 1995. Repellency of predator urine to woodchucks and meadow voles. In: REPELLENTS IN WILD LIFE MANAGEMENT. (Ed.) R.J. Mason. National Wildlife Research Centre, Fort Collins, Colorado, USA. pp.271-289.

SWEETMAN, H.L. 1958. THE PRINCIPLES OF BIOLOGICAL CONTROL. WM. C. Brown, Dubuque, Iowa, USA.

THEYLANDER, C.G. 1994. LIFE ON THE EDGE: A GUIDE TO CALFORNIA'S ENDANGERED NATURAL RESOURCES: WILDLIFE. SANTA CRUZ, CALIFORNIA. Biosystems Books, Berkley, USA.

TOOR, H.S & M. RAMZAN 1974. Seasonal food and feeding habits of the birds of the Punjab. *J. Res. Punjab Agric. Univ.* 11:191-196.

WALKER, S & S. MOLUR 2003. Summary of the Status of South Asian Chiroptera, *Zoo Print, XVIII.* No. 7: 5-8.

WARREN, G.L 1971. Introduction of the masked shrew to improve control of forest insects in Newfound land. *Proc. Tall Timbers Cont. Ecol. Animal Contr.* Habitat Management. 2:185-202.

WOODS, A. 1974. PEST CONTROL: A SURVEY. McGraw Hill CO., London.

VERGHESE, A & T.P. SRIHARAN. 1993. Insectivorous Birds and their use as Biological control agents. In: BIRD CONSERVATION (Eds.): Verghese, A., S. Sridhar & A.K. Chakravarthy. Ornithological Society of India, Bangalore, India. PP. 207-211.

ZHANG, ZHI-QIANG, 1992. The use of beneficial birds for biological pest control in China. *Biological News and Information.* 13(1): 11-16.

VOZNESSENSKAYA, V.V., C.J. WYSOCKI & E.P. ZINKEVICH. 1992. Regulation of rat oestrus cycle by predator odors: role of the vomeronasal organ. In: CHEMICAL SIGNALS IN VERTEBRATES (Eds.): Doty, R.L & D. Muller-Schwartze. Plenum Press, New York. pp. 281-284.

❑❑❑

23

TRADITIONAL AND UNCONVENTIONAL METHODS OF MANAGING VERTEBRATE PESTS

A.K. Chakravarthy and K. Srihari

Department of Entomology, University of Agricultural Sciences, G.K.V.K Campus, Bangalore – 560 065, Karnataka
E-mail: Chakravarthyakshay@yahoo.com

ABSTRACT

Effective traditional crop protection methods can facilitate sustainable agriculture. They can combine well with unconventional methods and can form important tools for Integrated Vertebrate Pest Management strategies. Traditional methods are often cheap, eco-friendly, practical and generally do not result in vertebrate mortality, thereby preserving ecological roles of the concerned species. Wrapping leaves around fruits and maize cobs, timely harvest, clean cultivation, covering portions of canopy with sprigs of foliage, trunk banding with locally available materials, wooden/snap trapping, trap/buffer cropping, etc. are proven traditional tools to reduce vertebrate damage to crops. Traditional methods, however, are not a panacea for all problems posed by vertebrate pests. Judicious use of chemicals and fabricated materials can facilitate more effective vertebrate pest management. A number of newer materials like fumigant poisons, second generation anticoagulants, sticky boards, electrocution traps, cellulose-based lethal feeds, plant based materials, solar-operated electric fence, UV light, strobe lights, etc. are being employed for reducing damage caused by vertebrate herbivores. In view of the scattered information there is a need to evaluate traditional crop protection methods against vertebrate pests, to identify the lacunae and priorities in addition to evolve policies for such management as most of the vertebrate pests also have conservation status.

INTRODUCTION

Cultivated crops are prone to heavy losses, sometimes almost complete crop failure as a result of feeding by vertebrate herbivores. In such cases every part of the plant is susceptible to specialized vertebrate herbivores. Historically, all methods of crop protection against vertebrates have been traditional or non-conventional. Conventional or traditional crop protection methods are adopted in traditional agricultural systems where no serious efforts are made to protect crops *per se* but maintain an 'Ecological Balance'. These traditional systems have emerged over centuries of cultural and biological evolution and represent the accumulated experiences of local farmers interacting with nature in situations or places where there was no access to external inputs, capital or modern scientific knowledge. Non-conventional crop protection methods are often deployed in modern agricultural systems where strict quality protocols and productivity are based on equipments and chemicals.

The social and economic pressures for development especially in the tropics are on the rise. The development schemes are often incompatible with environmental concerns, resulting in degradation and deforestation. It is from these situations that many vertebrates encroach into cultivated ecosystems, cause crop losses and come in conflict with human interests. There is an urgent need to evolve research-based management practices that will protect crops and yet conserve vertebrate crop raiders, their ecological functions and natural resources. Evolving practical management practices for vertebrate pests is becoming increasingly complex. The problems posed by vertebrates are often unique and ingenious methods for management usually involving locally available materials are desired (Chakravarthy & Srihari, 2000). It is more difficult to protect crops from vertebrate herbivores than invertebrates on account of their conservation status. The feeding habits of vertebrates vary widely and are area or location specific. This greatly affects the efficacy and efficiency of pest management tactic. So much so any level of standardization of a strategy would be partly effective. It is, therefore, quite imperative that a holistic approach is evolved. The information on the role, validity, characteristics of traditional and non-conventional (or both) methods of crop protection from vertebrate depredation under Indian conditions is reviewed in this chapter.

TRADITIONAL METHODS

Traditional approaches to crop protection for vertebrates are mentioned by many workers (Gee, 1951; Fitzwater & Prakash, 1973).

Many of these traditional crop protection approaches to vertebrate pests have not been evaluated. Replicated, multi location field trials are urgently needed to confirm the efficacy of certain promising traditional practices in an era of organic farming and sustainable agriculture of twenty first century. There are several successful, indigenous examples of vertebrate pest management practiced by farmers. These include: wrapping walnut-kernels in poplar leaves to protect them from rats, birds and insects (Raze Amber, 2000), placing wild onion (*Urginea indica*) around mulberry to repel rats (Orlob, 1973) and tying spiny branches of *Acacia nilotica* around the stem of mahua (*Madhuka indica*) to keep away troupes of monkeys (Nene *et al.*, 2002). Some farmers of Saurashtra have an interesting method of deterring rodent damage. They catch/trap one or two rats. A little quantity of 'Sindur' (vermillion) is mixed either in water or milk. A mixture of sindur and water or milk is poured on rats caught in the trap/cage. These 'coloured rats' on being released in fields where rodent damage to crops is observed are considered to repel pest rodents (Anon, 1996). Rapid deforestation and depletion of food in forests of Madhya Pradesh have caused *Sus scrofa* to become a major rice pest. Farmers in the village are now growing dwarf rice cultivars with long auricles such as *Kranti* and *Madhuri* to lessen damage to rice by wild boar (Chakravarthy, 1993; Thomas & Naidu, 1995).

NON-CONVENTIONAL METHODS

Increase in the commercial value for the cultivated crops in India and abroad has compelled growers to increasingly raise mono crops since early sixties of the twentieth century. Crop protection measures using chemicals against invertebrates and vertebrates have formed an integral part of the intensive cultivation of crops. However, these crop-protection measures (non-conventional) have resulted in several undesirable, harmful effects to environment and health of both man and animals. These include pesticide residue, pesticide resistance, adverse effects on non-target species, biomagnification and so on. Non-conventional methods include mostly chemicals whose applications may not always be desirable. For instance, 20% of the fruits in apple, pear, peach and apricot orchards in Himachal Pradesh are damaged by birds like Red billed Blue Magpie (*Cissa erythrorhyncha),* Red vented Bulbul (*Pycnonotus cafer*), White cheeked Bulbul (*Pycnonotus leucogenys*) and Slaty headed Parakeets (*Psittacula himalayana*). The orchards were successfully protected from depredations of birds by repeated sprays of landarin @ 1.5 kg.a.i./ha. However, fruits became unfit for consumption due to accumulation of toxic residues (Karkar *et al.*, 1986).

To protect ripening grapes from bird and bat damage, farmers adopt scaring methods which are traditional and in most cases ineffective. Erecting nylon netting around bower and covering the canopy gap on bower with briars and twigs gave almost complete control of bat damage in vineyards (Verghese, 1998). In Himachal Pradesh, apple orchards suffered 6.7–21.5% damage at nursery and 4.4% damage during storage due to rodents. Zinc phosphide poisoning resulted in 86% control in apple orchards. Burrow baiting proved superior to surface baiting (Anon, 1992-1993).

Chemicals used judiciously can help in protecting crops with desirable effects. Rats, for example, are serious pests of rice in Bangladesh. In flood-prone environments, *Bandicota bengalensis* and *Bandicota indica* are dominant and *R. rattus, Bandicota bengalensis* in non-flood prone areas (Catling & Islam, 1999). Damage is at tillering stage and intensifies during ripening (Islam *et al.,* 1993). Phostoxin pellets that produce lethal aluminium phosphide gas achieved 81.8% rat control, whereas chronic poison Lamirat (bromadiolone) achieved 42%. These measures revealed that targeting rats in the burrows with pesticides during ripening when most other traditional techniques are ineffective is a good option. Phostoxin was better than Lamirat in achieving immediate population reduction (Islam & Hussain, 2003). However, in hilly tracts of cultivation in Western Ghats (Mudigere) in Karnataka, phostoxin adversely affected rat snakes (Chakravarthy, 1993).

It is not always necessary to use chemicals to protect crops from vertebrates. The novel, non-conventional management tools include solar power operated fence, plant/animal products, provision of alternate or known natural foods for vertebrates in the vicinity of cultivated ecosystems, lacing repellent pastes made of easily available materials and mechanical barriers made of locally available, biodegradable and renewable materials. For instance, to protect coconut (*Cocos nucifera)* palms from vertebrates like rodents, monkeys, wild pigs, bats, bison, etc. trunk banding (thatch or metal), burrow baiting of rodents, clean basin cultivation and crown cleaning were effective. To prevent porcupine (*Hystrix indica*) damage, lacing base of grown-up tree trunks with waste oil or covering base of seedlings with thorny bamboo or a mat of dry sticks, timely harvests and proper disposal of fallen nuts have been found effective (Anon, 1997, 1998). It is however, important to distinguish damage caused by the major vertebrate pest species before adopting protection practices.

RODENTICIDES

Many different techniques or compounds are used to control rodents. First, there are ingested 'bait poisons' or 'contact powders' ingested while grooming or foraging. Accidental poisoning of non-target animals can occur, although this can be reduced to a minimum with well-designed bait stations, appropriate dyes and lures and bitter substances that deter non-rodents. Secondary poisoning can affect animals that eat dead or dying rodents. Anticoagulant poisoning is by far the most common means of rodent control (Pesticide Safety Directorate (PSD), 1997) as they are extremely effective and easy to use, cause little bait shyness and safe. Zinc phosphide is the most commonly used rodenticide worldwide in developing countries after the anticoagulants (Meehan, 1984). It is one of the more effective rodenticides, useful when a rapid population reduction is required and causes higher death rates than any of the second generation anticoagulants. The disadvantages include bait shyness, less effective if weathered or in damp conditions and generally less effective than anticoagulants.

Calciferol/Cholecalciferol, a form of Vitamin-D is another acute poison, formulated as a one-feed bait requiring no pre-baiting (Eason & Wickstrom, 2001). But it is relatively expensive. Alpha-chloralose is potentially highly effective and efficient and can give excellent, rapid results without pre-baiting, even in the presence of alternative food. However, it is expensive and rats and mice can develop tolerance (Meehan, 1984). Secondary poisoning is possible.

Fumigants offer the most efficient rodent control method for burrows or enclosed indoor sites or stores with several advantages viz. affected animals are not dangerous to predators, so that secondary poisoning risks are negligible and all animals within a burrow system are simultaneously poisoned. But fumigation is expensive and cannot be used in domestic settings, near livestock or when soil is sandy or loose for safety reasons. Aluminium phosphide is used largely as a fumigant. Sulphur dioxide and carbon dioxide are also used in indoor sites and cold stores.

CONVENTIONAL METHODS

(a) Trapping

Traps can be included under conventional methods. Traps are not practical for large fields because a large number of traps are required, they can be bulky to carry and require regular checking. However, many sectors of the food industry rely on non-toxic control as trapping is not only successful but also safe (Hughes, 1998). Trapping has the additional advantage of easy and safe disposal of captured rodents.

Live box traps may be baited or unbaited but should be placed on runways where the animals travel regularly. Efficient live traps can catch several animals at once and can be more effective than other traps (Islam & Karim, 1995). Trapped rodents can be released offsite or humanely destroyed.

Snap traps are spring-based devices which kill by means of a rapidly descending sharp metal bar. They can be baited with natural foods available in a locality namely -fruits, peanut butter, peanuts, cooked meals – all being effective baits. A practical disadvantage is that they need to be re-set each time they catch a rodent. But they are easier to monitor than live traps and in field they can be more effective.

Electrocution traps kill by electrocution and are a relatively recent innovation. These devices consist of an open-ended box baited with dry food. The practical advantages are that they can work outdoors or indoors, battery-driven, portable, can be remotely monitored and bodies can be rapidly disposed. The disadvantage is that they work best after a pre-baiting trial and have to be reset between kills.

(b) Non toxic lethal feeds

Cellulose based lethal feed pellets (non-toxic lethal feeds) are relatively new products. The pellets consist of plant based material, primarily cellulose flavoured for palatability. They are non-toxic and hence safe for non-target species and have no secondary poisoning risks.

(c) Habitat management

Deterring or proofing or removing rodents usually leaves a temporary void, soon occupied by immigrants and the rapidly proliferating descendants of surviving animals (Allen, 1999). In a study on an agricultural site, the rat population took only 2-8 weeks to recover after 70% were killed (Lu *et al.*, 1994). Therefore, for long term population reduction, unless control is frequently repeated, other techniques need to be employed to reduce carrying capacity of the site and surrounding areas to exclude or repel rodents. The most important and obvious tactic is to minimize available food, water and shelter by eliminating nest-sites and refuse areas. Finally, physical exclusion is also important. Sinking low foundations to prevent animals burrowing in, erecting surrounding walls topped with T-pieces, putting metal collars on pipes or trees to stop them from climbing, plugging gaps in the walls/floors with wire wool or wire mesh are all important.

INTEGRATED VERTEBRATE PEST MANAGEMENT

(a) Orchard crops

Orchards in India represent usually biologically diverse ecosystems in tropical and subtropical parts with a small, mixed crop. This implies that methods of pest management like chemical applications are not desirable. Much of the damage to orchards is as a result of conversion of, or encroachment into the natural habitats of wild animals. Wild animals use orchards for food, shelter, nesting and roosting causing often either temporary or permanent damage. The fruit-loss due to vertebrate pests is of great magnitude and little attention is paid to their management. This is because cultivation of edible fruits has come to fore in India only in the twentieth century. Also crop losses due to vertebrates are often accepted as part of the normal run of events or the crop itself is avoided.

Roseringed parakeets (*Psittacula krameri)* feed on seeds, berries, fruit, blossoms and nectar and are serious pests on guava (*Psidium guajava*) (Bhatnagar *et al.*, 1982). In Hyderabad, guava served as food throughout the year for the birds (Shivanarayan, 1982). In guava orchards, birds particularly parakeets caused on an average 23.5% fruit loss in Mudigere, Karnataka. Birds preferred fully ripened fruits most followed by partially ripened ones. Covering vulnerable portions of canopy during fruit-bearing period with thatch (locally available materials) saved 20% guava fruits (Chakravarthy, 1993). Shooting, trapping, scaring, destruction of bird roosts and nests and encouragement of natural enemies like owls helped in keeping birds away from orchards (Singh & Kumar, 1982).

Bank mynas (*Acridotheres gingianus*) and Indian mynas (*Acridotheres tristis*) damaged both immature and mature grape (*Vitis vinifera*) berries in Ludhiana causing a fruit loss of 19% (Toor, 1982). More damage occurred in grapevines trained with the Head System than in Bower System and birds preferred Beauty Seedless Variety of grapes having purple berries (Sandhu & Chakravarthy, 1982). This preference is exercised in the presence of other non-preferred cultivars or foods. Farmers in a patch can 'spread' or 'distribute' vertebrate feeding damage by growing different foods or varieties of a crop or stagger blooming or fruiting over time, so that the damage is not concentrated on a spot or a cultivar.

In Bangalore, Jungle crow (*Corvus macrorhynchos*), House crow (*Corvus splendens*), Barbet (*Megalaima viridis)* and three species of bats caused 30-36% loss to different varieties of grapes (Prasad & Verghese, 1985). Damage due to bird pests was offset by compensatory increase in number of grape bunches per vine. Birds also fed on packed berries which were prevented by covering bunches with muslin or cloth bags.

Damage to apple (*Malus domestica*) in Himachal Pradesh was mostly confined to the upper and middle canopy of the tree. Maximum damage to fruits (80%) was caused by Blossom headed Parakeet (*Psittacula cyanocephala*). The early maturing compared to late maturing cultivars suffered more bird damage (Narang & Chandel, 1995). The damaged fruits were subsequently fed by a number of insect pests and fungi adapted to decaying sweet resources. Such fruits became unfit for marketing.

Flying foxes (*Pteropus giganteus)* caused damage to apple, citrus, peach, guava and other fruits in India. The bat (*Cynopterus sphinx)* is a serious pest on grapes in Bangalore foraging maximally on mature fruits. Vines adjacent to open space had more bat damage (Verghese, 1998). Branches of thorny trees and old fish nets or wire netting protected the fruit bunches. Deterring bats from roosting near orchards was suggested to prevent bat damage to fruits (Toor & Ramzan, 1974; Toor & Sandhu, 1979, 1981).

In hill region of Karnataka, oranges (*Citrus sinensis*) are grown in coffee estates (*Coffea arabaica* and *C. robusta*). More than 90 bird species roosted/nested in shade trees planted or naturally present in estates. Fruit losses due to various species of vertebrates in Karnataka on oranges are given in Fig. 1. Jungle crow was the dominant species feeding on oranges. A single crow destroyed eight fruits/day and crows destroyed over 2000 fruits in a fortnight (Chakravarthy, 1993). In addition Monkey, bison, elephant and wild pig caused damage to citrus orchards (Ghosh, 1990). While monkeys fed on fruits, others dislodged/ trampled seedlings, broke branches/twigs or debarked while rubbing horns or body against the tree trunk. Porcupines (*Hystrix indica*) burrowed at the base of the tree and dislodged young trees in severe cases of attack. The peeled off branches were prone to termites or bark borer attack. By guarding orange orchards against vertebrate pests mainly frugivorous birds and monkeys 42% fruit losses were prevented (Chakravarthy, 1993).

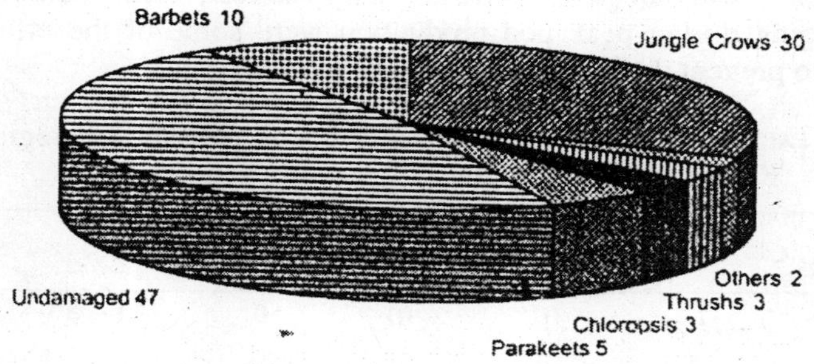

Fig. 1. Yield losses (per cent) in Oranges by birds in Mudigere

Snap trapping, sticky bands, covering fruits with polybags, cloth-bags laced with waste oils, etc. were the methods adopted to protect fruits in small orchards (< 1 ha) and backyards of houses in Bangalore. Ripening fruits of pomegranate and custard-apple laced with oils of neem (*Azadirachta indica*) and castor (*Ricinus communis)* protected the fruits from vertebrate depredation for a longer period (Chakravarthy, 1993).

In coastal Karnataka, three banana varieties were depredated by monkeys, barbets and bats. The barbets damaged ripening fruits. It was found that by retaining damaged fruits on the plant, bird damage to banana can be lessened because birds revisited the damaged fruits to feed on the sweetened fruit portions. There was varietal preference of banana by birds in the order of *Nendran* > *Nanjangud* > *Mysore*. Isolated banana orchards were prone to heavier vertebrate pest damage (Anon, 1997, 1998)

(b) Plantation crops

In South India, cocoa (*Theobroma cacao*) is cultivated with areca, coconut, coffee and cardamom or as a pure crop generally in small holdings (upto two acres). Jungle cat (*Felis chaus),* squirrel (*Funambulus palmarum),* rat *(Rattus rattus),* monkey (*Macaca radiata),* jackal (*Canis aureus),* bison *(Bos gaurus),* stag (*Cervus canis)* and birds constituted the vertebrate pest complex (Table 1). Squirrel damage commenced at the beginning of the fruiting season (July-August), that of jungle cat, monkey, jackal and birds during pod ripening stage (October-November) and rat damage throughout the year. Jungle cat is the most important pest (Table 1) incurring 20% out of 35% total loss in pod yields due to vertebrate pests (Chakravarthy, 1993; Thyagaraj *et al.*, 1996). Carbofuran 3% baiting in banana fruits or wet jaggery, intercropping with areca or coffee, timely harvest, clean cultivation, snap trapping during peak pod production were some of the effective methods to prevent damage to cocoa by vertebrate pests.

Table 1. Damage* by vertebrate pests to cocoa pods in hill region of Karnataka, South India

Location Jungle Cat	**Monkey**	**Squirrel**	**Rat**	**Jackal**	**Birds**	
Sringeri 29	1	01	01	0	1	
Koppa	24	0	02	01	0	0
Kalasa	07	1	13	07	1	1
T.halli	09	0	19	13	1	1
Sagar	18	0	15	04	1	5
Soraba	14	1	11	17	12	2
Sirsi	43	2	27	19	19	1
Total	144	5	88	72	34	11
Mean	20.5	0.71	12.57	10.28	4.85	1.57

* Actual pods damaged; Mean of two seasons' data
(Source : Thyagaraj *et al.*, 1996)

(c) Spice crops

Wild pig, monkey, rodents and birds are important vertebrate pests of cardamom (*Elettaria cardamomum*) in South India. *B. bengalensis, M. booduga, R. rattus, R. meltada* and squirrel (*F. palmarum)* constitute rodent pests incurring on an average ten per cent yield losses. Rodent damage to cardamom capsules is age related.

Clean cultivation, overlapping of panicles from August, timely harvests and mulching saved capsules considerably. Locally made wooden snap traps baited with germinating paddy seeds reduced capsule damage from 8.20 to 4.90%. These practices are cheap, practicable and eco-friendly and are additive in effect. Crude fish oil and neem oil (1-4%) and bingo, a sticky-rodent trap were ineffective (Srihari & Chakravarthy, 1998).

Studies to protect the cardamom nurseries from wild pig damage using a mixture of Thimet 10% G and sand 5 gm each and suspending the same in punched polybags (10 cm x 10 cm size) along the borders of cardamom nursery (at 3 m distance and 2 m height) minimized damage for 45 days. Providing good drainage also helped. Where feeding pressure of the animal is more, selective shooting would ameliorate the problem. Though wild pig is a protected animal, the Government of Karnataka has permitted shooting of wild pigs when the animals create pestilent problems.

Cultivation of cardamom is being severely affected in several districts of Karnataka due to monkey menace. Damage due to monkeys varied from 2.80 to 17.50 per cent suckers in cardamom. Peak damage to coffee and cardamom occurred during January to April when in general, there was a decline in the availability of natural foods. Two or three year old suckers were the most preferred. Methyl parathion and 10% G Thimet in fruit bait failed to repel monkeys. Selective debranching of shade trees and watch and ward during early morning (before 7.00am) and late evening (after 5.00pm) hours significantly reduced clump damage from 49 to 3.2% due to monkey menace in cardamom (Chakravarthy, 1993; Chakravarthy *et al.,* 1998; Anon, 1998).

(d) Palm crops

The coconut palm is depredated by vertebrates right from the time of sowing to maturity. Growers are currently using indigenous devices and tools to protect the palms from vertebrates. However, the efficacy of these traditional methods is low (Bhat *et al.*, 1990; 1995). Monkeys damaged the maximum number of nuts (150-200/15 days in March, 1997), rodents mainly *B. bengalensis* damaged 15-20 nuts/15 days and wild boar the least, 0-4 nuts/15 days. ANOVA revealed significant differences in nut damage caused by three species of vertebrates. Obviously wild pigs targeted only the fallen nuts.

Bison damage was by consumption of tender leaves of coconut seedlings which resulted in reduction in palm height, reduction being proportional to the extent of damage. Starting from July bison damage became severe during August to November.

Porcupine damage was mainly by debarking and burrowing at the base of the palm and thus damaging roots in coastal Karnataka. Upto July grasses were cut and basins of palms remained open and there was no porcupine damage. During the same period palms in the forest-tract whose base was covered with grasses were damaged (Anon, 1997; 1998), indicating that grass cover facilitates porcupine damage.

Coconut palms were damaged by elephants by trampling, browsing, uprooting and detopping of coconut palms in Coorg, Karnataka. Maximum damage (8%) occurred during July.

Trunk banding (metal), burrow baiting, clean basin cultivation, crown cleaning, lacing the base of mature tree trunks with waste oils or covering base of seedlings with thorny bamboo or a mat of dry sticks, timely harvests and pick-up of fallen nuts (against porcupine, rat and squirrel) proved effective in preventing vertebrate damage to coconut (Anon, 1997, 1998).

Comparative pattern of damage to areca by vertebrates in coastal region of Karnataka is given in Table 2 (Anon, 1997, 1998). Damage by vertebrates begins from October, i.e. at ripening stage. Monkeys, rats, wild pigs and bats damage nuts. Monkey damage on nuts was identified by the irregular peeling off fruit-epicarp, partially or completely. The unripe fruits were plucked-off and dropped. Bat damage was restricted to ripened or over ripened leaving the fruit surface damaged. Wild pigs fed only on the fallen ripened nuts. It chewed the whole nut, leaving the crushed nuts. Nuts damaged by monkeys are collected and sold by farmers as second quality nuts at 20% reduced rates.

Table 2. Comparative pattern of damage to areca by vertebrates in coastal region (Anon, 1997; 1998)

Plot	Month	% damage		
		Monkey	Bat	Wildpig
Ennakala	Dec.'96	100	00	00
Harpadi	Nov.	100	00	00
	Dec. '96	100	00	00
Eramadaka	Nov	100	00	00
	Dec. '96	100	00	00
Charmadi				
(1) Abdulla plot	Sep. '96	78.80	21.20	00
	Nov	61.50	24.80	13
(2) Anantha Rao's plot	Sep. '96	61.80	38.20	00
	Oct.	58.90	41.10	00
	Nov	84.00	16.00	00
	Dec.	73.80	26.20	00
Average damage (%)		83.52	15.23	1.25

Growers protected areca crop from vertebrate pests by patrolling with guns, Ibex fencing, bursting crackers, scaring by intense illumination and repellent odours like burning waste-tyre, bonfire, etc. Damage to areca seedlings from wild pig and rat could be avoided using locally available porcelain pipes. Monkey damage could be reduced by disfavouring the habitat like debranching shade trees, keeping dog for chasing monkeys, etc. Clean cultivation and burrow baiting with zinc phosphide (2%) reduced rodent damage. A rigorous and continuous guarding of the plantations proved costly and impracticable. However, Ibex fencing, solar fence, timely harvests and clean cultivation reduced vertebrate damage to areca nut (Chakravarthy, 1993; Anon, 1997, 1998).

(e) Vegetable crops

Cowpea (*Vigna unguiculata*) forms a staple vegetable in coastal region of Karnataka and is cultivated as a solo crop or intermixed with maize (*Zea mays*), ridge gourd (*Lufa acutangula*), cucumber (*Cucumis sativus*) or bottle gourd (*Lagenaria longisasima*). Small green barbet, Red vented bulbul and Roseringed parakeet, *B. bengalensis* and *F. palmarum* fed on cowpea flower buds and seeds. Intercropping with ridge gourd and erecting wooden platforms to a height of 0.5 m reduced damage, resulting in higher pod yields. Scare crows, reflective ribbon, mechanical barriers, pyrotechniques and bioacoustics are suggested to protect vegetables from bird damage at sowing and germination stages (Anon, 1992-97).

Cucumber and Ridge gourd are grown in small patches in coastal Karnataka. Damage to cucurbits was noticed at sowing, flowering and fruiting stages. At sowing, Blue rock pigeon (*Columba livia*), squirrel (*F. palmarum*) and hare (*Lepus nigricollis*) damaged the seedlings in the range of 17.2 to 30.5%. Birds even consumed petals, detached androecium and damaged the reproductive parts. They preferred flowers of cucumbers to that of snake gourd (*Trichosanthes anguinn*), ridgegourd (*Luffa acutangula*) and bitter gourd (*Momardica charantia*) that are less tasty and high in cucurbitacins (Srihari & Chakravarthy, 1998).

Damage by small green barbet to cucumber was 38% at fruiting stage. It was reduced to 8.5% when cucumber was grown on raised basins (2 m diameter, 0.3 m height) intercropped with pumpkin, bottle gourd and ridge gourd (Srihari & Chakravarthy, 1998) (Table 3). In general, for most vegetables, cultural practices that require no additional labour are effective fully or partly.

(f) Field Crops

(a) Paddy (*Oryza sativa)* is cultivated in small, fragmented plots in low-lying areas in the hilly regions of Karnataka and is attacked by vertebrates including bird, rodent, hare, wild pig, elephant and bison. In addition blue rock pigeon (*Columba livia*), Roseringed parakeet, rodents (*B. bengalensis*) and two species of *Mus* and hare (*Lepus nigricollis*) feed on sprouts or germinating paddy seeds. The damage caused by vertebrates begins within five days of sowing before seeds are germinated (Anon, 1992, 1997).

Table 3. Small green barbet damage to cucumber in relation to the Planting system

Plot	Cucumber fruits damaged (%) (n = 15)	
	Pendal	**Basin system**
1	30	10
2	35	12
3	28	8
4	24	9
Mean	29.25	9.75
Binomial expansion		
Test (m = 0.01)	2.964	

(Source: Chakravarthy, A.K. 1993)

Nine bird species mainly parakeets and finches in hill region of Karnataka (Chakravarthy, 1993), seven in Andhra Pradesh and eleven in Kerala (Anon, 1992-97) are implicated in damage at different phenophases of paddy. While the damage by parakeets is significant, that by finches is not (Chakravarthy *et al.,* 1994; Verghese & Chakravarthy, 1979). In Kole lands of Kerala due to saucer shaped topography of fields and difficulties in dewatering the sown fields, ducks devour the seeds and sprouts. Pintail, *Anas acuta*, garganey, *Anas querquedula* and lesser whistling teal, *Dendrocygna javanica* are the species involved. Farmers use floating banana leaves, kerosene lamps and decoys to scare away birds. Paddy fields at the vicinity of perching and roosting sites of birds suffer more damage (Anon, 1980-86). Baya weavers spend first 6 weeks of their life feeding on insects and cereals like rice.

Where rice is interspersed with sugarcane patches as in Mandya, Karnataka, the sugarcane patches serve as nesting and roosting sites and rice fields as the feeding sites for baya weavers (Chakravarthy & Gubbiah, 1986) (Table 4). Maintaining water level of 3 cm deep in paddy field during transplanting minimizes seedlings loss by crows (Chakravarthy, 1993).

Some rice varieties received higher damage from rodents; age of varieties to some extent influenced damage. Rats damaged clumps are susceptible to lodging. Practices chosen to prevent damage were bund clearing, trimming, weeding at the beginning of monsoon (June); baiting with 0.005% bromadiolone/roban cake; scaring at boot-leaf stage (September-October); use of decoys (coloured poly sheets) and use of waste cassette tapes at maturing stage (December) of the crop. The grain damage in the 'control' plot was 32% compared to 12% in the treated plots. The above instances of crop protection against vertebrates demonstrate that it is possible to protect crops by cultural and crop husbandry practices that also help in harvesting sustainable crop yield. The practices are cheap, practicable, ecofriendly and harmless to nontarget species including humans.

Table 4. Feeding rate, flock size and abundance of rice birds, VCT, Mandya, India

Bird	**Feeding rate$^{\alpha}$ (pecks/min) (x ± SD)**	**Flock size (max – min)**	**Population % e**	**Habitat preferenc**
Streaked weaver bird	8.2 ± 2.4	46-15	69	Wet
Baya weaver bird	6.0 + 0.9 17-4	8	Wet	
Black headed munia 6.0 + 1.7	33-14		19	Wet
White backed munia 3.0 + 1.0	3-2		<1	Dry
Spotted munia	2.5 + 1.2	5-2	<1	Dry
Red munia	3.6 + 0.8	9-3	3	Wet
CD at 5% by 'F' test	ns	ns	18	-

$^{\alpha}$Numbers in a column are + of 15 observations recorded from fields adjacent to patches of marsh and sugarcane.

(Source: Chakravarthy and Gubbiah, 1986)

But use of rodenticides at vulnerable stages of crops needs further research. Rice germplasms resistant/unattractive to rodents and wild pig damage have also been screened.

Wild pig is the most abundant quadruplet in hill region causing enormous losses in rice yield by trampling, lodging and directly feeding on ears/grains. The extra-depredative losses far exceed the depredative losses. The extent of damage depends on the number of animals in the herd and stage of the crop.

Five G Thimet (10g) mixed with 5g of sand when suspended in bags along borders of rice fields protected the crop from wild boar damage for 30 to 50 days. The bags were suspended at 3m distance and one meter height. Growing early maturing, dwarf rice varieties, timely harvest, scaring away wild pig during milky to grain maturation stages and synchronizing sowing and crop maturity, protected rice crop from wild boar and bird damage in hill region (Chakravarthy, *et al.,* 1992, 1994).

(b) Maize: Blue rock pigeon and rats together caused 72% seeds and seedlings losses to maize in coastal Karnataka. Jungle crows and mynas of *Acridotheres* species damaged maize from milky to grain maturing stages while parakeets damaged cobs extensively (Anon, 1998). Cobs growing at heights below 50 cm are relatively less attacked even when sown at borders or edges. Throughout India psittacids and corvids are the dominant pest species on maize.

Spathes completely covering the cobs and grains, tightness and compactness of inner and outer spathes of cobs reduce bird damage. There is a need for maize cultivars showing resistance to both insect and bird damage.

Crops cultivated in or near forests have high vertebrate pressure and face unique problems. This is because such agro ecosystems are found in the natural habitat of vertebrates concerned. In such situations, part of the food requirements of the vertebrates are met by cultivated crops. The Central Horticultural Experimental Station at Chetthalli in Coorg, Karnataka is a case in point. The station is surrounded by thick teak forests on one side which is the natural habitat of elephants, bisons, wild pigs, flying foxes and other animals. On the other side the station is surrounded by wetlands, coffee plantations, which imitate natural habitat of wild animals (Anon, 1998). One of the important functions of the station is to collect and maintain germplasms of tropical and subtropical fruit crops. But maintenance of the germplasms of coconut, jackfruit, mango, papaya and banana were abandoned due to 100% damage inflicted by vertebrates particularly elephants, bisons and wild pigs. This is despite rigorous night patrolling, burning of crackers, exploding dynamite, construction of elephant-proof trenches, lighting fires and use of high intensity lights. Lack of grass and natural food in the forest forced vertebrates to attack fruit crops at the research station.

LACUNAE, PRIORITIES AND SUGGESTIONS

An analysis of published literature on vertebrate pests shows that the information is largely fragmentary and preliminary. It is clear that barring a few species that are host-specific or have restricted geographical distribution, the same vertebrate species is implicated in causing damage to more than one crop. This is because of the preference and adaptiveness of the species to forage in a variety of ecosystems. Determination of economic status and ecological roles of vertebrates in their natural habitats and cultivated ecosystems is important. For instance, adult house sparrow feeds on fruits, nuts and grains but feeds its young ones with insects, some of which injure crop plants. Similarly rats, squirrels and porcupines which variously damage a variety of fruit crops are important regulators of surface and sub-surface vegetation, soil fertility and soil fauna. Perhaps with the exception of Roseringed parakeets and *B. bengalensis* and a few other species of rats, all other animals perform diversified roles in their habitats and cultivated ecosystems. Establishing the economic status of vertebrate pests is therefore difficult and complicated. Studies with multidisciplinary approach in this direction are urgently required.

The secretive and in some species, nocturnal habits of vertebrates and their mobility makes it highly impossible to visually observe crop damage (Thyagaraj *et al.,* 1996). There is a lack, in general, of trained personnel in vertebrate pest management. That apart, the few scientists working in agricultural universities and research centres are mostly entomologists and are handicapped by lack of basic knowledge on wildlife ecology and behaviour. No program of pest management in agro ecosystems can be effective unless it is based on the knowledge of the animal's natural history. Studies on habitat preferences should be combined with records of seasonal population changes so as to provide basic information for pest management. Estimates on crop losses are pre-requisite for implementation of management practices against all pests including vertebrate depredations. Varied methods are being used for assessing crop-losses due to vertebrate pests. Parameter used for assessing crop-losses differs from one investigator to another, thus rendering the data often incomparable. Thus there is an urgent need to standardize the methods of assessing losses due to vertebrate pests. The All India Coordinated Research Project on rodents and on economic ornithology has made significant progress in formulating pest management techniques. It is inevitable that a committee of experts be constituted to find gaps in research and set priorities for zone-wise research on vertebrate pests for the 21st century. The vertebrate pest managers should not aim at eliminating a specific community of animals to realize cent per cent crop protection. For instance, in cardamom plantation when in a patch of two and half acres squirrels were removed, rats occupied the niche. When rats were exterminated from the patch, birds devoured the capsules in a major way. When specific bird species (jungle fowl) were trapped and eliminated from the patch, ground birds like thrush and pitta, occupied the niche. Thus, a succession in the community of animals depredating cardamom capsules occurred. Parallely, in rice agro ecosystems due to the application of insecticides like thimet and zinc phosphide baiting, biomagnification linking rats-snakes-birds-of-pey occurred killing animals at all the three trophic levels. Thus, complete pest eradication may lead to consequences whose effects in the ecosystem are still little known and could be disastrous.

Some of the novel tools for vertebrate pest management in cultivated fields in India may include solar power operated fencing, plant/animal products, provision of alternate or known natural foods for vertebrates in the vicinity of agro ecosystem, trenches, cultural practices that do not involve much labour such as polyculture, timely harvest, lacing repellent pastes made of commonly and easily available materials and mechanical barriers made of locally available, biodegradable materials. Use of botanicals especially in the evergreen tropical forest tracts has considerable potential (Bhatnagar *et al.,* 1982; 1993). In fact, botanicals may serve as one of the ideal tools for integrated vertebrate pest management (IVPM). But little is done in this regard. Neem, cluster beans, agave, tree bark decoctions and species of *Acacia* are plants worthy of investigation (Bindra & Toor, 1972).

CONCLUSIONS

A compatible blend of conventional and unconventional methods is desired for managing vertebrate depredators. Farmers or farm managers should not aim at cent per cent crop protection against vertebrate pests and eliminate the species from the cultivated ecosystems. Botanicals may provide an alternate tool for protecting crops, but they are not fully explored. In tropical countries there is considerable scope for utilizing solar power too. Judicious use of chemicals, traps and fabricated structures can facilitate effective vertebrate pest management. The protected areas, forests and wild patches need to contain and sustain activities of wild animals so that wild animals do not feed in cultivated ecosystems.

ACKNOWLEDGEMENT

Authors are thankful to the Indian Council for Agricultural Research, New Delhi, and Department of Science and Technology, New Delhi for granting research projects through which research data could be generated and are included in this chapter. Thanks are due to the authorities of University of Agricultural Sciences, Bangalore, for support and encouragement. The authors also thank Mr. K. Ramakrishnan, for typing and preparing the outlay of this chapter.

REFERENCES

ALLEN, D. 1999. Rodents – least toxic control. *Pesticides News* 46: 20-21.

ALI, S & S.D. RIPLEY. 1983. HANDBOOK OF THE BIRDS OF INDIA AND PAKISTAN, Vol. 3. Oxford University Press, Delhi, pp.163-190.

ANON. 1980-1986. *Annual Reports of All India Coordinated Project on Economic Ornithology,* ICAR, Andhra Pradesh Agricultural University, Hyderabad, India.

ANON. 1992-1993. *Research highlights of AICRP on agricultural ornithology.* Andhra Pradesh Agricultural University, Hyderabad, India.

ANON. 1992-97. *Annual reports of All India network project on agricultural ornithology,* Andhra Pradesh Agricultural University, Hyderabad, India.

ANON. 1996. Rodent control and keeping away langurs. *Honey bee* 7:11 & 16.

ANON. 1997. *Annual report of integrated vertebrate pest management in hilly and coastal zones of Karnataka* (IVPM), ICAR ad-hoc project, University of Agricultural Sciences, Bangalore, India.

ANON. 1998. Annual report of Integrated vertebrate pest management in hilly and coastal zones of Karnataka (IVPM), ICAR ad-hoc project, University of Agricultural Sciences, Bangalore, India.

BHAT, S.K., P.S.V.P. VIDYASAGAR & A. SUJATHA. 1990. Record of *Rattus rattus* Wroughtoni Hinton as a pest of oil palm seedlings. *Trop. Pest. Manage.* 36:75-76.

BHAT, S.K., R.P. ADVANI, A.S. SUKUMARAN & P.S.P.V. VIDYA-SAGAR. 1995. Ecology, Biology, economic importance and control of economically important rodents in plantation crops. In: *Annual Report* 1994-95, Central Plantation Crops Research Institute, Kasargod, Kerala, India. pp.168-178.

BHATNAGAR, R.K., SWARUP, PRAKASH & K.N. MEHROTRA. 1982. Bird damage to maize cobs-relationship between susceptibility of maize germplasms and spathe configuration. In: MANAGEMENT OF PROBLEM BIRDS IN AVIATION AND AGRICULTURE (Eds.): R.A. Agarwal & R.K. Bhatnagar. Indian Agricultural Research Institute, New Delhi, India.

BHATNAGAR, RK., K.P. SRIVASTAVA & R.K. PALTA. 1993. Plastic bagging for controlling bird damage in sunflower heads. In: BIRD CONSERVATION (Eds.): Abraham Verghese, S. Sridhar & A.K. Chakravarthy. Ornithological Society of India, Bangalore, India.

BINDRA, O.S & H.S. TOOR. 1972. HARMFUL BIRDS AND THEIR CONTROL. Punjab Agricultural University, Ludhiana, India.

CATLING, H.D & Z. ISLAM. 1999. Pests of deepwater rice and their management. *Integr. Pest Manage. Rev.* 4:193-229.

CHAKRAVARTHY, A.K. 1993. *Vertebrates Pest Management. Final report of ICAR ad-hoc project,* University of Agricultural Sciences, GKVK, Bangalore, India.

CHAKRAVARTHY, A.K & GUBBAIAH. 1986. Bird damage to rice in Visveswaraiah canal tract. *IRRN* 11:2.

CHAKRAVARTHY, A. K. & K. SRIHARI. 2000. Vertebrate pest management in horticultural crops. In: INTEGRATED PEST MANAGEMENT IN HORTICULTURAL ECOSYSTEMS. (Eds.): P.R. Reddy, A. Verghese & Krishna Kumar, N.K. Capital Publishing Co., New Delhi. pp. 236-262.

CHAKRAVARTHY, A.K., V.G. SADAKSHARI & E. GANGAPPA. 1992. Rodent damage to rice germplasms in Mudigere, Karnataka, *Rod. Newsl.* 16:6

CHAKRAVARTHY, A.K., N.E. THYAGARAJ & J.B.NARENDRA KUMAR. 1994. Rice yield losses to finches in hill region, Karnataka, India. IRRN.19:23-24.

CHAKRAVARTHY, A.K & V.K.ABRAHAM. 1995. Bird pests of oilpalm in Karnataka. *Indian Oil Palm J.*5 (28):177-179.

CHAKRAVARTHY, A.K., K. SRIHARI., N.E. THYAGARAJ & S. NAGARAJU. 1998. Rodent pest management in cardamom with cultural practices. In: ADVANCES IN IPM FOR HORTICULTURAL CROPS (Eds.): P.P. Reddy, N.K. Krishna Kumar & Abraham Verghese. *Proc. 1st Nat. Symp. Pest Management in Horticultural crops,* Bangalore, India. pp.188-194.

EASON, C.T & M. WICKSTROM. 2001. VERTEBRATE PESTICIDE TOXICOLOGY MANUAL (POISONS): Information on Poisons used as Vertebrate Pesticides, 2 Ed. Department of Conservation: Wellington, New Zealand.

FITZWATER, W.D. & I. PRAKASH. 1973. HANDBOOK OF VERTEBRATE PEST CONTROL. Indian Council of Agricultural Research, New Delhi, India.

GEE, E.P. 1951. What is the best means of control and destruction of flying foxes? *Pterops giganteus* Brum. *J. Bombay nat. Hist. Soc.*50:17-20.

GHOSH, S.P. 1990. Citrus In : FRUITS: TROPICAL AND SUB-TROPICAL (Eds.): T.K. Bose and S.K. Mitra. Naya Prakash, Calcutta, India.

HUGHES, J. 1998. PEST CONTROL PROGRAMMES. American Institute of Baking, Manhattan, Kansas, USA.

ISLAM, Z & A.N.M.R. KARIM. 1995. Rat control by trapping in deepwater rice. *Intern. J. Pest Manage.* 41: 229-233.

ISLAM, Z & M.M. HUSSAIN. 2003. Control of rats at the ripening stage of a rice crop. IRRN, 28(1): 49-50.

ISLAM, Z., R.G. MORTON & B.P. JUPP. 1993. The effect of rat damage on deepwater rice yields in Bangladesh. *Int. J. Pest. Manage.* 39(2): 250-254.

JACOB, S.A. 1993. A simple device for scaring away wild boars, *Sus scrofa* in newly planted oil palm field. *The Planter*, Kuala Lumpur 69: 475-477.

KARKAR, K.L., AMITH NATH & G.S. DOGRA. 1986. Studies on the feeding behaviour of birds and their control under mid-hill conditions of Himachal Pradesh. In: ADVANCES IN RESEARCH ON TEMPERATE FRUITS (Eds.): Chadha, T.R., V.P. Bhutani & J.L. Kaul. Dr. Y.S. Parmar University of Hort. & Forestry, Solan, India. pp. 389-395.

LU K.H., L.H. LEE., S.K. YOUNG., T.S. LIN., S.K. HOU., J.T. HUNG., C.M. KUO & W.C. LIN. 1994. The strategy for improving efficacy of rodent control in public fields: using sustainable bait. *Plant Protection Bulletin Taichung* 36: 209-223 (in Chinese; English abstract).

MEEHAN, A.P. 1984. RATS AND MICE–THEIR BIOLOGY AND CONTROL. The Rentokil Library: East Grinstead, UK.

NARANG, M.L & J.S. CHANDEL. 1995. Preliminary studies on bird damage to apple orchards in Himachal Pradesh (Abstracts). ENVIRONMENT AND BIRDS, Secretariat Box 45, Srinagar, Garhwal, U.P, India.

NENE, Y.L., G.V. RANGA RAO & C.L.L. GOWDA. 2002. In: *Diseases and insect pest management practices in Ancient and Medieval india:* Relevance today (Eds.) Y.L. Nene & S.L. Choudhary. Agricultural Heritage of India, AAHF, Secunderabad, pp. 149-162.

ORLOB, G.B. 1973, ANCIENT AND MEDIEVAL PLANT PATHOLOGY PFLANZENSCHUTZ. Nachichten Bayer, Bayer A.G., Leverkusen, Germany, pp. 281.

PESTICIDE SAFETY DIRECTORATE (PSD) 1997. Assessment of Humaneness of vertebrate control agents – Evaluation of fully approved or provisionally approved products. No. 171, PSD, York, UK.

PRASAD, V.G & A. VERGHESE.1985. Birds as pests of horticultural crops. *Bull. of Entomology* 26(1): 94-96.

RAZE AMBER. 2000. *Nuskha Dar Fanni-Falahat (The Art of Agriculture). Agri-History Bulletin, No. 1.* Asian Agri History Foundation (AAHF), Secunderabad, India. pp. 94.

SANDHU, J.S & S.CHAKRAVARTHY. 1982. Orchard birds-a review. In: ENVIRONMENT AND BIRDS (Abstract), Secretariat Box 45, Srinagar (Garhwal), UP, India.

SHIVANARAYAN, N. 1982. The Roseringed parakeet, the problem bird in agriculture and its management. In : MANAGEMENT OF PROBLEM BIRDS IN AVIATION AND AGRICULTURE (Eds.): R.A. Agarwal & R.K. Bhatnagar. Indian Agricultural Research Institute, New Delhi, India.pp.151-158.

SINGH, A K & P. KUMAR. 1982. Damage to maize and guava crops by the parakeets. In: MANAGEMENT OF PROBLEM BIRDS (Eds.): R.A. Agarwal & R.K. Bhatnagar. Indian Agricultural Research Institute, New Delhi, India. pp.164-167.

SRIHARI, K & A.K. CHAKRAVARTHY. 1998. Ecological approaches for the management of vertebrate pests in western ghat region of Karnataka. *Indian J. Ecol.* 25: 55-61.

THOMAS, M. & A.K. NAIDU. 1995. Indian wild boar, *Sus scrofa* L.: a rice pest in Madhya Pradesh, INDIA. IRRN 20(1):28.

THYAGARAJ, N.E., A.K. CHAKRAVARTHY & K. SRIHARI 1996. Vertebrate pests of cocoa and their management practices in hill region of Karnataka. *IPM & Sustain. Agriculture* 6: 214-218.

TOOR, H.S. 1982. Problem birds and their management in Punjab. In: MANAGEMENT OF PROBLEM BIRDS IN AVIATION AND AGRICULTURE (Eds.): R.A. Agarwal. & R.K. Bhatnagar. Indian Agricultural Research Institute, New Delhi, India. pp.132-142.

TOOR, H.S & M. RAMZAN. 1974. A study on bird pests of grapes. *Punjab Hort. J.* 14: 46-48.

TOOR, H.S & P.S. SANDHU. 1979. Note on bird pests of grapes at Ludhiana. *Indian J. Agric. Sci.* 49: 817.

TOOR, H.S & P.S. SANDHU. 1981. An estimation of bird damage of peach. *Prunus persica* Batsch. *Indian J. Ecol.* 8:308-310.

VERGHESE, A. 1998. Non-destructive control of the bat, *Cynopterus sphinx* Vahl (Chiroptera: Pteropodidae) in grapes (*Vitis vinifera* Linn.) in India. *Intl. J. Pest Manage.* 44: 81-85.

VERGHESE, A & A.K. CHAKRAVARTHY. 1979. Preliminary studies on the granivorous munias (*Lonchura* spp.) of Bangalore. *J. Bombay nat. Hist. Soc.* 78: 384-385.

VERGHESE, A & P.L. TANDON. 1993. Enhanced number of grape bunches per vine offset damage due to bird pests. *Newsletter of Birdwatchers* 34:70-71.

❑❑❑

24

VERTEBRATE PESTICIDES

Shakunthala Sridhara

AINP on Rodent Control, University of Agricultural Sciences, G.K.V.K., Bangalore – 560 065

ABSTRACT

Management of vertebrate pests is a sensitive issue with no or minimum use of toxicants except while controlling serious grainivorous bird pests such as red-winged black birds and rodents. Mostly mammalian pests are scared, deterred or repelled from crops using auditory, visual scares, and chemical repellents. Killing is never reported to be successful in managing unconventional and sporadic mammalian pests such as elephants, blue bull, deer and antlers, bears, monkeys etc. A dozen species of rodents and a few species of birds are pests in true sense and effective chemicals are available to manage them. Hence this chapter deals mainly with rodenticides, avicides and chemical repellents. The bird pest management is by using avicides such as starlicide, DRC-1339 and PA-14 and mostly by using primary and secondary repellents, Methylanthranilate is the only successful primary repellent. Methiocarb, 4-aminopyridine and anthraquinone are the secondary repellents. But in India bird pests are managed by conventional auditory and visual repellents and use of chemicals is almost non-existent. Rodent control in the country is mostly chemical both in residential premises and in agricultural fields. Again the choice is limited to zinc phosphide amongst acute rodenticides and bromadiolone amongst chronic rodenticides. All other acute rodenticides described are of theoretical importance only. Zinc phosphide at 2% in the bait has been found effective for all cropping situations although prebaiting is necessary and bait shyness in surviving population is inherent. Bromadiolone is the only marketed second generation anticoagulant and is mainly used in residential areas. Aluminium phosphide is effective in irrigated fields. For optimum control of rodents, initial knockdown of rodents using zinc phosphide followed by aluminium phosphide fumigation of burrows of

surviving members in irrigated fields or followed by bromadiolone baiting in both irrigated and rained crops is suggested.

INTRODUCTION

Almost all pest species are killed using chemical compounds which are conveniently called pesticides. Depending on specific use, pesticides are variously termed as insecticides, fungicides, acaricides, nematicides, herbicides, weedicides, etc. Amongst vertebrates the most harmful pests are members of rodentia and some birds and the chemicals employed to reduce rodent and avian pests are termed rodenticides and avicides, respectively. Most of the unconventional pests viz. few species of ungulates, non-human primates etc and most bird pests are largely dealt with repellents which may be auditory, visual or chemical.

CHEMICALS FOR BIRD MANAGEMENT

Birds are managed usually employing lethal toxicants and repellents which can be auditory, visual and chemical. The lethal chemicals are restricted to only three products-Starlicide, DRC-1339 and CPT.

AVICIDES

(a) Starlicide (3-chloro-*p*-toulidene hydrochloride) is the only registered avicide. It is found effective against starlings, blackbirds and gulls. It is a slow acting and apparently painless toxicant causing death due to uraemia i.e. uric acid concentrated in blood leading to circulation impairment in the liver and kidney. Baits containing 1% starlicide are mixed with plain bait in 1: 9 ratio and broadcast. Six weeks of baiting gives nearly cent per cent kill (Brooks & Husain, 1990)

(b) DRC-1339 (3-chloro-4-methylaniline hydrochloride) is emerging as a successful avicide. Its LD_{50} value has been determined for 55 bird species, acute toxicity is known for six species and dietary tests have been conducted on 12 species. Death is believed to be due to irreversible necrosis in the kidney. Pest birds like blackbirds, corvids, rock doves, gulls and starlings are highly susceptible to the DRC-1339 in the laboratory (Eisamann *et al.*, 2003) and some field conditions (Linz *et al.,* 2003).

(c) CPT (3-chlor-4-methylbenzenamine, CAS 95-74-9) also termed PA-14 is a potential aerial toxicant to control roosts of blackbirds (Mott *et al.*, 1987). It lowers the surface tension of water, enhances wetting and death is due to excessive loss of heat. It effectively reduced large populations of starlings and blackbirds (Lefebure & Seubert, 1970).

BIRD REPELLENTS can be auditory, visual and chemical. Auditory and visual repellents are factually scares that frighten away birds from depredating on crops.

Auditory repellents: These include both sonic and ultrasonic devices. The sound producing devices are exploders, shell crackers and rockets. All three make loud noises, sometimes accompanied by light. Propane cannons become more effective when supplemented with shooting. For better results these devices should be operated from different locations in the field and the time interval should be changed.

Electronic noise systems, synthetic bird calls and pyrotechnics are also used along with exploders. In developing countries such as India playing amplified pre-recorded alarm and distress calls of birds and changing their location often is both affordable and practical as most of the farmers are poor and uneducated.

Ultrasonic devices are ineffective in repelling birds

Visual repellents include scare crows, balloons, kites, plastic flagging and mylar tapes. Functionally visual scares cause startle response which is similar to aposematic colours and predatory odours. Being one of the oldest bird control device throughout the world, the scare crow can be made much more effective by increasing the numbers and by changing their position every two to three days.

Reflecting tape has been found effective to deter birds. The reflected sunlight accompanied by the humming as well as thunder-like noise when the tape is stretched by wind makes it a powerful repellent.

Tactile repellents: Several oily, sticky substances are avoided by birds. Polybutane products repel birds. Compounds containing aliphatic petroleum compounds, polybutanes, polyisobutanes, mineral oil, zinc oxide and diphenylamine when applied to surface repel pest birds. Lithium stearate soap has similar effect.

Chemical repellents act in one or more of the following ways. They cause pain, illness or they scare the animal. The basic principal of repellent use is to generate aversive signals which motivate the target animal to avoid the food items. Aversive signals are associated with pain, discomfort, illness. Thus repellents may not be lethal but they definitely disturb and make the animal sick physiologically or behaviourally. Clark (1998) further classifies chemical repellents into primary and secondary chemical repellents.

(i) Primary chemical repellents are olfactorily scaring, distasteful and cause irritation or pain or both. These are avoided upon first exposure. Examples include avoidance of predator odour by prey and effect of alarm pheromones. Other primary repellents are sucrose and thiram. Cinnamamide, coniferyl benzoates and acetophenones also act as primary chemical repellents for birds.

Methylanthranilate (O-aminobenzoic acid methyl ester): It is the only effective and registered (USA) primary bird repellent. It is a chemical irritant to birds and birds usually avoid normally preferred foods when these foods are adulterated with methyl antranilate at concentrations ranging 0.5 to 1.005.

Secondary chemical repellents cause illness or an unpleasant experience leading to aversion learning and avoidance of associated sensory cues. The three widely used repellents of this category are methiocarb, 4-aminopyridine and anthraquinone. Most secondary chemical repellents are derived from insecticides and Methiocarb from carbamate. Derivatives from lindane, captan, thiram, thymol, denatonium succharide are all potential bird repellents.

(a) Methiocarb is derived from a carbonate insecticide. It inhibits acetylcholinesterase activity. Birds associate post ingestional distress with the foods ingested and avoid it. It has been found effective in protecting a wide variety of crops across the world (Mott *et al.*, 1976; Ruelle & Bruggers, 1979; Brooks *et al.*, 1982; Sandhu *et al.*, 1987)

(b) 4-aminipyridine (Avitrol) is used as a frightening agent to repel social birds that feed in flocks. Birds ingesting lethal doses die. Those consuming sub-lethal doses behave erratically, give warning calls which in turn frightens the rest of the flock compelling them to move out from the crop field. Sub lethal doses result in bait shyness. It is used at 3% in the bait which is mixed with untreated grain in the proportion of 1:29 and broadcast over the field at the rate of 1 kg/ha (Sultana *et al.*, 1986).

(c) Anthraquinone: Unlike methiocarb and 4-aminopyridine, anthraquinone does not affect the nervous system of birds. Poison mixed foods are not rejected. The chemical repelled Canada geese, Brown-headed cow birds and red winged blackbirds from turf and grain fields reducing damage. In view of decreased potency a week after initial application, additional treatments were found necessary at frequent intervals (Dolbeer *et al.*, 1998).

CHEMICALS FOR RODENT AND OTHER MAMMALIAN PESTS

The largest number of chemicals is used for rodent control. Based on their action rodenticides are classified into:

1. Acute rodenticides
2. Sub-acute rodenticides
3. Chronic rodenticides
4. Fumigants
5. Chemosterilants

(1) ACUTE RODENTICIDES (Table 1)

These rodenticides are highly toxic, required in very small quantities in comparatively lesser quantities of baits consequently reducing requirements of bait, labour and time giving a higher cost: benefit ratio. The major advantage of acute rodenticides is their rapid action which is of unparallel use in situations of severe outbreak of diseases and plagues of rodents in agriculture. The quick action, with target animals dying within 24 hours makes the rodenticide popular with public, both urban and rural. Although theoretically the list of acute rodenticides includes a large number of chemicals such as zinc phosphide, thallium phosphide, strychnine, red squill, alpha-napthyl thiourea (ANTU), Arsenic trioxide, Barium carbonate, norbormide, Sodium monofluoracotate (1080) and Fluoracetamide (1081), phosphorous, pyriminal etc. only few of them meet the criterion of an effective, safe, rodenticide which Gratz (1973) defined as a chemical with high degree of toxicity to rodents, readily accepted, does not induce bait aversion and has a greater degree of specificity to rodents. The following is an account of acute rodenticides:

Table 1. Toxicity of Acute rodenticides to Indian rodents

Sl. No	Rodenticide	Species	Mortality (%)	LD_{50} (mg/kg)	Reference
1.	Zinc phosphide	*T. indica*	-	35.0	Prakash *et al*'69
		T.i. cuveiri	-	2.21	Srivastava'65
		M. hurrianae	-	35	Prakash *et al*'69
		R. rattus	-	40-300	Paranjothy'39 Barnett *et al*' 75
		M. musculus	-	250	Rao & Pakash'81
		B. bengalensis	-	2.51 to 25	Srivastava'65 Htun & Brooks'79

2.	Barium carbonate 15% in bait	*R. rattus*	-	975	Kunhardt & Chitre' 21
		M. hurrianae	37.5	-	-
3.	Arsenic trioxide 5% in Bait	*B. bengalensis*	-	Male-450 Female-350	Srivastava, 68
		M. hurrianae	42.8	1080 mg/kg (lethal dose)	Prakash'64
			28.5	200mg/kg	
			31.2	600mg/kg	
4.	Strychnine	*M. hurrianae*	-	5	Prakash *et al*'69
		T. indica	-	5(LD_{60})	-do-
5.	Thallium sulphate 3% in bait	*M. hurrianae*	50	20mg/kg	Prakash'64
		T. indica	30	20mg/kg	-do-
		M. hurrianae	100	-	-do-
6.	Alphanaphthyl thiourea	*R. rattus*	-	100-400	Richter'45
7.	Norbormide 0.5% in Bait	*T. indica*	60	500mg/kg	Prakash & Jain'68
		M. hurrianae	40		-do-
		R. meltada	-	50	-do-
		G. gleadowii	100	200mg/kg	
		B. bengalensis	-	306mg/kg	
		M. musculus	-	2250mg/kg	Kapoor *et al*'65
		R. rattus	Male-46	-	
			Female-47	-	
8.	Red squill	*R. norvegicus*	found effective		Sharif'51
		R. rattus			
		B. bengalensis			Srivastava *et al*'80
		T. indica			
9.	Sodium fluoroacetate	*T. indica*	-	0.9	Prakash & Jain'70
		M. hurrianae	-	0.6	

Zinc phosphide (Zn_3p_2) was first used as a rodenticide in Italy in 1911. It is a greyish black, dense powder with a characteristic garlic-like odour which is attractive to rodents but repels other animals. The chemical is soluble in dilute hydrochloric and sulphuric acids. On being exposed to moisture it decomposes slowly releasing phosphine gas which is highly inflammable. The phosphine gas is lethal. However, zinc phosphide is quite stable when kept dry and sealed in air tight containers.

In dry conditions zinc phosphide is quite stable releasing very small amounts of phosphine gas rendering dry baits toxic indefinitely.

Mineral oil added to enhance bait acceptancy and also to make the poison adhere to the cereal bait protects against moisture and makes the poison bait stable. Under field conditions zinc phosphide baits remain toxic for several months until eroded by weather or ants eating the grain component. When zinc phosphide is dusted into wet baits such as meat, fish, fruits and vegetable, it loses its potency rapidly. In soil the chemical breaks down rapidly into phosphine, which is either released into atmosphere or quickly converted into harmless phosphates and zinc complexes.

Pharmocology: The ingested zinc phosphide on coming in contact with dilute hydrochloric acid in the stomach gets hydrolysed and releases phosphine gas which causes death. Death occurs overnight and symptoms include convulsions, paralysis and coma. Death is due to asphyxia. In the early stages of poisoning zinc phosphide produces nausea, vomiting characterized by black vomit smelling of phosphorous, abdominal pain, tightness in the chest and feeling of coldness. Fatal cases result in death anywhere between 12 to 30 hrs with liver, kidney and heart damaged severely. Breathing small amounts of phosphine produces symptoms similar to food poisoning like diarrhoea and stomach pain.

Zinc phosphide poisoning does not produce distress symptoms till the terminal stage of death. Rodents die with legs and tail outstretched. There is no secondary poisoning of zinc phosphide as it is not stored in the stomach but the toxicant remains active inside the stomach of dead animal which may cause death in animals eating lethal quantity of gut contents.

Toxicity: Zinc phosphide is probably the most used rodenticide worldwide. It is generally toxic to almost all animals. Care must be taken to make bait inaccessible to children and pets when baiting in and around residential premises. It is effective against a wide range of rodent pests of India with LD_{50} values ranging from 2.5mg to 40/mg/kg with the exception of 250 mg/kg for *Mus musculus* (Rao & Prakash, 1981) and 300mg/kg for *Rattus rattus* (Paranjothy, 1939). For best results zinc phosphide should be used at 1-2% concentration in the bait as concentrations above this render baits unacceptable and induce bait shyness (Idris & Prakash, 1992).

Extreme caution should be exercised while preparing and handling zinc phosphide baits. Preparing baits outdoors does not pose any hazard but if the poison bait is mixed indoors, adequate precautions against slow, poisoning of phosphine should be taken. As even zinc phosphide dust is hazardous indoor mixing should be done only if there is good ventilation and provision for exhausts. It is advisable to wear gloves while preparing baits as the phosphorous is likely to dissolve in small amounts in fatty acids and may be absorbed through the skin. Continuous absorption is toxic to the body immediately or later in life.

Strychnine (2, 4a, 5, 5a, 7, 8, 15, 15a, 15b, 15c-dehydro-4, 6-methano-6H, 14H-indolo (3, 2, 1, -ij) oxepino (2, 3, 4, -de) pyrrolo (2, 3-h) Quinolin-4-one): Strychnine is one of the oldest rodenticides being used since seventeenth century. It is derived from the dry, ripe seeds of *Strychnos nox vomica,* a small tree native to India, North Australia, Vietnam and Sri Lanka. The white crystalline powder is readily soluble in water and is very bitter in taste. It was commonly used to control rodents, pest birds, and nuisance mammals but in 1988 its use above the ground was banned except to control subterranean mammalian pests in USA.

Strychnine is the fastest acting of all rodenticides. It is neither stored in the body nor absorbed through the skin. The highly toxic chemical upon ingestion enters the blood and acts on central nervous system. It acts very fast in animals with an empty stomach. Symptoms which appear 5 to 30 minutes after ingestion are mainly in the form of convulsions with death occurring due to respiratory failure. Similar to zinc phosphide, the poison is not accumulated in tissues or bone but the remnants in the gastrointestinal tract become lethal to animals eating the alimentary canal. Birds are less susceptible to strychnine.

Being highly toxic to mammals the LD_{50} ranges from 0.7 to 1.5 mg/kg for many mammals and for rats it is reported to range from 4.8 to 6.0mg/kg (Gratz, 1966). Amongst Indian species the LD_{50} and LD_{60} for *Meriones hurrianae* and *Tatera indica* was 5.0 mg/kg (Prakash *et al.,* 1969). Its bitter taste and rapid action makes rodents to be cautious and ingest sub lethal doses.

Antidotes are effective if administered immediately after poisoning. Sodium pentobarbital and sodium amytal reduce convulsions. Emetics should be administered after the convulsions cease.

Thallium sulphate (Tl_2SO_4) is a white crystalline powder, soluble in water, with low levels of odour and taste. Because of these it is readily accepted by rodents, a fact which has made the rodenticide a popular one to control rodents and predators at 1.5% in the bait. But its extreme toxicity, profound metabolic disturbances in almost all the systems of the body and ability to penetrate unbroken skin make it a hazardous chemical to use on a large scale leading to the ban on its use by WHO in 1973. At 3% in the bait it killed both *M. hurrianae* and *T. indica* (Prakash, 1964).

Barium carbonate ($BaCo_3$) is a white crystalline solid, not readily soluble in water but reacts with stomach acids to produce barium chloride which is absorbed by the body and is lethal. It affects the central nervous system, respiratory and digestive systems, the symptoms being strong diarrhoea and diuresis. Although it was a popular rodenticide till the Second World War, its low and highly variable toxicity to rodents coupled with high toxicity to humans, livestock, pets and birds has gradually led to its replacement with safer and more toxic rodenticides.

Arsenic trioxide (AS_2O_3) is the earliest known rodenticide and is still in use. It is almost tasteless, insoluble in water and very stable in baits. Death is due to damage to gastrointestinal lining resulting in haemorrhage and shock, the symptoms being diarrhoea and thirst. Although suggested to be used at 1-10% in the bait, it is not encouraged and is restricted in its use because of highly erratic toxicity to different species of rodents and misuse of the poison for criminal acts of human poisoning.

Alphanapthyl Thiourea (ANTU) [$CloH_7NHCSNH_2(C_{11}H_{10}N_2S)$] is in use only from 1945. It is a fine bluish grey powder, stable in air and bait, insoluble in water and bitter in taste. It is an excellent poison for Norway rats but unsatisfactory for the control of roof rats and house mouse. It causes pulmonary oedema and pleural effusion with death occurring from 16 hours to 8 days. Suggested concentration for use is 1-3% but is banned in many countries because of toxicity to non-target animals and carcinogenic effects.

Norbormide (C5-(1-hydroxy-λ-Z-pyridyl-benzyl)-7(2-pyridyl-benzylidene) norborn-5-ene-2, 3-dicarboximide)]. It is an off-white crystal and specifically toxic to *Rattus* species and was found unsatisfactory to Indian rodent species (Prakash & Jain, 1968; Kapoor *et al.*, 1965). The poison acts as a vasoconstrictor, elevating blood pressure dangerously. Although it is not toxic to non-target animals, its selective action on *Rattus norvegicus* makes it unsuitable for use in situations where other species of rodent pests occur.

Red squill or scillirocide ($C_{32}H_{44}O_{12}$) is one of the oldest rodenticide still in use. It is extracted from the bulbs of *Urginea maritoma*. The poison is yellow-pink to dark brown in colour, is granular with a bitter taste. The poison contains a heart glycoside and a nerve glycoside, both are toxic. Due to its emetic properties, it is not hazardous to non-target animals. Death is due to paralysis of heart. Symptoms are severe diarrhoea and excessive urination. The pure chemical has LD_{50} at 1-2-mg/kg and 0.015 to 0.05% in the bait is lethal

(Fitzwater & Prakash, 1978; Srivastava *et al.,* 1980). Mice are not susceptible to red squill.

Sodium fluoroacetate is also known as sodium mono-fluoroacetate and compound 1080. The rodenticide is a water soluble granular powder and is the most toxic among rodenticides. It is a metabolic inhibitor by synthesising fluorocitrate which blocks a crucial step in the Krebs cycle leading to neurotoxicity ultimately resulting in convulsions, failure of respiratory and circulatory systems. The LD_{50} values are very low falling at 0.1 to 8.0 mg/kg for various species of rodents. Amongst Indian rodents the LD_{50} was 0.9mg/kg for *T. indica* and 0.6 mg/kg for *M. hurrianae* (Prakash & Jain, 1970). It is effective at very low concentration of 0.08 to 0.5% in solid, food baits and 0.25 to 0.5% in liquid baits. Not recommended because of extreme toxicity to non-target mammals.

Miscellaneous rodenticides

In late seventies (1974-1980), a new rodenticide RH-787 also called "vacor" was found quite effective against almost all Indian rodents (Prakash & Mathur, 1987) but detection of carcinogenic effects lead to withdrawal of the chemical. Other not well known rodenticides include gophacide, thiosemicarbamide, alphachloralose, silatrane, crimidine etc.

(2) SUB-ACUTE RODENTICIDES

The category of rodenticides which exhibit characters of acute rodenticides by requiring consumptions only for a day or two but resembling chronic rodenticides in producing lethal effect 4-10 days after ingestion of poison are termed sub-acute rodenticides (Rennison, 1974). Calciferol, cholecalciferol, flupropadine and bromethalin fall under this category.

Calciferol and Cholecalciferol: The vitamin D_2 activated ergo-sterol is considered moderately hazardous rodenticide with oral toxicity of 15-17mg/kg for *Mus musculus* and 40mg/kg for *R. norvegicus* (Greaves *et al.,* 1974). Consumption of baits containing 0.1% calciferol, 0.1% calciferol plus 0.025% warfarin and 0.025% cholecalciferol for 1-2 days resulted in 100% mortality of *R. rattus* (Arora *et al.,* 1982; Saini & Parshad, 1992).

Bromethalin is toxic at 2 to 8mg/kg for *M. musculus* and *R. norvegicus.* It is readily acceptable, hence no prebaiting is required and there is no poison shyness. Baits containing 0.005% bromethalin was found effective against *B. bengalensis* in Bangladesh (Mian *et al.*, 1993).

(3) CHRONIC RODENTICIDES (ANTICOAGULANT)

The anticoagulants successfully used at present are derivatives of either hydrocoumarin or inandione compounds. The coumarin derivatives are warfarin, fumarin, coumachlor and racumin (Coumatetralyl). The inandione derivatives are piral, pivalyn, diphacinone, chlorophacinone, PMP etc. The wide spread occurrence of warfarin resistance in Europe and America lead to the discovery and use of second generation anticoagulants namely difenacoum, bromadiolone, brodifacoum, flocoumafen and difethiolone.

Anticoagulants act by preventing blood clotting on one hand and causing damage to blood vessels on the other hand. They interfere with the synthesis of vitamin K in the liver, which is essential for production of prothrombin levels. When prothrombin levels fall, clotting cannot occur, the damaged blood vessels bleed continuously causing mortality.

First generation anticoagulants (Tables 2 & 3) also known as multiple dose anticoagulants are cumulative in their effect. They are consumed in small doses over a period of several days before a lethal dose is achieved, death occurs several days after the poison consumption. Of the several first generation anticoagulants only 0.025% warfarin and 0.025% fumarin have been evaluated extensively and used. Warfarin has been found to be effective against most Indian rodents at 0.025% fed over 8-14 days (Mathur *et al.*, 1992). In comparison fumarin was effective against *R. rattus*, less so in *B. bengalensis* (Sridhara, 1979), squirrels, house mice and not at all in *R. meltada* (Mathur *et al.*, 1992). Several factors like long duration of feeding with resultant higher requirements of baits, bait boxes, labour and lengthy periods for mortalities rendered warfarin and fumarin unpopular in India although warfarin resistance was an equally important reason for its discontinued use elsewhere in the world.

Racumin (coumatetralyl) at 0.075% tracking powder and 0.0375% in cereal bait has been found as effective as second generation anticoagulants against *B. bengalensis* and less so against other species (Chopra & Parshad, 1985; Parshad & Malhi, 1995; Mathur & Prakash, 1984). It appears to be a promising rodenticide for wider application in godowns in metropolitan cities and irrigated crop fields where *B. bengalensis* is the predominant species.

Table 2. Mortality of Indian rodents subjected first to generation anticoagulants during choice tests (Mathur *et al.*, 1992)

Species	% mortality					
	Chlorophacinone		Coumatetralyl		Warfarin	Fumarin
	0.075%	0.01%	0.0375%	0.05%	0.025%	0.025%
F. pennanti	50	83	83	83	17	8
T. indica	50	60	70	70	40	50
M. hurrianae	58	90	80	80	75% at 0.0125% 92% at 0.025%	60
R. rattus	-	-	55	-	42	-
B. bengalensis	-	-	90	-	-	-
M. booduga	-	-	-	-	34	-

Second generation anticoagulants (Tables 4 & 5): Difenacoum, brodifacoum, bromadiolone, flocoumafen and difethiolone belong to this group. Used at 0.005% in the bait, either loose grain, liquid form or wax formulations except difethiolone which is used at 0.025%, second generation anticoagulants have been found effective both in the laboratory and field conditions over a wide range of rodent species and crop conditions. Apart from lower concentrations, they are also more effective; require less time to cause mortality and effective in killing warfarin resistant rodents. Amongst these brodifacoum and flocoumafen take less time than bromadiolone to be effective. The laboratory efficacies are corroborated by field tests (Table 5).

Table 3. Efficacy of First generation anticoagulants in different crops, cropping systems and fields (Mathur *et al.*, 1992)

Sl. No	Crop/ Cropping system/ land	FIRST GENERATION ANTICOAGULANTS							
		Warfarin (0.0025%)		Fumarin (0.025%)		Coumatetralyl (0.0375%)		Chlorophacinone (0.0075%)	
		location/ species	% cont.	location/ Species	% Cont.	Location/ species	% Cont.	Location/ Species	% Cont.
1.	PADDY	1.Bangalore/Bb 2.Titupathi/Bb, Mm, Mb, Bi	79.6 66	1. Bang-alore/Bb 2. Banga-lore/Rr, Bb	79.55 93	-	-	-	-
2.	WHEAT	1. Ludhiana/Bb	97	-	-	-	-	-	
3.	SUGARCANE & GRAPES	1. Maharashtra/ Bb, Ti, Mb	100	-	-	-	-	-	-

4.	MAIZE	1. Maharashtra/ Bb, Ti, Mb	90.5	-	-	-	-	-	-
5.	COCO-NUT	1. Lakshwad-weep/Rr 2. Minicoy/Rr 3.Krishna Dt. (A.P)/Rr 4.Arasikere-(Karnataka)/Rr 5. Gujarat/Rr	100 68.4 84 100 61.9	-	-	-	-	-	-
6.	OIL PALM	1. Little Andaman/Rr	100	-	-	-	-	-	-
7.	GRASS-LAND	-	-	-		Bikaner/ Mh, Ti, Gg	66.9	Bikaner/ Mh, Ti, Gg	72.4
8.	POUL-TRY FARM	Ludhiana/Rr	57.1	-	-	Ludhiana/ Rr	55.5	-	-
9.	VILL-AGE/ HOUSES	Hapur, Jodhpur, Hyderabad, Ludhiana, Jaipur/Rr, Mm	66-100	Ludhiana/Rr	effe-ctive	Hapur/Rr, Mm	71.28	-	-

Bb-*Bandicota bengalensis*; Mm-*Millardia meltada*; Ti-*Tatera indica*; Mb-*Mus booduga*, Rr-*Rattus rattus*; Mh-*Meriones hurrianae*; Gg-*Gerbillus gleadowii*

Table 4. Mortality of Indian rodents subjected to second generation anticoagulants during choice tests (Mathur *et al.*, 1992)

S. No	species	SECOND GENERATION ANTICOAGULANTS (0.005%)								
		Bromadiolone			Broadifacoum			Flocoumafen		
		0.025 %	0.005 %	0.01%	0.002 %	0.0025%	0.005 %	0.0025%	0.005 %	0.01%
1.	*F. pennanti*	-	100	70	100	-	100	60	60	-
2.	*T. indica*	100	100	100	100	-	100	70	100	-
3.	*M. hurria-nae*	85	85	-	100	95	100	87.5	60	-
4.	*M. meltada*	-	100	-	100	-	100	-	100	-
5.	*R. rattus*	100	97	-	-	-	-	87.5	80	80
6.	*B. bengale-nsis*	-	90	-	-	-	100	100	-	-
7.	*B. indica*	-	90	-	-	-	-	-	-	-
8.	*M. musculus*	-	80	-	100	-	100	-	70	-
9.	*M. booduga*	92	89	-	-	92	100	-	-	-
10	*M. platythrix*	-	100	-	-	-	-	-	-	-
11.	*G. ellioti*	-	-	-	100	-	100	-	100	-

Table 5. Efficacy of second generation anticoagulants in different crops, cropping systems and field conditions (Mathur *et al.*, 1992)

Sl. No	Crop/ Cropping system/ Field condition	SECOND GENERATION ANTICOAGULANTS (0.005%)					
		Bromadiolone		Brodifacoum		Flocoumafen	
		Location/Sp.	% Control	Location/ Sp.	% Control	Location/ Sp.	% Control
1.	PADDY	Hyderabad/Bb, Mb,Rr	87-94	Hyderabad/Bb, Mb, Mm	68	Bangalore/ Bb,Mm	88 91
		Tirupathi/Mb, Bb, Bi, Mm	68-79	Tirupathi/Mb/Bb/ Mm	74	Jabalpur/ Bb, Rr	
		Bangalore/Bb	100	Ludhiana/Bb	60		
		Punjab/Bb, Mm, Mus Sp.	70-80				
		Manipur/Bb, Mm	100				
		Tamil Nadu/Bb, Mm	90-100				
2.	WHEAT	Jodhpur/Mh,Ti, Mm	95 80-100	Ludhiana/Bb,Ti, Rm, Mm, Mp	68-96	Ludhiana/ Bb	72
		Ludhiana/Mm,Bb	93-98				
		Maharashtra/Bb	94				
		Junagadh/Bb,Ti	62-64				
		Garhwal Himalayas					
3.	SUGARCA NE	Ludhiana/Mm,Bb	79-100			Ludhiana/ Mm	65-85
		Bangalore/Bb	70				
		Kanpur/Bb	97				
		Jabalpur/Bb	99				
4.	MILLET	Jodhpur/Mh	93				
5.	GROUND-NUT	1. Junagadh/Bb	95	Ludhiana/Bb,Ti, Mm, Mus sp	67		
				Bangalore/Bb/ Mm	79		
6.	GRAM	1. Nagpur/Bb	89-91				
7.	MIXED CROPS: WHEAT, GRAM, BARLEY, PULSES & VEGE-TABLES	New Delhi/Ni, Rr, Mh, Bb	99				
8.	FIELD CROPS: G.NUT, PADDY, JOWAR, S.CANE,	Maharashtra/ Bb, Bi	92-93				
		Tenali, A.P/ Bb, Rr	93				

	BAJRA, LUCERNE				
9.	JHUM CULTIVA-TION-MIXED CROPS	Arunachal Pradesh/ *R.r. brunnensculus*	100		
10	WHEAT-PADDY-SUGAR-CANE	Uttar Pradesh	89-93		
11.	COCONUT	AP, Karnataka, Gujarat, Lakshadweep, Minicoy islands/Rr	78-100	Minicoy islands/Rr	74-76
12	OIL PALM	Little Andamans/ *R.r. andamanensis*	100		
13	COCOA	Kerala/Ft, *R.r.wroughtonii*	84	Kerala/R.r. wroughtonii	76-95
14.	GROUND-NUT-COTTON-MOONG-RICE-ARHAR-S.CANE	W.Rajasthan/Mm, Mh, Ti,Bb,Mm	82-94		
15	FODDER/ GRASS-LANDS			Bikaner/Mh, Mm, Ti, Bb, Mm	91

Bb-Bandicota bengalensis; Mm-Millardia meltada; Ti-Tatera indica; Mb-Mus booduga, Rr-Rattus rattus; Mh-Meriones hurrianae; Gg-Gerbillus gleadowii, Ft-Funambulus tristriatus; Rra-Rattus rattus andamanensis; Ni-Nesokia indica; Mp-Mus platythrix; Rrb- Rattus rattus brunneusculus.

Being marketed in India since 1988 bromadiolone has been subjected to innumerable evaluations in almost all types of crop fields across the country infested by rodents. The percent control achieved was 68-108% in paddy, 62-100% in wheat, 70-100% in sugarcane, 93% in millet, 95% in groundnut, 89-91% in grain, 82-99% in mixed crops, 100% in *jhum* cultivation, 78-100% in coconut plantations, 100% in oil palm and 84% in cocoa (Table 5).

Broadifacoum has not been tested so widely. The limited evaluations show 60-74% control in paddy, 68-96% in wheat, 67-79% in groundnut, 73-74% in coconut, 76-85% in cocoa and 91% in fodder-grassland (Table 5).

Flocoumafen is least evaluated: only in paddy, wheat and sugarcane giving 88-91%, 72% and 65-85% population reduction respectively (Table 5).

The wide disparity in success with second generation anticoagulants in crop fields, ranging from 60-100% is attributed to several ecological, behavioural and baiting factors. The most important ones are the differences in pest species infesting, density, size of the baiting area, climatic and agroclimatic changes, harbourage, availability of alternate foods, behavioural responses to poison baits, method of bait application, and timing of poisoning, reinfestation and methods of evaluation (Parshad, 1999).

(4) FUMIGANTS

Fumigants are effective in killing rodents and other vertebrates which are inaccessible to control using poisons such as buildings, ships, burrows, caves etc. Generally fumigants cause death by asphyxiation which can be either simple or through chemical gases. It is advisable that fumigants are used by experienced technical personnel as the gases liberated are dangerous to the persons using, other humans and animals around the fumigating area. The fumigants employed for rodent control are calcium cyanide (releases hydrogen cyanide), methyl bromide, chloropicrin and aluminium phosphide (releases phosphine). Some conventional and unconventional methods are used to fumigate burrows with carbon-di-oxide, carbon monoxide and sulphur di oxide.

Calcium cyanide ($Ca\ (CN)_2$) is greyish white powder or in granular form which is blown or spooned into a burrow. Soon after blowing the chemical all the burrow holes have to be closed as the gas released, hydrogen cyanide being lighter than air accumulates in the upper part of the burrow system. Normally calcium cyanide is blown into the burrows using a special pump. Amyl nitrate is the antidote for hydrogen cyanide poisoning. Its ampules should be carried by the personnel involved in calcium cyanide fumigation. Due to its hazardous nature Calcium cyanide should not be used within 3 meter radius of buildings.

Methyl bromide, (CH_3Br) is an odourless, highly toxic gas, usually used to control insect pests of storage, and warehouses to kill rodents before grain is stored. Though effective when used to fumigate rodent burrows using a pump by a trained person, methyl bromide is quite hazardous and is not recommended for rodent control.

Chloropicrin (CCl_3No_2) also known as tear gas, is occasionally used to expel mice from grain storage areas. It is effective in killing mice in closed buildings and also as a burrow fumigant. As it is a chemical irritant, not being a common fumigant, information on its efficacy as a rodenticide is almost nil.

Aluminium phosphide, (Al_3P_2) is available in the form of a highly compressed tablet containing aluminium phosphide and ammonium carbonate. On being exposed to moisture, the compound decomposes into hydrogen phosphine, aluminium hydroxide, ammonia and carbon di oxide. A 3g tablet releases 1g of hydrogen phosphine. The odour is colourless, highly toxic, spontaneously inflammable, and heavier than air and smells like garlic. It has been found effective in paddy fields in India (Sridhara & Krishnamoorthy, 1979; Advani & Mathur, 1981) and lower Sind in Pakistan (Greaves *et al.,* 1977). However the efficacy is dependent on soil moisture and also the area, depth and ramification of the burrow rendering it selectively lethal to rodents of simple and shallow burrows, tight soil and preferably in irrigated fields. Simple method of using, lower cost, safety to non-target species, shorter duration to death and high percent of kill are the advantages of using aluminium phosphide tablets.

Carbon di oxide, (Co_2) has been successfully used to fumigate mice infested cold storage warehouses and corn-ricks. It is used as solid Co_2 (dry ice) in refrigerated warehouses where low temperature has to be maintained. After distributing the crushed dry ice, an electric fan is switched on to dispose the gas.

An indigenous method of fumigating rodent burrows in paddy fields by burning dry farm waste like paddy straw has been used successfully in Andhra Pradesh which is also cost-effective. The smoke generated is pushed into the burrow tunnel with the help of a blower. Within 10-15 minutes the burrow gets filled with smoke and the rodents inside die of suffocation (Reddy *et al.*, 2005).

Carbon monoxide (Co): The easiest way to fumigate rat burrows using carbon monoxide is to attach a flexible hose to the exhaust pipe of a petrol engine and after inserting the other end into a rodent burrow run the engine for about five minutes. Since there is a possibility of carbon mono oxide getting back into the exhaust system and gas leaking into the vehicle, care should be taken to ensure good ventilation of the vehicle.

Sulphur di oxide (So_2), a colourless, non-inflammable gas with a strong suffocating smell was used earlier to fumigate ships. Although smoke liberated by burning sulphur mixed with potassium nitrate and tallow is excellent to ferret out rats from their burrows, it is not advisable to use it as a burrow fumigant.

From the above account of rodenticides zinc phosphide and bromadiolone emerge as the only two effective and available rodenticides in India. However, singly used neither zinc phosphide nor bromadiolone can give optimum rodent control. The former because of bait shyness in animals that have survived sub-lethal doses and the later because of inherent genetic resistance. To make rodent control successful and economical, it is ideal to kill the population using zinc phoshide in the initial stage. If the fields are irrigated and small holdings, the surviving population can be fumigated with aluminium phosphide. However it is more convenient to kill the residual population with burrow or broadcast baiting using bromadiolone wax cakes or loose baits.

(5) CHEMOSTERILANTS

Chemosterilants are mostly organic compounds capable of sterilizing male or female rats either permanently or temporarily. Chemicals like clomiphene, tetradifon, furadantin, colchicine, glycopherol, ethyl methane sulphonate and alphachlorohydrin have been tested for antifertility effect in Indian rodents. Since rodents are prolific breeders where one fertile male can mate with several females, sterilizing a portion of males in the population is virtually useless. This narrows the selection to a chemical that sterilizes females effectively. An ideal chemosterilant should be one that should kill in lethal doses and render surviving populations sterile so that control programme results in very low pest populations whose recovery is suppressed, at least for considerable time. Only one chemical at present has these characters. Alphachlorohydrin (Epibloc) is toxic to rodents at lethal doses and exerts chemosterilant effects on rats ingesting sub lethal doses. The chemical has been widely tested both in the laboratory and fields in Punjab. At 0.5% in cereal bait it effectively killed *B. bengalensis* both in the laboratory and fields. In the later situation the remaining population after 82.9% control success became permanently sterile (Saini & Parshad, 1988, 1992). Although alphachlorohydrin is readily available outside the country, it is yet to be made available in India for rodent pest management.

MAMMALIAN REPELLENTS

The large mammalian herbivores like elephant, non-human primates and some ungulates which cause extensive damage regularly or during periods of drought or around degraded habitats are difficult and sensitive to manage. Almost all nuisance species are under conservation status and protected by Wildlife Act of India (1972). The two non-lethal approaches are repellents and exclusion. There is good amount of research on repellents both for mammals and birds in USA as well as Europe. Much of it is summarized below for adaption to Indian problems. Exclusion by several methods is the more prevalent practice in the country.

Repellents of late are receiving lot of attention in vertebrate pest management as they are non-lethal and environmentally safe. Repellents can be chemical, visual, acoustic or combination of selected characteristics of all these.

Mammal Repellents: Chemical repellents cause sensory irritation or gastric illness or mimic a semiochemical. Neophobia and aversion learning form the basis for visual and acoustic repellents. Although possibilities of forcing animals to abandon an area using chemical repellents exist, at present they are effective in reducing consumption when applied directly to foods. Abandonment is possible in rare instances in the presence of alternate, palatable foods at locations away from treated areas (Milunas *et al.,* 1994). However, when alternate foods are scarce or not palatable, animals come back to treated areas and resume eating the treated crop/crop produce (Elhani & Conover, 1998).

Chemical irritants are more effective in repelling mammalian depredators than semiochemicals or chemicals that cause post ingestional distress. Because not only avoidance is immediate, aversion and its persistence does not include learning and habituation is minimum. The immediate, fast response is attributed to the fact that irritants are chemically similar to endogenous substances released when tissue is damaged (Clark, 1998). The popular and effective mammalian irritants include *capsaicin*, the hot principle in capsicum, allyl isothiocyanate, the active principle in mustard, tear gas, ammonia, carbon di oxide and formaldehyde (Mason & Otis, 1990).

Irritants are broad spectrum chemicals, effective within a taxonomic group and the mammalian repellent, capsaicin rarely affects birds. Conversely the bird repellent, methyl anthranilate hardly acts on mammals at least at the concentration levels that are effective against birds.

Semiochemicals best exemplified by predator urine and their smell effectively repel herbivores. The presence of sulphur compounds and volatile fatty acids are causative agents (Mason, 1997), because sulphur signals presence of predators or plant toxicants. Unlike capsaicin, work on practical application of semiochemicals is not much, although there are instances of animals moving away from treated areas (Milunas *et al.,* 1994).

When gastro-intestinal distress is accompanied by ingestion of sub-lethal doses of a toxicant, conditioned taste aversion or simply poison shyness develops. The phenomenon is based strongly on taste. This behaviour has been successfully used to prevent coyote predation on sheep (Conover & Kessler, 1990), racoon predation on eggs (Nicolaus, 1987) and bird depredation on grain and fruits (Avery, 1989).

Visual repellents are more effective for birds because of their comparatively better vision and the presence of colour vision in them. In spite of possibility of repellent action of eyespots, predator effigies, Mylar ribbons etc against birds the only successful case of visual repellent in mammals is strobe light disrupting coyote predation in sheep (Linhart, 1984).

Auditory repellents are not so successful even to affect mammals although quite effective for birds. But this has not deterred farmers and villagers throughout the world from shouting, beating drums, and making all possible noises to drive away crop raiding elephants, monkeys and ungulates. However this sound production if accompanied by fire balls, lighted torches etc, the cumulative effect temporarily scares elephants from entering the field or makes them to retreat. In India at the time of grain maturity and hence vulnerable to bird depredation, women and children guard the crops. Every time a flock is sighted, they shout and simultaneously throw stones at them. The method is simple, effective and people are hired for this purpose. Similar approach is employed to keep away other crop raiding herbivores like monkeys and nilgai, where a group of professionals may also be employed to repel (Chhangani *et al.*, 2005). The claims of manufacturers that ultra sound emitting gadgets keep away insects and rodents lack evidence.

Mason (1998) emphasizes that repellents act best when they are a part of an integrated pest management with or without involving lethal methods. Acoustic signals like propane exploders used with guard dogs are one such example. Occasional shooting enhances acoustic repellent activity. The best strategy seems to be integrating non-lethal and lethal approaches.

DISCUSSION

The problem of vertebrate pests in India as elsewhere in tropical Asia and Africa has existed from time immemorial, much more severely than in Europe and America. However, their scientific management is receiving attention only from past three-four decades. Although rodenticides and avicides have been evaluated in the laboratory and field conditions elsewhere, the recommendations originating from these places cannot be adapted in India for the simple reason that the agro-ecological scenario and the pest complex is not only different but also the physiology, ecology and life history strategies of the pest species is also different. This makes it imperative to modify the already standardized methods to suit Indian conditions and evaluate it. Even toxicity studies have to be repeated for Indian species.

Amongst acute rodenticides, zinc phosphide is the only rodenticide commercially available. Laboratory evaluations have shown it to be quite effective against all the Indian rodents at 1 to 2% in the bait. Field tests have shown it to be quite satisfactory killing about 60-70% of pest population. The toxicant is economical, effective and popular. However two factors of poison application need elaboration. Since most species of rodents are neophobic (Sridhara, 1977) prebaiting for 2-3 days with unpoisoned bait is vital. Secondly with wide geo-and ecological conditions prevalent in India from alpine weather in Himalayas to tropical and sub-tropical in rest of the country with arid desert zone in the western corner, the rodent pest species are diverse. Being adaptive feeders and intelligent learners rodents across the country exhibit diverse feeding preferences which rules out the possible use of a common bait mixture. Regional tests are important to prepare attractive baits for field conditions as the rodent has an alternate, readily available and nutritious food in its habitat. Finally rodents surviving sub-lethal doses of zinc phosphide develop poison aversion (Sridhara, 1999) and refuse not only zinc phosphide but also all the components associated with the poison bait namely cereal and oil. However, used at 2% in preferred bait after prebaiting for two-three days zinc phosphide effectively knocks down a large proportion of infesting rodent population economically and quickly, the two qualities which make it a very popular rodenticide amongst farmers.

Aluminium phosphide though its sale is restricted, is selectively effective in irrigated cropping systems and is easy to apply.

The same problem of commercial availability is faced by anticoagulants. Only bromadiolone is available in loose and wax block formulations although coumatetralyl and difenacoum have been found effective both in the laboratory and field conditions. Warfarin resistance was vastly prevalent in USA, UK and Europe leading to formulations of second generation anticoagulants too. Of late there are reports of resistance to second generation anticoagulants. Both cases of resistance are undoubtedly due to prolonged use of anticoagulants, a situation which will take a long period to occur in India because rodent control except in factories, industries, star hotels, big poultry and meat industries, who can afford contract basis pest control, remains a remedial option for majority of farmers and people i.e. control measures are taken after the problem of rodents becomes visible that too at individual level. Although rodent control can be successful only when carried over entire villages/block of houses, neither the public nor the administration of state and country have a policy to facilitate such a programme on an annual basis at least. Apart from losses to agriculture, rodents are of considerable importance in public health. Government need to pay much more attention to motivate local authorities-civic, NGOs and others to make rodent control an annual event of public participation.

Bird pest management is comparatively simple mainly because damage to cereals is not alarming in India. A number of fruits are vulnerable to bird damage. The lesser degree of damage coupled with conservation status of birds restricts bird management techniques to non lethal approaches. These are mainly netting, block plantation, scare crows, reflector ribbons, pyrotechnics and playing of distress and alarm cells. Still the average Indian farmer with small holdings and bigger problems of crop loss to insects, diseases and rodents is content with hiring or employing family members to guard fields during vulnerable stages of damage by birds. The problem is severe for sunflower growers who are trying block plantation and netting to alleviate the problem. Sometimes poverty, religion and lack of education are a blessing for conservation purpose because an average Indian cannot imagine using avicides, pyrotechniques, shooting or any of the sophisticated methods used by his counterparts in Western world. To that extent our avifauna is saved.

REFERENCES

ADVANI, R & R.P. MATHUR. 1982. Experimental reduction of rodent damage to vegetable crops in Indian villages. *Agro ecosystems* 8:39-45

ARORA, K.K., S.S. LAL & S. RANI. 1982. Studies on newer anticoagulant rodenticides for the control of black rat (*Rattus rattus* Linn) *Pestology* 6: 5-7

AVERY, M.L. 1989. Experimental evaluation of partial repellent treatment for reducing bird damage to crops. *J. Appl. Ecol.* 26: 433-439

BARNETT, S.A., P.E. COWAN, G.G. REDFORD & I. PRAKASH. 1975. Circadian rhythm of movements of the house rat, *Rattus rattus. Indian J. exp. Biol.* 13: 153-155.

BROOKS, J.E. & I. HUSSAIN. 1990. Chemicals for bird control. In: VERTEBRATE PEST MANAGEMENT – A TRAINING MANUAL. (Eds.) J.E. Brooks, E. Ahmed, I. Hussain, S. Munir & A.A. Khan. Pakistan Agricultural Research Council, Islamabad, Pakistan. Pp. 193-195.

BROOKS, J.E., P.T. HTUN & H. NAING 1980. The susceptibility of *Bandicota bengalensis* from Rangoon, Burma to several anticoagulant rodenticides. *J. Hyg. Camb.* 84: 127-135.

CHOPRA, G & V.R. PARSHAD. 1985. Evaluation of coumatetralyl against two predominant murid species. *J. Hyg. Camb.* 94: 327-330.

CLARK, L. 1998. Physiological, ecological and evolutionary bases for the avoidance of chemical irritants by birds. In: CURRENT ORNITHOLOGY. V. Vol. 14 (Eds.): Nolan, V. & E. Ketterson. Plenum Press, New York. pp.1-137.

CONOVER, M.C & K.K. KESSLER 1990. Diminished producer participation in an aversive conditioning program to reduce coyote predation on sheep. *Wildl. Soc. Bull.* 22: 229-233.

COSLICK, J.W & D.J. DECKER 1979. Economic feasibility of a deer-proof fence for apple orchards. *Wildl. Soc. Bull.* 7: 173-175.

EISEMANN, J.D., P.A. PIPAS & J.L. CUMMINGS 2003. Acute and chronic toxicity of compound DRC-1339 (3-chloro-4-methyl-aniline hydrochloride) to birds. In: MANAGEMENT OF NORTH AMERICAN BLACKBIRDS *Proc. Special Symp. of the Wildl. Soc. 9th Ann. Conf.* (Ed.) G.M. Linz. Bismarck, North Dakota, USA. pp. 49-63.

ELHANI, A & M.R. CONOVER 1998. Comparative analysis of deer repellents. In: REPELLENTS IN WILDLIFE MANAGEMENT (Eds.): Mason, J.R. Colorado State University, Fort Collins, Colorado, USA. pp. 147-155.

FITZWATER, W.D & I. PRAKASH 1978. HANDBOOK OF VERTEBRATE PEST CONTROL. Indian Council of Agricultural Research, New Delhi, 92p.

GRATZ, N.G. 1966. A critical review of the currently used acute rodenticides. *Proc. Sem. on Rodents and Rodent Ectoparasites.* Geneva 99: 79-88.

GREAVES, J.H., M.A. CHOUDHURY & A.A. KHAN 1977. Pilot rodent control studies in Sind, using five rodenticides. *Agroecosystems.* 3: 119-130

HTUN, P.T. & S.E. BROOKS 1979. Laboratory evaluation of the toxicity and efficacy of zinc phosphide used as a rodenticide against the lesser bandicoot rat, bandicoot rat, *Bandicota bengalensis* PANS 25(3): 245-250.

IBID. 1970. Eco-toxicology and control of Indian desert gerbil, *Meriones hurrianae* Jerdon. 6. one shot baiting technique. *J. Bombay nat. Hist. Soc.* 67: 274-278.

IBID. 1979. Rodenticidal action of four poisons in *Bandicota bengalensis* and *Tatera indica. Pest control.* 47(9): 30-31.

IBID. 1988. Evaluation of toxicant–sterilant: alpha-cholohydrin, for the control of Indian mole rat *Bandicota bengalensis. Ann. appl. Biol.* 113: 307-312.

IBID. 1999. Quarter century of rodent research on rodent control (1973-98) in the University of Agricultural Sciences, Bangalore-high-lights and recommendations. University of Agricultural Sciences, Bangalore. 126p.

IDRIS, M & I. PRAKASH 1992. Shyness behaviour In: RODENTS IN INDIAN AGRICULTURE (Eds.) Prakash, I & P.K. Ghosh, Scientific Publishers (India), Jodhpur, Pp. 433-444.

KAPOOR, I.P., T. RAMASIVAN & K. KRISHNAMURTHY 1965. Evaluation of "Shoxin" as rodenticide. *Bull. Grain Technol.* 3: 160

KUNHARDT, J.C.G & G.O. CHITRE 1921. Further experiments in plague prevention carried out at Poona: Second report (for the period 1916-1918) *Indian J. Med. Res.* 8: 446-489.

LEFEBVRE, P.W & J.L .SEUBERT 1970. Surfactants as blackbird stressing agents. *Proc. Vertebr. Pest Control. Conf.* West Sacramento, California, USA. 4: 156-161.

LINHART, S.B. 1984. Strobe light and siren devices for protecting fenced pasture and range sheep from coyote predation. *Proc. Vertebr. Pest Conf.* 11: 154-156.

LINZ, G.M., G.A. KNUTSEN, H.J. HOMAN & W.J. BLEIER. 2003. Baiting blackbird (Icteridae) in stubble grain fields during spring migration in South Dakota. *Crop protection* 22: 261-264.

MASON, J.R & D.L. OTIS 1990. Effectiveness of six potential irritants on consumption by red-winged blackbirds (*Agelacius phoeniceus)* and starlings (*Sturnus vulgavis)*. In: CHEMICAL SENSES: Vol.2, IRRITATION. (Eds.): Green, B.G., J.R. Mason & M.R. Kare. Marcel Dekker, New York. pp. 309-324.

MASON, J.R. 1997. REPELLENTS IN WILDLIFE MANAGEMENT, Colorado State University Press, Ft. Collins, Co, USA. 447p.

MASON, J.R. 1998. Mammal repellents: options and considerations for development. *Proc. 18th Vertebrate pest conference.* (Eds.): Baker, R.O & A.C. Crabb. University of California, Davis. USA. pp. 325-329.

MATHUR, R.P & I. PRAKASH 1984. Reduction in the population of Indian desert rodents with anticoagulant rodenticides. *Proc. Indian, Acad. Sci.* 93: 585-589.

MATHUR, R.P., N. KASHYAP & M. MATHUR 1992. Bioefficacy of anticoagulant rodenticides. In: RODENTS IN INDIAN AGRICULTURE (Eds.): Prakash, I. & P.K. Ghosh. Scientific publishers, Jodhpur, India. pp. 517-583.

MIAN, Y., E. HAQUE, J.E. BROOKS & P.J. SAVARIE 1993. Laboratory evaluation of bromethalin and a rodenticide against *Bandicota bengalensis* Gray. *Univ. J. Zool.*, Rajasthan University. 12: 15-19.

MILUNAS, M.C., A.F. RHOADS & J.R. MASON 1994. Effectiveness of odour repellents for protecting ornamental shrubs from browsing by white tailed deer. *Crop Protection* 13: 393-397.

MOTT, D.F., G.M. LINZ & E.W. SCHAFER, Jr. 1987. An evaluation of CPT-avicide aerial sprays for control of roosting blackbirds and starling. Denver Wildlife Research Centre, Colorado, USA. 46pp.

MOTT, D.F., J.L. GUARINO, E.W. SCHAFER, Jr., & D.J. CUNNINGHAM 1976. Methiocarb for preventing blackbird damage to sprouting rice. *Proc. 7th Vertebr. Pest. Conf.* 7: 22-25.

NICHOLAUS, L.K. 1987. Conditioned aversions in a guild of egg predators: implications for aposematism and prey defence mimicry. *Am. Midland Naturalist* 117: 405-419.

PARANJYOTHY, J.T. 1939. Rat virus enquiry report including a note on the raticidal effect of certain commercial and other chemical poisons. *Bull. Inst. Med. Res.* F.M.S. 1:23.

PARSHAD, V.R & C.S. MALHI 1995. Comparative efficacy of two methods of delivering an anticoagulant rodenticide to three species of South Asian rodents. *Int. Biodet. Biodgr.* 36: 89-102.

PARSHAD, V.R. 1999. Rodent control in India. *Integrated pest manage. Review*. 4: 97-126.

PRAKASH, I & A.P. JAIN 1968. Toxicity of norbormide for desert rodents. *Proc. International Symp. on Bionomics and control of rodents.* Kanpur, India. pp.146-149.

PRAKASH, I & R.P. MATHUR 1987. MANAGEMENT OF RODENT PESTS. Indian Council of Agricultural Research, New Delhi, India. 133p.

PRAKASH, I. 1964. Ecotoxicology and control of Indian desert gerbil, *Meriones hurrianae* Jerdon. 4. Toxicity, relative acceptability and efficiency of certain poisons. *Indian Forester.* 90: 519-528.

PRAKASH, I., W.D. FITZWATER & A.P. JAIN. 1969. Toxic chemicals and baits for the control of the gerbils, *Meriones hurrianae* Jerdon and *Tatera indica* Hardwicke. *J. Bombay nat. Hist. Soc.* 66: 500-509.

RANA, B.D & R.S. TRIPATHI 1999. *Recent advances in co-ordinated research on Rodent Control.* Central Arid Zone Research Institute, Jodhpur, India. 48p

RAO, A.M.K.M & I. PRAKASH 1981. Rodenticidal evaluation of zinc phosphide against house mouse, *Mus musculus bactrianus*. *Indian J. Pt. Prot.* 13: 61-63.

REDDY, A.R., K. VASANTHA BANU, S.M. ZAHARUDDIN & P. RAGHAVA REDDY 2005. *Burrow fumigator – An ecofriendly device for rodent management*. Technical bulletin. All India Network Project on Rodent Control, Jodhpur, India.

RENNISON, B.D. 1974. Field trials of calciferol against warfarin resistant infestations of the Norway rat, *Rattus norvegicus*, Berk, *J. Hyg., Camb.* 73: 361-67.

RICHTER, C.P. 1945. The development and use of alphanapthylthiourea (ANTU) as a rat poison. *J. Amer. Med. Ass.* 129: 927-931.

RUELLE, P & R.L. BRUGGERS 1979. Evaluating bird protection to mechanically sown rice seed treated with methiocarb at Nijanga, Senegal, West Africa. In: VERTEBRATE PEST CONTROL AND MANAGEMENT MATERIALS. (Ed.) J.R. Beck. ASTM STP 680. American Society for testing and materials, Philadephia. Pp: 211-216.

SAINI, M.S & V.R. PARSHAD 1992. Control of *Rattus rattus* with cholecalciferol: Laboratory acceptance of freshly prepared and ready-to-use bait formulations. *Int. Biodet. Biodgr.* 30: 87-96.

SANDHU, P.S., M.S. DHINDSA & H.S. TOOR 1987. Evaluation of methiocarb and thiram as seed treatments for protecting sprouting maize from birds in Punjab (India). *Tropical pest Manage.* 33 (4): 370-372.

SHARIF, M. 1951. Spread of plague in Southern and Central division of Bombay province and plague endemic centre in Indo-Pakistan subcontinent. *Bull. Wild. Health. Org.* 4: 75-109.

SRIDHARA, S & R.V. KRISHNAMOORHTY 1979. Efficacies of three rodenticides in paddy fields. *Z. angew. Zool.* 66: 233-242.

SRIDHARA, S. 1977. New food reaction in four Indian wild rodents. *Z. angew. Zool.* 64: 449-452.

SRIVASTAVA, A.S. 1965. Annual Progress Report 1963-64. Co-ordinated scheme for Research on the study of field rats in Uttar Pradesh, Section of the Entomologist to Government of U.P., Kanpur, India. 28p.

SRIVASTAVA, A.S. 1968. RODENT CONTROL FOR INCREASED FOOD PRODUCTION. Rotary club (West), Kanpur, India. 152p.

SRIVASTAVA, A.S., Y.K. MATHUR, K.D. UPADHAY, A.S BHADURIA & J.P. SRIVASTAVA 1980. Evaluation of "Silmurin" (Scillurocide) against field and domestic rodents. *Rod. Newsl.* 4:19.

SUKUMAR, R. 1990. Ecology of the Asian elephant in Southern India. II. Feeding habits and crop raiding patterns. *J. Tropical Ecol.* 6:33-53.

SULTANA, P., J.E. BROOKS & R.L. BRUGGERS. 1986. Repellency and toxicity of bird control chemicals to pest birds in Bangladesh. *Tropical Pest Manage.* 32(3): 246-248.

WADE, D.A. 1982. The use of fences for predator damage control. *Proc. Vert. Pest. Conf.* 10: 24-33.

25

PERSPECTIVES OF VERTEBRATE PEST MANAGEMENT

Shakunthala Sridhara

AINP on Rodent Control, University of Agricultural Sciences, G.K.V.K., Bangalore - 560 065

ABSTRACT

Vertebrate pests cause considerable damage to environment, agriculture and biodiversity apart from transmitting diseases. The problem is more pronounced in tropical Asia and Africa with non-human primates, elephants, several species of ungulates, rodents and granivorous birds causing agricultural losses. The damage in Europe and America is due to carnivore predation on livestock, bird damage in agriculture and rodents in both urban and rural situations. The four essential factors of vertebrate pest management (VPM) are: recognizing damage patterns, identification and understanding the biology of pest species, selection of appropriate strategy and awareness about laws pertaining to vertebrate pests. The management has several options like long term control, sustained management, cultural practices, proofing and netting, frightening devices, repellents, shooting, guarding, trapping and toxicants. None of these are effective if applied singly. An integrated approach using several methods depending on the pest problem, area affected and species involved will yield optimum results. Consulting and involving affected people in planning and implementing management measures is vital. Reduction of damage, not reduction of pest numbers and humane and painless killing if that becomes the only solution is the foundations of VPM. Public pressure should never influence a professional biologist's advice and judgement regarding implementation of VPM which should be based on principles of population ecology alone. As most problems of vertebrate pests are due to unplanned and uncontrolled human population increase, they can be solved by modifying our own behaviour and by regulating our own numbers (Dolbeer, 1998).

INTRODUCTION

All agro-ecosystems are artificial ecosystems, created by converting land ecology. They are highly simplified and highly unstable systems. Of all the agro ecosystems mono-culture especially involving cereal, pulse and oil seed cultivation is the most unstable one. It is a haven of easily available and plentiful food and shelter to pests, plant and soil microbes. These plant devouring organisms when coupled with weather extremes, lead to total crop failure. On the other hand polycultures with diverse and intermixture of crop varieties with well segregated planting dates represent a more stable and less susceptible eco system to climatic vagaries and pest attacks. But this very diversity of food crops cultivated in a stagnated manner becomes an attractive source of constant food availability to various vertebrate pests.

Although diverse groups of vertebrates such as amphibians, reptiles, birds and mammals may cause losses, factually only few species of rodents and a far lesser number of bird species become pests. Amongst vertebrate pest rodents are most devastating followed by birds. Rodent menace is worldwide while granivorous birds reduce agricultural yield substantially in America, Europe and Africa but comparatively less in Asia. Other vertebrates become pests in extreme cases of habitat shrinkage and degradation. They directly damage biodiversity and productivity by competition, predation, habitat degradation, damage to crops, pastures and structures, harbouring parasites and pathogens.

The major mammalian crop depredators in tropical Asia and Africa are few species of non-human primates, bats, elephants and some ungulates causing losses to food and plantation crops. In contrast predation on livestock by carnivores such as coyotes, wolves, lions, leopards etc is significant in America and Sub-Saharan Africa.

THE DILEMMA OF VERTEBRATE PEST MANAGEMENT

Almost all wild animals are potential pests of crops and predators of livestock, especially in shrunken and degraded habitats. Wild animals compete with resources either in natural or cultivated fields and become predators or pests. Summarily many reptiles are poisonous and scary, granivorous birds damage crops and frugivorous cause yield losses in orchards, predatory birds prey on chicken, carnivorous mammals on livestock and herbivores become crop pests. Earlier pests were usually 'controlled' which was literally attempting to eradicate a species. But it is highly impossible to eliminate an animal species entirely be it a mammoth elephant or the smaller but fast reproducing species of insects and rodents. Eventually the economic impact of pest control in terms of cost : benefit analysis is the sole criterion to take up management of pest animals with due consideration for animal welfare and humane killing while planning management strategies. If pests also happen to be endangered or threatened the problem becomes complicated with a section of society insisting on conservation at any cost and the affected farmers demanding immediate damage prevention. Thus managing endangered animals that become pests is not only a technical problem but a social and ethical one.

With the exception of few instances where culture and religion dictate man's attitude of being compassionate and protective about animals, farmers fear and resent wildlife damage to crops, livestock and environment. Further, the problem of vertebrate pests is an emotional one to both the affected people and public in general. The government spends enormous amount of money on their conservation, preservation and improvement of their habitat. On the other hand when these animals particularly elephants, primates, some ungulates raid crops causing huge losses in a very short period, the farmer is helpless and cannot do anything because of the protected status of these animals. Either he expects hefty monetary compensation or wants government under whose patronage these animals thrive in protected areas to take suitable action to ameliorate the problem. The anger and helplessness escalates into public protest and acts of violence when an injured elephant or a carnivore attacks and kills people or livestock.

TRADITIONAL METHODS OF VERTEBRATE PEST MANAGEMENT

Reducing animal damage is a mean to achieve an objective but not an end itself. Raising tolerance to damage is considered as important as reducing damage itself in situations of man-animal conflict. Several approaches are integrated to manage the problem of vertebrate pests such as compensating the farmers for crop loss, repelling the damage causing animal and educating the communities to live with the animal or to adapt non-lethal approaches to deter them. Managing crop raiding vertebrate pests is not only an agricultural problem but a social responsibility too.

Amongst vertebrate pests lethal approach is limited to rodent and some bird pests. As this aspect is discussed in detail elsewhere in this book, it will not be dealt with in this chapter. Most other pests are managed with non-lethal approaches. Of all the non-lethal approaches to vertebrate pest management (VPM), the most economical and wide spread is patrolling fields and chasing the intruders (Hill, 2000;

Chhangani *et al.*, 2005). The farmers in affected fields either hire or the family members specially women and children chase monkeys (Chhangani *et al.*, 2005), nilgai (Rajpurohit *et al.*, 2005) and birds (Verghese, 2005; Rao & Dubey, 2005). The nocturnal and powerful elephants are chased using light and sound by groups of men folk (Sridhara, 2005).

Other non lethal methods of preventing vertebrate damage are repellents and deterrents. Basically repellents are of three types- visual, acoustic and chemical. *Visual* repellents are mostly effective in preventing bird damage in crop fields. Scare crows of different types keep bird pests temporarily away from crops. *Acoustic repellents* are also effective in bird management. These include propane cannons which explode noisily, pre-recorded alarm and distress calls broadcast over a speaker system, pyrotechniques etc (Bomford & O'Brien, 1990). *Chemical repellents* are used successfully to protect crops from insects, birds and mammals. The non-lethal, bird specific repellents used effectively in crop protection include methiocarb and Avitrol (Avery, 1989; Mason, 1989; Nolte *et al.*, 1993). Consumption of Avitrol leads to erratic flight and the bird emits distress calls before dying; rest of the birds in the flock get frightened and keep away from crops (Dolbeer *et al.*, 1994). Predator urine has been found to deter damage by several mammals like bears, hares, voles, rabbits, deer etc (Lindgren *et al.*, 1995; Sullivan *et al.*, 1985; Morgan & Woolhouse, 1995).

Shooting is resorted to very rarely either illegally or after obtaining permission from the local forest officials specially to kill wild boar and to scare elephants.

VERTEBRATE PEST MANAGEMENT: PRINCIPLES AND OBJECTIVES

Vertebrate pest damage is only one of the several factors that is involved in practices and priorities of environmental management of land. Management of vertebrate pests should be based on a wholistic approach taking into consideration the social and economic implications of pest management as well as environmental safety. With this approach the primary objective of vertebrate pest management (VPM) is to alleviate the damage caused, not to destroy the pest. Many problems of vertebrate damage can be prevented or solved with non-lethal approaches such as exclusion and habitat management. Most of the following discussion on VPM is based on the document of Australian Capital Territory (ACT) which deals with almost all aspects of managing vertebrate pests (Anon, 2002). Indian work and context are included wherever data is available.

A successful VPM programme should be based on definite principles, have well defined objectives and viable strategies. The basic principle of such a programme is the involvement of affected people in the early stages of planning and during implementation. The later should be under the supervision of a responsible and knowledgeable team.

Broadly speaking the three objectives of VPM are:

- Identification of different aspects of vertebrate pest management.
- Planning the management of vertebrate pests after appropriate consultation and liaison with affected people.
- Entrusting the implementation of the programme to people affected.

The four basic and essential steps of VPM are:

- Recognizing the damage patterns and the species responsible.
- Understanding the biology of pest species causing damage.
- Knowledge about strategies of management.
- Awareness about local, state and central government laws about pest control and wildlife protection.

1. Recognising damage patterns and species causing damage

Not all members of a vertebrate pest population cause damage as in insects and other invertebrates. Only few individuals in a population cause losses as in an elephant group or most of them as in rodents. The only exception is some granivorous birds where the entire flock as in blackbirds cause damage. Granivorous bird pests like baya weaver, parakeets, doves, peacock, various species of cranes and munias are visible and diurnal. For instance ripening corn, sunflower, groundnut and fruits are susceptible to bird damage which is conspicuous to the naked eye. However, rodent and ungulate damage to sprouting crops and orchard trees occurs underground or at night. Such damages go unnoticed or underestimated. In contrast damage to ripening grains, fruits, nuts, seeds and vegetables is clearly recognisable and methods are available to estimate them. Species damaging can be identified by signs of activity such as tracks, foot marks, faeces and burrows which also serve as indices of density.

Damage characteristics also help in species identification. Debarking of stem is by porcupines. Roots of trees and shrubs are damaged by burrowing species like *Bandicota bengalensis* and *Nesokia indica.* Rodent damage to sugarcane is at internodes with a characteristic, semicircular cut while damage by rabbit results in severing of internodes. Ungulates damage agriculture, forestry and ornamental plants by feeding, trampling and antler rubbing. Their damage is rough with a shredded appearance as they lack incisors, which characteristically cuts the plant parts neatly. However, objective estimates of losses due to vertebrate pests are difficult to obtain specially quantifying damages caused by monkeys, wild boar and elephants are almost impossible. Thus there is an urgent need to standardise methods of damage estimates for these animals to fine tune the management strategies (Imam & Malik, 2005; Sridhara, 2005; Sridhara & Chakravarthy, 2005).

An important point in vertebrate pest management is detection of damage. An early detection may make control measures more effective and less expensive. As less amount of pesticides or less amount of time and energy is spent on less number of animals, environmental deterioration and threat to non-target animals is also very less. Delayed control in addition to becoming costly may result in habitat conditioning wherein the sites become more attractive to additional animals and hence increased damage.

2. Understanding the biology of pest in relation to damage

(a) Pest biology and ecology

(i) ***Biological factors of pest species affecting damage*****:** The biological factors of vertebrate pests to consider while managing the problem are: habitat, home range, nesting/ burrowing/ resting activities and foraging behaviour of pest species, regulators of population growth such as breeding season, natality, mortality, dispersion, migration, competition, predators, pathogens and reproductive strategy. Management of pests will be more effective with more information on the resource requirements of pest species. There is a direct relationship between numbers and damage levels. Further removal of animals by lethal or non-lethal method has predictable effects on birth rates, death rates and dispersal. All parameters of population regulation are species specific and have a bearing on method selected to manage.

(ii) ***Seasonal and daily activity patterns*** of the pest are important in planning the management strategies for vertebrate pest problems. Seasonal migrations of several mammals and birds influence the extent of damage. Migrating elephants, birds cause greatest damage to ripening cereals. Knowledge about stages of crop growth viz. sown seeds, seedlings, vegetative stages, ripening grains, seeds, fruits, nuts, vegetables and the time of damage i.e. day or night depending on diurnal or nocturnal activity pattern of pest animals are crucial for timing vertebrate pest control.

Seasonal breeding makes it more economical and efficient to control the vertebrate pests during non-breeding season, especially before the onset of breeding.

(iii) ***Pest damage*****:** Although it is unquestionably true that vertebrate pests severally damage environment and cause considerable yield losses in agriculture, evidence for the link between density of pest population and the degree of damage have not been documented accurately in many cases. This leads to two strategies. It is more important that the emphasis be on reducing damage than on reducing pest numbers. If the option is to reduce pest numbers, then the fundamental issue will be to determine what level of pest reduction will result in optimum and sustainable reduction in damage levels. These decisions are area, pest and crop specific and will vary from location to location.

(iv) ***Pest ecology*****:** Pests are in dynamic equilibrium with the environment in which they live. This equilibrium will change once the pest numbers are drastically reduced. A very important question is whether there will be unwanted and harmful consequences of drastically reducing the pest numbers. The answer can come only from specific research on the pest animal involved. In the absence of such information if pest management decisions have to be made, it is advisable to test the logical assumptions made in a precontrol evaluation. Based on the results subsequent programmes should be modified / refined.

(b) Environmental factors affecting VPM

(i) ***Weather*** affects indirectly by its effect on vegetation. During rainy and post rainy season when food is plentifully available damage to crops is less compared to dry season when damage is very high. At the same time, dry seasons with characteristic low populations are less expensive to control. The absence or scarcity of food during this period facilitates ready acceptance of poison bait.

3. Strategies for VPM

The basic strategy of all VPM should be reduction of damage which should be cost-effective, long lasting and least damaging to environment.

Three major aspects greatly influence the approach to and the success of VPM. These are: Ecological, Economical and Social in nature. Ecological aspects of pest management takes into account the relationship between animals and the environment, more so the interrelationship between plant and animal communities, and the various environmental factors. Economic factors pertain to cost- benefit aspects of VPM while social factors are multiple, varied and complex. They range from the attitude of people to political consideration, ethical issues of animal killing and animal welfare.

The strategy of VPM includes defining the problem, prioritization of tasks, technique selection, grades of approach to manage and implementing management.

(a) Definition of problem

The first step in management is defining the pest problem. This involves description of the extent of damage, its impact and the people affected. Mapping damaged areas, correctly identifying the pest species causing damage and collecting information on the biotic and abiotic factors influencing pest biology are important. Broadly speaking definition of the problem involves collection of information necessary to understand the problem and identifying the practical solutions to ameliorate the damage.

(b) Prioritisation of tasks

Prior to finalizing the strategies, priorities should be identified because the resources available for VPM are often inadequate and hence budgeting should be based on priority. Areas where these pests cause considerable production losses, affect conservation programmes adversely and cause damage to cultivation should be tackled first. This prioritization is very important in planning the strategy for VPM as well as in implementing because VPM is a complex strategy requiring considerations of conservation, aesthetics and economic losses.

Priorities can be determined by dividing the affected area into manageable units, each unit is an area of known dimensions preferably with well defined boundaries. These should be categorised into units with vertebrate pests and units without vertebrate pests. The later category should be treated with preventive measures, which in spite of not being cost-effective at the time of execution of IPM will yield long lasting results. The areas with established pest populations are difficult and expensive to treat. Even here before management practices are implemented, each unit has to be ranked as to conservation or pest control needs. The final scores will help to identify areas to be treated either for conservation or for pest management. Even at this juncture control programme may not be feasible or desirable where conservation of pests, livestock and non-target wildlife becomes more important than reducing pest numbers. This evaluation of economic and conservation considerations should be based on discussions and consultations with individuals, agencies and local people who are knowledgeable about the species and damage.

An issue of equal importance is to determine whether the problem is real or perceived. It is likely that the damage may not be significant in terms of economics and ecology but the localised, sudden and unexpected damages make the losses appear spectacular. It is essential that the problem is evaluated with open mind and is agreeable to everyone concerned.

(c) Selection of Technique

Selection of appropriate technique depends on local circumstances, the resource under threat, the nature of land, available techniques, and availability of resources and attitude of the local population. The following are the various methods of alleviating damage by vertebrate pests:

Habitat management and cultural practices: Measures to make habitat less attractive, access to food difficult, area repellents, diversion of feeding site etc deter vertebrate pests from attacking crops. Some of the successful examples of making habitat less attractive are given below.

Thinning marshy vegetation to reduce habitat to baya weaver birds which forage considerably in adjoining paddy fields is an instance of making habitat less attractive.

Frightening agents in conjunction with lure crops which provide alternate food can protect grain and sunflower (Cummings *et al.*, 1987). Grain scattered on ground reduced damage to corn seedlings in no tillage fields and damage to apple trees by rodents (Johnson, 1986). Clean cultivation and weed control reduced pine vole damage in apple orchards significantly (Davis, 1976).

Bird resistant varieties of corn, sunflower and sorghum effectively reduced bird damage. The characteristics responsible were mostly those which prevented easy access to cobs and seeds.

Proofing and netting: Plastic netting is effective in excluding birds from individual fruit trees of high value crops such as grapes.

Many types of fences including electric ones are used to exclude elephants, deer and other ungulates. Wire netting and electric fences kept away rabbits from crop fields in England (Mckillop & Wilson, 1987). Rabbit and rodent damage in orchards can be prevented by wrapping trees with hardware cloth or metal sheets (0.5 mm) and also by burying the same 10 cm deep around the tree base (Tubin & Richmond, 1993; Anon, 2004).

Frightening devices such as propane cannon, shell crackers, rockets, whistle bombs, alarm and distress calls, repel pests from crop fields.

Repellents: Odour and taste repellents reduced deer and rodent damage with variable effect but were not cost effective. Methyl anthranilate acts as a sensory irritant while Methiocarb induces post ingestional distress in birds. Cinnamamide, thiram and capsaicin are reported to protect seedlings from rodent and ungulate damage.

Shooting and Trapping: Shooting is effective when the bird numbers are low but has little effect on large flocks. Shooting is one of the best ways to reduce deer population but needs government permission.

Trapping removes pest population satisfactorily when bird populations are small. Rodents can be trapped using live and snap traps but the method is successful only in small plots.

Use of toxicants in VPM is greatly restricted. Between 1974-1992 using the wetting agent PA-14, 38 million blackbirds were killed in USA (Dolbeer *et al.*, 1997). But the chemical is banned since. Two bird toxicants that are still used are DRC-1339 and Avitrol.

(d) Graded approach to management: Timely harvest, resistant cultivars, frightening agents, barriers, repellents, trap crops, and finally toxicants after considering conservation status of the pest in an integrated way will yield sustainable damage reductions. Once the management technique is chosen or a combination of techniques is chosen, the management option has to be defined. The degree of pest reduction has to be determined by the value of the resource being affected and the cost of control. Often the value of resource to be protected justifies the cost of control. The effective management of vertebrate pest damage has several graded approaches. Mainly these are local eradication, strategic management which includes long term control, sustained management, targeted management and no management.

(i) Local eradication of a vertebrate species i.e. permanent removal of every individual is rarely feasible or economically and ecologically sensible except at a local level. Such local eradications can succeed only in islands or when sources of immigration can be controlled by erecting barriers. Eradications are expensive and can be achieved when pests can be killed at a faster rate than they can reproduce or when immigration is prevented or when monitoring of very low populations is possible and the socio-economic environment supports the eradication.

(ii) Strategic management is the popular option and is of three types namely long term or permanent reduction, sustained management and targeted management,

Long term or permanent reduction in the damage caused by a vertebrate pest is possible with a single or a set of actions such as fencing or modifying the habitat and making it least suitable to pests.

In sustained management, pest density is reduced to minimal level and maintained at threshold density when no further benefit from control measures is derived. This strategy involves two steps: An initial knock down of a high proportion of the population followed by periodic control to slow down or prevent recovery.

In targeted management reductional measures are directed at the individual/group of individuals that are responsible for maximum damage or applied when damage is maximum E.g. Displacing 'rogue' and habitual crop raiding elephants. However, this approach is difficult to implement as it requires a thorough understanding of the behaviour and biology of the pest to identify the individuals to be targeted for control. Targeted management will be more effective if control is carried out at critical times.

No management: In some cases the most cost-effective solution to vertebrate pest problem may be to allow the pest to continue to exist without taking any action. Such a step is taken when the continued presence of the pest will not lead to any significant or long term deterioration of resources. Or it can be the option if the ecological processes involved are not understood completely or acceptable techniques for control are not available.

(e) Implementing management

Once the strategy is selected the details of management are to be worked out with a feasible time table and resources committed. In other words what equipment is needed, who does what and when. A very important aspect of management planning is consulting the affected people while developing the programme. This fastens understanding, mutual trust, confidence and co-operation resulting in easier and successful implementation of programme.

Monitoring the effectiveness and evaluation are essential elements of a pest management programme. They provide inputs as to the practically, success/failure of the programme which in turn will be helpful in improving the efficiency and effectiveness of the strategy and to modify the objectives if necessary.

4. Awareness about laws and alternatives

With awareness and concern for wildlife percolating to all levels of society, more and more regulations and restrictions are being brought into vertebrate pest management. Actually the conservation status of pest animals, the local, state and central laws that protect them should be thoroughly understood before initiating any VPM.

Most mammalian pests with the exception of rodents are protected from killing under the provisions of Wildlife Protection Act, 1972 of India which is amended from time to time. The Indian elephant, *Elephas maximus* is included under Schedule I, the Hanuman langur, *Semnopithecus entellus*; and Rhesus macaque, *Macaca mulatta* in schedule II, Nilgai, *Boselaphus tragocamelus* and wildboar, *Sus scrofa* under schedule III and hares, *Lepus nigricollis* under schedule IV. All these animals are protected and hence cannot be subjected to lethal control of any type that results in their death. All birds are similarly protected. However the fruit bats are considered as vermin.

A second legal aspect of VPM is compliance of regulations regarding the use, storage, disposal and record keeping of vertebrate pesticides. Not following the instructions diligently may lead to not only ineffective control but risky to non-target wildlife and environment, bad publicity and public disapproval and protests.

It becomes easier to comply with legal restrictions if the available management options are understood thoroughly. Management options generally include lethal chemicals, trapping, modification of habitat and behaviour of the pest animal. Options are dependent on effectiveness, cost, advantages and disadvantages. Some methods work well in certain conditions and others in some other context. But every option should be carefully evaluated before finalizing the VPM strategy. Irrespective of choice, the single most criterions should be minimum use of pesticides.

ANIMAL WELFARE AND HUMANE CONTROL IN VPM

While managing vertebrate pests it is crucial to remember that they require the same level of consideration for their welfare as animals in other contexts irrespective of the harm/damage they cause. Many a times selective killing of vertebrates becomes inevitable but such a control should be justified and carried out in the most humane way possible. Humane killing can be interpreted as death without pain, suffering or distress perceptible to the animal. This can be achieved by rendering the animal unconscious and ensuring that the animal stays so till death. Every step towards minimising pain, suffering or distress is crucial and ensures humane control of vertebrate pests. The six guiding principles of humane vertebrate pest control are: Justification for control, lethal or non-lethal control, probability of success, co-ordinated and staggered approach, target specificity and humaneness. Some of the instances of human control are avoidance of steel-jawed leg hold traps, painful/distress causing poisons and baiting strategies, suffocating burrow fumigation etc. Additionally vertebrate pest control programmes must be designed to minimise the number of animals targeted for control. Finally as and when more humane methods are developed, they should be immediately adapted.

SOCIAL, POLITICAL AND ETHICAL ASPECTS OF VPM

Along with unprecedented growth of human population several successful vertebrate pests are also proliferating rapidly reaching very high density. This has lead to competition for resources and conflict of interest between vertebrate pests and humans. The management of problem animals mostly mammalian herbivores is besotted with diverse opinions and pressures. Those suffering damage insist that the traditional methods of hunting, trapping and poisoning are adequate and satisfactory methods. But changing social perceptions and emerging animal rights/welfare activism looks down on these age old practices as cruel and inhuman. However, it is important to bear in mind that usually these protesters are not victims of vertebrate damage directly or indirectly.

In many situations, methods to deal with vertebrate pest problems are limited. Local, national or international regulations, laws and treatises interfere with implementation. However, research and committed concern has brought out huge improvements. Application procedures are changing so also the options for managing vertebrate damage. Still the choice of method is riddled with controversy and susceptible to political pressures. At the end, reviews of lethal and non-lethal approaches to wildlife damage reduction indicate Integrated Vertebrate Pest Management as the best approach for resolving the problems.

Dolbeer (1998) draws attention to the following issues while formulating vertebrate pest management:

(a) Since management of wildlife and vertebrates is a complex issue, VPM should be based on detailed information on behaviour and population dynamics. The exact economic and health impact of pest species is a prerequisite.

(b) Vertebrate pests are highly intelligent, adaptable and secretive animals. Management techniques which are mainly techniques of census, monitoring and measuring should precede intense studies of pest species.

(c) The sociological aspects of VPM are not only diverse but also emotional when it comes to killing to reduce vertebrate damage because the animals are large, visible, attractive, awe inspiring and capable of thinking and complex behaviour patterns, and much like human beings. Their response to management either by repelling or killing makes it a battle of wits, somewhere kindling the feelings of compassion as well as guilt.

(d) Lastly vertebrate pest management is overwhelmingly complicated by the legal status of these animals, conservation issues and registration of lethal chemicals.

There has been a growing public sentiment in many counties, particularly in the past quarter century, opposing the management of wildlife damage including that of vertebrate pests. This has led to new legislations and restrictions on the methods and strategies to mitigate human-wildlife conflict. There is also an increased emphasis on non-lethal methods such as managing the behaviour of individuals or local group of animals instead of attempting to manage populations. Thus, there is a pressing need for the development of new technologies for resolving human-wildlife conflicts because of increased restrictions on traditional methods as well as emergence of new problems in new environments involving new species.

Ultimately we have to remember eternally that most human-wildlife problems are direct outcome of human activities and can be easily solved by modifying our own behaviour or controlling our own population. Problems arise because of unprecedented increase in human population and human thirst for high standards of life which depletes natural resources at a faster rate than their regeneration. Decrease in resources resulting in degraded habitat as well as its shrinkage places immense pressure on vertebrate herbivores. The second controversial issue is about killing. For most vertebrate pest species lethal control is more effective than reproductive control in reducing population levels. But public opinion favours non-lethal methods, particularly reproductive control to contain population. Dolbeer (1998) argues vehemently that such public pressure should not influence a professional biologist who should advise and implement pest reduction measures based on principles of population dynamics alone and not by emotional blackmail. No doubt, reproductive control has a definite role in population management but it is more effective in species with a short life span like rodents and insects. Such a strategy will not work satisfactorily for species with long life like deer, elephants, nilgai etc. for several reasons. Finally lethal methods should be employed only if they are based on scientific analysis, when alternate methods are less effective or impractical, that too when results can be monitored, and only when it is certain that killing indeed will reduce damage.

CONCLUSIONS

"Tolerance" continues to be the most widely used strategy for wildlife conflict resolution. The earlier management strategy was based on game management philosophy, relying on habitat improvement and harvest management to manipulate wildlife populations. This approach is no longer the panacea. There is a critical need to evolve newer, effective, eco-friendly, humane and economic methods of vertebrate pest management. Some potential areas of research include behaviour modifying drug therapy fertility control, computer-assisted management, capture technology, barrier technology, surveillance technology and behaviour contingent technology. So far wildlife damage management had been reactive and labour intensive with unsatisfactory results. It is time to evolve a practical, preventive, environmentally safe and mechanised approach to manage vertebrate pests.

REFERENCES

ANON, 2002. *ACT VERTEBRATE PEST MANAGEMENT STRATEGY*. Conservation series No. 19. Australian Capital Territory, Canberra, Australia. 73pp.

ANON. 2004. *Annual progress report of AICRP on Rodent Control*, University of Agricultural Sciences, Bangalore, India.

AVERY, M.L. 1989. Experimental evaluation of partial repellent treatment for reducing bird damage to crops. *J. Appl. Ecol.* 26: 433-439.

BOMFORD, M & P.H. O'BRIEN. 1990. Sonic deterrents in animal damage control: a review of devise tests and effectiveness. *Wildlife Soc. Bull.* 1: 411-422.

CHHANGANI, A.K., L.S. RAJPUROHIT & S.M. MOHNOT. 2005. Crop damage by hanuman langur, *Semnopithecus entellus* and its management In: VERTEBRATE PESTS IN AGRICULTURE (Ed.) Shakunthala Sridhara, Scientific Publishers (India), Jodhpur, (this book).

CUMMINGS, J.L., J.L. GUARINO, C.E. KNITTLE & W.C. ROYALL, Jr. 1987. Decoy plantings for reducing blackbird damage to ripening sunflower. *Proc. Verteb. Pest. Conf.* 12: 286-291.

DAVIS, D.E. 1976. Management of pine voles. *Proc. Verteb. Pest. Conf.* 7: 270-275.

DOLBEER, R.A. 1998. Overview and management of vertebrate pests. In: HANDBOOK OF PEST MANAGEMENT (Eds.) J.R. Roberson. Marcel Dekker, Inc., New York. pp. 863-691.

DOLBEER, R.A., D.F. MOLT & J.L. BELANT. 1997. Blackbirds and starlings killed at winter roosts from PA-14 applications. Implications for regional population management. *Proc. East. Wildl. Damage. Mgmt. Conf.* 7: 77-86.

DOLBEER, R.A., N.R. HOLLER & D.W. HAWTHORNE. 1994. Identification and control of wildlife damage. RESEARCH AND MANAGEMENT TECHNIQUES FOR WILDLIFE AND HABITATS. (Eds.) T.A. Bookhout. The Wildlife Society, Bethesda, Maryland, USA.

HILL, C.M. 2000. Conflict of interest between people and baboons: crop raiding in Uganda. *International J. Primatol.* 21: 299-315

IMAM, I. & IQBAL MALIK. 2005 Rhesus monkey (*Macaca mulatta*) problems in India and their management In: VERTEBRATE PESTS IN AGRICULTURE (Ed.) Shakunthala Sridhara, Scientific Publishers, (India), Jodhpur (this book).

JOHNSON, R.J. 1986. Wildlife damage in conservation tillage agriculture: a new challenge. *Proc. Verteb. Pest Conf.* 12: 127-132.

LAWS, R.M., I.S.C. PARKER & R.C.B. JOHNSTONE. 1975. ELEPHANTS AND THEIR HABITATS. THE ECOLOGY OF ELEPHANTS IN NORTH BUNYORO, UGANDA. Colorado Press, Oxford.

LINDGREN, P.M.F, T.P. SULLIVAN & D.R. CRUMP. 1995. Review of synthetic odour semiochemicals as repellents for wildlife management in the pacific North West. In REPELLENTS IN WILDLIFE MANAGEMENT (Ed.) J.R. Mason. Colorado State University, Fort Collins, Colorado, USA. pp. 217-230.

MASON, J.R. 1989. Avoidance of methiocarb-poisoned apples by red-winged blackbirds. *J. Wildl. Mgmt.* 53: 836-846.

McKILLOS, I.G & C.J. WILSON. 1987. Effectiveness of fences to exclude European rabbits from crops. *Wildl. Soc. Bull.* 15: 394-401.

MORGAN, D.R & A.D. WOOLHOUSE. 1995. Predator odours as repellents to Bush tail possums and rabbits. In: REPELLENTS IN WILDLIFE

MANAGEMENT (Ed.) J.R. Mason. Colorado State University, Fort Collins, Colorado, USA. pp. 241-252.

NOLTE, D.L., J.R. MASON & L. CLARK 1993. Non lethal rodent repellents; differences in chemical structure and efficacy from non-lethal bird repellents. *J. Chem. Ecol.* 19: 2017-2019.

RAJPUROHIT, L.S., A.K. CHHANGANI & S.M. MOHNOT 2005. Crop damage by blue-bull, *Boselaphus tragocamelus* and its management In: VERTEBRATE PESTS IN AGRICULTURE. (Ed.) Shakunthala Sridhara, Scientific Publishers (India), Jodhpur (this book).

RAO, V & O.P. DUBEY 2005. Granivorous bird pests and their management In: VERTEBRATE PESTS IN AGRICULTURE. (Ed.) Shakunthala Sridhara, Scientific Publishers (India), Jodhpur, India (this book).

SRIDHARA, S. 2005. Management of crop raiding elephants In: VERTEBRATE PESTS IN AGRICULTURE. (Ed.) Shakunthala Sridhara, Scientific Publishers (India), Jodhpur, India (this book).

SRIDHARA, S & A.K. CHAKRAVARTHY. 2005. Wild boar, *Sus scrofa,* an underestimated crop pest of India In: VERTEBRATE PESTS IN AGRICULTURE. (Ed.) Shakunthala Sridhara, Scientific Publishers (India), Jodhpur, India (this book).

SULLIVAN, T.P., L.O. NORDSTROM & D.S. SULLIVAN. 1985. The use of predator odours as repellents to reduce feeding damage by herbivores. II Black tailed deer (*Odocoileus hemonius columbianus*). *J. Chem. Ecol.* 11:921-935.

TUBIN, M.E & M.E. RICHMOND. 1993. Vole management in fruit orchards. *U.S. Fish. Wildl. Serv. Biol. Repl.* 5.

VERGHESE, A. 2005. Frugivorous birds and their management In: VERTEBRATE PESTS IN AGRICULTURE. (Ed.) Shakunthala Sridhara, Scientific Publishers (India), Jodhpur, India (this book).

❑❑❑

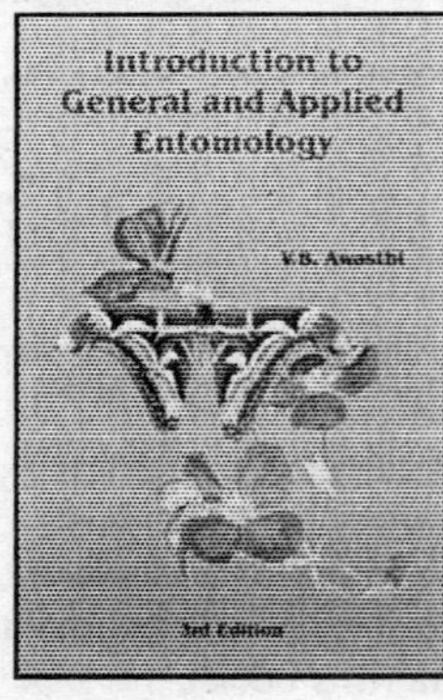

Introduction to General and Applied Entomology
3rd Edition

V.B. Awasthi

This text book aims at to provide an up to date, general and applied account of Entomology which is written in a simple and lucid language keeping in view the syllabi of Entomology in M.Sc. Zoology & M.Sc. (Ag.) of various Indian Universities. The book will also be useful to those who are preparing for competitive examinations conducted by various central & state Government agencies for recruitment. The book is equally useful to those dealing with pests and pest control. It is hoped the book will be useful not only to students but also to agriculturist, horticulturist, foresters, veterinary and clinical doctors and laymen dealing with pest control.

978-81-7233-597-7 | 499 pp.

Agricultural Insect Pests and their Control

V.B. Awasthi

This book has been mainly written keeping in view the requirements of the students of Agricultural Entomology of various Indian Universities. The book provides a brief account on the structure, function, development and metamorphosis, and a comprehensive account on almost all the methods of pest control. The nature of damage caused by the important insect pests of crops, their life histories and control have been described. It is hopped that the students of Agricultural Entomology, will find the book extremely useful. The book is also useful to the Plant Protection Departments and similar faculties. The book will also be useful to those who are preparing for competitive examinations conducted by various central and state Government agencies for recruitment. Every attempt has been made to provide necessary information from the point of view of students of Agricultural Entomology, which is not available in other books.

978-81-7233-712-4 | 267 pp.

Other Related Books

Principles of Insect Behaviour
2nd Edition

V.B. Awasthi

The aim of writing this book has been to provide an account of the principles of insect behaviour for the post graduate students of Entomology and general Zoology of all the Universities and Research Institutions as well as to the research scholars engaged in the study of different aspects of Insect behaviour. It is also useful to general readers who have basic knowledge of Biology and have curiosity to know why the insects are vividly coloured, have peculiar shapes, produce sound and light; how they communicate and why bees and wasps sting when disturbed. Significance of insects in pollination of plants associated with the production of seeds, gains and fruits, the role of hormones, pheromones and other chemicals controlling various vital events of the insect behaviour has also been described. The importance of insect migration, control of insect pests by behavioural means has also been discussed. Thus, this book deals with almost all major aspects of insect behaviour. It is hoped that this book will be useful not only to the students of Entomology-Zoology and research scholars of insect behaviour, but also to naturalists who wish to know about some of the mysteries of these small creatures and their lifestyle.

978-81-7233-825-1 | 299 pp.

A Manual of Practical Entomology
3rd Edition

M.M. Trigunayat

A Manual of Practical Entomology (Field and Laboratory guide) is written to provide text material on different aspects of the practical syllabi of M.Sc. Entomology. All chapters are illustrative and well explained. Easy text will help students understanding the exercises. Dissections of the individual insect have been described so as to understand and follow the finest details of the anatomy of the insect concerned. Physiological and Behavioural Exercises are well presented usually in the style followed by the students. Individual Orders have been dealt with relevant necessary text and vital aspects of biology of the insect concerned. Chapters on Natural History and Rearing will inculcate interest among the students for their quest on these tiny creatures. Profusely illustrated with simple figures presentation and style gives the reader an insight to make it unique. Primarily intended for use by the post-graduate and graduate students of Entomology of the Universities in Rajasthan and elsewhere in India, it can also be used by agriculture departments, naturalists and workers in other related fields.

978-81-7233-983-8 | 358 pp.

Other Related Books

ISBN	Title	Author
978-81-7233-430-3	Animal Biodiversity: Patterns and Processes	Ananthakrishnan, T.N.
978-81-7233-492-3	Biodiversity of Ranthambhore Tiger Reserve Rajasthan	Singh, V.
978-81-7233-923-4	Biological and Molecular Approaches in Pest Management	Singh, Balwinder
978-81-7233-797-1	Biopesticides in Environment and Food Security: Issues and Strategies	Kaul, Opender
978-81-7233-865-7	Biopesticides in Sustainable Agriculture Progress and Potential	Koul, Opender
978-81-7233-402-8	Changing Faunal Ecology in the Thar Desert	Tyagi, B.K.
978-93-8369-228-6	Comparative Anatomy of the Vertebrates	Theodore H. Eaton, Jr.
978-81-7233-794-0	Consciousness, Entropy and Evolution : The Laws of Biology	Sekhar, DMR
978-81-7233-740-7	Dynamics of Insect Behaviour	Ananthakrishnan, T.N.
978-81-7233-599-1	Ecodynamics of Insect Communities	Ananthakrishnan,T.N.
978-81-7233-509-0	Ecological Entomology: Insect Life in Odd Environments	Ananthakrishnan,T.N.
978-81-7233-401-7	Ecology of Small Mammals of Desert & Montane Ecosystems	Prakash, I.
978-81-7233-805-3	Ecology Workbook	Misra, R.
978-81-7233-727-8	Entomology Ecology & Biodiversity	Tyagi, B.K.
978-93-8369-225-5	Evolution of the Vertebrates, 3rd Edition	Colbert , H. Edwin
978-93-8369-204-0	Freshwater Fish Pond Culture and Management	Chakroff , Marilyn
978-81-7233-724-7	Genopsych : A Coinage in the Foundry of Biology	Sekhar, DMR
978-81-7233-680-6	Handbook of Practical Nematology	Bajaj , K. Harisha
978-81-7233-628-8	Insect, Mite and Vertebrate Pests and their Management in Horticultural Crops	Reddy, P.P.
978-81-7233-523-6	Integrated Nematode Management in Horticultural Crops	Reddy, P.P.
978-81-7233-850-3	Integrated Pest Management	Dhawan, A.K
978-93-8369-232-3	Introductory Insect Physiology	Potton, L. Robert
978-81-7233-766-7	Laboratory Manual of Microbiology, Biochemistry and Molecular Biology	Saxena, J.
978-81-7233-782-7	Microbial Diversity : Exploration & Bioprospecting	Reddy, S.R.
978-81-7233-521-2	Objective Type Questions and Answers in Veterinary Immunology	Kannaki, T.R.
978-93-8369-211-8	Pest and Vector Management in the Tropics	Youdeowei , Anthony
978-81-7233-532-8	Pests of Forest Importance and their Management	Tyagi, B.K.
978-81-7233-763-6	Principles of Animal Taxonomy	Simpson, G.G.
978-93-8369-227-9	Principles of Systematic Zoology	Mayr , Ernst
978-81-7233-583-0	Scientist : A Unique Species & the Backbone of a Nations Economic &Intellectual Prosperity	Tyagi, B.K.
978-81-7233-819-0	The Chemistry of Pesticides their Metabolism, Mode of Action and Uses in Crop Protection	Hassall, K.A.
978-93-8369-231-6	The Structure and Life of Bryophytes	Watson, E.V.
978-81-7233-760-5	Theory and Practice of Integrated Pest Management	Arora, R.